Manual of Physical Status and Performance in Childhood

Volume 2: Physical Performance

Manual of Physical Status and Performance in Childhood

Volume 2: Physical Performance

Robert M. Malina
University of Texas
Austin, Texas

and

Alex F. Roche
Wright State University School of Medicine
Yellow Springs, Ohio

PLENUM PRESS • NEW YORK AND LONDON

Library of Congress Cataloging in Publication Data

Roche, Alex F., 1921—
    Manual of physical status and performance in childhood.

    Author's names in reverse order in v. 2.
    Includes bibliographies and indexes.
    Contents: v. 1. Physical status (2 v.)—v. 2. Physical performance.
    1. Children—North America—Growth. 2. Children—North America—Anthropometry. I. Malina, Robert M. II. Title. (DNLM: 1. Child Development—Tables. 2. Motor skills—In infancy and childhood—Tables. WS16 R637m)
R637m)

RJ131.R592 1982   612'.65'0973   82-16515
ISBN 0-306-41136-9 (v. 1)
ISBN 0-306-41137-7 (v. 2)

© 1983 Plenum Press, New York
A Division of Plenum Publishing Corporation
233 Spring Street, New York, N.Y. 10013

Printed in the United States of America

Dedicated to

WILTON MARION ("BILL") KROGMAN

A great teacher, leader, and scholar

PREFACE

The inspiration for the present work was the outstanding compilation of data published by "Bill" Krogman in 1941. At that time, he included data from all available sources. We could not match this achievement partly because of the tremendous recent increase in the volume of scientific material published. Therefore, we restricted the coverage to data published after 1940 for the age range birth to 18 years. Almost all the data included are for normal North American children; those for other children have been included only if the variables are not covered adequately in reports for North American children.

Great pains have been taken to ensure that the tables and references are accurate but some errors may be present. Sample sizes, ranges, correlations, and results of significance tests have been omitted when making these extracts from published tables and data have been omitted if the number in an age group is less than 10. The bibliography includes notes regarding the size and nature of the samples studied and the variables for which data are included in this volume.

The captions refer to "U.S. children" or "Canadian children" only if the data are for a nationally representative sample. Captions that refer to the children of a state, province or city do not infer that a representative sample was examined. Footnotes give the source of the data and some define unusual measurements. The format for the footnotes varies because of copyright requirements.

This volume contains data relating to performance, functional measurements, body composition, some aspects of maturation and a wide array of physiological variables. The data complement those in Volume 1 which contains tables relating to body size and proportions, measurements of the skeleton and of maturity and stature prediction. Some tables include many variables; the position of each of these tables is determined by the first variable listed.

We are most grateful for the assistance given by members of our advisory board (Claude Bouchard, Francis E. Johnston, Buford L. Nichols, Andrew K. Poznanski and G. Lawrence Rarick). Also we wish to express our gratitude to those who released their copyright privileges to allow the inclusion of published data. Unfortunately, copyright permission could not be obtained for the inclusion of some material published by Harper and Row. Financial demands made it impossible to include some other data published by The American Dietetic Association, The American Society for Clinical Nutrition, Prentice-Hall, Inc., W. B. Saunders Co., and some publications of the Williams and Wilkins Company. Almost all authors of published work have been extremely helpful. References to the material that had to be excluded from this volume for the above reasons are listed after the tables, almost as a *salon des refusés*. We, of course, did not constitute the jury.

The preparation of this volume, and of Volume 1, has been a large task. It could not have been accomplished without the capable and enthusiastic help given by Kathleen Blamey, Anna Gregor, Nancy Harvey, Joan Hunter, Betty Mullins, Linda Shook and Louise Wechsler of The Fels Research Institute, and Peter Buschang of The University of Texas at Austin. We are most appreciative also for the advice and technical help provided by Mr. James Busis and his associates at the Plenum Publishing Corporation. It is hoped that this compilation will prove useful to physical educators, pediatricians, nutritionists, human biologists, physical anthropologists, and design engineers.

Robert M. Malina
Alex F. Roche

CONTENTS

VOLUME 2

INTRODUCTION

The first volume of this work considers measurements of the body as a whole (e.g., stature) and specific body parts (e.g., limb lengths, tooth dimensions, etc.), and indicators of maturity. This second volume presents complementary data dealing with functional measurements, body composition, organ dimensions, physiological data, and physical performance, in addition to reference findings for other less easily grouped developmental characteristics.

Reference data relevant to anthropometric considerations in the design of workspace, clothing, furniture and toys are included in this volume. Subject comfort, safety, and function are of basic importance, yet a broad range of normal variability must be accommodated (Burdi et al., 1969). For example, seating for elementary school children must be wide enough and long enough to accommodate larger children in the age group but not uncomfortable for a small child. To solve such problems, designers and manufacturers need data on anthropometric dimensions. The data provided here are both static, i.e., measurements made with the subject in a fixed, standardized position, and functional or dynamic, i.e., measurements made while the subject is in the position required for a specific task or while the body is in motion (e.g., functional arm or leg reach). Measurement errors are greater for dynamic dimensions, and proper interpretation of the data is dependent upon close attention to the precise methodology used (Damon et al., 1966).

Data concerning the location of the center of gravity are included among the data for functional measurements. The center of gravity is important in maintaining balance and in the design of some equipment, e.g., tricycles. Cooper and Glassow (1968) offer a good discussion of locating the center of gravity in man and the application of the concepts involved to human movement.

Measurement of body composition is more complex than the measurement of body size and dimensions. Although studies of stature and weight provide information that is valuable in understanding growth processes and in many practical applications, these external dimensions represent many tissues. Stature is the sum of a large number of bone lengths, plus the intervening cartilages less an amount due to the fact that the long axes of the bones that contribute to stature are not in the same straight line. Similarly, body weight is the sum of the weights of muscle, fat, the skeleton and numerous organs. These global measurements are important because they are highly reliable, they can be made easily, and excellent reference data are available. However, the information they provide about body composition is very incomplete. Hence, much effort has been expended to quantify body composition _in vivo_, i.e., in the intact living individual (Brožek and Henschel, 1961; Garn, 1961, 1963; Brožek, 1963, 1965; Bergner and Lushbaugh, 1967; National Academy of Sciences, 1968; Malina, 1969, 1980; Forbes, 1978; Lohman, 1981).

The biochemical analysis of body composition is obviously limited to cadavers (carcasses) and deals with composition _in vitro_, i.e., in glass. Usually, in these studies the body is reduced to four basic chemical constituents: water, protein, mineral and fat (Brožek, 1966). This has been done in only a few human cadavers; consequently, the data are inadequate statistically. Nevertheless, they provide the benchmarks for estimating body composition _in vivo_, since all other methods are indirect and their validity must be established by comparison with _in vitro_ studies in man or animals. Direct chemical analyses of the human body, in conjunction with indirect measurements, have led to development of the concept of the "reference man," i.e., an average individual free of disease. The

1

most widely used "reference man" is the young adult reference male; a male "reference infant" at birth is available also (Brožek et al., 1963; Fomon, 1966). There are no corresponding reference data for females at any age or for the chemical composition of the male body between infancy and adulthood.

Indirect estimates of body composition are usually based upon a two compartment model, i.e., body weight is partitioned into a fat component and a lean component. The weight of one component is measured indirectly, while the weight of the other is derived by subtraction from total body weight. However, definition of the lean component varies. It is referred to as either lean body mass (LBM), lean body weight (LBW), fat free mass (FFM) or fat free weight (FFW). Some treat these terms as synonymous, but there are basic differences between them. LBM and LBW are in vivo concepts, while FFM and FFW are based on in vitro concepts. The two differ in that essential lipids, variously estimated from 2 to 10% of FFW, are included with LBM and LBW but not with FFM or FFW (Behnke, 1961; Brožek, 1961).

The two compartment model describes body composition in an holistic or gross manner. It provides little information about specific tissue components except fat. For example, LBM fails to distinguish between muscle, bone and various viscera. Additionally, this approach does not provide information about the regional distribution of the compositional components. The model does not distinguish between two individuals of the same weight, each of whom has 10 kg of body fat, in one of whom most of the fat is deposited on the trunk while in the other most is on the extremities. Such differences are important relative to specific compositional changes, sites of change, the genesis of sex differences during growth and development, and perhaps later disease outcomes.

The three most common methods for estimating body composition within the two compartment model are densitometry (body density by underwater weighing), gamma ray spectrometry (potassium^{40}) and hydrometry (total body water). These procedures, their reproducibility and the errors in the body composition estimates have been described in considerable detail (Malina, 1980).

Tissue-specific and regional approaches to body composition include skinfold thicknesses with "constant-pressure" calipers, creatinine excretion as an estimate of skeletal muscle mass, radiographic widths and areas of bone, muscle and fat, ultrasonic impulses as estimates of subcutaneous fat thickness, and photon absorptiometric scans as estimates of bone mineral. The measurement of the triceps skinfold is often used, in combination with the relaxed arm circumference at the same level, to estimate the width, circumference or area of arm muscle. This indirect measurement of arm muscle is interpreted as a guide to total body muscle or to total lean body mass. The underlying assumptions involved, the methods of measurement, their reproducibility, and the associated errors of the estimates have been described (Malina, 1980).

The dimensions of specific organs and levels of physiological function are related to estimates of body composition. Organ size, for example, has been measured directly in the living, e.g., tooth size, and in dissection studies; or indirectly with radiographic or ultrasonic procedures or biochemical tests. More complicated methods utilizing radioactive substances provide additional data.

Not all physiological functions can be covered in a single data manual. A selection has been made based on our view of the utility of the measure and the nature of the samples from which data have been reported. Some of the physiological data result from rather routine measurements, e.g., blood pressure, electrocardiographs, while some result from more specialized procedures, e.g., endocrine assays, and the measurement of maximal oxygen consumption. Measurement procedures, quality control, and underlying assumptions are described in many reports. For example, Consolazio et al. (1963) and McHardy et al. (1967) present the measurement of respiratory functions including the estimation of the volume and content of respiratory gases, the analysis of blood gases, and the computation of metabolic balance. Consolazio et al. (1963) also consider physical fitness testing. Weiner and Lourie (1969) describe many physiological measurements that can be applied in field studies.

Other works are more specific, for example, Cotes (1968) provides a detailed discussion of the assessment of respiratory function while Polgar and Promadhat (1971) discuss the measurement of respiratory function in children.

Data relative to circulating hormones have been included. In this rapidly developing area, new techniques have become available and the complex interrelationships among hormones and nutrition are being actively studied. A comprehensive account is presented by Gardner (1969) while much recent material can be found in La Cauza and Root (1980). These works provide much of the background necessary to interpret data but, in addition, reported values for an individual must always be compared with the reported distributions of values from the same laboratory.

In the hematology area, data are presented for cell size and number and for hemoglobin concentration. These are derived from well-established techniques that are described in many sources such as Davidsohn and Henry (1974). A selection had to be made among the many aspects of blood chemistry reported. Emphasis has been placed on lipids and lipoproteins because of their health-related significance and because better techniques are being applied now than was the case a decade or more ago. The basic reference to this methodology is the Manual of Laboratory Operations (1974) prepared for use in the U.S. Lipid Research Centers.

Age at menarche is the most commonly reported developmental milestone in relation to sexual maturation. Much recent data are based upon the status quo method; this involves asking girls, within the appropriate age range, whether they have or have not reached menarche and applying probit analysis to the recorded data (Finney, 1971; Marubini, 1980). Most earlier data were obtained with the retrospective method, which is limited in accuracy because of recall errors and a tendency to round to the nearest whole year of age. Ratings of pubic hair development in each sex, breast development in girls and genital development in boys are commonly used to grade sexual maturation. The development of these secondary sex characteristics has been summarized in five-stage scales for each trait. The ratings are made from standardized photographs or direct inspection during clinical examinations. The scales used most commonly in recent studies for pubic hair, breast and genital development are those of Tanner (1962). It should be noted that the development of secondary sex characteristics is a continuous process; the division to stages is arbitrary, but convenient.

The measurement of physical performance is perhaps as varied as any set of procedures considered thus far. Physical performance is a general concept that can be viewed in many ways. For convenience, a primary twofold subdivision to motor and organic components of performance is used. The motor component relates to the development and performance of motor or movement skills. The organic component relates to energy production, work output, and cardiovascular fitness. The two subdivisions are not mutually exclusive. Two additional components of physical performance should be considered also. Muscular strength is common to both the motor and organic components of performance, while flexibility, not strictly a measure of physical performance, significantly influences both motor and organic performance.

The motor domain can be divided into motor development and motor performance. The former is the process through which a child develops and/or learns fundamental movement patterns and skills (e.g., walk, run, jump). Motor development is oriented to documenting the sequence of changes leading to the attainment of "mature" movement patterns. Motor development during the first few years of life has been documented using scales such as those of Bayley (1935, 1969), Gesell and Amatruda (1947), Knobloch and Pasamanick (1974) and Frankenburg and Dodds (1967). Wickstrom (1977) has consolidated information on the development and refinement of running, jumping, throwing, catching, striking and kicking movement patterns. Methods used in the study of motor development are largely observational, although cinematographic procedures permit finer examination of the component movements, e.g., hip rotation, arm action and leg action in throwing, and specific mechanical elements, e.g., angle of take-off in a jump, lengths of lever arms. Because of their complex nature, cinematographic studies are usually limited to small samples.

In contrast to motor development, motor performance is oriented to production of a specific activity, e.g., the distance jumped, the time elapsed. Thus motor performance is viewed in terms of particular tasks amenable to rather precise measurements and performed under specified conditions. Despite considerable inter- and intra-individual variability in motor performance, by five or six years of age children tend to perform consistently enough for group comparisons on tests that are extensions of fundamental motor skills, e.g., running, jumping. Between-day and within-day reliabilities of motor performance vary from task to task. Children tend to perform more consistently on motor items requiring an all-out effort, e.g., throw for distance, standing long jump. Performance on items which require a carefully directed movement of either the entire body or body part, or the movement of some object toward a specific goal, are generally more variable, e.g., throw for accuracy, cable jump. Reliability data for a variety of motor tasks are provided by several authors (Espenschade, 1940; Seils, 1951; Glassow and Kruse, 1960; Glassow et al., 1965; Hanson, 1965; Keogh, 1965, 1969; Klesius, 1968; Malina, 1968a,b; Rarick et al., 1976).

Maximal aerobic power is generally considered the best single indicator of cardiorespiratory fitness or the efficiency of the oxygen transporting system. Direct measurement of aerobic power is made during, and/or towards the end of, maximal or exhausting physical work. Treadmill running, cycling on an ergometer and step tests are the most commonly used modes of exercise. Of course, it is necessary that the subject performs work to the maximum of his or her capacity. Procedures for testing maximal aerobic power have been described in detail (Consolazio et al., 1963; Larson, 1974; Bouchard et al., 1975; Åstrand and Rodahl, 1977; Shephard, 1978).

Repeated determinations of maximal oxygen uptake on the same children are quite reliable (r about +.7 to +.8; Wilmore and Sigerseth, 1967; Shephard et al., 1969; Katch et al., 1973; Rowell, 1974; Cunningham et al., 1977). However, there are some problems in testing maximal aerobic power in children. One primary criterion for the establishment of maximal effort in a subject is a plateau in oxygen uptake. Not all children reach such a plateau with an increase in work load; some stop their physical effort before reaching their maximal oxygen uptake.

Submaximal work testing has also received attention as an indicator of cardiorespiratory performance. Oxygen uptake at submaximal work levels (e.g., at a heart rate of 130 beats per minute) can be measured directly. Also oxygen uptake can be estimated indirectly from heart rate during work at a specified submaximal work load. These estimates are based upon the linear relationship between work load and measured oxygen consumption within a wide range of submaximal intensities. The relationship is not linear, however, at higher work loads (Åstrand and Rodahl, 1977).

Another method for measuring the capacity of an individual for prolonged work was developed by Sjöstrand (1947) and modified by Wahlund (1948). The test measures the work load that produces a steady heart rate of 170 beats/min during exercise on a bicycle ergometer. Hence, it is ordinarily indicated as PWC_{170}. Test-retest reliability of PWC_{170} testing in children is quite high (r>.90; Watson and O'Donovan, 1976).

A variety of distance running events are also used to indicate organic performance or cardiorespiratory fitness. The tests take two forms: (1) the time required to run a specified distance, and (2) the distance covered in a specified time. Although motivation is important in distance running, the tests are highly reliable in children and youth (Askew, 1966; Willgoose et al., 1961; Doolittle and Bigbee, 1968; Klesius, 1968; Maksud and Coutts, 1971).

Muscular strength is common to motor performance and the functional capacity of the cardiorespiratory system. Strength is the capacity of an individual to exert force against an external resistance. Studies of muscular strength during childhood and adolescence have generally measured static strength, i.e., force exerted without change in muscle length. Pull-ups, sit-ups, bar-dips and the flexed arm hang are common indicators of dynamic strength, i.e., shortening and lengthening of muscles (Asmussen, 1968; Clarke, 1967). Jumping and throwing tests, in addition to being measures of motor performance, also indicate explosive strength, i.e., ability of muscles to release maximal force in the shortest possible time.

Static strength tests used in growth studies relate to a variety of muscle groups, using primarily dynamometers and cable tensiometers. The most common strength measurement is gripping strength. Reliability of strength tests is generally good during childhood and adolescence (Metheny, 1941; Jones, 1949; Fleishman, 1964; Keogh, 1965; Clarke, 1966, 1971; Malina, 1968a; Simons et al., 1969) and, with proper motivation, strength tests can be used reliably during preschool ages (Metheny, 1940, 1941; Krogman, 1972; Parískova et al., 1977). Reliabilities for pull-ups, sit-ups, bar-dips and the flexed arm hang are more variable. Pull-ups have good reliability among boys 10 years of age and older (Clarke, 1967; Klesius, 1968); the test is less reliable in girls and young children. As a result, modified pull-up tests and the flexed arm hang are often used for boys under 10 years and for girls at all ages. Nevertheless, young children, especially at 5, 6 and 7 years of age, have difficulty with such dynamic strength tests.

Flexibility, like muscular strength, is common to both the motor and organic components of physical performance. Flexibility is the range of motion of different body segments at various joints. Commonly it is viewed as the maximum range of joint motion, and is joint-specific. Goniometers and flexometers are used to measure the range of motion at different joints (Leighton, 1955, 1960; Clarke, 1967; Sigerseth, 1970).

In any compilation of data for a variety of dimensions and functions, there are several that are difficult to classify, e.g., bedwetting, thumbsucking, meal preferences. Some sets of data for such variables are included in the present volume.

Robert M. Malina
Alex F. Roche

References

Askew, N. R., 1966, Reliability of the 600-yard run-walk test at the secondary school level, Res. Q. Am. Assoc. Health Phys. Educ., 37:451.

Asmussen, E., 1968, The neuromuscular system and exercise, in: "Exercise Physiology," H. B. Falls, ed., Academic Press, New York.

Åstrand, P. O., and Rodahl, K., 1977, "Textbook of Work Physiology," 2nd edition, McGraw-Hill, New York.

Bayley, N., 1935, The development of motor abilities during the first three years, Monogr. Soc. Res. Child Dev., 1 (1).

Bayley, N., 1969, "Manual for the Bayley Scales of Infant Development," Psychological Corporation, Berkeley.

Behnke, A. R., 1961, Comment on the determination of whole body density and a resumé of body composition data, in: "Techniques for Measuring Body Composition," J. Brožek and A. Henschel, eds., National Academy of Science-National Research Council, Washington.

Bergner, P. E., and Lushbaugh, C. C. (eds.), 1967, "Compartments, Pools, and Spaces in Medical Physiology," U.S. Atomic Energy Commission, Washington.

Bouchard, C., Carrier, R., Boulay, M., Poirier, M. C. T., and Dulac, S., 1975, "Le Développement du Système de Transport de l'Oxygène chez les Jeunes Adultes," Pélican, Québec.

Brožek, J., 1961, Editor's comment, in: "Techniques for Measuring Body Composition," J. Brožek and A. Henschel, eds., National Academy of Science-National Research Council, Washington.

Brožek, J. (ed.), 1963, Body composition, Ann. NY Acad. Sci., 110:1.

Brožek, J. (ed.), 1965, "Human Body Composition," Pergamon Press, Oxford.

Brožek, J., 1966, Body composition: Models and estimation equations, Am. J. Phys. Anthropol., 24:239.

Brožek, J., and Henschel, A. (eds.), 1961, "Techniques for Measuring Body Composition," National Academy of Science-National Research Council, Washington.

Brožek, J., Grande, F., Anderson, J. T., and Keys, A., 1963, Densitometric analysis of body composition: revision of some quantitative assumptions, Ann. NY Acad. Sci., 110:113.

Burdi, A. R., Huelke, D. F., Snyder, R. G., and Lowery, G. H., 1969, Infants and children in the adult world of automobile design: pediatric and anatomical considerations for design of child restraints, J. Biomech., 2:267.

Clarke, H. H., 1966, "Muscular Strength and Endurance in Man," Prentice-Hall, Englewood Cliffs.

Clarke, H. H., 1967, "Application of Measurement to Health and Physical Education," 4th edition, Prentice-Hall, Englewood Cliffs.

Clarke, H. H., 1971, "Physical and Motor Tests in the Medford Boys' Growth Study," Prentice-Hall, Englewood Cliffs.

Consolazio, C. F., Johnson, R. E., and Pecora, L. J., 1963, "Physiological Measurements of Metabolic Functions in Man," McGraw-Hill, New York.

Cooper, J. M., and Glassow, R. B., 1968, "Kinesiology," 2nd edition, Mosby, St. Louis.

Cotes, J. E., 1968, "Lung Function: Assessment and Application in Medicine," 2nd edition, Blackwell, Oxford.

Cunningham, D. A., Van Waterschoot, B. M., Paterson, D. H., Lefcoe, M., and Sangal, S. P., 1977, Reliability and reproducibility of maximal oxygen uptake measurement in children, Med. Sci. Sports, 9:104.

Damon, A., Stoudt, H. W., and McFarland, R. A., 1966, "The Human Body in Equipment Design," Harvard University Press, Cambridge.

Davidsohn, I., and Henry, J. B. (eds.), 1974, "Todd-Sanford Clinical Diagnosis by Laboratory Methods," 15th edition, W. B. Saunders Co., Philadelphia.

Doolittle, T. L., and Bigbee, R., 1968, The twelve-minute run-walk: a test of cardiorespiratory fitness of adolescent boys, Res. Q. Am. Assoc. Health Phys. Educ., 39:491.

Espenschade, A. S., 1940, Motor performance in adolescence, Monogr. Soc. Res. Child Dev., 5(1).

Finney, D. J., 1971, "Probit Analysis," The University Press, Cambridge.

Fleishman, E. A., 1964, "The Structure and Measurement of Physical Fitness," Prentice-Hall, Englewood Cliffs.

Fomon, S. J., 1966, Body composition of the infant. Part I: The male "reference infant," in: "Human Development," F. Falkner, ed., W. B. Saunders Co., Philadelphia.

Forbes, G. B., 1978, Body composition in adolescence, in: "Human Growth, Volume 2," F. Falkner and J. M. Tanner, eds., Plenum, New York.

Frankenburg, W. K., and Dodds, J. B., 1967, The Denver Developmental Screening Test, J. Pediatr., 71:181.

Gardner, L. I. (ed.), 1969, "Endocrine and Genetic Diseases of Childhood," W. B. Saunders Co., Philadelphia.

Garn, S. M., 1961, Radiographic analysis of body composition, in: "Techniques for Measuring Body Composition," J. Brožek and A. Henschel, eds., National Academy of Science-National Research Council, Washington.

Garn, S. M., 1963, Human biology and research in body composition, Ann. NY Acad. Sci., 110:429.

Gesell, A., and Amatruda, C. S., 1947, "Developmental Diagnosis," 2nd edition, Harper and Row, New York.

Glassow, R. B., and Kruse, P., 1960, Motor performance of girls age 6 to 14 years, Res. Q. Am. Assoc. Health Phys. Educ., 31:426.

Glassow, R. B., Halverson, L. E., and Rarick, G. L., 1965, "Improvement of Motor Development and Physical Fitness in Elementary School Children," Cooperative Research Project, no. 696, University of Wisconsin, Dept. Phys. Educ. and Dance, Madison.

Hanson, M. R., 1965, "Motor Performance Testing of Elementary School Age Children," Ph.D. dissertation, University of Washington.

Jones, H. E., 1949, "Motor Performance and Growth," University of California Press, Berkeley.

Katch, F. I., McArdle, W. D., Czula, R., and Pechar, G. S., 1973, Maximal oxygen uptake, endurance running performance, and body composition in college women, Res. Q. Am. Assoc. Health Phys. Educ., 44:301.

Keogh, J., 1965, "Motor Performance of Elementary School Children," University of California, Dept. Phys. Educ., Los Angeles.

Keogh, J., 1969, "Change in Motor Performance During Early School Years," University of California, Dept. Phys. Educ., Los Angeles.

Klesius, S. E., 1968, Reliability of the AAHPER youth fitness test items and relative efficiency of the performance measures, Res. Q. Am. Assoc. Health Phys. Educ., 39:809.

Knobloch, H., and Pasamanick, B. (eds.), 1974, "Gesell and Amatruda's Developmental Diagnosis," 3rd edition, Harper and Row, New York.

Krogman, W. M., 1972, The Manual and Oral Strengths of American White and Negro Children,
Ages 3-6 Years, in Proc. 58th Mid-Winter Meeting, May 14-17, 1972, Chem. Specialties
Mfg. Assoc., New York.

La Cauza, C., and Root, A. W. (eds.), 1980, "Problems in Pediatric Endocrinology," Academic
Press, New York.

Larson, L. A. (ed.), 1974, "Fitness, Health, and Work Capacity: International Standards
for Assessment," Macmillan, New York.

Leighton, J. R., 1955, An instrument and technic for the measurement of range of joint
motion, Arch. Phys. Med. Rehabil., 36:571.

Leighton, J. R., 1960, On the significance of flexibility for physical educators, J. Health
Phys. Educ. Rec., 31:27.

Lohman, T. G., 1981, Skinfolds and body density and their relation to body fatness: a
review, Hum. Biol., 53:181.

Maksud, M. G., and Coutts, K. D., 1971, Application of the Cooper twelve-minute run-walk
test to young males, Res. Q. Am. Assoc. Health Phys. Educ., 42:54.

Malina, R. M., 1968a, "Growth, Maturation and Performance of Philadelphia Negro and White
Elementary School Children," Ph.D. dissertation, University of Pennsylvania.

Malina, R. M., 1968b, Reliability of different methods of scoring throwing accuracy, Res.
Q. Am. Assoc. Health Phys. Educ., 39:149.

Malina, R. M., 1969, Quantification of fat, muscle and bone in man, Clin. Orthop., 65:9.

Malina, R. M., 1980, The measurement of body composition, in: "Human Physical Growth and
Maturation: Methodologies and Factors," F. E. Johnston, A. F. Roche, and C. Susanne,
eds., Plenum Press, New York.

Marubini, E., 1980, General statistical considerations in analysis, in: "Human Physical
Growth and Maturation: Methodologies and Factors," F. E. Johnston, A. F. Roche, and
C. Susanne, eds., Plenum Press, New York.

McHardy, G. J. R., Shirling, D., and Passmore, R., 1967, "Basic Techniques in Human
Metabolism and Respiration," Blackwell, Oxford.

Metheny, E., 1940, Breathing capacity and grip strength in preschool children, University
of Iowa Studies in Child Welfare, 18(2).

Metheny, E., 1941, The present status of strength testing for children of elementary school
and preschool age, Res. Q. Am. Assoc. Health Phys. Educ., 12:115.

National Academy of Sciences, 1968, "Body Composition in Animals and Man," National Academy
of Sciences, Washington.

National Institutes of Health, 1974, "Manual of Laboratory Operations, Vol. 1, Lipid and
Lipoprotein Analysis," DHEW Publication (NIH) 75-628, U.S. Government Printing Office,
Washington.

Parízková, J., Cermak, J., and Horna, J., 1977, Sex differences in somatic and functional
characteristics of preschool children, Hum. Biol., 49:437.

Polgar, G., and Promadhat, V., 1971, "Pulmonary Function Testing in Children: Techniques
and Standards," W. B. Saunders Co., Philadelphia.

Rarick, G. L., Dobbins, D. A., and Broadhead, G. D., 1976, "The Motor Domain and Its
Correlates in Educationally Handicapped Children," Prentice-Hall, Englewood Cliffs.

Rowell, L. B., 1974, Human cardiovascular adjustments to exercise and thermal stress,
Physiol. Rev., 54:75.

Seils, L. G., 1951, The relationship between measures of physical growth and gross motor
performance of primary-grade school children, Res. Q. Am. Assoc. Health Phys. Educ.,
22:244.

Shephard, R. J., 1978, "Human Physiological Work Capacity," Cambridge University Press,
Cambridge.

Shephard, R. J., Allen, C., Bar-Or, O., Davies, C. T. M., Degre, S., Heldman, R., Ishii, K.,
Kaneko, M., Lacour, J. R., di Prampero, P. E., and Seliger, V., 1969, The working
capacity of Toronto schoolchildren. Part 1, Can. Med. Assoc. J., 100:560.

Sigerseth, P. O., 1970, Flexibility, in: "An Introduction to Measurement in Physical Education, Volume 4, Physical Fitness," H. J. Montoye, ed., Phi Epsilon Kappa Fraternity,
Indianapolis.

Simons, J., Beunen, G., Ostyn, M., Renson, R., Swalus, P., Van Gerven, D., and Willems, E., 1969, Construction d'une batterie de tests d'aptitude motrice pour garcons de 12 a 19 ans, par la methode de l'analyse factorielle. Kinanthropologie, 1:323.

Sjöstrand, T., 1947, Changes in respiratory organs of workmen at an ore smelting works, Acta Med. Scand., 198:687.

Tanner, J. M., 1962, "Growth at Adolescence," 2nd edition, Blackwell, Oxford.

Wahlund, H. G., 1948, Determination of the physical working capacity: a physiological and clinical study with special reference to standardization of cardio-pulmonary functional tests, Acta Med. Scand., Suppl. 215.

Watson, A. W. S., and O'Donovan, D. H., 1976, The physical working capacity of male adolescents in Ireland, Ir. J. Med. Sci., 145:383.

Weiner, J. S., and Lourie, J. A., 1969, "Human Biology: A Guide to Field Methods," F. A. Davis, Philadelphia.

Wickstrom, R. L., 1977, "Fundamental Motor Patterns," 2nd edition, Lea & Febiger, Philadelphia.

Willgoose, C. E., Askew, N. R., and Askew, M. P., 1961, Reliability of the 600-yard run-walk test at the junior high school level, Res. Q. Am. Assoc. Health Phys. Educ., 32: 264.

Wilmore, J. H., and Sigerseth, P. O., 1967, Physical work capacity of young girls, 7-13 years of age, J. Appl. Physiol., 22:923.

PERFORMANCE

TABLE 1

AGES (months) AT WHICH THE MOTOR ITEMS OF
THE BAYLEY SCALE OF INFANT DEVELOPMENT ARE PASSED

Item		Percentiles		
		50	5	95
1.	Lifts head when held at shoulder	0.1	--	--
2.	Postural adjustment when held at shoulder	0.1	--	--
3.	Lateral head movements	0.1	--	--
4.	Crawling movements	0.4	0.1	3.0
5.	Retains red ring	0.8	0.3	3.0
6.	Arm thrusts in play	0.8	0.3	2.0
7.	Leg thrusts in play	0.8	0.3	2.0
8.	Head erect: vertical	0.8	0.3	3.0
9.	Head erect and steady	1.6	0.7	4.0
10.	Lifts head: dorsal suspension	1.7	0.7	4.0
11.	Turns from side to back	1.8	0.7	5.0
12.	Elevates self by arms: prone	2.1	0.7	5.0
13.	Sits with support	2.3	1.0	5.0
14.	Holds head steady	2.5	1.0	5.0
15.	Hands predominantly open	2.7	0.7	6.0
16.	Cube: ulnar-palmar prehension	3.7	2.0	7.0
17.	Sits with slight support	3.8	2.0	6.0
18.	Head balanced	4.2	2.0	6.0
19.	Turns from back to side	4.4	2.0	7.0
20.	Effort to sit	4.8	3.0	8.0
21.	Cube: partial thumb opposition (radial-palmar)	4.9	4.0	8.0
22.	Pulls to sitting position	5.3	4.0	8.0
23.	Sits alone momentarily	5.3	4.0	8.0
24.	Unilateral reaching	5.4	4.0	8.0
25.	Attempts to secure pellet	5.6	4.0	8.0
26.	Rotates wrist	5.7	4.0	8.0
27.	Sits alone 30 seconds or more	6.0	5.0	8.0
28.	Rolls from back to stomach	6.4	4.0	10.0
29.	Sits alone, steadily	6.6	5.0	9.0
30.	Scoops pellet	6.8	5.0	9.0
31.	Sits alone, good coordination	6.9	5.0	10.0
32.	Cube: complete thumb opposition (radial-digital)	6.9	5.0	9.0
33.	Prewalking progession	7.1	5.0	11.0
34.	Early stepping movements	7.4	5.0	11.0
35.	Pellet: partial finger prehension (inferior pincer)	7.4	6.0	10.0
36.	Pulls to standing position	8.1	5.0	12.0
37.	Raises self to sitting position	8.3	6.0	11.0
38.	Stands up by furniture	8.6	6.0	12.0
39.	Combines spoons or cubes: midline	8.6	6.0	12.0
40.	Stepping movements	8.8	6.0	12.0
41.	Pellet: fine prehension (neat pincer)	8.9	7.0	12.0
42.	Walks with help	9.6	7.0	12.0
43.	Sits down	9.6	7.0	14.0

TABLE 1 (continued)

AGES (months) AT WHICH THE MOTOR ITEMS OF
THE BAYLEY SCALE OF INFANT DEVELOPMENT ARE PASSED

| Item | | Percentiles | | |
		50	5	95
44.	Pat-a-cake: midline skill	9.7	7.0	15.0
45.	Stands alone	11.0	9.0	16.0
46.	Walks alone	11.7	9.0	17.0
47.	Stands up I	12.6	9.0	18.0
48.	Throws ball	13.3	9.0	18.0
49.	Walks sideways	14.1	10.0	20.0
50.	Walks backward	14.6	11.0	20.0
51.	Stands on right foot with help	15.9	12.0	21.0
52.	Stands on left foot with help	16.1	12.0	23.0
53.	Walks up stairs with help	16.1	12.0	23.0
54.	Walks down stairs with help	16.4	13.0	23.0
55.	Tries to stand on walking board	17.8	13.0	26.0
56.	Walks with one foot on walking board	20.6	15.0	29.0
57.	Stands up II	21.9	11.0	30+
58.	Stands on left foot alone	22.7	15.0	30+
59.	Jumps off floor, both feet	23.4	17.0	30+
60.	Stands on right foot along	23.5	16.0	30+
61.	Walks on line, general direction	23.9	18.0	30+
62.	Walking board, stands with both feet	24.5	17.0	30+
63.	Jumps from bottom step	24.8	19.0	30+
64.	Walks up stairs alone: both feet on each step	25.1	18.0	30+
65.	Walks on tiptoe, few steps	25.7	16.0	30+
66.	Walks down stairs alone: both feet on each step	25.8	19.0	30+
67.	Walking board: attempts step	27.6	19.0	30+
68.	Walks backward, 10 feet	27.8	20.0	30+
69.	Jumps from second step	28.1	21.0	30+
70.	Distance jump: 4 to 14 inches	29.1	22.0	30+
71.	Stands up III	30+	22.0	30+
72.	Walks up stairs: alternating foot forward	30+	23.0	30+
73.	Walks on tiptoe, 10 feet	30+	20.0	30+
74.	Walking board: alternate steps part way	30+	24.0	30+
75.	Keeps feet on line, 10 feet	30+	23.0	30+
76.	Distance jump: 14 to 24 inches	30+	25.0	30+
77.	Jumps over string 2 inches high	30+	24.0	30+
78.	Distance jump: 24 to 34 inches	30+	28.0	30+
79.	Hops on one foot, 2 or more hops	30+	--	--
80.	Walks down stairs: alternating forward foot	30+	--	--
81.	Jumps over string 8 inches high	30+	28.0	30+

TABLE   2

POINT SCORES ON THE BAYLEY INFANT SCALES
OF MENTAL AND MOTOR DEVELOPMENT BY AGE FOR
INFANTS IN 10 METROPOLITAN CENTERS

Age	Mental Total Infants		Motor Total Infants	
(months)	Mean	S.D.	Mean	S.D.
1	13.83	4.57	6.40	2.03
2	22.78	6.21	9.69	2.28
3	33.56	8.55	12.75	2.72
4	43.10	8.03	15.16	3.07
5	56.13	8.32	20.08	3.52
6	69.61	9.00	25.71	4.64
7	74.87	7.06	29.58	4.93
8	80.74	6.35	35.04	5.41
9	84.30	5.94	37.69	4.35
10	90.66	6.02	41.31	3.73
11	94.46	5.68	43.70	2.92
12	99.65	6.86	45.41	3.77
13	103.68	6.13	46.21	5.15
14	107.39	6.93	48.42	3.23
15	110.68	7.47	49.18	3.18

(data from Bayley, 1965)

TABLE   3

BAYLEY SCALES OF MOTOR DEVELOPMENT: POINT SCORES
BY SEX FOR INFANTS IN 10 METROPOLITAN CENTERS

Age	Boys		Girls	
(months)	Mean	S.D.	Mean	S.D.
1	6.07	1.87	6.86	2.74
2	10.05	2.30	9.41	2.20
3	12.76	3.03	12.81	2.41
4	15.60	3.22	14.72	2.86
5	19.67	3.34	20.59	3.58
6	26.27	4.52	24.88	4.75
7	29.34	4.75	29.69	5.01
8	36.57	4.61	34.14	5.63
9	38.14	3.96	37.64	4.65
10	40.25	3.98	41.28	3.63
11	43.02	3.05	43.79	2.72
12	45.65	3.16	44.63	4.29
13	46.67	2.99	46.38	3.35
14	48.37	3.60	48.36	2.78
15	49.59	3.23	49.37	3.09

(data from Bayley, 1965)

TABLE 4

POINT SCORES OF THE BAYLEY INFANT SCALE
OF MOTOR DEVELOPMENT FOR FIRST-BORN AND
LATER-BORN BABIES IN 10 METROPOLITAN CENTERS

Age (months)	First-Born		Later-Born	
	Mean	S.D.	Mean	S.D.
1	6.83	1.99	5.98	2.18
2	9.65	1.88	10.25	2.22
3	12.52	2.68	12.62	3.04
4	14.74	3.21	15.00	3.07
5	20.50	3.22	19.84	3.84
6	26.81	3.29	25.65	5.12
7	30.55	5.17	28.50	4.65
8	35.14	5.65	35.00	5.44
9	37.92	3.43	37.82	4.39
10	41.37	3.72	40.46	3.99
11	43.19	2.93	43.73	2.55
12	46.42	2.59	44.19	4.36
13	47.09	2.57	46.41	3.30
14	49.31	2.44	47.61	3.28
15	50.12	1.99	48.59	3.56

(data from Bayley, 1965)

TABLE 5

POINT SCORES OF THE BAYLEY INFANT SCALE OF MOTOR DEVELOPMENT
FOR BABIES BY RACE IN 10 METROPOLITAN CENTERS

Age (months)	White Babies		Negro Babies		Puerto Rican Babies	
	Mean	S.D.	Mean	S.D.	Mean	S.D.
1	6.34	2.03	6.39	2.98	7.00	1.00
2	9.31	2.20	9.89	2.22	11.17	2.40
3	12.12	2.57	13.39	2.82	13.20	1.79
4	14.57	3.20	16.29	2.92	14.88	1.36
5	18.83	3.32	21.25	3.46	20.75	1.50
6	25.73	4.40	25.76	4.78	27.80	3.77
7	28.47	4.88	30.46	4.64	30.33	5.50
8	34.41	5.27	35.67	5.02	39.00	3.00
9	37.13	4.06	38.95	4.17	37.25	8.18
10	40.11	3.62	41.32	3.93	42.80	1.30
11	42.84	3.06	44.00	2.99	44.13	1.36
12	44.22	3.16	45.88	4.42	46.20	1.64
13	46.45	6.49	47.08	3.27	46.14	2.41
14	48.33	3.01	48.68	4.03	49.00	1.73
15	49.35	3.08	48.39	3.42	50.00	.82

(data from Bayley, 1965)

TABLE 6

PERFORMANCE IN THE DENVER DEVELOPMENTAL SCREENING TEST
BY CHILDREN IN DENVER (CO) AGES AT
WHICH GIVEN PERCENTAGES OF THE POPULATION PASS

Item	25 percent	50 percent	75 percent	90 percent
Gross motor				
Prone, lifts head	--	--	--	0.7 mo.
Prone, head up 45 degrees	--	--	1.9 mo.	2.6 mo.
Prone, head up 90 degrees	1.3 mo.	2.2 mo.	2.6 mo.	3.2 mo.
Prone, chest up, arm support	2.0 mo.	3.0 mo.	3.5 mo.	4.3 mo.
Sits-head steady	1.5 mo.	2.9 mo.	3.6 mo.	4.2 mo.
Rolls over	2.3 mo.	2.8 mo.	3.8 mo.	4.7 mo.
Bears some weight on legs	3.4 mo.	4.2 mo.	5.0 mo.	6.3 mo.
Pulls to sit, no head lag	3.0 mo.	4.2 mo.	5.2 mo.	7.7 mo.
Sits without support	4.8 mo.	5.5 mo.	6.5 mo.	7.8 mo.
Stands holding on	5.0 mo.	5.8 mo.	8.5 mo.	10.0 mo.
Pulls self to stand	6.0 mo.	7.6 mo.	9.5 mo.	10.0 mo.
Gets to sitting	6.1 mo.	7.6 mo.	9.3 mo.	11.0 mo.
Stands momentarily	9.1 mo.	9.8 mo.	12.1 mo.	13.0 mo.
Walks holding on furniture	7.3 mo.	9.2 mo.	10.2 mo.	12.7 mo.
Stands alone well	9.8 mo.	11.5 mo.	13.2 mo.	13.9 mo.
Stoops and recovers	10.4 mo.	11.6 mo.	13.2 mo.	14.3 mo.
Walks well	11.3 mo.	12.1 mo.	13.5 mo.	14.3 mo.
Walks backwards	12.4 mo.	14.3 mo.	18.2 mo.	21.5 mo.
Walks up steps	14.0 mo.	17.0 mo.	21.0 mo.	22.0 mo.
Kicks ball forward	15.0 mo.	20.0 mo.	22.3 mo.	2.0 yr.
Throws ball overhand	14.9 mo.	19.8 mo.	22.8 mo.	2.6 yr.
Balances on 1 foot 1 second	21.7 mo.	2.5 yr.	3.0 yr.	3.2 yr.
Jumps in place	20.5 mo.	22.3 mo.	2.5 yr.	3.0 yr.
Pedals trike	21.0 mo.	23.9 mo.	2.8 yr.	3.0 yr.
Broad jump	2.0 yr.	2.8 yr.	3.0 yr.	3.2 yr.
Balances on 1 foot 5 seconds	2.6 yr.	3.2 yr.	3.9 yr.	4.3 yr.
Balances on 1 foot 10 seconds	3.0 yr.	4.5 yr.	5.0 yr.	5.9 yr.
Hops on 1 foot	3.0 yr.	3.4 yr.	4.0 yr.	4.9 yr.
Catches bounced ball	3.5 yr.	3.9 yr.	4.9 yr.	5.5 yr.
Heel-to-toe walk	3.3 yr.	3.6 yr.	4.2 yr.	5.0 yr.
Backward heel-toe	3.9 yr.	4.7 yr.	5.6 yr.	6.3 yr.
Fine motor-adaptive				
Follows to midline	--	--	0.7 mo.	1.3 mo.
Symmetrical movements	--	--	--	--
Follows past midline	--	1.3 mo.	1.9 mo.	2.5 mo.
Follows 180 degrees	1.8 mo.	2.4 mo.	3.2 mo.	4.0 mo.
Hands together	1.3 mo.	2.2 mo.	3.0 mo.	3.7 mo.
Grasps rattle	2.5 mo.	3.3 mo.	3.9 mo.	4.2 mo.
Regards raisin	2.5 mo.	3.3 mo.	4.2 mo.	5.0 mo.
Reaches for object	2.9 mo.	3.6 mo.	4.5 mo.	5.0 mo.

TABLE 6   (continued)

PERFORMANCE IN THE DENVER DEVELOPMENTAL SCREENING TEST
BY CHILDREN IN DENVER (CO) AGES AT
WHICH GIVEN PERCENTAGES OF THE POPULATION PASS

Item	25 percent	50 percent	75 percent	90 percent
Sits, looks for yarn	4.8 mo.	5.6 mo.	6.9 mo.	7.5 mo.
Sits, takes 2 cubes	5.1 mo.	6.1 mo.	7.0 mo.	7.5 mo.
Rakes raisin, attains	5.0 mo.	5.6 mo.	6.2 mo.	7.8 mo.
Transfers cube hand to hand	4.7 mo.	5.6 mo.	6.6 mo.	7.5 mo.
Bangs 2 cubes held in hands	7.0 mo.	8.4 mo.	9.8 mo.	12.3 mo.
Thumb-finger grasp	7.1 mo.	8.3 mo.	9.1 mo.	10.6 mo.
Neat pincer grasp of raisin	9.4 mo.	10.7 mo.	12.3 mo.	14.7 mo.
Scribbles spontaneously	11.9 mo.	13.3 mo.	15.8 mo.	2.1 yr.
Tower of 2 cubes	12.1 mo.	14.1 mo.	17.0 mo.	20.0 mo.
Dumps raisin from bottle-spontaneous	12.7 mo.	13.4 mo.	16.4 mo.	2.0 yr.
Dumps raisin from bottle-demonstrative	13.7 mo.	14.8 mo.	2.1 yr.	3.0 yr.
Tower of 4 cubes	15.5 mo.	17.9 mo.	20.5 mo.	2.2 yr.
Imitates vertical line within 30 degrees	18.4 mo.	21.7 mo.	2.2 yr.	3.0 yr.
Tower of 8 cubes	21.0 mo.	23.8 mo.	2.4 yr.	3.4 yr.
Copies circle	2.2 yr.	2.6 yr.	2.9 yr.	3.3 yr.
Imitates bridge	2.3 yr.	2.7 yr.	3.1 yr.	3.4 yr.
Copies +	2.9 yr.	3.4 yr.	3.8 yr.	4.4 yr.
Copies square	4.1 yr.	4.7 yr.	5.5 yr.	6.0 yr.
Imitates square, demonstrative	3.5 yr.	4.1 yr.	4.7 yr.	5.7 yr.
Draws man, 3 parts	3.3 yr.	4.0 yr.	4.7 yr.	5.2 yr.
Draws man, 6 parts	4.6 yr.	4.8 yr.	5.4 yr.	6.0 yr.
Picks longer line, 3 of 3	2.6 yr.	2.9 yr.	3.4 yr.	4.4 yr.

(data from Frankenburg and Dodds, 1967, Journal of Pediatrics 71:181-191)

TABLE 7

AGE (weeks) OF ACHIEVEMENT OF TWELVE BEHAVIOR PATTERNS
IN THE NEUROMUSCULAR DEVELOPMENT OF NEGRO INFANTS
IN WASHINGTON (D.C.)

Neuromuscular Steps	"Private" Infants		"Clinic" Infants		Combined "Private & Clinic"	
	Mean	S.D.	Mean	S.D.	Mean	S.D.
Smile	6.60	3.07	5.27	2.17	5.88	2.70
Vocal	7.38	3.08	7.06	3.24	7.19	3.18
Head Control	12.55	3.61	11.76	3.71	12.12	3.67
Hand Control	17.59	3.65	15.20	4.06	16.32	4.06
Roll	22.91	4.81	17.99	4.89	20.05	5.42
Sit	23.43	3.46	22.52	4.43	24.51	4.42
Crawl	29.98	6.07	27.25	7.28	28.41	6.90
Prehension	33.20	5.76	31.48	5.96	32.21	5.97
Pull up	34.30	5.30	33.09	6.82	33.69	6.16
Walk with support	37.04	5.77	37.97	6.47	37.48	6.10
Stand alone	44.41	7.14	43.97	6.35	44.18	6.77
Walk alone	49.70	6.60	50.23	6.56	49.95	6.62

(data from Scott, Ferguson, Jenkins, et al., 1955, Pediatrics 16:24-30. Copyright
American Academy of Pediatrics 1955)

TABLE 8

MEAN AGE (weeks) OF ACHIEVEMENT
FOR FOUR REPRESENTATIVE BEHAVIORAL PATTERNS IN NEGRO INFANTS
IN WASHINGTON (D.C.)

Sex	Head Control		Sit		Pull Up		Walk Alone	
	Mean	S.D.	Mean	S.D.	Mean	S.D.	Mean	S.D.
Male	12.2	3.35	23.8	4.64	32.9	6.46	50.2	7.0
Female	12.0	3.66	23.8	4.50	33.8	6.39	49.8	6.6

(data from Scott, Ferguson, Jenkins, et al., 1955, Pediatrics 16:24-30. Copyright
American Academy of Pediatrics 1955)

TABLE 9

FLEXIBILITY (degrees) IN GIRLS

	Age (years)				
Percentiles	6	9	12	15	18

NECK FLEXION-EXTENSION

95	153	145	167	153	145
70	130	135	145	135	125
50	123	128	134	123	120
30	118	124	126	116	114
5	103	107	107	100	97

NECK ROTATION

95	184	191	190	184	181
70	175	183	177	170	173
50	170	175	172	163	167
30	164	168	168	158	155
5	151	152	152	131	135

LEFT ARM FLEXION-EXTENSION

95	248	242	241	230	232
70	240	227	227	221	221
50	231	223	219	213	214
30	223	220	211	208	206
5	206	210	192	196	195

RIGHT ARM FLEXION-EXTENSION

95	244	239	236	237	232
70	236	224	224	218	221
50	231	220	214	212	214
30	225	214	208	206	207
5	202	203	197	195	192

LEFT FOREARM FLEXION-EXTENSION

95	172	171	174	169	165
70	162	165	163	162	157
50	157	162	160	157	153
30	151	158	155	152	150
5	146	148	142	140	140

TABLE 9 (continued)

FLEXIBILITY (degrees) IN GIRLS

	Age (years)				
Percentiles	6	9	12	15	18

RIGHT FOREARM FLEXION-EXTENSION

Percentiles	6	9	12	15	18
95	168	168	173	167	164
70	160	161	162	158	156
50	156	158	158	155	152
30	152	155	154	152	148
5	146	143	143	138	138

LEFT HAND FLEXION-EXTENSION

Percentiles	6	9	12	15	18
95	168	176	181	177	175
70	153	162	163	161	159
50	147	156	157	155	154
30	142	150	151	151	148
5	130	135	141	133	131

RIGHT HAND FLEXION-EXTENSION

Percentiles	6	9	12	15	18
95	167	175	180	177	170
70	154	158	163	160	159
50	148	152	155	152	151
30	141	146	147	145	147
5	127	128	133	126	132

LEFT THIGH ABDUCTION-ADDUCTION

Percentiles	6	9	12	15	18
95	64	67	66	66	63
70	53	59	58	57	56
50	50	56	53	53	52
30	45	53	50	50	49
5	38	47	42	41	36

RIGHT THIGH ABDUCTION-ADDUCTION

Percentiles	6	9	12	15	18
95	64	67	68	69	66
70	51	59	59	56	58
50	45	53	54	51	54
30	42	44	47	46	48
5	37	37	40	40	40

TABLE   9   (continued)

FLEXIBILITY (degrees) IN GIRLS

Percentiles	Age (years)				
	6	9	12	15	18

LEFT LEG FLEXION-EXTENSION

95	151	143	139	135	128
70	135	134	131	127	121
50	130	128	126	123	117
30	125	123	122	117	113
5	113	113	110	110	101

RIGHT LEG FLEXION-EXTENSION

95	145	143	140	134	132
70	138	135	128	126	121
50	132	128	124	122	116
30	125	122	122	117	112
5	116	111	117	111	103

LEFT FOOT FLEXION-EXTENSION

95	90	91	89	87	84
70	82	80	81	77	76
50	77	74	76	72	72
30	71	70	72	67	69
5	61	59	59	56	60

RIGHT FOOT FLEXION-EXTENSION

95	89	95	93	93	91
70	81	81	82	80	78
50	75	76	78	73	74
30	71	72	73	68	70
5	61	61	64	60	61

Leighton flexometer used

(data from Sigerseth, 1970)

TABLE 10

FLEXIBILITY (degrees) IN BOYS

			Age (years)			
Percentiles	10	12	14	16	18	College

RIGHT ARM FLEXION-EXTENSION

95	269	269	275	275	266	243
70	250	251	255	264	252	226
50	239	245	249	258	236	220
30	230	238	243	251	224	214
5	213	225	223	236	203	201

LEFT ARM ADDUCTION-ABDUCTION

95	199	196	201	195	212	209
70	192	186	188	177	193	191
50	187	182	183	173	186	183
30	182	177	177	168	180	173
5	165	167	154	156	166	161

RIGHT ARM ADDUCTION-ABDUCTION

95	200	196	198	187	205	199
70	190	186	188	177	182	183
50	185	181	183	173	175	175
30	180	176	176	170	170	166
5	166	166	160	156	153	158

LEFT ARM ROTATION

95	205	208	189	212	198	238
70	187	184	179	177	180	209
50	181	177	173	170	169	203
30	174	171	167	161	160	195
5	163	160	155	147	142	180

RIGHT ARM ROTATION

95	205	212	195	212	197	229
70	187	180	181	177	180	211
50	180	175	176	168	171	199
30	173	171	171	161	161	188
5	161	161	153	146	141	172

LEFT FOREARM FLEXION-EXTENSION

95	164	162	170	153	167	177
70	154	150	156	145	154	163
50	150	144	147	143	150	158
30	145	140	142	141	144	153
5	136	130	128	131	133	146

TABLE 10   (continued)

FLEXIBILITY (degrees) IN BOYS

			Age (years)			
Percentiles	10	12	14	16	18	College

RIGHT FOREARM FLEXION-EXTENSION

Percentiles	10	12	14	16	18	College
95	164	159	169	152	167	177
70	155	149	155	144	153	163
50	150	144	149	142	148	158
30	145	140	143	140	143	153
5	130	132	128	122	125	146

LEFT HAND SUPINATION-PRONATION

Percentiles	10	12	14	16	18	College
95	216	209	203	179	200	217
70	205	194	187	166	179	200
50	198	187	177	161	167	185
30	187	181	167	154	161	175
5	163	168	153	142	138	148

RIGHT HAND ABDUCTION-ADDUCTION

Percentiles	10	12	14	16	18	College
95	116	104	97	90	103	103
70	97	91	81	80	92	83
50	89	87	77	76	85	74
30	82	83	73	72	77	71
5	75	69	67	56	62	60

THIGH ADDUCTION-ABDUCTION

Percentiles	10	12	14	16	18	College
95	70	76	76	76	72	64
70	56	63	68	65	61	55
50	51	58	60	61	55	51
30	46	53	50	49	50	47
5	36	42	41	39	40	36

LEFT THIGH ROTATION

Percentiles	10	12	14	16	18	College
95	135	111	84	99	135	132
70	117	84	74	74	100	113
50	108	80	72	69	87	102
30	101	74	67	64	78	92
5	79	57	42	46	60	73

RIGHT THIGH ROTATION

Percentiles	10	12	14	16	18	College
95	140	108	86	92	135	124
70	117	86	74	73	101	103
50	108	79	72	70	87	94
30	99	75	69	65	75	91
5	80	60	42	43	53	80

TABLE 10   (continued)

FLEXIBILITY (degrees) IN BOYS

| Percentiles | Age (years) | | | | | |
	10	12	14	16	18	College

LEFT LEG FLEXION-EXTENSION

Percentiles	10	12	14	16	18	College
95	165	158	152	154	148	165
70	153	151	143	143	137	155
50	146	147	139	138	131	150
30	138	144	135	129	124	145
5	118	135	119	118	115	135

RIGHT LEG FLEXION-EXTENSION

Percentiles	10	12	14	16	18	College
95	166	158	158	157	153	165
70	152	151	148	145	142	155
50	145	147	143	140	137	151
30	138	144	138	128	132	145
5	120	136	121	116	122	136

LEFT FOOT FLEXION-EXTENSION

Percentiles	10	12	14	16	18	College
95	100	75	77	82	80	77
70	87	62	64	64	70	67
50	80	58	60	60	64	62
30	74	55	55	58	60	54
5	61	47	50	51	46	46

RIGHT FOOT FLEXION-EXTENSION

Percentiles	10	12	14	16	18	College
95	99	77	90	85	81	79
70	88	64	60	65	70	66
50	80	59	56	60	64	62
30	73	54	52	57	59	56
5	60	46	45	51	50	47

LEFT FOOT INVERSION-EVERSION

Percentiles	10	12	14	16	18	College
95	64	63	63	59	70	73
70	55	51	51	48	53	63
50	48	46	46	44	45	58
30	44	41	40	41	37	52
5	36	30	23	32	23	37

RIGHT FOOT INVERSION-EVERSION

Percentiles	10	12	14	16	18	College
95	64	63	64	59	72	70
70	54	49	53	48	52	61
50	48	45	43	44	42	55
30	44	40	38	39	37	48
5	35	30	30	32	23	34

TABLE 10   (continued)

FLEXIBILITY (degrees) IN BOYS

			Age (years)			
Percentiles	10	12	14	16	18	College

TRUNK LATERAL FLEXION

Percentiles	10	12	14	16	18	College
95	121	103	105	123	129	133
70	108	90	91	105	107	116
50	102	84	87	92	99	110
30	96	80	82	87	91	100
5	83	70	61	76	68	83

TRUNK ROTATION

Percentiles	10	12	14	16	18	College
95	158	163	170	154	170	153
70	149	154	158	143	144	131
50	145	148	148	137	134	117
30	141	143	140	132	127	102
5	130	135	97	121	115	71

RIGHT HAND SUPINATION-PRONATION

Percentiles	10	12	14	16	18	College
95	213	207	201	179	202	225
70	204	193	186	168	179	197
50	199	186	175	162	166	184
30	185	180	166	153	158	174
5	165	168	153	143	136	157

LEFT HAND FLEXION-EXTENSION

Percentiles	10	12	14	16	18	College
95	186	175	167	157	151	188
70	175	149	149	140	134	173
50	169	143	132	127	126	165
30	163	135	121	123	118	156
5	135	111	105	111	98	146

RIGHT HAND FLEXION-EXTENSION

Percentiles	10	12	14	16	18	College
95	184	176	167	159	153	181
70	175	148	149	136	133	172
50	171	142	131	126	125	166
30	165	135	120	120	119	160
5	140	92	107	106	110	145

LEFT HAND ABDUCTION-ADDDUCTION

Percentiles	10	12	14	16	18	College
95	112	104	97	92	106	107
70	98	93	81	80	90	87
50	89	86	78	77	83	80
30	83	81	74	73	77	72
5	73	70	69	53	63	58

TABLE 10   (continued)

FLEXIBILITY (degrees) IN BOYS

Percentiles	Age (years)					
	10	12	14	16	18	College
NECK FLEXION-EXTENSION						
95	150	153	150	150	148	177
70	134	145	143	128	138	157
50	126	141	136	123	129	151
30	119	135	127	118	123	144
5	90	111	119	108	101	117
NECK LATERAL FLEXION						
95	117	116	109	104	127	155
70	103	101	96	93	110	125
50	98	97	91	89	94	116
30	91	93	86	84	86	111
5	73	80	66	71	72	85
NECK ROTATION						
95	209	182	179	176	188	208
70	191	172	165	164	172	194
50	183	163	160	158	163	188
30	173	154	154	153	151	182
5	135	127	140	138	110	166
LEFT ARM FLEXION-EXTENSION						
95	268	270	274	275	266	245
70	250	252	256	264	248	230
50	239	244	247	260	236	224
30	231	236	240	253	225	215
5	216	225	222	239	203	201

Leighton flexometer used

(data from Sigerseth, 1970)

TABLE 11

GRIP STRENGTH (kg) IN CALIFORNIA CHILDREN

Age (years)	Boys		Girls	
	Mean (kg)	S.D. (kg)	Mean (kg)	S.D. (kg)
9	68.91	14.49	60.40	14.95
10	83.32	14.18	75.74	14.76
11	94.05	16.29	86.59	16.14
12	104.82	18.82	98.63	18.71
13	120.41	24.35	110.29	20.87
14	137.27	28.96	112.60	19.43
15	153.00	30.69	111.49	19.50
16	173.91	28.45	112.13	19.00
17	198.36	28.71	121.51	19.20
18	212.09	28.34	124.60	17.46

(data from Tuddenham and Snyder, 1954)

TABLE 12

GRIP STRENGTH (kg) IN CALIFORNIA CHILDREN

Age	Boys		Girls	
(years)	Mean	S.D.	Mean	S.D.
Right Hand				
11.0	25.14	4.09	21.04	3.86
11.5	26.28	3.89	22.62	4.82
12.0	27.62	3.71	24.15	4.89
12.5	29.37	4.42	26.36	5.03
13.0	30.96	4.60	27.72	5.20
13.5	33.39	5.68	28.72	4.97
14.0	36.33	6.96	29.19	5.21
14.5	39.55	7.24	30.34	5.60
15.0	43.40	7.15	32.50	5.32
15.5	46.62	7.35	33.08	5.62
16.0	49.10	7.09	33.69	5.59
16.5	51.74	6.82	34.61	5.19
17.0	54.50	7.06	35.15	5.47
17.5	56.26	7.25	35.79	5.05
Left Hand				
11.0	23.46	3.93	19.73	3.52
11.5	24.91	3.63	20.16	4.78
12.0	26.29	3.69	21.41	4.61
12.5	27.69	4.06	23.48	4.43
13.0	28.77	4.58	24.92	5.69
13.5	31.50	5.15	25.90	5.05
14.0	33.82	6.11	26.41	5.21
14.5	37.06	6.06	27.11	5.90
15.0	40.48	7.03	28.26	5.65
15.5	43.61	7.25	29.82	4.74
16.0	45.65	6.77	30.78	5.19
16.5	48.73	6.48	31.42	5.75
17.0	50.08	7.03	31.78	4.93
17.5	52.28	6.94	31.81	5.51

Ages are $\pm$ 0.25 years.

(data from Jones, 1949)

TABLE 13

GRIP STRENGTH (lb. pressure; before demonstration) IN CHILDREN
OF PHILADELPHIA (PA) AND BERKELEY-OAKLAND (CA)

Age (years)	Philadelphia				Berkeley-Oakland			
	MN	FN	MW	FW	MN	FN	MW	FW
RIGHT HAND								
3	11.1	9.2	12.6	10.5	10.6	10.0	8.9	8.3
4	12.6	9.9	14.7	12.5	13.4	11.3	11.6	11.1
5	15.7	17.8	16.1	16.0	18.0	14.0	16.1	14.0
6	23.3	19.5	20.7	18.2	21.9	19.8	20.0	18.5
LEFT HAND								
3	11.5	9.8	11.8	10.3	11.0	9.8	10.1	8.4
4	13.5	9.7	14.1	12.8	13.8	12.5	12.9	11.8
5	16.5	17.3	16.5	15.5	20.2	15.5	17.5	15.0
6	22.6	19.7	19.5	17.6	25.1	22.5	21.0	20.3

MN = male Negro; FN = female Negro; MW = male white; FW = female white.

(data from Krogman, 1971)

TABLE 14

GRIP STRENGTH (lb. pressure; after demonstration) IN CHILDREN
OF PHILADELPHIA (PA) AND BERKELEY-OAKLAND (CA)

Age (years)	Philadelphia					Berkeley-Oakland			
	MN	FN	MW	FW		MN	FN	MW	FW
				RIGHT HAND					
3	10.9	10.0	13.5	11.8		9.8	9.9	9.1	8.0
4	13.1	12.5	13.4	13.7		13.6	11.0	11.8	10.3
5	17.6	19.7	17.9	15.3		17.1	14.0	15.8	14.0
6	22.8	20.0	20.7	18.9		23.1	19.6	19.0	17.4
				LEFT HAND					
3	10.3	10.5	12.7	10.6		9.8	10.5	10.1	8.9
4	13.2	12.4	13.9	12.5		13.6	12.0	13.1	11.8
5	17.4	18.4	16.4	16.2		18.9	15.7	16.0	14.1
6	23.0	18.4	20.1	17.8		24.1	21.0	19.9	19.5

MN = male Negro; FN = female Negro; MW = male white; FW = female white.

(data from Krogman, 1971)

TABLE 15

HAND GRIP (lb pressure) FOR URBAN
CHILDREN IN 7 AREAS OF THE U.S.

BOYS Age (years)						Percentile	GIRLS Age (years)					
13	14	15	16	17	18		13	14	15	16	17	18
95	115	121	140	144	144	95th	60	67	74	79	86	91
75	93	106	118	120	125	75th	53	54	60	67	72	76
65	78	93	106	109	114	50th	42	43	55	59	63	67
50	59	81	93	98	101	25th	30	37	47	51	56	57
39	41	61	76	82	86	5th	20	25	36	38	43	46

Using a Narragansett hand dynamometer.

(data from Fleishman, 1964)

TABLE 16

RIGHT HAND GRIP (lb) USING NARRAGANSETT
DYNAMOMETER FOR PHILADELPHIA CHILDREN

Age (years)	White Boys		Black Boys	
	Mean	S.D.	Mean	S.D.
6	28.11	8.08	29.55	6.75
7	32.65	8.23	33.76	7.90
8	37.53	7.63	38.32	6.95
9	42.86	8.42	44.67	8.80
10	48.51	10.89	49.11	11.07
11	52.18	10.72	53.52	10.96
12	56.47	11.11	58.58	12.81
13	--	--	61.22	12.21

Age (years)	White Girls		Black Girls	
	Mean	S.D.	Mean	S.D.
6	25.45	7.99	27.55	7.26
7	29.75	7.71	31.51	7.50
8	32.64	7.85	35.58	7.79
9	37.02	8.32	37.17	6.70
10	40.15	7.67	42.52	8.73
11	43.05	8.37	48.22	12.60
12	44.63	7.71	55.97	11.33
13	--	--	61.18	12.09

Age is for completed years

(data from Malina, unpublished)

TABLE 17

LEFT HAND GRIP (lb) USING NARRAGANSETT
DYNAMOMETER FOR PHILADELPHIA CHILDREN

Age (years)	White Boys		Black Boys	
	Mean	S.D.	Mean	S.D.
6	27.25	7.69	28.17	6.60
7	31.57	8.25	31.80	7.43
8	35.65	7.80	35.65	6.93
9	40.05	8.37	41.30	8.96
10	44.78	10.79	45.62	11.07
11	47.34	9.84	50.18	10.54
12	52.08	8.35	53.19	11.87
13	--	--	58.85	12.67

Age (years)	White Girls		Black Girls	
	Mean	S.D.	Mean	S.D.
6	23.97	7.50	26.84	6.61
7	28.04	8.20	29.26	6.46
8	30.73	7.65	32.69	7.12
9	35.70	7.76	34.34	7.60
10	37.86	7.44	39.46	7.47
11	39.75	8.58	44.72	11.46
12	42.44	7.39	50.25	11.36
13	--	--	56.36	13.84

Age is for completed years

(data from Malina, unpublished)

TABLE   18

GRIP STRENGTH (kg) IN MICHIGAN CHILDREN

					GRADE					
	Second		Third		Fourth		Fifth		Sixth	
Sex	Fall	Spring	Fall	Spring	Fall	Spring	Fall	Spring	Fall	Spring
						Right				
Boys										
Mean	12.59	12.91	13.94	14.96	16.46	17.52	19.70	20.60	22.37	24.48
S.D.	2.48	3.04	3.26	3.07	3.27	3.54	4.14	4.04	5.07	5.00
Girls										
Mean	12.32	12.62	13.17	14.24	15.26	16.00	17.76	19.26	21.70	22.60
S.D.	2.42	2.09	2.42	2.48	2.49	2.89	3.03	3.25	3.24	3.61
						Left				
Boys										
Mean	11.32	11.23	12.30	13.62	14.61	16.61	16.93	18.50	19.00	21.67
S.D.	2.64	3.15	2.79	3.33	3.15	3.25	3.56	3.86	4.19	4.21
Girls										
Mean	11.32	11.31	11.85	13.44	13.36	15.06	15.58	17.68	19.60	20.54
S.D.	2.75	2.09	2.45	2.81	3.35	3.01	3.49	3.76	4.27	3.45

(data from Govatos, 1966)

TABLE 19

STATURE, WEIGHT AND GRIP STRENGTH IN CALIFORNIA CHILDREN

| | Age (years) | | Stature (in.) | | Weight (lbs.) | | Initial Strength (kg.) | |
	Boys	Girls	Boys	Girls	Boys	Girls	Boys	Girls
Mean	8.0	8.0	51.2	49.6	61.6	57.7	8.9	7.3
S.D.	0.32	0.24	3.1	2.2	8.6	10.2	1.8	2.3
Mean	9.0	8.8	54.0	52.8	68.5	63.6	12.6	10.1
S.D.	0.31	0.24	2.1	1.4	11.1	8.1	2.9	3.4
Mean	9.8	10.0	55.8	55.6	80.3	73.3	14.6	11.1
S.D.	0.23	0.11	3.2	1.9	15.6	4.2	1.5	2.5
Mean	11.2	11.1	57.2	58.6	93.4	92.2	18.2	15.4
S.D.	0.14	0.24	3.8	2.7	13.9	16.8	1.9	4.5
Mean	12.0	11.9	60.8	59.7	101.8	105.6	21.6	19.1
S.D.	0.33	0.32	3.3	3.5	21.7	19.1	4.5	3.8
Mean	13.1	13.1	61.2	62.4	119.4	120.2	22.5	21.0
S.D.	0.19	0.20	3.1	2.8	15.5	9.7	3.9	4.6
Mean	14.0	13.8	63.7	64.0	117.8	125.7	29.3	23.0
S.D.	0.30	0.26	2.5	4.1	20.1	17.3	5.8	3.5
Mean	15.0	15.0	63.9	65.0	121.0	128.9	29.6	22.3
S.D.	0.27	0.19	3.0	2.5	18.4	9.1	7.1	2.9
Mean	16.1	16.0	68.6	64.7	136.7	127.9	37.6	22.8
S.D.	0.32	0.27	2.9	1.9	14.6	11.2	3.1	3.4
Mean	17.0	17.0	67.8	65.3	148.0	134.8	38.6	25.2
S.D.	0.28	0.28	2.7	2.2	16.6	22.0	6.1	3.8

(data from Rich, 1957, Research Quarterly 31:485-498, by permission of the American Alliance for Health, Physical Education, Recreation and Dance, 1900 Association Drive, Reston, VA 22091)

TABLE 20

RESULTS OF SUM OF GRIP STRENGTH TESTS (kg)
IN MICHIGAN CHILDREN

Percentile	Age (years)								
	10	11	12	13	14	15	16	17	18
MALES									
90	34	42	52	69	89	96	106	111	117
80	30	37	47	60	80	90	99	105	106
70	26	34	41	53	72	84	95	99	101
60	24	32	38	48	66	80	91	93	98
50	22	29	34	44	61	76	87	89	96
40	20	26	31	42	58	73	84	85	93
30	18	23	30	39	54	69	78	81	90
20	15	21	27	34	49	64	74	76	86
10	11	16	23	28	39	55	68	70	81
Mean	23.6	30.2	37.4	47.5	64.3	76.6	87.6	91.5	97.1
S.D.	8.8	9.9	11.9	14.4	18.1	15.1	15.5	18.2	15.5
FEMALES									
90	30	37	44	49	65	60	58	61	59
80	25	33	40	44	50	54	53	54	55
70	22	30	36	41	48	49	49	50	52
60	20	27	33	38	44	45	48	47	49
50	18	25	31	36	41	43	43	44	46
40	17	23	28	34	39	41	41	42	43
30	15	20	26	32	36	38	39	39	39
20	14	17	22	30	32	36	36	36	36
10	10	12	18	26	27	31	33	31	31
Mean	19.9	26.2	32.5	37.8	43.6	45.7	45.8	46.8	47.1
S.D.	7.5	9.7	10.9	9.5	13.4	10.8	10.1	12.1	9.7

(data from Montoye and Lamphear, 1977, Research Quarterly for Exercise and Sport 48:109-120, by permission of the American Alliance for Health, Physical Education, Recreation and Dance, 1900 Association Drive, Reston VA 22091)

TABLE   21

LATERAL DIFFERENCES IN GRIP STRENGTH
IN CALIFORNIA CHILDREN

Age	Differences in kg		Differences in S.D. Units	
(years)	Boys	Girls	Boys	Girls
11.5	1.37	2.46	.35	.51
12.5	1.68	2.88	.38	.57
13.5	1.89	2.82	.33	.57
14.5	2.49	3.33	.34	.59
15.5	3.01	3.24	.41	.58
16.5	3.01	3.19	.44	.61

The ages are $\pm$ 0.25 years

(data from Jones, 1949)

TABLE   22

STRENGTH (kg) IN CALIFORNIA GIRLS
IN RELATION TO MENARCHE

Age	Premenarcheal		Postmenarcheal	
(years)	Mean	S.D.	Mean	S.D.
	Right grip			
12.5	44.5	9.4	54.1	8.4
13.0	43.8	10.0	50.7	8.7
13.5	38.7	8.9	49.0	8.2
	Total Strength			
12.25	86.0	16.30	100.6	15.55
12.75	92.0	16.35	105.6	15.30
13.25	93.0	16.20	108.8	16.35
13.75	97.1	16.35	105.6	16.50
14.25	100.7	19.50	107.9	18.00

The ages are $\pm$ 0.25 years.

(data from Jones, 1949)

TABLE 23

PHYSICAL FITNESS NORMS FOR COLLEGE FRESHMEN
IN KANSAS AGED 17.6 TO 19.5 YEARS

Percentile	Stature (cm)	Weight (kg)	Right Grip (kg)	Left Grip (kg)	Strength Leg (kg)	Strength Back (kg)	Trunk Flexibility (cm)	EVO_2 l/min	EVO_2 ml/kg/min	% Fat	Triceps S.F. (mm)
						MEN					
100	200.7	110.9	74.1	70.0	277	295	61.0	4.80	65.0	3.0	3.0
90	186.4	84.5	60.0	56.4	215	214	54.9	3.60	51.5	6.5	5.8
80	183.1	79.5	56.5	52.7	200	195	51.8	3.33	46.7	7.7	7.0
70	181.4	76.9	53.5	50.0	182	181	49.8	3.13	43.6	8.6	8.1
60	179.3	74.1	51.5	48.0	178	170	47.5	2.95	41.0	9.6	9.1
50	177.5	71.7	49.6	45.6	168	159	45.2	2.79	38.8	10.8	10.3
40	176.0	69.5	46.8	43.6	156	150	43.4	2.65	36.7	12.0	11.5
30	174.2	67.7	44.5	41.1	146	136	40.1	2.45	34.4	13.5	12.9
20	172.0	65.0	41.4	39.1	132	126	36.3	2.33	32.3	15.3	15.2
10	168.4	61.4	36.8	34.8	110	107	29.7	2.13	29.3	19.0	18.5
0	147.3	51.4	21.4	19.1	50	52	15.2	1.50	21.0	35.0	33.0
						WOMEN					
100	188.0	92.3	58.2	53.6	209	200	63.5	3.80	65.0	11.0	5.0
90	173.2	69.1	34.8	31.8	118	114	56.4	2.90	49.5	18.3	11.0
80	170.2	65.0	31.4	28.9	110	101	53.1	2.60	45.0	19.8	12.7
70	167.9	62.0	29.7	26.7	100	91	50.5	2.41	41.8	20.9	14.3
60	166.4	59.7	28.0	25.0	91	86	48.3	2.35	40.3	22.0	15.7
50	164.6	58.0	26.5	24.0	90	80	46.5	2.16	37.2	23.2	17.1
40	162.6	56.2	25.0	22.4	79	73	44.4	2.05	35.0	24.1	18.9
30	160.8	54.2	23.9	28.8	73	67	41.4	1.93	32.9	25.3	20.5
20	158.8	52.1	21.4	19.9	66	59	37.6	1.70	30.5	26.8	22.4
10	156.0	49.5	19.9	17.3	55	50	30.5	1.60	27.7	29.3	25.0
0	132.0	37.7	10.0	8.6	36	36	10.2	1.10	18.0	40.0	35.0

EVO_2 = estimated VO_2 maximum uptake.
S.F. = skinfold

(data from Zuti and Corbin, 1977, by permission of the American Alliance for Health,
Physical Education, Recreation and Dance, 1900 Association Drive, Reston VA 22091)

TABLE 24

STRENGTH (lb) IN ILLINOIS BOYS

Percentiles	Right Hand	Left Hand	Back	Legs	Total	Strength/ Weight
			Age 7 years			
95	46	40	157	200	402	7.01
75	35	31	128	157	331	5.77
50	23	20	91	103	243	4.22
25	10	9	55	50	154	2.67
5	0	0	25	7	83	1.43
			Age 8 years			
95	48	37	195	248	484	7.49
75	38	31	156	203	393	6.13
50	25	23	108	134	278	4.43
25	12	15	60	65	164	2.73
5	2	8	22	9	73	1.37
			Age 9 years			
95	55	54	216	284	555	7.67
75	45	43	176	226	457	6.31
50	32	29	126	153	334	4.61
25	20	14	76	80	211	2.91
5	10	3	36	22	112	1.55
			Age 10 years			
95	62	58	225	302	603	8.57
75	51	46	188	247	504	7.05
50	38	31	141	177	381	5.15
25	24	16	95	108	257	3.25
5	13	4	58	53	158	1.73
			Age 11 years			
95	70	62	238	311	658	8.07
75	57	49	200	256	546	6.59
50	41	34	151	188	406	4.74
25	25	18	103	119	266	2.89
5	13	5	64	64	154	1.41
			Age 12 years			
95	129	108	363	465	1009	9.19
75	99	83	287	367	798	7.35
50	61	51	192	244	533	5.05
25	24	20	97	121	269	2.75
5	0	0	21	23	57	0.91

TABLE   24   (continued)

STRENGTH (lb) IN ILLINOIS BOYS

Percentiles	Right Hand	Left Hand	Back	Legs	Total	Strength/ Weight
			Age 13 years			
95	121	102	385	510	1042	8.20
75	99	84	312	407	855	6.92
50	70	62	221	278	621	5.32
25	42	39	131	148	387	3.72
5	20	21	58	45	200	2.44

Hand data from Narragansett dynamometer; back and leg data from back lift and leg lift dynamometer.

(data from Cureton and Barry, 1964)

TABLE 25

ANTHROPOMETRIC AND BODY COMPOSITION MEASURES
FOR 7-YEAR-OLD WISCONSIN CHILDREN

Measures	Boys		Girls	
	Mean	S.D.	Mean	S.D.
Strength (lb.)	86.74	15.43	76.86	12.24
Muscle area,* A-P view (sq. cm.)	133.72	17.02	126.94	8.29
Total area,* A-P view (sq. cm.)	162.41	23.33	160.44	14.29
Total area,* lateral view (sq. cm.)	174.06	24.97	169.61	14.26
Muscle area,* lateral view (sq. cm.)	148.32	20.10	140.85	9.77
Bone area, A-P view (sq. cm.)	69.61	10.13	65.45	5.53
Muscle breadth,* A-P view (cm.)	7.01	0.56	6.69	0.41
Muscle breadth,* lateral view (cm.)	7.31	0.73	6.96	0.36
Calf girth (cm.)	25.14	2.15	24.87	1.53
Stature (cm.)	121.58	5.83	121.37	3.97
Weight (lb.)	52.65	8.23	50.83	5.60

*Includes bone tissue; calf.

(data from Rarick and Thompson, 1956, Research Quarterly 27:321-332, by permission of the American Alliance for Health, Physical Education, Recreation and Dance, 1900 Association Drive, Reston VA 22091)

TABLE 26

WRIST-TURNING STRENGTH (lb pressure) IN
CHILDREN OF PHILADELPHIA (PA)   AND BERKELEY-OAKLAND (CA)

Age (years)	Philadelphia				Berkeley-Oakland			
	MN	FN	MW	FW	MN	FN	MW	FW
BEFORE DEMONSTRATION								
3	28.7	26.3	19.5	20.6	24.9	22.7	25.9	23.1
4	40.7	31.9	45.6	38.7	32.9	28.7	36.9	31.5
5	44.4	43.4	51.4	45.5	43.7	31.6	47.1	42.4
6	53.9	46.9	61.1	55.5	44.4	40.6	47.4	47.0
AFTER DEMONSTRATION								
3	31.4	26.6	27.7	21.5	26.8	21.1	22.6	24.9
4	40.2	36.6	46.1	41.5	32.8	29.8	36.8	32.4
5	49.4	49.0	58.4	50.5	44.7	33.1	46.5	43.2
6	54.3	51.2	69.8	58.5	48.1	37.9	47.4	46.5

MN = male Negro; FN = female Negro; MW = male white; FW = female white.

(data from Krogman, 1971)

TABLE 27

THUMB-OPPOSABILITY STRENGTH (lb pressure) IN CHILDREN
OF PHILADELPHIA (PA)   AND BERKELEY-OAKLAND (CA)

Age (years)	Philadelphia				Berkeley-Oakland			
	MN	FN	MW	FW	MN	FN	MW	FW
BEFORE DEMONSTRATION								
3	13.5	12.5	14.0	11.8	12.3	10.9	12.3	11.8
4	18.4	16.3	17.5	16.0	16.3	14.4	17.2	15.4
5	24.8	24.7	24.4	20.5	25.0	17.8	25.3	20.9
6	30.7	24.4	30.7	24.7	24.7	23.4	26.0	22.5
AFTER DEMONSTRATION								
3	13.8	12.4	15.3	13.1	13.6	11.8	13.2	12.0
4	15.8	14.3	17.6	18.3	16.8	13.5	18.3	16.2
5	26.1	26.4	25.8	21.8	24.5	19.3	26.2	21.5
6	28.4	24.8	32.6	26.2	26.1	24.2	27.0	24.5

MN = male Negro; FN = female Negro; MW = male white; FW = female white.

(data from Krogman, 1971)

TABLE 28

PERFORMANCE DATA FOR ILLINOIS BOYS

Measures		Age (years)									
		6	7	8	9	10	11	12	13	14	15
18 item motor fitness test	Mean	4.20	4.73	4.99	5.62	6.86	8.02	9.03	9.79	11.13	11.79
	S.D.	2.89	2.50	2.51	2.97	2.80	2.81	3.54	3.01	3.02	2.76
Dynamometer strength (R. hand; lb)	Mean	--	22.9	25.2	32.5	37.6	41.2	61.3	70.4	69.8	77.7
	S.D.	--	8.4	8.5	8.4	8.9	10.5	25.0	18.8	19.7	17.3
Dynamometer strength (L. hand; lb)	Mean	--	19.9	22.6	28.6	31.0	33.6	51.5	61.7	63.6	59.7
	S.D.	--	7.5	5.4	9.5	10.2	10.5	20.9	14.9	20.0	16.6
Dynamometer strength (back lift; lb)	Mean	--	91.23	108.33	125.78	141.38	151.29	192.39	221.41	211.51	241.83
	S.D.	--	24.4	32.0	33.2	31.1	32.2	81.8	60.4	63.0	61.5
Dynamometer strength (leg extension; lb)	Mean	--	103.40	133.59	153.12	177.43	187.50	244.22	277.61	264.15	293.12
	S.D.	--	35.6	46.0	48.6	46.2	45.8	81.8	86.2	99.4	118.8
Dynamometer strength (sum of 4 items)	Mean	--	242.83	278.46	333.60	380.60	405.95	533.26	620.77	598.58	667.09
	S.D.	--	59.1	76.2	81.9	82.4	93.2	176.3	155.8	186.6	191.4
Dynamometer strength/lb of body wt.	Mean	--	4.22	4.43	4.61	5.15	4.74	5.05	5.32	5.08	5.37
	S.D.	--	1.0	1.1	1.1	1.3	1.2	1.5	1.1	1.2	1.8
60 yard dash (sec)	Mean	--	12.2	11.9	10.8	10.7	10.4	9.85	9.85	--	--
	S.D.	--	1.19	1.87	1.08	.62	.95	1.32	.49	--	--
Running long jump (in)	Mean	--	68	76	85	96	102	114	123	--	--
	S.D.	--	15.8	19.4	14.7	15.2	20.3	18.1	15.5	--	--
Running high jump (in)	Mean	--	27	28	31	34	36	37	38	--	--
	S.D.	--	4.05	3.95	3.61	3.25	5.41	4.50	6.12	--	--
Standing long jump (in)	Mean	--	49	50	53	57	60	61	64	--	--
	S.D.	--	6.72	7.18	3.71	7.76	6.80	9.78	5.55	--	--
Shot put (3 lb; in)	Mean	--	98	111	130	160	206	237	--	--	--
	S.D.	--	12.4	23.6	22.9	25.4	17.3	37.6	--	--	--

TABLE 28   (continued)

PERFORMANCE DATA FOR ILLINOIS BOYS

Measures		6	7	8	9	Age (years) 10	11	12	13	14	15
600 yard	Mean	--	163	169	166	158	149	149	137	--	--
run (min)	S.D.	--	17.80	24.89	23.85	21.37	22.97	22.69	9.57	--	--
Chins (no)	Mean	--	1.0	1.5	1.2	1.5	1.1	1.1	--	--	--
	S.D.	--	1.18	.39	1.65	.43	1.29	1.34	--	--	--
Dips (no)	Mean	--	1.9	1.8	1.8	1.8	2.1	2.8	--	--	--
	S.D.	--	1.90	1.91	2.13	2.37	2.54	2.75	--	--	--
Floor push-	Mean	--	--	7	8	8	9	9	10	--	--
ups (no)	S.D.	--	--	4.3	5.7	5.6	5.1	6.1	6.6	--	--
Vertical	Mean	--	8.0	9.5	10.4	11.0	11.3	12.7	--	--	--
jump (no)	S.D.	--	.15	2.09	1.43	1.58	1.80	2.11	--	--	--
Visual reaction	Mean	.51	.46	.45	.42	.41	.39	.39	.38	--	--
time (sec)	S.D.	.10	.09	.09	.07	.08	.09	.05	.08	--	--
Auditory reaction	Mean	.48	.45	.43	.39	.39	.38	.36	.35	--	--
time (sec)	S.D.	.08	.09	.08	.06	.06	.08	.06	.09	--	--
Combined (visual + auditory) reaction	Mean	.47	.45	.43	.39	.39	.37	.36	.36	--	--
time (sec)	S.D.	.09	.08	.09	.07	.06	.09	.05	.09	--	--
Vital capacity (in^3)	Mean	--	89.2	99.4	125.7	139.6	142.8	183.3	197.3	235.0	--

(data from Cureton and Barry, 1964)

TABLE 29

RATIO OF SUM OF GRIP STRENGTHS TO BODY WEIGHT (kg/kg body wt.)
IN MICHIGAN CHILDREN

	AGE (years)								
Percentile	10	11	12	13	14	15	16	17	18
Boys									
90	.89	1.04	1.10	1.20	1.40	1.45	1.56	1.62	1.62
80	.83	.94	.97	1.13	1.28	1.36	1.49	1.48	1.50
70	.76	.85	.93	1.03	1.23	1.29	1.43	1.44	1.47
60	.69	.80	.86	.98	1.15	1.25	1.39	1.39	1.44
50	.65	.77	.80	.93	1.09	1.22	1.30	1.35	1.37
40	.62	.71	.74	.86	1.02	1.17	1.19	1.28	1.33
30	.52	.66	.70	.80	.96	1.14	1.14	1.21	1.31
20	.45	.58	.66	.77	.90	1.06	1.09	1.16	1.22
10	.38	.48	.61	.65	.81	.95	.98	1.01	1.16
Mean	.65	.76	.83	.93	1.11	1.21	1.28	1.32	1.37
S.D.	.20	.22	.20	.22	.24	.21	.23	.24	.19
Girls									
90	.71	.91	.92	.96	1.08	1.04	.98	1.12	1.02
80	.65	.80	.80	.86	.96	.92	.94	1.02	.95
70	.62	.72	.75	.81	.86	.89	.84	.96	.90
60	.57	.68	.72	.79	.81	.84	.79	.89	.82
50	.54	.64	.66	.73	.76	.79	.76	.83	.78
40	.50	.57	.61	.68	.71	.76	.73	.78	.72
30	.46	.53	.57	.65	.67	.69	.71	.71	.69
20	.42	.45	.54	.63	.65	.65	.65	.66	.65
10	.32	.33	.47	.52	.58	.57	.57	.52	.58
Mean	.53	.62	.67	.74	.79	.79	.79	.83	.79
S.D.	.16	.21	.17	.18	.21	.19	.19	.22	.18

(data from Montoye and Lamphear, 1977, Research Quarterly for Exercise and Sport 48:109-
120, by permission of the American Alliance for Health, Physical Education, Recreation
and Dance, 1900 Association Drive, Reston VA 22091)

TABLE 30

OUTSIDE GRIP DIAMETER (cm)
FOR CHILDREN IN 8 STATES

Age (months)	Boys		Girls	
	Mean	S.D.	Mean	S.D.
0 - 3	3.7	0.5	3.6	0.4
4 - 6	4.1	0.4	4.0	0.4
7 - 9	4.4	0.4	4.4	0.4
10 - 12	4.4	0.3	4.5	0.3
13 - 18	4.6	0.3	4.5	0.3
19 - 24	5.0	0.3	4.9	0.3
25 - 30	5.2	0.4	5.0	0.3
31 - 36	5.4	0.4	5.3	0.4
37 - 42	5.5	0.4	5.4	0.3
43 - 48	5.6	0.4	5.5	0.4
49 - 54	5.8	0.3	5.6	0.3
55 - 60	5.9	0.4	5.8	0.4
61 - 66	6.2	0.4	6.0	0.4
67 - 72	6.3	0.4	6.2	0.4
73 - 78	6.5	0.4	6.3	0.4
79 - 84	6.6	0.4	6.4	0.4
85 - 96	6.9	0.5	6.7	0.4
97 - 108	7.1	0.5	7.0	0.5
109 - 120	7.3	0.4	7.2	0.5
121 - 132	7.6	0.5	7.6	0.6
133 - 144	8.0	0.5	7.7	0.5
145 - 156	8.0	0.7	8.2	0.5

The child or infant grips a cone with the right hand so that the
middle finger first touches the thumb. The measurement is from
the joint between the proximal and middle phalanges of the middle
finger and the metacarpophalangeal joint of the thumb.

(data from Snyder et al., 1975)

TABLE   31

MEASURES OF PERFORMANCE IN CANADIAN BOYS
AGED 18 AND 19 YEARS

Percentiles	Grip strength (kg)		Flexibility (in)	Standing Long Jump (in)	Speed sit-ups (no./1 min)
	Right Hand	Left Hand			
95	64	62	18	100	55
75	55	52	15	92	48
50	50	47	13	86	41
25	45	43	11	81	36
5	38	37	8	69	22

Age is at last completed year. Flexibility based on the sit and reach test.

(data from The Canadian Association for Health, Physical Education and
Recreation, n.d.)

TABLE   32

MEASURES OF PERFORMANCE IN CANADIAN GIRLS
AGED 18 AND 19 YEARS

Percentiles	Grip strength (kg)		Flexibility (in)	Standing Long Jump (in)	Speed sit-ups (no./1 min)
	Right Hand	Left Hand			
95	37	35	18	78	47
75	33	31	16	73	37
50	29	28	14	67	28
25	26	22	12	60	22
5	20	17	6	48	12

Age is at last completed year.   Flexibility based on the sit and reach test.

(data from The Canadian Association for Health, Physical Education and
Recreation, n.d.)

TABLE 33

MEASUREMENT OF GRIP STRENGTH AND AEROBIC POWER
IN CANADIAN YOUTHS AGED 16-19 YEARS

	Mean	S.D.
Grip strength (kg)	54.2	9.2
Forced vital capacity (ℓBTPS)	5.21	0.68
One-second forced expiratory volume (ℓBTPS)	4.49	0.63
FEV%	86.2	--
Aerobic power (ℓO_2 STPD per min)	3.31	0.60
Relative aerobic power (ml. O_2STPD per kg min)	47.9	7.9

FEV = forced expiratory volume

(data from Shephard, Jones and Brown,
originally published in Canadian Medical
Association Journal May 25, 1968, Vol. 98)

TABLE 34

GRIP STRENGTH (right hand, kg) IN CALIFORNIA CHILDREN
IN RELATION TO RATE OF MATURING

Age	Boys			Girls		
(years)	Early	Average	Late	Early	Average	Late
11.0	27.1	24.0	22.7	21.1	20.9	20.6
11.5	29.3	25.9	25.2	24.4	23.2	21.2
12.0	29.3	26.9	26.0	26.1	25.8	22.5
12.5	31.3	28.4	27.0	29.1	26.8	23.7
13.0	33.3	30.4	28.1	30.3	28.8	25.7
13.5	37.6	32.5	30.0	29.3	30.3	26.8
14.0	44.2	34.3	30.2	29.7	30.7	26.4
14.5	47.1	38.6	33.3	31.0	32.2	28.4
15.0	50.0	43.0	36.3	32.5	33.3	31.4
15.5	52.2	47.6	41.1	33.4	35.2	32.7
16.0	54.3	49.0	43.9	33.4	35.8	32.4
16.5	55.9	50.9	48.4	34.7	36.1	34.4
17.0	57.2	53.5	51.3	34.3	36.5	34.8
17.5	--	55.8	54.3	33.9	37.8	35.3

The boys are classified on the basis of skeletal maturity the girls on the
basis of age at menarche.

The ages are ± 0.25 years.

(data from Jones, 1949)

TABLE 35

RIGHT GRIP STRENGTH AND WEIGHT IN TERMS
OF PERCENTAGE OF TERMINAL STATUS IN CALIFORNIA
GIRLS GROUPED BY RATE OF MATURING

Age (years)	Early-Maturing		Late-Maturing	
	Strength	Weight	Strength	Weight
11.5	79.8	72.0	64.3	60.1
12.5	89.1	85.8	69.3	67.1
13.5	93.3	86.4	76.8	75.9
14.5	95.6	91.5	86.5	80.4
15.5	97.0	98.5	93.5	92.6
16.5	99.5	102.4	97.7	97.4
17.5	100.0	100.0	100.0	100.0

(data from "Sex differences in physical abilities," Human Biology
19:12-25, 1947, by H. E. Jones, by permission of the Wayne State
University Press. Copyright 1947, Wayne State University Press,
Detroit, Michigan 48202)

TABLE 36

MEAN GRIP STRENGTH (kg) FOR MEXICAN-
AMERICAN BOYS IN AUSTIN (TX)

Age (years)	Mean	S.D.
9	15.6	2.6
10	17.8	2.9
11	21.1	3.9
12	22.8	3.6
13	26.9	5.1
14	31.8	7.3

Measured on right side with a Stoelting
dynamometer.

(data from Zavaleta, 1976)

TABLE 37

MONTHLY GAINS IN TOTAL STRENGTH (kg; right-grip,
left-grip, pull and thrust) IN CALIFORNIAN CHILDREN

Month	Boys		Girls	
	Mean	S.D.	Mean	S.D.
January	1.82	1.94	0.30	2.16
February	1.82	1.81	0.47	2.12
March	1.69	1.82	0.71	2.28
April	2.26	1.71	1.33	1.98
May	2.02	1.72	0.55	2.47
June	1.98	1.68	0.36	2.41
July	1.88	1.75	0.36	2.41
August	1.93	1.68	0.18	2.42
September	1.61	1.70	-0.06	2.51
October	1.47	1.61	-0.85	2.33
November	1.72	1.90	-0.10	2.01
December	1.82	1.84	0.17	2.11
Monthly mean	1.83		0.29	

(data from Jones, 1949)

TABLE   38

HAND PINCH STRENGTHS (1b) IN RELATION TO LATERALITY IN IOWA GIRLS

Later-ality	Grade	Right Hand					Left Hand				
		Index	Long	Ring	Little	Lateral Pinch	Index	Long	Ring	Little	Lateral Pinch
Right	K	4.328	4.008	2.924	1.933	6.908	4.109	3.832	2.790	1.933	6.630
		1.201	1.238	1.035	0.778	2.042	1.072	1.068	1.040	0.745	1.881
	1	5.098	4.677	3.511	2.504	8.263	4.511	4.278	3.226	2.293	7.752
		1.547	1.406	1.152	0.958	2.177	1.229	1.269	1.098	0.903	1.811
	2	5.293	4.667	2.960	1.973	7.667	4.320	3.920	2.747	1.920	7.253
		1.600	1.638	1.132	0.838	2.570	1.286	1.323	1.187	0.850	2.641
	3	6.535	5.724	3.874	2.945	10.220	5.811	5.079	3.614	2.394	9.268
		2.088	1.979	1.345	4.592	2.845	2.054	1.712	1.322	0.993	2.632
	4	6.574	5.787	3.889	2.759	11.074	5.611	5.148	3.713	2.556	10.167
		2.171	1.835	1.537	1.191	2.781	1.617	1.707	1.421	1.053	2.563
	5	7.651	7.205	4.867	3.494	12.398	6.795	6.590	4.675	3.157	11.530
		1.953	1.898	1.429	1.130	2.627	1.744	1.858	1.654	1.204	2.183
	6	4.185	3.761	2.598	1.761	6.239	3.750	3.435	2.435	1.641	5.696
		1.533	1.270	1.038	0.894	1.738	1.315	1.170	0.941	0.622	1.784
Left	K	4.267	3.667	2.467	1.667	7.067	4.133	3.800	2.733	1.600	6.533
		1.033	0.724	1.125	0.617	1.981	1.125	1.424	0.961	0.828	1.506
	1	5.353	4.647	3.706	2.412	8.412	4.941	4.294	3.176	2.353	7.882
		1.730	1.455	1.047	1.004	2.785	1.600	1.359	1.185	1.169	1.691
	2	5.500	5.000	3.200	2.100	8.500	4.800	4.100	3.000	2.000	8.200
		1.650	2.211	1.135	0.568	2.224	1.229	0.738	0.816	0.667	2.348
	6	4.133	3.533	2.600	1.867	6.067	4.333	3.667	2.867	1.867	5.733
		1.356	1.060	0.828	0.915	1.907	1.234	1.291	1.060	0.915	1.534

(data from Burmeister et al., 1974)

TABLE 39

RESULTS OF ARM STRENGTH TESTS (kg)
IN MICHIGAN CHILDREN

					Age (years)				
Percentile	10	11	12	13	14	15	16	17	18
Boys									
90	48	52	58	66	86	90	98	100	106
80	42	47	52	60	77	84	91	93	98
70	38	44	48	56	71	80	86	88	94
60	36	42	46	53	67	77	83	86	90
50	35	40	44	50	64	74	80	83	87
40	34	38	43	49	61	71	77	81	84
30	32	36	41	47	58	67	74	78	80
20	29	33	38	44	53	63	72	72	74
10	25	29	34	41	46	58	67	62	65
Mean	36.1	40.9	46.1	52.8	65.3	74.5	81.5	83.4	88.0
S.D.	7.0	8.0	8.8	10.6	14.7	13.4	12.6	14.3	15.2
Girls									
90	40	43	49	50	56	54	53	57	55
80	36	41	45	48	52	51	48	52	53
70	33	38	43	45	48	48	47	49	51
60	31	36	41	43	46	45	45	47	49
50	30	34	38	41	44	43	43	46	47
40	29	32	36	39	42	41	42	44	45
30	27	30	34	37	40	39	40	43	44
20	25	27	32	34	36	36	37	39	40
10	23	25	29	30	33	33	33	33	36
Mean	31.3	34.9	39.3	42.0	45.2	44.8	44.0	47.0	47.3
S.D.	6.2	6.8	7.9	8.9	9.8	8.2	7.4	8.8	7.6

(data from Montoye and Lamphear, 1977, Research Quarterly for Exercise and Sport 48:109-120, by permission of the American Alliance for Health, Physical Education, Recreation and Dance, 1900 Association Drive, Reston VA 22091)

TABLE 40

RATIO OF ARM STRENGTH TO BODY WEIGHT (kg/kg body wt.)
IN MICHIGAN CHILDREN

	Age (years)								
Percentiles	10	11	12	13	14	15	16	17	18
Boys									
90	1.28	1.33	1.26	1.30	1.39	1.40	1.43	1.44	1.52
80	1.18	1.19	1.20	1.20	1.28	1.33	1.36	1.38	1.45
70	1.10	1.14	1.15	1.13	1.18	1.29	1.30	1.32	1.38
60	1.03	1.10	1.10	1.09	1.15	1.24	1.24	1.27	1.30
50	.97	1.04	1.07	1.07	1.11	1.18	1.20	1.23	1.20
40	.94	.97	1.00	1.04	1.09	1.12	1.14	1.20	1.18
30	.91	.94	.94	.98	1.02	1.08	1.09	1.10	1.13
20	.88	.88	.86	.88	.96	.97	1.04	1.03	1.04
10	.79	.82	.80	.79	.88	.91	.95	.90	1.00
Mean	1.01	1.04	1.04	1.05	1.12	1.17	1.20	1.20	1.24
S.D.	.18	.19	.17	.18	.19	.19	.20	.22	.21
Girls									
90	1.11	1.03	1.00	1.02	.96	.96	.97	1.04	1.00
80	.97	.98	.95	.93	.91	.91	.87	.98	.95
70	.92	.93	.90	.90	.89	.88	.82	.92	.88
60	.88	.87	.87	.86	.87	.86	.79	.90	.83
50	.85	.84	.85	.83	.83	.83	.75	.85	.82
40	.81	.82	.80	.80	.79	.76	.71	.82	.76
30	.77	.77	.76	.76	.75	.72	.69	.74	.71
20	.73	.73	.70	.69	.72	.62	.65	.70	.66
10	.69	.71	.65	.64	.66	.54	.59	.63	.63
Mean	.86	.86	.83	.83	.83	.79	.76	.84	.80
S.D.	.17	.15	.16	.16	.15	.17	.15	.17	.14

(data from Montoye and Lamphear, 1977, Research Quarterly for Exercise and Sport 48:109-120, by permission of the American Alliance for Health, Physical Education, Recreation and Dance, 1900 Association Drive, Reston VA 22091)

TABLE 41

STATURE (cm), WEIGHT (kg), BICEPS MUSCLE STRENGTH (kg) AND MANUAL
DEXTERITY (sec) IN PHILADELPHIA CHILDREN

| Age | Stature | | Weight | | Biceps Muscle Strength | | | | Manual Dexterity | |
| (years) | | | | | Right | | Left | | | |
	Mean	S.D.	Mean	S.D.	Mean	S.D.	Mean	S.D.	Mean	S.D.
					BOYS					
8	130.6	6.5	29.1	5.6	9.2	2.0	8.8	2.0	12.4	2.7
10	140.8	7.6	35.8	8.4	13.6	3.6	13.0	3.5	10.4	1.8
12	152.3	8.5	44.8	9.9	16.2	4.7	14.8	3.7	9.0	1.0
14	164.7	8.5	56.1	9.4	21.4	5.6	20.0	5.1	8.5	0.9
16	172.4	6.8	63.5	9.4	30.8	6.8	30.0	6.3	8.6	0.9
18	175.8	5.7	66.7	9.5	34.1	7.4	32.5	7.2	8.4	1.0
20	177.2	6.4	73.2	10.8	31.9	5.4	31.2	5.3	8.2	1.1
					GIRLS					
8	128.8	5.4	28.0	5.8	8.5	1.7	8.3	1.9	12.1	1.8
10	140.3	7.0	35.3	7.4	11.1	2.7	10.5	2.3	10.1	1.5
12	154.2	7.4	45.3	8.0	14.4	2.6	13.8	3.2	8.8	1.1
14	158.7	6.2	49.3	7.1	16.2	3.3	15.7	3.0	8.1	0.8
16	163.3	5.7	55.0	7.4	18.7	3.2	18.0	3.6	8.1	1.1
18	162.0	6.6	54.3	6.7	17.9	3.7	17.2	3.6	8.0	1.3
20	164.9	6.3	57.9	6.2	17.6	2.8	17.3	2.4	7.6	0.7

[a]Age groups defined as 7.5 to 8.5, 9.5 to 10.5, 17.5 to 18.5
[b]Biceps strength: modification of Hettinger and Müller Method--subject seated, shoulders
and upper arms fixed so that angle between arm and forearm was 90°; loop placed over wrist
and connected to strain gauge and a Wheatstone's bridge; muscle contraction was insometric.
[c]Manual dexterity based on O'Connor Test: time required to fill a board of 100 holes with
3 pins in each hole; recorded with stopwatch.

(data from Malina, unpublished)

TABLE 42

PERFORMANCE DIFFERENCES OF OREGON BOYS
CLASSIFIED INTO PUBESCENT DEVELOPMENT GROUPS

Age (years)	Means					
	1	2	3	4	4 + 5	5

Mean Cable-Tension Strength Test Scores

10	54.25	58.89	--	--	--	--
13	--	75.34	81.40	--	--	--
13	--	75.34	--	--	94.50	--
13	--	--	81.40	--	94.50	--
16	--	--	--	112.42	--	128.01

Mean Lung Capacities (in^3)

10	119.28	124.86	--	--	--	--
13	--	158.90	173.73	--	--	--
13	--	158.90	--	--	207.84	--
13	--	--	173.73	--	207.84	--
16	--	--	--	245.33	--	261.93

Mean Standing Long Jump Distances (in)

10	57.33	49.07	--	--	--	--
13	--	69.84	67.72	--	--	--
13	--	69.84	--	--	76.00	--
13	--	--	67.72	--	76.00	--
16	--	--	--	79.00	--	83.39

(data from Clarke and Degutis, 1962, Research Quarterly, 33:356-368, by permission of The American Alliance for Health, Physical Education, Recreation and Dance, 1900 Association Drive, Reston VA 22091)

TABLE 43

MUSCLE STRENGTH (thrust; kg) IN CALIFORNIAN CHILDREN

Age	Boys		Girls	
(years)	Mean	S.D.	Mean	S.D.
11.0	21.86	4.78	21.30	5.78
11.5	22.14	5.00	22.86	6.34
12.0	24.30	5.26	24.39	6.36
12.5	26.14	5.56	27.11	6.10
13.0	27.46	6.22	28.82	6.76
13.5	30.49	7.08	29.62	6.28
14.0	32.51	9.34	30.02	6.52
14.5	35.75	8.65	29.78	6.06
15.0	39.61	10.61	29.39	6.62
15.5	42.97	10.58	30.10	6.75
16.0	47.70	10.87	31.44	6.27
16.5	52.45	10.00	32.53	6.33
17.0	56.04	10.40	32.19	6.16
17.5	58.20	10.49	31.37	6.07

The ages are $\pm$ 0.25 years.

(data from Jones, 1949)

TABLE 44

PUSH STRENGTH (lb) FOR PHILADELPHIA CHILDREN

Age	White Boys		Black Boys	
(years)	Mean	S.D.	Mean	S.D.
6	22.13	9.65	21.07	8.76
7	28.62	9.90	26.73	9.10
8	32.70	10.09	34.62	10.05
9	39.94	12.00	43.83	11.62
10	46.80	12.15	49.77	12.02
11	51.69	12.13	53.83	12.81
12	56.55	12.77	60.79	13.59
13	--	--	61.16	9.56

Age	White Girls		Black Girls	
(years)	Mean	S.D.	Mean	S.D.
6	19.94	8.91	16.85	7.47
7	24.13	9.27	21.97	8.11
8	27.74	9.94	27.76	8.36
9	33.74	9.76	31.75	8.92
10	38.07	10.53	38.72	10.69
11	39.84	11.00	42.45	13.34
12	42.56	13.97	46.82	15.00
13	--	--	54.39	12.77

Age is for completed years

(data from Malina, unpublished)

TABLE   45

PALM-PUSH STRENGTH (lb pressure) FOR CHILDREN
IN PHILADELPHIA (PA)   AND BERKELEY-OAKLAND (CA)

Age (years)	Philadelphia				Berkeley-Oakland			
	MN	FN	MW	FW	MN	FN	MW	FW
BEFORE DEMONSTRATION								
3	10.6	10.5	12.8	10.3	13.7	11.0	12.4	11.5
4	17.6	15.9	20.9	16.0	18.5	14.4	19.2	17.2
5	21.0	24.6	24.7	21.3	24.2	19.8	27.0	22.3
6	30.3	23.3	32.5	26.1	31.1	26.2	29.0	26.9
AFTER DEMONSTRATION								
3	10.3	12.1	13.7	11.6	13.7	11.9	13.2	12.7
4	19.0	17.5	19.6	18.2	20.7	17.2	20.4	16.3
5	22.9	27.6	29.6	23.8	27.3	18.2	29.2	21.7
6	30.0	24.4	33.2	29.1	31.8	24.5	32.7	27.3

MN = male Negro; FN = female Negro; MW = male white; FW = female white.

(data from Krogman, 1971)

TABLE 46

MUSCLE STRENGTH (pull; kg) IN CALIFORNIAN CHILDREN

Age (years)	Mean	S.D.
	BOYS	
11.0	18.41	3.71
11.5	19.16	4.31
12.0	20.72	4.42
12.5	22.24	5.42
13.0	23.26	6.26
13.5	25.69	6.63
14.0	28.79	7.33
14.5	31.28	7.53
15.0	34.71	7.98
15.5	38.82	8.63
16.0	43.10	9.53
16.5	45.02	8.74
17.0	49.25	9.17
17.5	50.42	9.30
	GIRLS	
11.0	16.51	3.86
11.5	17.40	4.74
12.0	17.60	4.87
12.5	18.84	5.04
13.0	19.20	5.61
13.5	20.05	5.36
14.0	20.25	5.40
14.5	21.15	5.99
15.0	21.68	6.29
15.5	23.42	6.39
16.0	24.90	6.37
16.5	25.74	5.59
17.0	26.53	6.31
17.5	25.41	5.99

Ages are $\pm$ 0.25 years.

(data from Jones, 1949)

TABLE 47

CABLE-TENSION STRENGTH TEST PERFORMANCE IN
IN OREGON BOYS

Test	7 years		9 years		12 years		15 years	
	Mean	S.D.	Mean	S.D.	Mean	S.D.	Mean	S.D.
Elbow flexors	30.58	5.79	40.95	8.50	57.24	11.54	101.10	20.40
Shoulder flexors	39.25	10.80	53.06	12.67	71.91	20.02	116.72	28.42
Shoulder inward rotators	22.58	4.92	29.27	7.36	34.84	8.72	57.63	12.78
Trunk flexors	32.62	6.73	42.00	10.26	68.43	16.07	104.97	31.63
Trunk extensors	33.00	7.67	43.21	11.26	71.17	18.06	104.50	32.12
Hip flexors	34.83	8.05	44.53	10.31	62.21	14.14	105.43	24.19
Hip extensors	29.26	8.12	39.90	10.32	64.93	16.14	119.26	31.26
Hip inward rotators	17.56	5.59	21.96	7.63	34.72	11.23	50.92	15.00
Knee flexors	33.86	8.84	51.25	16.76	67.02	15.39	100.71	31.14
Knee extensors	48.16	9.97	66.71	16.24	101.91	20.23	175.68	47.82
Ankle dorsal flexors	22.84	5.53	31.58	9.43	41.45	9.94	63.38	14.94
Ankle plantar flexors	47.09	10.11	67.62	16.12	106.85	20.92	166.20	43.44

(data from Harrison, 1958)

TABLE   48

PULL STRENGTH (lb) FOR PHILADELPHIA CHILDREN

Age (years)	White Boys		Black Boys	
	Mean	S.D.	Mean	S.D.
6	22.06	8.99	21.13	7.45
7	26.81	8.47	26.04	7.81
8	30.06	8.42	30.57	9.09
9	36.20	10.50	37.71	9.83
10	39.72	10.25	42.84	12.84
11	42.45	9.65	46.52	13.08
12	49.01	9.67	49.74	12.27
13	--	--	49.64	9.24

Age (years)	White Girls		Black Girls	
	Mean	S.D.	Mean	S.D.
6	19.80	8.96	17.40	6.36
7	22.90	7.97	21.51	8.25
8	24.19	9.10	26.56	7.44
9	29.53	8.12	28.23	7.76
10	33.68	8.16	32.25	9.17
11	36.16	8.95	35.57	11.25
12	37.50	8.63	39.99	11.39
13	--	--	45.82	10.35

Age is for completed years

(data from Malina, unpublished)

TABLE 49

STRENGTH INDEX (kg)
IN MICHIGAN CHILDREN

Percentiles	Age (years)								
	10	11	12	13	14	15	16	17	18
Boys									
90	81	91	110	125	175	180	199	203	210
80	72	83	96	119	154	173	190	195	202
70	66	78	87	113	141	166	182	186	195
60	62	75	84	106	132	158	175	180	190
50	57	72	82	98	126	152	168	174	185
40	54	67	77	92	119	146	163	169	179
30	51	61	73	85	113	138	156	161	171
20	46	56	68	79	105	128	149	151	164
10	39	50	64	73	86	115	139	136	155
Mean	59.7	71.1	83.5	100.4	129.6	151.1	169.0	174.8	185.2
S.D.	14.3	16.2	18.7	22.8	31.1	25.7	25.0	29.4	27.5
Girls									
90	68	79	92	99	118	111	108	114	112
80	60	70	85	93	105	104	102	106	108
70	56	67	80	88	95	98	96	101	103
60	53	65	75	82	90	93	91	97	98
50	50	62	71	76	86	90	88	93 ·	92
40	47	58	67	71	82	85	84	89	87
30	44	54	62	68	77	81	83	84	83
20	40	48	57	65	72	75	79	80	79
10	34	40	49	61	65	68	73	65	75
Mean	51.2	61.1	71.9	79.8	88.8	90.5	98.8	93.7	94.4
S.D.	12.1	14.9	17.2	16.2	21.5	17.1	15.4	18.9	14.9

(data from Montoye and Lamphear, 1977, Research Quarterly for Exercise and Sport 48:109-120, by permission of the American Alliance for Health, Physical Education, Recreation and Dance, 1900 Association Drive, Reston VA 22091)

TABLE 50

RATIO OF STRENGTH INDEX TO BODY WEIGHT (kg/kg body wt.)
IN MICHIGAN CHILDREN

	Age (years)								
Percentile	10	11	12	13	14	15	16	17	18
Boys									
90	2.09	2.22	2.30	2.38	2.79	2.75	2.98	3.00	3.02
80	1.97	2.09	2.18	2.26	2.53	2.66	2.84	2.84	2.92
70	1.85	2.00	2.04	2.14	2.39	2.55	2.70	2.71	2.82
60	1.77	1.94	1.95	2.06	2.27	2.50	2.61	2.64	2.73
50	1.68	1.83	1.85	2.00	2.18	2.42	2.51	2.55	2.66
40	1.58	1.75	1.77	1.88	2.11	2.35	2.38	2.48	2.53
30	1.46	1.60	1.70	1.92	2.03	2.19	2.25	2.33	2.40
20	1.33	1.52	1.60	1.70	1.95	2.07	2.16	2.24	2.35
10	1.24	1.37	1.47	1.54	1.73	1.97	1.99	1.95	2.24
Mean	1.67	1.81	1.87	1.98	2.22	2.37	2.48	2.52	2.61
S.D.	.33	.35	.31	.35	.39	.34	.39	.42	.35
Girls									
90	1.72	1.92	1.88	2.04	2.00	1.92	1.97	2.06	1.97
80	1.53	1.79	1.74	1.82	1.85	1.79	1.75	1.98	1.85
70	1.48	1.62	1.64	1.67	1.74	1.73	1.65	1.88	1.72
60	1.44	1.54	1.57	1.60	1.67	1.69	1.60	1.79	1.68
50	1.38	1.47	1.50	1.57	1.61	1.66	1.53	1.70	1.61
40	1.32	1.41	1.45	1.49	1.52	1.53	1.46	1.58	1.50
30	1.26	1.34	1.38	1.42	1.44	1.47	1.37	1.47	1.48
20	1.23	1.24	1.26	1.38	1.38	1.32	1.31	1.41	1.42
10	1.08	1.10	1.17	1.27	1.25	1.14	1.17	1.25	1.21
Mean	1.38	1.49	1.51	1.58	1.62	1.59	1.55	1.67	1.60
S.D.	.26	.30	.27	.29	.30	.32	.30	.34	.29

(data from Montoye and Lamphear, 1977, Research Quarterly for Exercise and Sport 48:109-120, by permission of the American Alliance for Health, Physical Education, Recreation and Dance, 1900 Association Drive, Reston VA 22091)

TABLE 51

RELATIVE STRENGTH INDEX *
IN MICHIGAN CHILDREN

				Age (years)					
Percentiles	10	11	12	13	14	15	16	17	18
Boys									
90	122	127	116	117	117	119	113	115	114
80	114	118	110	111	109	115	107	112	109
70	106	112	106	108	103	108	105	109	103
60	102	106	103	104	98	104	103	105	101
50	96	104	100	101	95	101	100	101	98
40	93	98	96	98	93	99	98	99	97
30	87	95	92	96	89	97	95	95	93
20	80	90	87	90	86	92	90	90	91
10	75	82	82	82	81	84	86	85	87
Mean	97.2	102.5	99.1	101.2	97.0	103.0	99.7	100.4	99.0
S.D.	17.8	17.6	13.0	13.5	13.6	14.7	10.1	13.2	11.7
Girls									
90	116	122	118	123	121	123	119	125	120
80	109	117	109	112	115	112	109	112	112
70	105	111	104	108	106	106	104	109	109
60	100	105	102	104	102	104	101	105	103
50	97	100	98	100	97	101	97	102	96
40	92	95	94	97	94	96	95	99	92
30	89	93	91	94	90	93	91	95	91
20	85	90	87	88	87	87	86	91	85
10	79	81	80	86	81	76	82	80	81
Mean	97.5	101.8	98.7	101.7	99.9	100.1	98.6	101.9	98.6
S.D.	15.0	16.6	15.7	16.0	18.7	17.4	15.1	18.1	13.9

$$* = 100 \left(\frac{\text{observed strength} + \text{grip strength}}{\text{estimated arm strength} + \text{grip strength}} \right)$$

The estimates were derived from regressions based on body size and fatness.

(data from Montoye and Lamphear, 1977, Research Quarterly for Exercise and Sport 48:109-120, by permission of the American Alliance for Health, Physical Education, Recreation and Dance, 1900 Association Drive, Reston VA 22091)

TABLE   52

DISTANCE AND INCREMENT DATA FOR COMPOSITE, UPPER, AND LOWER
STRENGTH/STATURE AND STRENGTH/WEIGHT FOR THE BOYS TESTED
IN THE SASKATCHEWAN CHILD GROWTH AND DEVELOPMENT STUDY

Parameter and Age (years)	Composite Strength			Upper Strength			Lower Strength		
	Mean	S.D.	% In-crease	Mean	S.D.	% In-crease	Mean	S.D.	% In-crease
Strength (lb)/ stature (cm):									
10	0.225	.048		0.178	0.039		.347	.081	
			32.4			24.3			30.3
11	0.298	.048		0.239	0.035		.452	.089	
			9.7			7.1			13.3
12	0.327	.057		0.256	0.039		.512	.112	
			15.9			19.5			11.5
13	0.379	.073		0.306	0.051		.571	.132	
			22.2			23.5			19.8
14	0.463	.095		0.378	0.073		.684	.165	
			17.1			19.0			12.0
15	0.542	.120		0.450	0.095		.766	.195	
			12.4			16.4			8.0
16	0.609	.114		0.524	0.097		.827	.193	
Strength (lb)/ weight (kg):									
10	0.972	.193		0.773	0.167		1.498	.314	
			25.5			26.9			23.2
11	1.220	.203		0.981	0.156		1.846	.357	
			2.9			0.5			6.0
12	1.255	.207		0.986	0.163		1.956	.387	
			7.8			11.3			4.4
13	1.353	.244		1.097	0.184		2.042	.436	
			12.0			12.9			9.0
14	1.516	.270		1.289	0.216		2.225	.442	
			7.1			10.0			3.9
15	1.624	.293		1.363	0.230		2.312	.469	
			5.5			8.2			0.6
16	1.714	.293		1.475	0.254		2.326	.492	

(data from Carron and Bailey, 1974)

TABLE 53

DISTANCE AND INCREMENT DATA FOR COMPOSITE, UPPER, AND LOWER RAW STRENGTHS FOR
BOYS TESTED IN THE SASKATCHEWAN CHILD GROWTH AND DEVELOPMENT STUDY

Age (years)	COMPOSITE STRENGTH			
	Mean	Increase	S.D.	% Increase
10	36.11		6.89	
		6.63		37.4
11	42.74		7.64	
		5.93		13.9
12	48.67		9.52	
		10.14		20.8
13	58.81		12.96	
		16.49		28.0
14	75.30		18.09	
		15.96		21.2
15	91.26		21.79	
		13.77		15.1
16	105.03		20.73	

Age (years)	UPPER STRENGTH				LOWER STRENGTH			
	Mean	Increase	S.D.	% Increase	Mean	Increase	S.D.	% Increase
10	24.64		5.45		48.08		11.84	
		9.56		38.8		16.69		34.7
11	34.20		5.67		64.77		13.83	
		3.87		11.3		11.38		17.6
12	38.07		6.56		76.15		16.52	
		9.33		24.5		12.45		16.4
13	47.40		9.56		88.60		22.72	
		14.12		29.8		22.60		25.5
14	61.52		13.87		111.20		30.23	
		14.40		23.4		18.40		16.6
15	75.92		17.27		129.60		36.05	
		14.34		18.9		13.7		10.6
16	90.26		17.22		143.30		35.14	

(data from Carron and Bailey, 1974)

TABLE 54

DIFFERENCES BETWEEN MEAN CABLE TENSION STRENGTH TEST
SCORES (1b.) OF OREGON BOYS 10, 13, AND 16 YEARS OF AGE CLASSIFIED INTO
PUBESCENT DEVELOPMENT GROUPS

Age (years)	Groups					
	1	2	3	4	4+5	5
10	54.25	58.89	--	--	--	--
13	--	75.34	81.40	--	--	--
13	--	75.34	--	--	94.50	--
13	--	--	81.40	--	94.50	--
16	--	--	--	112.42	--	128.01

(data from Clarke and Degutis, 1962, Research Quarterly 33:356-368, by permission
of The American Alliance for Health, Physical Education, Recreation and Dance,
1900 Association Drive, Reston VA 22091)

TABLE 55

STRENGTH AND PHYSICAL MEASUREMENTS IN CALIFORNIA
BOYS AGED 17.3 TO 17.7 YEARS

Variable	Mean	S.D.
Total strength (kg)	216.3	26.2
Endomorphy	2.6	1.0
Mesomorphy	4.2	1.1
Ectomorphy	3.8	1.1
Weight (kg)	65.5	7.2
Stature (cm)	177.4	6.5

Somatotypes made according to the method of
Sheldon (1940)

(data from Jones, 1949)

TABLE 56

50-YARD DASH (sec) PERFORMANCE IN U.S. YOUTH

Percen-				Age (years)				
tiles	10	11	12	13	14	15	16	17
				BOYS				
95	7.3	7.1	6.8	6.5	6.2	6.0	6.0	5.9
75	7.8	7.6	7.4	7.0	6.8	6.5	6.5	6.3
50	8.2	8.0	7.8	7.5	7.2	6.9	6.7	6.6
25	8.9	8.6	8.3	8.0	7.7	7.3	7.0	7.0
5	9.9	9.5	9.5	9.0	8.8	8.0	7.7	7.9
				GIRLS				
95	7.4	7.3	7.0	6.9	6.8	6.9	7.0	6.8
75	8.0	7.9	7.6	7.4	7.3	7.4	7.5	7.4
50	8.6	8.3	8.1	8.0	7.8	7.8	7.9	7.9
25	9.1	9.0	8.7	8.5	8.3	8.2	8.3	8.4
5	10.3	10.0	10.0	10.0	9.6	9.2	9.3	9.5

(data from AAHPER Youth Fitness Test Manual, 1976, by permission of the American Alliance for Health, Physical Education, Recreation and Dance, 1900 Association Drive, Reston VA 22091)

TABLE 57

9-MINUTE (yards) AND 1-MILE (min-sec) RUN PERFORMANCE IN U.S. YOUTHS

Percen-tiles	9-Minute Run Age (years)			1-Mile Run Age (years)		
	10	11	12	10	11	12
		BOYS				
95	2294	2356	2418	5:55	5:32	5:09
75	1952	2014	2076	7:49	7:26	7:03
50	1717	1779	1841	9:07	8:44	8:21
25	1482	1544	1606	10:25	10:02	9:39
5	1140	1202	1264	12:19	11:56	11:33
		GIRLS				
95	1969	1992	2015	7:28	6:57	6:23
75	1702	1725	1748	9:16	8:45	8:11
50	1514	1537	1560	10:29	9:58	9:24
25	1326	1349	1372	11:42	11:11	10:37
5	1069	1082	1105	13:30	12:59	12:24

(data from AAHPER Youth Fitness Test Manual, 1976, by permission of the American Alliance for Health, Physical Education, Recreation and Dance, 1900 Association Drive, Reston VA 22091)

TABLE   58

12-MINUTE (yards) AND 1.5 MILE (min-sec)
RUN PERFORMANCE IN U.S. YOUTH AGED 13 YEARS AND OVER

Percentiles	12-Minute Run	1.5 Mile Run
	BOYS	
95	3297	8:37
75	2879	10:19
50	2592	11:29
25	2305	12:39
5	1888	14:20
	GIRLS	
95	2448	12:17
75	2100	15:03
50	1861	16:57
25	1622	18:50
5	1274	21:36

(data from AAHPER Youth Fitness Test Manual, 1976, by
permission of the American Alliance for Health, Physical
Education, Recreation and Dance, 1900 Association Drive,
Reston VA 22091)

TABLE 59

600-YARD RUN (sec) PERFORMANCE IN U.S. YOUTH

Percen-tiles	Age (years)							
	10	11	12	13	14	15	16	17
				BOYS				
95	2' 5"	2' 2"	1'52"	1'45"	1'39"	1'36"	1'34"	1'32"
75	2'17"	2'15"	2' 6"	1'59"	1'52"	1'46"	1'44"	1'43"
50	2'33"	2'27"	2'19"	2'10"	2' 3"	1'56"	1'52"	1'52"
25	2'53"	2'47"	2'37"	2'27"	2'16"	2' 8"	2' 1"	2' 2"
5	3'22"	3'29"	3' 6"	3' 0"	2'51"	2'30"	2'31"	2'38"
				GIRLS				
95	2'20"	2'14"	2' 6"	2' 4"	2' 2"	2' 0"	2' 8"	2' 2"
75	2'39"	2'35"	2'26"	2'23"	2'19"	2'22"	2'26"	2'24"
50	2'56"	2'53"	2'47"	2'41"	2'40"	2'37"	2'43"	2'41"
25	3'15"	3'16"	3'13"	3' 6"	3' 1"	3' 0"	3' 3"	3' 2"
5	4' 0"	4'15"	3'59"	3'49"	3'49"	3'28"	3'49"	3'45"

(data from AAHPER Youth Fitness Test Manual, 1976, by permission of the American Alliance for Health, Physical Education, Recreation and Dance, 1900 Association Drive, Reston VA 22091)

TABLE 60

SHUTTLE RUN PERFORMANCE (sec) IN U.S. YOUTH

Percen-tiles	Age (years)							
	10	11	12	13	14	15	16	17
				BOYS				
95	10.0	9.7	9.6	9.3	8.9	8.9	8.6	8.6
75	10.6	10.4	10.2	10.0	9.6	9.4	9.3	9.2
50	11.2	10.9	10.7	10.4	10.1	9.9	9.9	9.8
25	12.0	11.5	11.4	11.0	10.7	10.4	10.5	10.4
5	13.1	12.9	12.4	12.4	11.9	11.7	11.9	11.7
				GIRLS				
95	10.2	10.0	9.9	9.9	9.7	9.9	10.0	9.6
75	11.1	10.8	10.8	10.5	10.3	10.4	10.6	10.4
50	11.8	11.5	11.4	11.2	11.0	11.0	11.2	11.1
25	12.5	12.1	12.0	12.0	12.0	11.8	12.0	12.0
5	14.3	14.0	13.3	13.2	13.1	13.3	13.7	14.0

(data from AAHPER Youth Fitness Test Manual, 1976, by permission of the American Alliance for Health, Physical Education, Recreation and Dance, 1900 Association Drive, Reston VA 22091)

TABLE 61

SOFTBALL THROW (in) FOR U.S. CHILDREN

Percen- tiles	Age (years)							
	10	11	12	13	14	15	16	17
BOYS								
95	138	151	165	195	208	221	238	249
75	114	126	141	163	176	192	201	213
50	96	111	120	140	155	171	180	190
25	81	94	106	120	133	152	160	163
5	60	70	76	88	102	123	127	117
GIRLS								
95	84	95	103	111	114	120	123	120
75	65	74	80	86	90	95	92	93
50	50	59	64	70	75	78	75	75
25	40	46	50	57	61	64	63	62
5	21	32	37	36	45	45	45	40

(data from AAHPER Youth Fitness Test Manual, 1965, by permission of the
American Alliance for Health, Physical Education, Recreation and Dance, 1900
Association Drive, Reston VA 22091)

TABLE   62

SIT-UPS (no. completed: maximum 100 for
boys and 50 for girls) IN U.S. CHILDREN

Percen-				Age (years)				
tiles	10	11	12	13	14	15	16	17
				BOYS				
95	100	100	100	100	100	100	100	100
75	65	73	93	100	100	100	100	100
50	41	46	50	60	70	80	76	70
25	25	26	30	38	45	49	50	45
5	11	12	15	20	24	27	28	23
				GIRLS				
95	50	50	50	50	50	50	50	50
75	50	50	50	50	42	39	38	40
50	31	30	32	31	30	26	26	27
25	20	20	20	20	20	19	18	18
5	8	10	7	10	10	8	7	9

(data from AAHPER Youth Fitness Test Manual, 1965, by permission of the
American Alliance for Health, Physical Education, Recreation and Dance, 1900
Association Drive, Reston VA   22091)

TABLE 63

SIT-UP PERFORMANCE (N/min); WITH FLEXED LEG IN U.S. YOUTH

Percen-tiles	Age (years)							
	10	11	12	13	14	15	16	17
				BOYS				
95	47	48	50	53	55	57	55	54
75	38	40	42	45	47	48	47	46
50	31	34	35	38	41	42	41	41
25	25	26	30	30	34	37	35	35
5	13	15	18	20	24	28	28	36
				GIRLS				
95	45	43	44	45	45	45	43	45
75	34	35	36	36	37	36	35	35
50	27	29	29	30	30	31	30	30
25	21	22	24	23	24	25	24	25
5	10	9	13	15	16	15	15	18

(data from AAHPER Youth Fitness Test Manual, 1976, by permission of the American Alliance for Health, Physical Education, Recreation and Dance, 1900 Association Drive, Reston VA 22091)

TABLE   64

PULL-UPS PERFORMANCE (N/completed) FOR U.S. BOYS

Percen-				Age (years)				
tiles	10	11	12	13	14	15	16	17
95	9	8	9	10	12	15	14	15
75	3	4	4	5	7	9	10	10
50	1	2	2	3	4	6	7	7
25	0	0	0	1	2	3	4	4
5	0	0	0	0	0	0	1	0

(data from AAHPER Youth Fitness Test Manual, 1976, by permission of the
American Alliance for Health, Physical Education, Recreation and Dance, 1900
Association Drive, Reston VA 22091)

TABLE   65

FLEXED-ARM HANG PERFORMANCE (sec) FOR U.S. GIRLS

Percen-				Age (years)				
tiles	10	11	12	13	14	15	16	17
95	42	39	33	34	35	36	31	34
75	18	20	18	16	21	18	15	17
50	9	10	9	8	9	9	7	8
25	3	3	3	3	3	4	3	3
5	0	0	0	0	0	0	0	0

(data from AAHPER Youth Fitness Test Manual, 1976, by permission of the American
Alliance for Health, Physical Education, Recreation and Dance, 1900 Association
Drive, Reston VA 22091)

TABLE 66

STANDING LONG JUMP (feet, inches) PERFORMANCE IN U.S. YOUTH

Percen-tiles	Age (years)							
	10	11	12	13	14	15	16	17
				BOYS				
95	6' 0"	6' 2"	6' 6"	7' 1"	7' 6"	8' 0"	8' 2"	8' 5"
75	5' 4"	5' 7"	5'11"	6' 3"	6' 8"	7' 2"	7' 6"	7' 9"
50	4'11"	5' 2"	5' 5"	5' 9"	6' 2"	6'11"	7' 0"	7' 2"
25	4' 6"	4' 8'	5' 0"	5' 2"	5' 6"	6' 1"	6' 6"	6' 6"
5	3'10"	4' 0"	4' 2"	4' 4"	4' 8"	5' 2"	5' 5"	5' 3"
				GIRLS				
95	5'10"	6' 0"	6' 2"	6' 5"	6' 8"	6' 7"	6' 6"	6' 9"
75	5' 2"	5' 4"	5' 6"	5' 9"	5'11"	5'10"	5' 9"	6' 0"
50	4' 8"	4'11"	5' 0"	5' 3"	5' 4"	5' 5"	5' 3"	5' 5"
25	4' 1"	4' 4"	4' 6"	4' 9"	4'10"	4'11"	4' 9"	4'11"
5	3' 5"	3' 8"	3'10"	4' 0"	4' 0"	4' 2"	4' 0"	4' 1"

(data from AAHPER Youth Fitness Test Manual, 1976, by permission of the American Alliance for Health, Physical Education, Recreation and Dance, 1900 Association Drive, Reston VA 22091)

TABLE 67

PERFORMANCE DATA FOR U.S. BOYS

| Test | | Age (years) | | | | | | | |
		10	11	12	13	14	15	16	17
Pull-ups	Mean	2.3	2.6	2.8	3.6	5.0	6.5	7.1	7.2
(no. completed)	S.D.	3.0	2.8	3.2	3.4	4.1	4.4	4.1	4.4
Sit-ups	Mean	31.6	32.9	35.1	37.4	40.1	42.1	41.4	40.6
(no. completed)	S.D.	10.5	10.5	9.8	10.2	9.8	9.0	9.2	8.8
Shuttle Run	Mean	11.4	11.0	10.8	10.6	10.2	10.0	10.0	9.9
(sec.)	S.D.	1.1	1.1	1.1	1.0	1.0	1.0	1.0	1.0
Standing Long Jump	Mean	59.1	61.9	64.9	68.6	73.2	79.0	83.0	84.9
(in.)	S.D.	7.8	8.6	8.5	10.3	10.4	10.6	10.2	11.3
50-Yard Dash	Mean	8.4	8.2	7.9	7.6	7.3	6.9	6.8	6.7
(sec.)	S.D.	0.8	0.8	0.8	0.9	0.8	0.6	0.6	0.7
600-Yard Run	Mean	158.2	155.3	143.0	135.2	127.6	118.8	115.3	116.3
(sec.)	S.D.	27.3	32.3	23.5	24.2	24.6	19.4	18.4	22.5

(data from Hunsicker and Reiff, 1980

TABLE 68

PERFORMANCE DATA FOR U.S. GIRLS

| Test | | Age (years) | | | | | | | |
		10	11	12	13	14	15	16	17
Flexed Arm	Mean	12.7	13.0	11.9	11.2	13.0	12.6	10.4	11.6
(sec.)	S.D.	13.4	12.4	11.4	11.1	11.9	12.4	11.0	11.7
Sit-up	Mean	27.2	28.0	29.1	29.5	30.5	30.7	29.4	30.0
(no. completed)	S.D.	10.4	10.2	9.4	9.2	9.2	9.2	9.2	9.6
Shuttle Run	Mean	11.9	11.6	11.4	11.3	11.2	11.2	11.5	11.3
(sec.)	S.D.	1.3	1.3	1.1	1.1	1.3	1.2	1.4	1.4
Standing Long Jump	Mean	56.0	58.3	60.4	63.1	64.3	64.4	63.1	65.4
(in.)	S.D.	9.4	8.5	8.7	8.9	9.6	8.9	9.2	9.5
50-Yard Dash	Mean	8.7	8.4	8.3	8.1	7.9	7.9	8.0	7.9
(sec.)	S.D.	0.9	0.9	1.0	1.1	0.8	0.9	0.8	0.9
600-Yard Run	Mean	180.2	178.6	173.0	166.8	164.5	162.7	167.4	166.0
(sec.)	S.D.	32.4	35.2	36.1	32.5	33.8	31.4	30.7	34.5

(data from Hunsicker and Reiff, 1980

TABLE   69

PERFORMANCE SCORES FOR CHILDREN IN SANTA MONICA (CA)

		Age (years)						
		5	6	7	8	9	10	11

Side Step Mean Scores (lines crossed)

		5	6	7	8	9	10	11
BOYS	Mean	--	8.8	10.7	11.8	13.0	--	--
	S.D.	--	2.6	2.2	1.9	2.2	--	--
GIRLS	Mean	--	8.2	11.1	11.0	12.3	--	--
	S.D.	--	2.2	2.3	2.2	1.7	--	--

Number of shuffles from midline to lines 4' away in 10 seconds

Cable Jump Mean Scores (jumps)

		5	6	7	8	9	10	11
BOYS	Mean	--	1.8	3.0	4.9	5.5	--	--
	S.D.	--	2.3	2.4	2.6	2.6	--	--
GIRLS	Mean	--	3.5	5.2	5.2	6.2	--	--
	S.D.	--	2.7	2.2	2.7	1.8	--	--

Jump made with 24" rope

Grip Right Mean Scores (lb)

		5	6	7	8	9	10	11
BOYS	Mean	18.3	23.5	26.3	31.3	37.3	40.1	45.2
	S.D.	4.7	4.6	7.4	6.2	7.4	7.3	9.3
GIRLS	Mean	16.6	19.3	22.9	28.0	30.8	35.3	45.8
	S.D.	4.8	4.8	5.4	7.6	7.9	7.4	10.4

Beam Balance Mean Scores (seconds)

		5	6	7	8	9	10	11
BOYS	Mean	23.2	30.1	35.8	37.3	40.1	42.9	45.5
	S.D.	10.1	10.8	10.4	10.9	11.6	12.7	11.2
GIRLS	Mean	22.1	33.7	33.3	41.6	39.9	44.5	45.1
	S.D.	9.5	10.2	11.2	10.2	11.8	11.2	9.1

Sum of balancing on beams 1", 1 1/2", and 2" wide; maximum score per beam = 20 seconds

Beam Walk Mean Scores (steps)

		5	6	7	8	9	10	11
BOYS	Mean	15.7	17.8	21.0	21.9	20.9	24.3	25.9
	S.D.	5.7	6.6	4.7	5.9	6.0	4.4	4.4
GIRLS	Mean	15.7	20.6	20.7	24.2	24.3	24.4	25.8
	S.D.	5.7	5.9	7.2	4.2	4.6	4.3	3.4

Sum of steps walking on 3 beams 1", 1 1/2", and 2" wide

50-Foot Hop Median Scores (seconds)

		5	6	7	8	9	10	11
BOYS	Mean	10.6	7.4	7.0	6.1	6.0	4.9	4.8
GIRLS	Mean	10.2	7.3	5.9	5.7	5.2	4.7	4.5

TABLE   69   (continued)

PERFORMANCE SCORES FOR CHILDREN IN SANTA MONICA (CA)

					Age (years)			
		5	6	7	8	9	10	11

Mat Hop Test Mean Scores (points)

		5	6	7	8	9	10	11
BOYS	Mean	--	18.9	23.3	31.9	35.9	--	--
	S.D.	--	9.0	11.0	10.3	10.0	--	--
GIRLS	Mean	--	24.6	33.2	40.1	43.2	--	--
	S.D.	--	10.9	9.1	10.9	8.8	--	--

Continuous hopping from square to square; each square is 15" x 15"

Ball Throw Mean Scores (foot)

		5	6	7	8	9	10	11
BOYS	Mean	--	34.1	45.2	59.0	70.7	94.0	105.9
	S.D.	--	11.9	12.7	13.3	14.0	21.0	20.7
GIRLS	Mean	--	19.0	25.8	33.8	41.3	49.0	57.6
	S.D.	--	7.4	9.0	10.9	14.0	16.3	17.0

Overhand with 12" softball measured to point of landing

Accuracy Throw Mean Scores (points)

		5	6	7	8	9	10	11
BOYS	Mean	--	--	9.0	11.6	15.6	--	--
	S.D.	--	--	6.3	5.5	4.6	--	--
GIRLS	Mean	--	--	7.7	10.5	13.2	--	--
	S.D.	--	--	5.1	4.8	5.2	--	--

Overhand throws at a target; distance depends on sex and age

Standing Long Jump Mean Scores (inches)

		5	6	7	8	9	10	11
BOYS	Mean	35.9	42.9	49.1	55.2	56.6	61.4	66.5
	S.D.	6.7	7.2	6.2	6.8	7.4	6.2	7.7
GIRLS	Mean	33.1	41.2	48.6	49.9	52.7	57.3	61.9
	S.D.	6.1	5.0	5.9	6.7	6.6	6.8	9.4

Hurdle Standing Jump Mean Scores (inches)

		5	6	7	8	9	10	11
BOYS	Mean	14.4	17.8	19.0	21.5	22.9	--	--
	S.D.	2.93	2.8	2.2	2.6	2.5	--	--
GIRLS	Mean	14.4	17.8	20.1	21.2	22.1	--	--
	S.D.	2.70	2.4	2.6	2.8	2.5	--	--

30-Yard Dash Mean Scores (seconds)

		5	6	7	8	9	10	11
BOYS	Mean	7.47	6.78	6.28	5.94	5.66	--	--
	S.D.	0.64	0.60	0.41	0.38	0.41		
GIRLS	Mean	7.67	6.69	6.17	6.10	5.81		
	S.D.	0.71	0.50	0.50	0.52	0.32	--	--

TABLE 69   (continued)

PERFORMANCE SCORES FOR CHILDREN IN SANTA MONICA (CA)

		Age (years)						
		5	6	7	8	9	10	11

Shuttle Run Mean Scores (seconds)

		5	6	7	8	9	10	11
BOYS	Mean	--	13.83	13.12	12.17	11.99	--	--
	S.D.	--	1.13	0.87	0.70	0.76	--	--
GIRLS	Mean	--	13.83	12.99	12.92	12.62	--	--
	S.D.	--	1.02	0.80	1.04	0.88	--	--

(data from Keogh, 1965)

TABLE 70

MEANS FOR MOTOR PERFORMANCE IN GEORGIA GIRLS

Age (years)	Ball Bounce (sec)	Jump Rope (N/30 sec)	Jump-height (in)	Wall Ball (N catches /30 secs)	Accuracy throw (N correct/ 3 trials)	Side Step (N /30 sec)	Distance throw (ft)	Base run (sec)
12.3	8.88	46.98	12.24	28.07	2.33	21.72	33.93	9.60
13.5	8.41	46.53	11.21	28.12	2.44	21.81	34.23	9.83
14.4	8.02	46.43	12.61	31.04	2.69	21.71	38.16	9.55
15.3	7.48	52.84	12.15	33.65	2.89	24.54	43.85	10.46
16.4	7.62	49.07	11.58	32.67	2.77	26.17	42.82	10.38
17.2	7.63	52.07	11.95	32.20	2.63	27.27	43.73	10.37
18.4	7.66	58.92	11.75	33.35	2.85	33.67	33.77	10.02

Ball bounce is time taken to bounce a ball around a series of obstacles; jump height is recorded against a wall.

(data from Vincent, 1968, Research Quarterly 39:1094-1100, by permission of the American Alliance for Health, Physical Education, Recreation and Dance, 1900 Association Drive, Reston VA 22091)

TABLE 71

PERFORMANCE DATA FOR ILLINOIS BOYS

Percentiles	Age (years)						
	7	8	9	10	11	12	13

440-Yard Run (seconds)

Percentiles	7	8	9	10	11	12	13
95	60	73	73	53	67	--	--
75	84	97	87	79	84	--	--
50	134	125	105	112	106	--	--
25	184	154	123	145	127	--	--
5	224	176	138	171	144	--	--

600-Yard Run (seconds)

Percentiles	7	8	9	10	11	12	13
95	115	102	102	101	87	88	--
75	136	132	131	126	114	115	--
50	163	169	166	158	149	149	--
25	190	206	202	190	183	183	--
5	211	236	231	216	211	210	--

Vertical Jump (inches)

Percentiles	7	8	9	10	11	12	13
95	12.0	14.8	14.2	15.7	16.2	18.4	16.6
75	10.2	12.4	12.5	13.6	14.0	15.9	14.7
50	8.0	9.5	10.4	11.0	11.3	12.7	12.3
25	5.8	6.5	8.2	8.3	8.6	9.6	10.0
5	4.0	4.2	6.5	6.2	6.4	7.0	8.1

60-Yard Dash (seconds)

Percentiles	7	8	9	10	11	12	13
95	8.9	6.8	8.0	9.0	7.8	6.3	8.5
75	10.4	9.0	9.3	9.7	8.9	7.9	9.1
50	12.2	11.9	11.0	10.7	10.4	9.9	9.9
25	13.9	14.7	12.7	11.6	11.8	11.8	10.6
5	15.4	16.9	14.0	12.3	12.9	13.4	11.2

Agility Run (seconds)

Percentiles	7	8	9	10	11	12	13
95	18.7	18.0	18.3	17.3	17.0	15.2	16.1
75	20.8	20.4	20.3	19.3	19.0	17.6	18.1
50	23.3	23.4	22.8	21.8	21.5	20.4	20.6
25	25.8	26.4	25.3	24.3	24.0	23.4	23.1
5	27.9	28.8	27.3	26.3	26.0	25.7	25.1

Running High Jump (inches)

Percentiles	7	8	9	10	11	12	13
95	38	39	47	42	51	--	--
75	33	34	40	38	44	--	--
50	27	28	31	33.5	36	--	--
25	21	22	22	29	28	--	--
5	16	18	14	25	22	--	--

TABLE 71   (continued)

PERFORMANCE DATA FOR ILLINOIS BOYS

				Age (years)			
Percentiles	7	8	9	10	11	12	13

Balance Beam (score)

95	25	29	29	29	29	29	29
75	19	24	23	27	27	28	27
50	10	13	14	17	18	18	19
25	2	3	4	7	9	9	7
5	--	--	--	0	2	1	--

8-Pound Shot-Put (inches)

95	--	175	191	228	--	--	--
75	--	147	164	198	--	--	--
50	--	111	130	160	--	--	--
25	--	76	95	121	--	--	--
5	--	47	68	91	--	--	--

Standing Long Jump (inches)

95	67	70	63	78	78	87	79
75	59	62	59	68	70	75	72
50	49	50	53	57	60	61	64
25	39	41	48	45	50	47	55
5	31	32	43	36	42	35	49

Running Long Jump (inches)

95	111	138	123	137	157	--	--
75	92	111	106	119	132	--	--
50	68	77	85	96	102	--	--
25	44	42	65	73	71	--	--
5	26	15	48	55	47	--	--

Floor Push-Ups (number)

95	--	19	23	23	24	24	28
75	--	14	17	17	18	17	20
50	--	7	8	8	9	9	10
25	--	1	0	0	2	0	0

Endurance Hops (number)

95	855	876	901	1043	948	1034	1134
75	607	622	653	775	732	774	882
50	297	304	343	440	462	449	567
25	0	0	33	105	192	124	252

(data from Cureton and Barry, 1964)

TABLE 72

MEAN PERFORMANCE IN LOUISIANA GIRLS

Variable	Age (years)			Race	
	6	7	8	Caucasian	Black
Vertical Jump and Reach (in)	6.68	6.88	7.78	7.21	7.02
Standing Long Jump (in)	36.67	39.85	46.70	41.43	40.71
Modified Pull-Ups (total)	12.05	10.95	14.42	12.72	12.22
Modified Push-Ups (total)	10.63	9.48	8.33	9.44	9.52
Bent Arm Hang (secs)	9.96	12.42	12.92	12.14	11.39
Grip Strength (lbs)	18.03	21.20	25.53	23.55	19.62
Leg Lift (lbs)	145.33	214.99	234.33	211.56	184.89
Shoulder Extension Strength (lbs)	10.13	12.97	15.20	12.36	13.18
Hip Extension Strength (lbs)	12.23	17.63	19.57	16.04	16.91
Trunk Flexion Strength (lbs)	15.20	15.97	21.40	17.29	17.76
Wrist Flex.-Ext. Flexibility (degrees)	143.37	137.10	143.33	137.84	144.69
Trunk-Hip Flex.-Ext. Flexibility (degrees)	168.17	165.73	172.43	167.78	169.78
Leg Flex.-Ext. Flexibility (degrees)	146.90	144.47	148.57	145.38	147.91
Neck Flex.-Ext. Flexibility (degrees)	150.80	147.97	151.87	147.47	152.96
Arm Flex. on Back Flexibility (degrees)	93.80	92.77	92.80	90.89	95.36
Well's Sit and Reach (in)	12.55	12.37	11.83	12.38	12.12
Bass Lengthwise Balance (secs)	11.36	15.71	17.02	14.73	14.67
Bass Crosswise Balance (secs)	4.67	6.06	7.09	6.00	5.88
Railwalk (no. segments crossed)	3.33	4.23	4.90	3.87	4.44
6-Second Run (yards)	29.13	29.87	31.67	30.49	29.96
10-Yard Dash (secs)	3.35	3.18	3.05	3.17	3.22
50-Yard Dash (secs)	11.23	10.70	10.08	10.74	10.61
Time-Limit Shuttle Run (no. crossed zones)	26.67	28.83	29.37	27.71	28.87
Dodging Run (secs)	24.45	25.41	24.35	26.94	25.21
Scott Obstacle Race (secs)	26.63	25.13	23.94	25.47	24.99
Illinois Agility Run (secs)	28.51	26.57	25.16	27.60	25.80
300-Yard Run (secs)	95.67	88.51	84.64	89.28	89.93
600-Yard Run (secs)	213.15	206.46	198.30	203.29	208.65

Grip strength and leg lift measured with a manuometer dynamometer; shoulder, hip and trunk flexion strength measured with a cable tensiometer; flexibility measured with a Leighton flexometer.

(data from Dinucci and Shows, 1977, Research Quarterly 48:680-683, by permission of the American Alliance for Health, Physical Education, Recreation and Dance, 1900 Association Drive, Reston VA 22091)

TABLE 73

PHYSICAL GROWTH AND MOTOR SKILL PERFORMANCE IN
MASSACHUSETTS BOYS SEPARATED BY CHRONOLOGICAL AGE

Age Interval (months)	Stature		Weight	
	Mean	S.D.	Mean	S.D.
73-75	47.09	4.03	51.89	7.82
76-78	47.47	1.96	51.61	5.74
79-81	48.21	2.2	52.88	6.57
82-84	48.66	2.67	53.41	6.56
85-87	49.55	2.62	57.13	6.63
88-90	51.0	2.52	60.19	8.49
91-93	49.95	2.35	57.78	6.14
94-96	51.63	1.84	63.23	11.1
97-99	52.71	2.18	63.38	6.57
100-102	52.78	2.54	64.84	9.03
103-105	52.51	2.61	69.0	13.95

Age Interval (months)	Running (sec)		Balance (sec)		Agility (no)		Jumping (in)		Throwing (sec a)	
	Mean	S.D.	Mean	S.D.	Mean	S.D.	Mean	S.D.	Mean	S.D.
73-75	9.99	1.45	3.56	3.63	8.94	.84	35.5	2.62	135.81	38.06
76-78	9.36	1.09	5.12	3.40	9.00	1.48	35.82	5.76	148.66	45.25
79-81	8.26	.77	6.15	4.34	9.77	1.39	35.31	8.06	143.88	52.04
82-84	9.09	1.26	6.96	5.31	9.00	1.00	33.66	5.95	141.11	40.75
85-87	8.5	.67	6.21	4.62	9.35	1.24	38.35	5.12	194.28	42.39
88-90	8.39	1.02	8.69	7.69	9.33	1.74	36.79	7.33	153.67	44.54
91-93	8.19	.67	5.48	3.29	10.38	1.23	38.66	6.76	167.69	45.33
94-96	8.28	.47	12.03	10.79	10.58	1.68	42.17	4.69	213.04	45.61
97-99	8.1	1.82	7.32	6.75	10.29	1.52	42.1	6.67	196.33	46.29
100-102	7.82	.53	8.52	5.42	11.13	1.46	43.0	7.2	193.81	96.56
103-105	7.77	.63	11.66	10.47	10.94	1.25	43.06	6.22	224.18	37.22

TABLE   73   (continued)

PHYSICAL GROWTH AND MOTOR SKILL PERFORMANCE IN
MASSACHUSETTS BOYS SEPARATED BY CHRONOLOGICAL AGE

Age Interval (months)	Striking (sec b)		Catching (sec c)	
	Mean	S.D.	Mean	S.D.
73-75	4.31	1.9	6.38	2.16
76-78	4.18	2.09	6.55	2.61
79-81	5.15	1.89	7.46	1.99
82-84	4.59	1.8	6.82	2.10
85-87	5.45	2.25	8.25	1.81
88-90	5.79	1.82	8.17	1.58
91-93	4.69	2.19	8.56	1.08
94-96	5.17	1.5	8.67	1.68
97-99	5.52	2.14	8.48	1.48
100-102	4.94	2.01	8.5	1.77
103-105	5.82	2.21	9.41	.93

a = total of 3 throws to nearest half foot.
b = pendulum-controlled striking: number of
    successful attempts in 10 tries.
c = hoop-controlled catching tennis ball:
    number of successful catches in 10 tries.

(data from Seils, 1951, Research Quarterly 22:244-
264, by permission of the American Alliance for Health,
Physical Education, Recreation and Dance, 1900 Association
Drive, Reston, VA 22091)

TABLE 74

PHYSICAL GROWTH AND MOTOR SKILL PERFORMANCE IN
MASSACHUSETTS GIRLS SEPARATED BY CHRONOLOGICAL AGE

Age Interval	Stature		Weight	
(months)	Mean	S.D.	Mean	S.D.
71-73	45.59	1.69	46.64	2.57
74-76	46.85	1.33	49.31	8.71
77-79	46.84	2.02	51.0	5.04
80-82	47.73	1.93	50.6	6.0
83-85	48.68	1.49	54.03	9.02
87-89	48.29	2.78	50.06	9.2
90-91	50.14	1.98	61.0	7.57
92-94	51.12	.62	65.81	12.71
95-97	51.07	2.71	60.17	13.19
98-100	51.47	2.27	59.16	7.31
101-103	51.96	2.60	64.68	12.0
104-106	52.61	2.22	63.19	13.96

Age Interval (months)	Running (sec)		Balance (sec)		Agility (no)		Jumping (in)		Throwing (sec a)	
	Mean	S.D.	Mean	S.D.	Mean	S.D.	Mean	S.D.	Mean	S.D.
71-73	9.49	.93	3.1	2.18	8.45	1.27	32.27	4.94	57.14	25.67
74-76	9.9	1.19	3.68	3.61	9.29	1.26	34.06	4.66	71.97	25.39
77-79	9.6	1.08	4.16	2.26	8.71	1.41	36.36	5.54	74.14	35.4
80-82	9.18	1.15	5.85	7.1	8.85	1.24	35.75	5.56	76.55	17.0
83-85	9.01	.64	4.67	6.48	9.48	1.58	37.65	7.84	79.85	40.45
86-88	9.02	.82	4.14	3.28	9.15	1.46	35.65	5.3	85.1	24.9
89-91	9.23	1.04	7.63	8.36	9.7	1.08	34.83	6.72	76.04	13.09
92-94	8.65	.84	7.19	6.1	9.19	1.53	35.5	5.87	89.36	29.25
95-97	8.72	.66	4.55	2.62	10.17	1.58	39.67	7.05	92.70	28.37
98-100	8.95	.79	9.01	6.92	10.24	1.5	38.36	6.57	94.14	23.14
101-103	8.34	.5	8.93	3.11	10.71	1.77	44.71	7.13	121.21	36.87
104-106	8.46	.63	13.86	10.33	10.72	1.21	43.39	7.97	130.5	48.64

TABLE 74   (continued)

PHYSICAL GROWTH AND MOTOR SKILL PERFORMANCE IN
MASSACHUSETTS GIRLS SEPARATED BY CHRONOLOGICAL AGE

Age Interval (months)	Striking (sec b)		Catching (sec c)	
	Mean	S.D.	Mean	S.D.
71-73	3.36	1.5	5.27	3.05
74-76	4.59	1.57	5.88	2.2
77-79	4.29	1.7	6.86	2.19
80-82	4.65	2.01	6.75	2.43
83-85	5.09	1.37	7.13	2.46
86-88	3.8	1.93	7.9	1.93
89-91	4.65	1.87	7.65	2.04
92-94	4.76	1.57	8.67	1.40
95-97	5.44	2.13	7.72	1.97
98-100	4.76	1.75	8.84	1.46
101-103	4.64	.98	8.86	1.28
104-106	6.22	1.32	8.78	1.74

a = total of 3 throws to nearest half foot
b = pendulum-controlled striking: number of successful
    attempts in 10 tries
c = hoop-controlled catching a tennis ball: number of
    successful catches in 10 tries

(data from Seils, 1951, Research Quarterly 22:244-264,
by permission of the American Alliance for Health, Physical
Education, Recreation and Dance, 1900 Association Drive,
Reston VA 22091)

TABLE 75

PHYSICAL GROWTH AND MOTOR SKILL PERFORMANCE IN
MASSACHUSETTS CHILDREN SEPARATED BY PRIMARY SCHOOL GRADE

	Grade I		Grade II		Grade III	
	Mean	S.D.	Mean	S.D.	Mean	S.D.
BOYS						
Physical Growth						
Age (months)	78.93	4.98	91.03	6.71	103.40	7.16
Stature (cm)	47.75	2.30	50.90	2.16	53.38	2.34
Weight (lb)	51.82	6.50	58.94	6.92	67.43	10.95
Motor performance						
Dash (sec)	9.30	1.20	8.28	.66	7.97	.60
Stick test (sec)	5.02	4.04	7.59	6.69	9.19	8.22
Sidestep (no)	9.10	1.50	9.60	1.31	11.00	1.52
Jump (in)	34.97	6.60	38.45	5.92	44.00	6.12
Ball throw (sec a)	137.86	50.40	176.04	51.94	216.06	57.34
Striking (sec b)	4.70	2.60	5.40	2.24	5.60	2.01
Catching (sec c)	6.60	2.50	8.40	1.70	9.00	1.50
GIRLS						
Physical Growth						
Age (months)	78.35	4.74	89.88	5.14	101.57	5.60
Stature (cm)	46.95	1.79	49.80	2.46	52.10	2.42
Weight (lb)	49.19	7.90	58.32	12.87	61.67	9.28
Motor performance						
Dash (sec)	9.51	1.12	9.0	.89	8.64	.69
Stick test (sec)	5.17	5.61	5.07	5.42	10.50	9.42
Sidestep (no)	8.90	1.44	9.79	2.66	10.59	1.52
Jump (in)	35.62	5.45	35.22	6.51	41.96	6.93
Ball throw (sec a)	71.66	27.09	82.14	30.78	108.01	33.68
Striking (sec b)	4.33	1.84	4.46	1.69	5.33	1.73
Catching (sec c)	6.43	2.48	7.61	2.08	9.81	1.58

a = total of 3 throws to nearest half foot.
b = pendulum-controlled striking: number of successful attempts in
    10 tries.
c = hoop-controlled catching tennis ball: number of successful catches
    in 10 tries.

(data from Seils, 1951, Research Quarterly 22:244-264, by permission of
the American Alliance for Health, Physical Education, Recreation and
Dance, 1900 Association Drive, Reston, VA 22091)

TABLE   76

PERFORMANCE DATA FOR MICHIGAN CHILDREN

	GRADE									
	Second		Third		Fourth		Fifth		Sixth	
Sex	Fall	Spring	Fall	Spring	Fall	Spring	Fall	Spring	Fall	Spring
Standing Long Jump (Feet)										
Boys										
Mean	3.59	3.68	4.02	4.19	4.07	4.28	4.40	4.48	4.75	5.04
S.D.	0.49	0.45	0.48	0.48	0.39	0.42	0.41	0.53	0.47	0.58
Girls										
Mean	3.42	3.40	3.68	3.78	3.86	4.00	4.14	4.26	4.48	4.32
S.D.	0.51	0.51	0.53	0.50	0.58	0.51	0.56	0.59	0.54	0.58
Basketball Throw (Feet)										
Boys										
Mean	19.43	21.79	25.08	28.49	32.03	32.78	35.28	39.90	44.17	48.38
S.D.	5.32	5.57	6.13	6.08	6.37	6.17	5.32	8.27	11.38	12.74
Girls										
Mean	11.81	12.84	15.42	17.02	18.66	21.70	23.91	29.71	26.99	29.82
S.D.	3.05	3.31	4.13	5.01	4.88	5.46	5.43	11.20	4.92	5.32
Horizontal Ladder (Distance in feet)										
Boys										
Mean	24.38	26.00	28.33	29.31	28.98	29.55	31.22	29.20	33.40	35.57
S.D.	10.54	9.68	9.36	10.11	10.38	10.02	9.32	9.68	9.13	6.96
Girls										
Mean	22.21	21.38	27.84	25.29	28.11	25.94	29.67	30.44	31.64	29.50
S.D.	10.52	9.80	9.38	10.15	9.85	10.60	10.81	10.56	9.88	9.71
Modified Shuttle Run (Sec)										
Boys										
Mean	21.97	22.05	21.98	20.11	20.03	19.70	19.70	18.53	17.70	17.51
S.D.	1.42	1.65	1.57	1.00	1.79	1.40	1.39	1.46	1.34	1.58
Girls										
Mean	23.08	22.37	22.18	20.79	20.31	20.01	19.90	18.93	18.17	18.52
S.D.	1.80	1.76	1.96	1.28	1.39	1.30	1.64	1.31	1.32	1.65

TABLE 76   (continued)

PERFORMANCE DATA FOR MICHIGAN CHILDREN

	GRADE									
	Second		Third		Fourth		Fifth		Sixth	
Sex	Fall	Spring	Fall	Spring	Fall	Spring	Fall	Spring	Fall	Spring

Balance Beam (ft)

Boys										
Mean	8.84	11.84	14.74	16.59	17.37	17.82	19.35	19.51	21.23	20.91
S.D.	4.73	6.02	4.51	5.05	4.32	4.22	3.53	3.9	3.12	3.12
Girls										
Mean	12.43	14.40	18.15	19.36	19.80	21.19	20.32	20.26	21.02	19.22
S.D.	5.48	6.26	4.57	4.11	4.13	5.68	4.47	4.65	2.91	5.10

Turn Under Bar (No. in 30 Sec)

Boys										
Mean	3.57	4.63	5.33	5.85	5.67	6.07	6.39	6.91	7.67	7.54
S.D.	2.30	2.16	2.42	2.36	2.08	2.47	2.79	2.85	3.14	3.12
Girls										
Mean	4.92	5.60	6.44	6.88	7.30	7.54	7.44	8.08	8.52	8.14
S.D.	2.27	1.71	2.26	1.76	2.67	2.31	2.19	2.51	2.45	2.26

(data from Govatos, 1966)

TABLE 77

PHYSICAL PERFORMANCE IN CHILDREN OF MILWAUKEE (WI)

GIRLS

	Age in years					
	9		10		11	
Measures	Mean	S.D.	Mean	S.D.	Mean	S.D.
Run (50 yd.; sec)	9.87	1.03	9.14	1.03	8.84	.89
Standing Jump (in)	44.54	6.67	49.97	6.71	53.31	7.74
Softball Throw (sec)	32.94	10.30	43.58	13.12	46.41	12.50
Shuttle Run (sec)	13.58	.85	12.59	1.10	12.31	1.00
600-Yard Run-Walk (sec)	205.04	38.43	202.40	44.39	192.86	46.03
Flexed Arm-Hang (girls; sec) Pull-ups (boys; no)	5.83	3.94	6.18	3.98	7.93	5.92
Sit-ups (no)	25.39	13.50	20.40	12.00	29.00	16.23

BOYS

	Age in years					
	9		10		11	
Measures	Mean	S.D.	Mean	S.D.	Mean	S.D.
Run (50 yd.; sec)	8.56	.67	8.47	1.50	8.28	.74
Standing Jump (in)	49.90	6.27	52.89	6.40	56.55	7.56
Softball Throw (sec)	61.85	15.06	79.37	16.41	88.60	20.94
Shuttle Run (sec)	167.75	30.68	178.32	35.34	162.95	35.49
600-Yard Run-Walk (sec)	13.03	1.13	12.21	.87	11.73	.88
Flexed Arm-Hand (girls; sec) Pull-ups (boys; no)	3.05	2.56	2.84	2.08	2.80	2.91
Sit-ups (no)	71.20	32.27	83.16	28.26	64.80	32.57

(data from Safrit, unpublished)

TABLE 78

PITCHING ACCURACY FOR CHILDREN IN
GRADES ONE THROUGH SIX IN MINNESOTA

Percentile	GRADE						GRADE					
	1	2	3	4	5	6	1	2	3	4	5	6
	BOYS						GIRLS					
95	56	64	76	81	86	89	30	39	55	66	67	73
90	45	57	69	74	79	85	26	34	47	58	62	69
75	33	44	58	63	68	75	17	26	37	45	51	59
50	21	30	43	53	58	64	9	17	29	31	42	48
25	12	23	33	41	48	55	2	8	18	21	33	37
10	2	17	23	33	39	48	1	3	9	17	22	28
5	1	15	17	25	35	44	1	2	5	8	17	21

Using softball and a target 25 feet away with circles 1, 2, 3, 4 and 5 feet in diameter.
Scores are 5, 4, 3, 2 or 1 depending on the circle hit.

(data from Hanson, 1965)

TABLE 79

VOLLEYBALL SERVE (ft) FOR DISTANCE ON SCORES OBTAINED FOR
CHILDREN IN GRADES ONE THROUGH SIX IN MINNESOTA

Percentiles	GRADES						GRADES					
	1	2	3	4	5	6	1	2	3	4	5	6
	BOYS						GIRLS					
95	131	171	207	242	277	322	94	118	135	181	210	233
90	120	154	189	223	258	296	83	107	122	154	189	214
75	99	131	160	188	224	259	60	88	100	132	161	194
50	74	100	128	153	184	225	47	69	79	109	131	156
25	59	78	101	123	157	179	36	54	62	85	99	130
10	49	62	84	98	131	155	27	42	57	67	80	111
5	41	52	68	91	121	139	21	35	38	58	70	99

Distance ball is hit.

(data from Hanson, 1965)

TABLE 80

FULL-HANG PULL-UP SCORES FOR CHILDREN IN
GRADES ONE THROUGH SIX IN MINNESOTA

Percentile	GRADES						GRADES					
	1	2	3	4	5	6	1	2	3	4	5	6
	BOYS						GIRLS					
95	5	7	7	7	8	9	5	4	5	4	4	4
90	4	6	6	5	7	7	3	3	3	3	3	3
75	3	4	3	3	4	4	2	1	2	2	1	2
50	1	2	2	1	2	2	1	1	1	1	1	1
25	1	1	1	1	1	1	1	1	1	1	1	1
10	1	1	1	1	1	1	1	1	1	1	1	1
5	1	1	1	1	1	1	1	1	1	1	1	1

Scored as number of times.

(data from Hanson, 1965)

TABLE 81

POTATO RACE SCORES FOR CHILDREN IN GRADES ONE
THROUGH SIX IN MINNESOTA

Percentile	GRADE						GRADE					
	1	2	3	4	5	6	1	2	3	4	5	6
	BOYS						GIRLS					
95	29	30	32	32	33	34	28	29	29	31	31	33
90	28	29	30	31	32	33	27	28	28	30	31	32
75	26	27	28	30	30	31	25	26	27	28	29	30
50	24	25	27	28	28	30	24	25	25	27	28	28
25	22	23	25	26	27	27	22	23	24	25	26	27
10	21	21	24	24	25	26	21	21	23	24	24	25
5	20	21	22	24	24	25	19	20	21	22	23	24

Tested with lines 12 feet apart for 15 seconds.

(data from Hanson, 1965)

TABLE 82

SOCCER WALL VOLLEY ON SCORES OBTAINED FOR CHILDREN IN GRADES ONE THROUGH SIX IN MINNESOTA

	GRADES						GRADES					
Percentile	1	2	3	4	5	6	1	2	3	4	5	6
			BOYS						GIRLS			
95	48	51	57	60	65	74	42	47	51	54	56	65
90	44	47	53	55	60	68	39	44	48	50	54	60
75	38	41	45	49	54	60	34	39	41	44	48	53
50	33	36	39	43	46	52	29	34	35	39	42	46
25	27	30	33	37	40	45	24	29	30	34	37	40
10	23	26	28	32	34	38	21	25	25	30	32	34
5	20	23	25	30	31	34	18	22	21	26	30	32

Number of times ball kicked against wall in 30 secs.

(data from Hanson, 1965)

TABLE 83

SOCCER PUNT (ft) FOR DISTANCE FOR CHILDREN IN GRADES ONE THROUGH SIX IN MINNESOTA

	GRADE						GRADE					
Percentile	1	2	3	4	5	6	1	2	3	4	5	6
			BOYS						GIRLS			
95	42	54	67	74	85	96	24	33	40	51	63	70
90	39	51	61	71	80	92	22	30	36	47	56	63
75	29	40	42	64	72	80	18	24	30	39	46	53
50	21	32	42	52	61	70	14	17	22	29	36	43
25	15	24	32	41	51	57	9	13	15	22	27	34
10	8	18	23	33	42	47	6	9	10	17	20	24
5	4	15	19	26	35	42	4	6	7	15	14	19

(data from Hanson, 1965)

TABLE 84

WALL PASS FOR CHILDREN IN GRADES ONE THROUGH SIX IN MINNESOTA

Percentile	GRADES						GRADES					
	1	2	3	4	5	6	1	2	3	4	5	6
	BOYS						GIRLS					
95	65	68	83	88	97	116	52	66	76	81	87	98
90	59	66	78	84	93	107	49	61	72	78	84	95
75	50	60	69	78	87	97	44	53	62	71	79	89
50	42	52	60	71	80	87	38	46	53	63	73	80
25	35	43	52	62	71	79	32	39	44	55	64	73
10	28	37	43	54	62	70	27	32	37	49	56	65
5	23	33	39	49	55	61	24	29	31	45	53	60

Number of hits with soccer ball against wall caught in 30 secs.

(data from Hanson, 1965)

TABLE 85

OVERHAND THROW (ft) FOR DISTANCE FOR CHILDREN
IN GRADES ONE THROUGH SIX IN MINNESOTA

Percentile	GRADES						GRADES					
	1	2	3	4	5	6	1	2	3	4	5	6
	BOYS						GIRLS					
95	71	90	101	122	143	152	39	53	60	69	85	100
90	65	84	94	115	131	144	34	48	52	63	77	92
75	56	73	83	102	118	130	29	38	42	52	62	74
50	47	61	73	87	102	114	24	31	35	44	54	62
25	38	53	63	77	86	100	19	25	29	35	45	51
10	29	42	54	67	76	86	16	21	25	29	36	43
5	25	35	49	63	65	81	14	20	20	25	32	38

Using a 12" softball.

(data from Hanson, 1965)

TABLE 86

HANSON SHOULDER TEST ON SCORES OBTAINED FROM CHILDREN IN
GRADES ONE THROUGH SIX IN THE STATE OF MINNESOTA

Percentiles	GRADES						GRADES					
	1	2	3	4	5	6	1	2	3	4	5	6
	BOYS						GIRLS					
95	25	32	39	42	51	50	25	29	33	39	42	47
90	23	29	34	39	46	47	23	26	27	35	40	44
75	22	24	29	33	39	42	20	23	24	27	32	39
50	19	21	24	26	30	36	16	20	20	21	23	32
25	15	18	20	21	23	28	14	16	17	18	19	22
10	12	14	16	17	17	21	11	12	14	14	16	18
5	9	12	14	15	15	18	8	11	12	12	15	16

Based on shifting hand to left and to right when doing push-ups; number in 30 seconds.

(data from Hanson, 1965)

TABLE 87

SIT-UPS IN 60 SECONDS FOR CHILDREN IN GRADES ONE
THROUGH SIX IN MINNESOTA

Percentile	GRADES						GRADES					
	1	2	3	4	5	6	1	2	3	4	5	6
	BOYS						GIRLS					
95	33	35	42	41	42	45	30	35	36	37	35	39
90	30	32	38	37	39	42	27	32	31	32	32	36
75	24	27	29	30	33	36	22	25	26	27	27	29
50	19	21	24	24	27	29	16	20	21	21	22	24
25	13	16	19	19	21	23	11	14	16	16	16	18
10	7	10	14	13	12	17	5	10	11	10	11	12
5	4	6	10	9	6	13	2	6	8	4	8	8

(data from Hanson, 1965)

TABLE 88

ROPE SKIPPING SCORES (30 seconds) FOR CHILDREN IN GRADES
ONE THROUGH SIX IN MINNESOTA

	GRADES						GRADES					
Percentile	1	2	3	4	5	6	1	2	3	4	5	6
	BOYS						GIRLS					
95	42	54	58	55	61	66	57	69	76	79	79	79
90	38	46	51	51	57	62	51	66	69	71	72	76
75	29	38	41	45	48	53	46	56	62	63	65	69
50	20	27	34	37	40	43	40	48	52	55	56	60
25	15	18	23	28	31	34	32	42	42	47	49	50
10	12	13	17	20	22	27	18	36	37	40	43	43
5	10	11	14	16	18	24	13	32	32	36	39	39

(data from Hanson, 1965)

TABLE 89

JUMP AND REACH ON SCORES OBTAINED FROM CHILDREN IN GRADES
ONE THROUGH SIX IN THE STATE OF MINNESOTA

	GRADES						GRADES					
Percentile	1	2	3	4	5	6	1	2	3	4	5	6
	BOYS						GIRLS					
95	11.5	12.6	14.0	15.5	17.0	18.0	11.0	12.5	13.0	14.5	16.0	18.0
90	11.0	12.0	13.5	14.5	16.5	17.5	10.5	12.0	12.0	13.5	15.0	16.5
75	10.0	11.0	12.0	13.0	15.0	16.0	9.5	10.5	11.0	12.0	13.5	14.5
50	9.0	9.5	11.0	11.5	13.0	14.0	8.5	9.0	9.5	11.0	11.5	12.5
25	7.5	8.5	9.5	10.5	11.5	12.0	7.0	7.5	8.5	9.5	10.5	11.5
10	6.5	7.5	8.5	9.0	10.0	10.5	5.5	7.0	7.5	8.5	9.0	10.0
5	5.5	6.5	8.0	8.5	9.0	9.5	5.0	6.5	7.0	8.0	8.5	9.5

Measured for a jump reaching evenly with both hands; recorded in feet and inches.

(data from Hanson, 1965)

TABLE 90

50 YARD DASH TIMES (secs.) FOR CHILDREN IN GRADES ONE
THROUGH SIX IN MINNESOTA

Percentile	GRADES						GRADES					
	1	2	3	4	5	6	1	2	3	4	5	6
	BOYS						GIRLS					
95	8.4	8.1	7.8	7.5	7.4	7.1	8.9	8.2	8.0	7.6	7.7	7.2
75	9.2	8.7	8.3	8.1	7.8	7.7	9.5	8.9	8.7	8.3	8.2	7.8
50	9.9	9.2	8.8	8.6	8.2	8.0	10.2	9.3	9.2	8.7	8.6	8.3
25	10.6	9.9	9.1	9.0	8.6	8.5	11.0	9.9	9.6	9.1	9.1	8.8
10	11.3	10.5	9.6	9.5	9.1	9.0	11.6	10.6	10.3	9.7	9.6	9.2
5	11.6	11.0	9.7	9.6	9.5	9.3	12.0	11.1	10.8	10.1	9.8	9.5

(data from Hanson, 1965)

TABLE 91

MODIFIED BASS BALANCE TEST SCORES FOR CHILDREN
IN GRADES ONE THROUGH SIX IN MINNESOTA

Percentile	GRADES						GRADES					
	1	2	3	4	5	6	1	2	3	4	5	6
	BOYS						GIRLS					
95	70	108	161	229	237	297	88	112	163	200	210	223
90	49	68	96	166	180	227	62	79	108	160	156	171
75	31	40	57	88	87	138	41	53	65	81	87	101
50	18	22	31	42	40	69	19	24	37	48	44	53
25	10	13	17	23	22	29	10	12	17	25	20	25
10	7	8	10	13	13	16	8	10	12	14	14	14
5	6	7	8	10	11	12	6	8	10	11	11	10

Scored as the number of seconds a child can balance on a stick.

(data from Hanson, 1965)

TABLE 92

600-YARD WALK-RUN (minutes and secs) FOR CHILDREN IN
GRADES ONE THROUGH SIX IN MINNESOTA

	GRADES						GRADES					
Percentile	1	2	3	4	5	6	1	2	3	4	5	6
	BOYS						GIRLS					
95	29"	19"	17"	12"	5"	3"	42"	35"	27"	23"	22"	11"
90	38"	26"	22"	16"	12"	7"	50"	46"	36"	30"	27"	19"
75	52"	40"	35"	25"	19"	15"	7"	4"	47"	43"	37"	29"
50	13"	3'2"	49"	39"	32"	28"	22"	16"	8	3'0"	52"	41"
25	34"	18"	3'2"	54"	50"	47"	42"	36"	23"	18"	10"	3'3"
10	56"	37"	20"	11"	3'4"	11"	4'12"	52"	44"	43"	30"	27"
5	4'14"	59"	39"	21"	18"	30"	29"	4'12"	58"	48"	41"	40"

' - minutes; " - seconds.

(data from Hanson, 1965)

TABLE 93

STANDING LONG JUMP (in) FOR CHILDREN IN
GRADES ONE THROUGH SIX IN MINNESOTA

	GRADES						GRADES					
Percentile	1	2	3	4	5	6	1	2	3	4	5	6
	BOYS						GIRLS					
95	58	61	63	66	71	75	54	59	61	65	68	74
90	56	59	61	65	69	73	52	55	58	63	66	71
75	50	55	59	61	66	69	47	51	55	58	62	66
75	46	50	54	56	61	64	42	47	50	53	57	61
25	42	46	50	52	56	59	39	43	45	49	53	56
10	39	42	46	49	52	54	35	39	41	44	48	51
5	36	39	43	46	48	50	32	37	38	42	46	50

(data from Hanson, 1965)

TABLE 94

KICK, PASS-AND-CATCH, JUMP-AND-REACH, ZIG-ZAG RUN, BATTING
PERFORMANCE IN MINNESOTA CHILDREN

Percentiles	Kick (pts)		Pass-and-Catch (pts)		Jump-and-Reach (in)		Zig-Zag Run (sec)		Batting (No. of hits)	
	Boys	Girls	Boys	Girls	Boys	Girls	Boys	Girls	Boys	Girls
				GRADE 1						
95	28	27	26	23	9.0	8.5	9.2	9.4	5	4
75	25	24	21	18	8.0	7.5	10.0	10.8	--	--
50	--	--	17	--	6.5	--	10.9	11.6	2	--
25	--	19	13	10	--	4.5	11.6	12.2	--	--
5	14	10	9	5	3.5	3.0	12.8	13.4	0	--
				GRADE 2						
95	33	33	38	31	10.0	9.5	8.0	8.2	6	5
75	--	28	30	25	8.5	--	9.0	9.4	--	--
50	25	24	25	--	7.5	--	9.8	9.9	--	--
25	20	--	21	16	6.0	--	10.1	10.8	--	--
5	14	14	13	10	4.5	4.0	11.2	11.8	0	0
				GRADE 3						
95	37	34	40	34	11.5	10.0	7.8	8.0	7	--
75	--	--	36	29	9.5	8.5	8.6	8.9	--	--
50	29	25	32	25	--	7.5	--	9.5	3	--
25	26	21	--	21	7.5	6.5	9.8	10.0	2	--
5	20	17	21	16	5.0	5.0	10.6	11.0	--	--
				GRADE 4						
95	38	37	47	40	13.0	11.0	7.6	7.8	8	--
75	--	32	41	36	10.5	9.5	8.4	8.8	--	--
50	32	29	--	32	--	--	9.0	9.5	--	--
25	--	26	33	28	8.0	--	9.6	10.2	--	2
5	23	20	27	21	6.5	5.0	10.6	11.2	1	--

TABLE 94   (continued)

KICK, PASS-AND-CATCH, JUMP-AND-REACH, ZIG-ZAG RUN, BATTING
PERFORMANCE IN MINNESOTA CHILDREN

Percentiles	Kick (pts)		Pass-and-Catch (pts)		Jump-and-Reach (in)		Zig-Zag Run (sec)		Batting (No. of hits)	
	Boys	Girls	Boys	Girls	Boys	Girls	Boys	Girls	Boys	Girls
GRADE 5										
95	40	38	54	50	14.0	13.0	7.0	7.2	9	7
75	36	33	47	42	--	11.0	7.5	--	--	5
50	33	--	43	38	10.0	9.5	8.0	8.4	5	--
25	--	27	38	34	--	8.0	8.5	9.0	4	--
5	26	20	33	29	7.0	6.0	9.2	10.0	--	1
GRADE 6										
95	41	40	56	51	16.0	14.0	6.8	7.0	--	8
75	--	34	51	45	13.0	11.5	7.4	7.6	--	--
50	34	--	46	41	--	10.0	7.9	8.1	6	--
25	31	29	42	37	10.0	--	8.4	8.8	--	--
5	26	20	37	31	8.5	7.0	9.2	10.0	3	2

Kicking - points for each time ball kicked onto wall; Pass-and-Catch - points for throwing
ball onto target and catching it.

(data from R. D. Johnson, 1962, Research Quarterly 33:94-103, by permission of the American
Alliance for Health, Physical Education, Recreation and Dance, 1900 Association Drive,
Reston VA 22091)

TABLE 95

MEAN PERFORMANCE SCORES ON FUNDAMENTAL SKILL TESTS
FOR MINNESOTA BOYS AND GIRLS IN GRADES 1 TO 6

					Grades		
Tests		1	2	3	4	5	6
Kicking	Boys	22.56	26.58	29.62	33.25	34.40	32.88
	Girls	22.39	25.86	25.08	29.72	30.92	32.73
Throw-and-catch	Boys	16.38	26.64	30.75	38.55	44.14	46.50
	Girls	13.86	21.48	25.08	32.93	39.60	41.65
Zig-Zag Run	Boys	11.28	10.06	9.62	8.51	8.45	8.51
	Girls	12.05	10.44	9.83	9.88	9.17	9.23
Jump-and-Reach	Boys	6.07	7.02	8.77	9.90	11.10	11.70
	Girls	5.67	7.05	7.62	8.52	10.22	10.51
Batting	Boys	2.67	3.12	3.07	4.91	5.69	6.24
	Girls	2.08	2.12	2.92	4.21	4.75	5.57

Scores based on kicking (points for kicking ball onto wall longest), throw and
catch (points for throwing onto target and catching), zig-zag run (secs), jump
and reach (in) and batting (hits).

(data from R. D. Johnson, 1962, Research Quarterly 33:94-103, by permission
of the American Alliance for Health, Physical Education, Recreation and Dance,
1900 Association Drive, Reston VA 22091)

TABLE   96

PERFORMANCE DATA FROM CHILDREN IN NEW YORK CITY

Age	Boys Percentiles			Girls Percentiles		
(years)	25	50	75	25	50	75

Kraus Weber Test (Number Passed)

Age	25	50	75	25	50	75
7	4	5	6	4	5	6
9	5	6	6	5	6	6
11	5	6	6	5	6	6
13	5	6	6	5	6	6
15	5	6	6	5	6	6

Push-Ups (number completed)

Age	25	50	75	25	50	75
7	0	4	8	0	2	7
9	0	4	10	0	6	13
11	1	5	10	2	8	16
13	2	7	12	5	12	19
15	7	14	20	4	9	15

Standing Long Jump (inches)

Age	25	50	75	25	50	75
7	37	42	48	33	37	42
9	46	52	58	42	47	54
11	52	58	64	47	54	61
13	57	63	67	53	60	64
15	67	76	82	54	60	65

Agility Run (seconds)

Age	25	50	75	25	50	75
7	32	30	26	35	31	29
9	30	28	25	33	30	27
11	28	26	24	30	28	26
13	25	24	23	29	27	26
15	24	23	21	28	26	25

Chins (number passed)

Age	25	50	75	25	50	75
7	0	0	1	0	0	0
9	0	0	2	0	0	0
11	0	0	2	0	0	1
13	0	0	3	0	0	0
15	1	3	7	0	0	0

(data from Brown, unpublished)

TABLE 97

35-YARD DASH (sec) FOR PHILADELPHIA CHILDREN

Age (years)	WHITE BOYS Mean	S.D.	BLACK BOYS Mean	S.D.
6	7.38	0.79	7.00	0.64
7	6.87	0.55	6.47	0.57
8	6.50	0.56	6.22	0.55
9	6.30	0.52	6.02	0.41
10	6.03	0.48	5.93	0.44
11	5.86	0.41	5.80	0.43
12	5.82	0.43	5.60	0.64
13	--	--	5.51	0.43

Age (years)	WHITE GIRLS Mean	S.D.	BLACK GIRLS Mean	S.D.
6	7.84	0.98	7.30	0.70
7	7.31	0.85	6.97	0.66
8	6.85	0.54	6.52	0.53
9	6.58	0.49	6.38	0.57
10	6.33	0.52	6.28	0.53
11	6.07	0.45	6.00	0.66
12	--	--	5.74	0.40
13	--	--	5.69	0.49

Age is for completed years.

(data from Malina, unpublished a)

TABLE 98

STANDING LONG JUMP (in) FOR PHILADELPHIA CHILDREN

Age	WHITE BOYS		BLACK BOYS	
(years)	Mean	S.D.	Mean	S.D.
6	40.33	7.35	41.89	8.81
7	46.24	5.83	47.21	8.48
8	50.26	5.82	50.35	7.79
9	53.96	6.73	55.92	7.72
10	55.88	7.11	60.07	7.80
11	59.03	6.64	63.63	7.51
12	61.06	7.93	68.20	8.67
13	--	--	70.08	8.93

Age	WHITE GIRLS		BLACK GIRLS	
(years)	Mean	S.D.	Mean	S.D.
6	38.05	6.20	37.39	7.65
7	43.84	5.34	42.15	7.68
8	46.88	5.57	45.34	7.28
9	50.23	5.33	47.30	7.77
10	51.73	6.57	52.30	7.30
11	56.13	6.14	57.22	7.47
12	--	--	62.00	9.42
13	--	--	61.31	10.87

Age is for completed years.

(data from Malina, unpublished a)

TABLE 99

SOFTBALL THROW FOR DISTANCE
(ft) FOR PHILADELPHIA CHILDREN

| Age | WHITE BOYS | | BLACK BOYS | |
(years)	Mean	S.D.	Mean	S.D.
6	29.17	9.35	37.80	13.01
7	41.22	12.12	47.46	15.95
8	55.05	14.02	60.78	17.45
9	72.34	18.10	78.64	18.90
10	88.63	20.85	90.48	18.99
11	101.39	21.47	99.79	21.35
12	100.86	22.27	112.08	24.55
13	--	--	117.33	22.18

| Age | WHITE GIRLS | | BLACK GIRLS | |
(years)	Mean	S.D.	Mean	S.D.
6	18.80	8.20	22.17	7.49
7	23.66	8.62	27.80	9.47
8	26.22	6.90	31.80	9.82
9	33.08	11.05	37.99	13.10
10	41.79	14.93	46.51	14.35
11	54.08	15.90	56.51	19.33
12	--	--	56.75	18.17
13	--	--	69.38	19.75

Age is for completed years.

(data from Malina, unpublished a)

TABLE 100

PERFORMANCE DATA FROM TEXAS CHILDREN

Percentile	Sex	6	7	Age (years) 8	9	10	11
				50-Yard Dash (sec)			
90	Boys	8.8	8.6	7.9	7.8	7.6	7.7
	Girls	9.4	8.9	8.4	8.0	7.8	7.6
50	Boys	9.5	9.5	8.7	8.4	8.3	8.4
	Girls	10.2	9.9	9.3	9.0	8.8	8.5
				Standing Long Jump Norms (in)			
90	Boys	47	50	56	60	61	64
	Girls	44	48	52	56	57	61
50	Boys	40	42	49	51	54	55
	Girls	34	38	42	47	49	50
10	Boys	35	34	39	43	46	45
	Girls	24	30	37	37	40	46
				Softball Throw Norms (ft)			
90	Boys	60	74	88	101	110	115
	Girls	38	41	49	66	79	86
50	Boys	40	54	65	81	88	95
	Girls	23	27	36	45	51	62
10	Boys	23	36	46	54	66	65
	Girls	13	17	25	26	36	41

(data from Hardin and Ramirez, 1972)

TABLE 101

BALANCE BOARD SCORES FOR CHILDREN IN WICHITA (KA)

Age Groups (years)	Boys								Girls							
	0	1	2	3	4	5	6	Ave.	0	1	2	3	4	5	6	Ave.
13-15	3	2	2	8	11	2	7	3.7	1	7	4	5	2	0	0	2.0
12-13	2	1	7	4	12	6	4	3.6	2	6	2	7	3	0	1	2.3
11-12	0	3	4	8	7	6	6	3.8	2	3	3	5	3	2	0	2.5
10-11	3	9	6	5	7	8	3	3.0	2	5	4	6	0	4	0	2.4
9-10	2	8	7	7	5	4	3	2.8	5	4	7	7	4	1	0	2.1
8-9	7	11	8	9	11	2	2	2.5	7	6	6	5	4	2	0	2.0
7-8	12	10	9	7	0	4	0	1.6	5	3	3	3	3	2	0	2.1
6-7	17	11	7	1	0	1	0	0.9	3	2	3	2	1	0	0	1.6
4-6	10	1	0	0	0	0	0	0.1	9	1	2	0	0	0	0	0.5

Method of scoring: Each child was given three trials. A miss, stepping off the board, scored 0; Walking the length of the board before stepping off scored 1; Walking length of board, turning and walking back ("round trip") scored 2; A perfect score was 6.

(data from Cron and Pronko, 1957)

TABLE 102

PERFORMANCE ON THE 35-YARD DASH (sec)
FOR CHILDREN IN KANSAS CITY (MO)

	Grade 4		Grade 5		Grade 6	
	Boys	Girls	Boys	Girls	Boys	Girls
Negro	5.76	6.28	5.72	5.98	5.62	5.68
White	6.37	6.58	6.11	6.40	5.82	6.12

(data from Hutinger, 1959, Research Quarterly 30:366-368, by permission of the American Alliance for Health, Physical Education, Recreation and Dance, 1900 Association Drive, Reston VA 22091)

TABLE 103

PERFORMANCE SCORES OF WISCONSIN GIRLS

Age* (years)	Run 30 yd. (sec)		Long Jump (in)		Throw (ft/sec)	
	Mean	S.D.	Mean	S.D.	Mean	S.D.
6	6.37	.70	40.5	7.1	29.1	7.3
7	5.85	.58	43.5	6.6	30.5	6.2
8	5.56	.50	47.7	5.8	34.7	6.5
9	5.24	.41	52.9	7.6	36.4	6.9
10	5.02	.44	57.6	7.4	40.7	7.1
11	4.79	.61	61.5	7.4	44.0	8.3
12	4.60	.42	63.9	6.0	48.6	7.6
13	4.42	.48	68.0	6.2	51.9	10.3
14	4.25	.5	69.7	6.2	58.7	11.9

*Each age represents a 12-month span beginning with 67-78 months.

(data from Glassow and Kruse, 1960, Research Quarterly 31:426-433, by permission of the American Alliance for Health, Physical Education, Recreation and Dance, 1900 Association Drive, Reston VA 22091)

TABLE 104

50-YARD DASH (sec) FOR URBAN
CHILDREN IN 7 AREAS OF THE U.S.

BOYS Age (years)						Percentile	GIRLS Age (years)				
13	14	15	16	17	18		14	15	16	17	18
6.7	6.5	6.4	6.2	--	6.1	95th	7.4	7.5	7.3	--	--
7.4	7.0	6.8	6.6	--	--	75th	8.3	8.1	7.9	7.9	--
7.8	7.7	7.2	6.9	6.8	6.7	50th	8.6	8.5	8.4	8.3	8.6
8.3	8.2	7.6	7.3	7.1	7.0	25th	9.3	9.1	8.9	8.9	9.2
8.8	8.6	8.5	8.1	7.7	7.5	5th	10.2	10.2	10.2	10.0	10.1

(data from Fleishman, 1964)

TABLE 105

600 YARD RUN-WALK (min - sec) FOR URBAN
CHILDREN IN 7 AREAS OF THE U.S.

BOYS Age (years)				Percentile	GIRLS Age (years)
13	14	15	16-18		12-18
1'57	1'49	1'41	1'33	95th	2'25
2'18	2'06	2'01	1'52	75th	2'55
2'30	2'20	2'12	2'04	50th	3'12
2'48	2'38	2'30	2'21	25th	3'40
3'48	3'32	3'10	3'02	5th	4'30

(data from Fleishman, 1964)

TABLE 106

HOLD HALF SIT-UP (sec) FOR URBAN
CHILDREN IN 7 AREAS OF THE U.S.

| | Age (years) | | | GIRLS | | | | | |
| | | | Percentile | | | Age (years) | | | |
14-16	17	18		13	14	15	16	17	18
165	138	124	95th	222	159	92	92	79	63
90	85	87	75th	82	66	41	40	39	36
61	61	60	50th	56	39	24	23	21	17
44	45	42	25th	29	19	11	10	9	7
22	23	23	5th	7	6	2	2	2	2

(data from Fleishman, 1964)

TABLE 107

SHUTTLE RUN (secs) FOR URBAN
CHILDREN IN 7 AREAS OF THE U.S.

| BOYS | | | | | | | GIRLS | | | | | | |
| | Age (years) | | | | | Percentile | | | Age (years) | | | | |
12	13	14	15	16-17	18		12	13	14	15	16	17	18
20.7	20.0	19.0	19.0	18.5	18.3	95th	21.0	21.4	21.2	21.5	21.1	20.8	20.6
22.1	21.3	20.1	19.9	19.5	19.3	75th	22.6	22.2	22.2	22.8	22.6	22.3	22.5
23.1	22.5	20.8	20.5	20.3	20.1	50th	23.5	22.8	23.1	23.9	23.7	23.4	23.5
24.6	23.5	21.6	21.4	21.3	21.1	25th	24.8	23.9	24.0	25.1	24.8	24.5	25.1
29.0	25.5	23.0	23.2	23.7	23.4	5th	26.5	25.4	25.5	27.2	26.5	26.6	26.9

(data from Fleishman, 1964)

TABLE 108

DODGE RUN (sec) FOR URBAN
CHILDREN IN 7 AREAS OF THE U.S.

BOYS Age (years)						Percentile	GIRLS Age (years)				
13	14	15	16	17	18		14	15	16	17	18
16.3	15.8	15.4	15.2	15.3	15.6	95th	15.7	16.4	17.1	17.2	17.4
17.2	16.8	16.6	16.4	16.4	16.7	75th	17.8	18.2	18.6	18.3	18.9
17.9	17.7	17.5	17.4	17.4	17.4	50th	19.5	19.5	19.6	19.4	19.9
18.7	18.5	18.6	18.4	18.6	19.1	25th	21.6	21.1	20.6	20.5	21.1
20.5	20.5	20.3	20.4	20.4	20.8	5th	24.7	24.3	22.7	22.5	23.3

(data from Fleishman, 1964)

TABLE 109

LONG JUMP (ft - in) FOR URBAN
CHILDREN IN 7 AREAS OF THE U.S.

BOYS Age (years)						Percentile	GIRLS Age (years)			
13	14	15	16	17	18		13-15	16	17	18
6'10	7'6	8'0	8'2	8'3	8'4	95th	6'2	6'3	6'8	6'11
6'2	6'8	7'3	7'6	7'7	7'8	75th	5'6	5'7	5'9	6'1
5'8	6'2	6'10	7'1	7'1	7'1	50th	5'0	5'1	5'3	5'6
5'2	5'5	6'1	6'5	6'6	6'7	25th	4'7	4'8	4'10	4'11
4'6	4'7	5'1	5'6	5'9	5'10	5th	3'8	3'11	4'1	4'3

(data from Fleishman, 1964)

TABLE 110

LEG LIFTS (No.) FOR URBAN CHILDREN
IN 7 AREAS OF THE U.S.

| BOYS | | | | | | GIRLS | | | | | |
| Age (years) | | | | | | Age (years) | | | | | |
14	15	16	17	18	Percentile	13	14	15	16	17	18
26	27	28	28	28	95th	19	18	22	22	23	23
--	22	--	24	24	75th	13	--	17	18	18	18
20	20	21	22	22	50th	10	12	14	15	15	14
16	17	--	19	19	25th	7	9	10	10	10	9
11	13	15	15	15	5th	3	5	4	5	4	4

(data from Fleishman, 1964)

TABLE 111

PULL-UPS (No.) FOR URBAN CHILDREN
IN 7 AREAS OF THE U.S.

| BOYS | | | | | | | GIRLS | | | | | |
| Age (years) | | | | | | | Age (years) | | | | | |
12	13	14	15	16	17	18	Percentile	13	14	15	16	17	18
7	11	12	14	16	16	17	95th	2	--	--	--	2	2
3	5	7	--	11	11	11	75th	--	--	1	1	--	1
--	3	5	7	8	9	--	50th	--	--	0	--	--	--
0	1	2	4	5	5	6	25th	--	--	--	--	--	--
--	--	0	1	1	2	1	5th	--	--	--	--	--	--

(data from Fleishman, 1964)

TABLE 112

SOFTBALL THROW (ft) FOR URBAN
CHILDREN IN 7 AREAS OF THE U.S.

BOYS Age (years)							GIRLS Age (years)		
12	13	14	15	16-17	18	Percentile	13	14-16	17-18
130	160	170	199	204	205	95th	92	100	105
108	134	152	172	176	182	75th	63	74	78
94	109	134	150	156	164	50th	52	58	63
83	92	117	134	136	144	25th	44	48	53
68	70	95	105	107	111	5th	33	36	38

Standard 12" softball.

(data from Fleishman, 1964)

TABLE 113

DIFFERENCES BETWEEN MEAN STANDING LONG JUMP
DISTANCES (in.) OF BOYS 10, 13, AND 16 YEARS OF AGE CLASSIFIED
INTO PUBESCENT DEVELOPMENT GROUPS

Age (years)	Groups					
	1	2	3	4	4+5	5
10	57.33	59.07	--	--	--	--
13	--	69.84	67.72	--	--	--
13	--	69.84	--	--	76.00	--
13	--	--	67.72	--	76.00	--
16	--	--	--	79.00	--	83.39

The pubertal stages were assigned using the method of Greulich et al., (1942).

(data from Clarke and Degutis, 1962, Research Quarterly 33:356-368, by permission
of The American Alliance for Health, Physical Education, Recreation and Dance,
1900 Association Drive, Reston VA 22091)

TABLE   114

BEAM WALKING SCORES
FOR CALIFORNIAN BOYS

Age (years)	Mean	S.D.
By Chronological Age		
11.5	49.5	8.8
12.5	52.5	8.9
13.5	55.6	9.0
14.5	56.2	10.1
15.5	59.9	13.5
By Maturity Class		
Prepubescent I	52.2	9.3
Pubescent { II	55.6	9.6
III	54.7	9.8
IV	56.7	10.1
Postpubescent	57.6	11.8

Maximum score = 80

(data from Espenschade, Dable, and Schoendube,
1953, Research Quarterly 24:270-275, by
permission of the American Alliance for
Health, Physical Education, Recreation
and Dance, 1900 Association Drive, Reston
VA   22091)

TABLE 115

MOTOR PERFORMANCE IN MICHIGAN CHILDREN

		White				Black			
		BOYS		GIRLS		BOYS		GIRLS	
	Grade	Mean	S.D.	Mean	S.D.	Mean	S.D.	Mean	S.D.
Agility Shuttle Run	K	14.79	1.71	15.60	1.71	14.89	1.19	15.45	1.16
(sec)	1	13.42	1.23	14.06	1.41	13.58	.76	14.20	1.55
	2	13.00	1.36	13.76	1.27	12.97	1.24	13.53	1.75
Standing Long Jump	K	34.78	7.10	33.73	6.51	34.86	6.76	29.57	6.98
(in)	1	41.37	7.23	38.05	7.02	42.02	5.63	38.31	5.81
	2	46.41	7.71	41.42	5.62	47.05	6.54	40.73	8.61
30-Yd Dash (sec)	K	6.47	0.67	7.00	1.06	6.11	.74	6.63	0.84
	1	5.57	0.57	5.99	.59	5.51	.40	5.70	0.74
	2	5.36	0.64	5.74	.53	4.93	.35	5.54	0.73
Well's Sit and Reach	K	5.71	1.91	6.30	1.56	6.11	2.02	6.61	1.61
Flexibility Test (in)	1	5.35	1.92	6.24	1.99	5.70	1.93	5.96	1.53
	2	5.96	2.12	5.77	1.93	5.00	2.11	4.70	2.27
400-Ft Run (sec)	K	50.68	4.00	52.88	7.00	50.85	4.29	53.24	4.50
	1	46.90	8.20	47.35	4.56	47.59	3.25	50.19	5.30
	2	44.78	5.73	46.59	4.02	43.37	4.08	45.70	5.71

(data from Milne, Seefeldt, and Reuschlein, 1976, Research Quarterly 47:726-730, by permission of the American Alliance for Health, Physical Education, Recreation and Dance, 1900 Association Drive, Reston VA 22091)

TABLE 116

PHYSICAL PERFORMANCE IN ANGLO AND
SPANISH-AMERICAN BOYS IN NEW MEXICO

Event	Race	Mean	Critical Ratio
Baseball throw	Anglo	103.3	7.2
	Sp. Am.	124.1	
Base running	Anglo	12.389	3.2
	Sp. Am.	12.001	
Chinning	Anglo	3.71	3.89
	Sp. Am.	5.528	
60-yard dash	Anglo	9.795	3.5
	Sp. Am.	9.365	
Jump and reach	Anglo	11.44	4.2
	Sp. Am.	13.04	
Shot-put	Anglo	19.00	3.8
	Sp. Am.	21.80	

(data from Thompson and Dove, 1942, Research Quarterly 13:
341-346, by permission of the American Alliance for Health,
Physical Education, Recreation and Dance, 1900 Association
Drive, Reston VA 22091)

TABLE 117

PERFORMANCE AT STANDING LONG JUMP, FLEXED ARM HANG AND BENT KNEE SIT UPS
OF BOYS TESTED FROM 10 THROUGH 16 YEARS OF AGE
IN THE SASKATCHEWAN CHILD GROWTH AND DEVELOPMENT STUDY

| Age (years) | Standing Long Jump | | |
	Mean ± S.D. (cm)	Increase	Percent Increase
10	164.08 ± 14.22		
		5.34	3.3
11	169.42 ± 14.99		
		8.64	5.1
12	178.05 ± 15.49		
		9.14	5.1
13	187.20 ± 17.27		
		7.62	4.1
14	194.82 ± 19.56		
		14.22	7.3
15	209.04 ± 19.56		
		9.14	4.4
16	218.19 ± 28.19		

| Age (years) | Flexed Arm Hang | | | Bent Knee Sit Ups | | |
	Mean ± S.D. (sec)	Increase	Percent Increase	Mean ± S.D. no./min.	Increase	Percent Increase
10	30.6 ± 19.9			37.8 ± 10.2		
		5.5	17.9		1.6	4.2
11	36.1 ± 22.9			39.4 ± 10.5		
		10.3	28.5		2.5	6.3
12	46.4 ± 22.8			41.9 ± 8.4		
		4.5	9.7		1.4	3.3
13	50.9 ± 23.5			43.3 ± 8.1		
		5.2	10.2		1.8	4.2
14	56.1 ± 23.1			45.1 ± 8.3		
		9.6	17.1		2.7	6.0
15	65.7 ± 25.1			47.8 ± 8.7		
		- 1.2	- 1.8		1.1	2.3
16	64.5 ± 22.3			48.9 ± 8.9		

(data from "Physical performance in boys from 10 through 16 years," Human Biology 47:263-
281, 1975, by Ellis, Carron, and Bailey, by permission of the Wayne State University Press.
Copyright 1975, Wayne State University Press, Detroit, Michigan 48202)

TABLE   118

PERFORMANCE DATA FOR CANADIAN CHILDREN AGED 7 YEARS

Percentiles	Speed Sit Up (No.)	Stand Long Jump (Ft. In.)		Shuttle Run (Sec)	Flexed Arm Hang (Sec)	50 Yard Run (Sec)	300 Yard Run (Sec)
				BOYS			
95	33	4'	7"	12.0	55	8.7	73
75	27	4'	0"	13.2	29	9.4	79
50	20	3'	9"	14.0	18	10.0	84
25	12	3'	3"	14.9	10	10.7	89
5	3	2'	8"	16.3	4	12.1	101
				GIRLS			
95	33	4'	5"	12.7	54	9.1	75
75	22	3'	11"	13.7	22	9.8	80
50	17	3'	6"	14.7	11	10.5	85
25	10	3'	2"	15.7	6	11.3	92
5	0	2'	8"	17.6	1	12.8	106

Age is for completed years.

(data from the CAHPER Fitness Performance Test Manual for boys and girls aged
7 to 17 years of age, 1966).

TABLE 119

PERFORMANCE DATA FOR CANADIAN CHILDREN AGED 8 YEARS

Percentiles	Speed Sit Up (No.)	Stand Long Jump (Ft. In.)	Shuttle Run (Sec)	Flexed Arm Hang (Sec)	50 Yard Run (Sec)	300 Yard Run (Sec)
		BOYS				
95	37	4' 11"	11.6	63	8.3	69
75	29	4' 5"	12.6	37	9.0	74
50	24	3' 11"	13.3	23	9.5	78
25	19	3' 7"	14.2	13	10.2	83
5	5	2' 10"	15.8	6	11.5	96
		GIRLS				
95	34	4' 11"	12.3	45	8.7	71
75	24	4' 3"	13.2	23	9.3	76
50	19	3' 10"	14.0	13	9.9	80
25	12	3' 4"	14.8	7	10.5	85
5	2	2' 10"	16.2	2	11.5	94

Age is for completed years.

(data from the CAHPER Fitness Performance Test Manual for boys and girls aged 7 to 17 years of age, 1966)

TABLE   120

PERFORMANCE DATA FOR CANADIAN CHILDREN AGED 9 YEARS

Percentiles	Speed Sit Up (No.)	Stand Long Jump (Ft. In.)	Shuttle Run (Sec)	Flexed Arm Hang (Sec)	50 Yard Run (Sec)	300 Yard Run (Sec)
		BOYS				
95	40	5' 3"	11.3	64	8.0	66
75	33	4' 8"	12.1	43	8.7	70
50	26	4' 4"	12.7	27	9.1	75
25	19	3' 11"	13.6	16	9.6	80
5	10	3' 4"	15.2	5	10.6	91
		GIRLS				
95	36	5" 1"	12.0	52	8.3	69
75	26	4' 5"	12.8	26	9.0	73
50	20	4' 0"	13.5	14	9.5	77
25	13	3' 7"	14.5	7	10.1	83
5	3	2' 10"	16.5	2	11.4	91

Age is for completed years.

(data from the CAHPER Fitness Performance Test Manual for boys and girls 7 to 17 years of age, 1966)

TABLE   121

PERFORMANCE DATA FOR CANADIAN CHILDREN AGED 10 YEARS

Percen-tiles	Speed Sit up (No.)	Stand Long Jump (Ft. In.)	Shuttle Run (Sec.)	Flexed Arm Hang (Sec.)	50 Yard Run (Sec.)	300 Yard Run (Sec.)
			BOYS			
95	42	5' 5"	11.0	71	7.8	64
75	34	4'10"	11.9	44	8.3	68
50	27	4' 6"	12.6	27	8.8	72
25	21	4' 1"	13.4	18	9.3	77
5	10	3' 6"	15.0	7	10.2	85
			GIRLS			
95	39	5' 3"	11.6	56	7.9	65
75	30	4' 8"	12.3	27	8.5	70
50	22	4' 3"	13.0	17	9.0	75
25	16	3'10"	13.9	9	9.7	79
5	3	3' 4"	15.5	1	10.9	90

Age is for completed years.

(data from the CAHPER Fitness Performance Test Manual for boys and girls 7 to 17 years of age, 1966)

TABLE   122

PERFORMANCE DATA FOR CANADIAN CHILDREN AGED 11 YEARS

Percen- tiles	Speed Sit up (No.)	Stand Long Jump (Ft. In.)	Shuttle Run (Sec.)	Flexed Arm Hang (Sec.)	50 Yard Run (Sec.)	300 Yard Run (Sec.)
		BOYS				
95	46	5'10"	10.8	70	7.4	60
75	36	5' 3"	11.5	46	8.0	66
50	29	4'10"	12.2	31	8.4	70
25	24	4' 4"	13.1	20	8.9	74
5	14	3' 8"	14.7	6	10.0	85
		GIRLS				
95	41	5' 7"	11.3	59	7.5	62
75	31	5' 1"	12.1	28	8.2	68
50	25	4' 7"	12.8	16	8.7	72
25	18	4' 1"	13.6	8	9.4	77
5	6	3' 5"	15.2	1	10.7	89

Age is for completed years.

(data from CAHPER Fitness Performance Test Manual for boys and girls 7 to
 17 years of age, 1966)

TABLE   123

PERFORMANCE DATA FOR CANADIAN CHILDREN AGED 12 YEARS

Percen- tiles	Speed Sit up (No.)	Stand Long Jump (Ft. In.)	Shuttle Run (Sec.)	Flexed Arm Hang (Sec.)	50 Yard Run (Sec.)	300 Yard Run (Sec.)
			BOYS			
95	45	6' 1"	10.8	72	7.2	59
75	36	5' 5"	11.4	51	7.9	64
50	30	5' 0"	12.0	35	8.3	67
25	23	4' 7"	12.9	22	8.9	72
5	13	3'10"	14.7	7	9.9	83
			GIRLS			
95	39	5'10"	11.0	46	7.5	62
75	30	5' 1"	12.0	26	8.0	67
50	22	4' 8"	12.8	14	8.5	71
25	15	4' 2"	13.9	7	9.2	76
5	5	3' 6"	15.1	1	10.4	86

Age is for completed years.

(data from CAHPER Fitness Performance Test Manual for boys and girls 7 to
  17 years of age, 1966).

TABLE 124

PERFORMANCE DATA FOR CANADIAN CHILDREN AGED 13 YEARS

Percen- tiles	Speed Sit up (No.)	Stand Long Jump (Ft. In.)	Shuttle Run (Sec.)	Flexed Arm Hang (Sec.)	50 Yard Run (Sec.)	300 Yard Run (Sec.)
			BOYS			
95	48	6' 6"	10.3	75	6.9	56
75	39	5'10"	11.1	57	7.5	61
50	33	5' 3"	11.8	40	8.0	65
25	25	4' 9"	12.5	23	8.5	69
5	15	4' 0"	13.9	9	9.5	77
			GIRLS			
95	37	6' 0"	11.0	47	7.2	61
75	29	5' 3"	12.0	26	7.9	66
50	23	4'10"	12.6	13	8.6	70
25	16	4' 4"	13.5	6	9.2	75
5	5	3' 6"	14.9	0	10.2	85

Age is for completed years.

(data from CAHPER Fitness Performance Test Manual for boys and girls 7 to
17 years of age, 1966)

TABLE 125

PERFORMANCE DATA FOR CANADIAN CHILDREN AGED 14 YEARS

Percen-tiles	Speed Sit up (No.)	Stand Long Jump (Ft. In.)	Shuttle Run (Sec.)	Flexed Arm Hang (Sec.)	50 Yard Run (Sec.)	300 Yard Run (Sec.)
			BOYS			
95	49	7' 2"	10.1	84	6.5	53
75	39	6' 4"	10.8	60	7.1	57
50	32	5' 9"	11.3	45	7.5	62
25	25	5' 2"	12.0	32	8.0	66
5	18	4' 4"	13.6	14	9.0	74
			GIRLS			
95	36	6' 2"	10.9	43	7.4	62
75	26	5' 6"	11.8	23	8.0	67
50	20	4'11"	12.4	12	8.4	72
25	13	4' 4"	13.4	5	9.1	76
5	0	3' 6"	14.8	0	10.2	88

Age is for completed years.

(data from CAHPER Fitness Performance Test Manual for boys and girls 7 to 17 years of age, 1966).

TABLE 126

PERFORMANCE DATA FOR CANADIAN CHILDREN AGED 15 YEARS

Percentiles	Speed Sit up (No.)	Stand Long Jump (Ft. In.)	Shuttle Run (Sec.)	Flexed Arm Hang (Sec.)	50 Yard Run (Sec.)	300 Yard Run (Sec.)
			BOYS			
95	50	7' 5"	9.9	82	6.3	52
75	41	6' 8"	10.5	62	6.9	56
50	33	6' 2"	11.0	50	7.1	58
25	25	5' 8"	11.6	35	7.6	62
5	16	4'10"	13.2	15	8.5	69
			GIRLS			
95	39	6' 4"	10.9	43	7.2	61
75	28	5' 6"	11.7	21	7.9	67
50	22	5' 0"	12.5	13	8.3	71
25	15	4' 5"	13.1	6	8.9	75
5	6	3'10"	14.5	0	10.1	86

Age is for completed years.

(data from CAHPER Fitness Performance Test Manual for boys and girls 7 to 17 years of age, 1966).

TABLE 127

PERFORMANCE DATA FOR CANADIAN CHILDREN AGED 16 YEARS

Percen-tiles	Speed Sit up (No.)	Stand Long Jump (Ft. In.)	Shuttle Run (Sec.)	Flexed Arm Hang (Sec.)	50 Yard Run (Sec.)	300 Yard Run (Sec.)
		BOYS				
95	50	7'10"	9.7	85	6.2	51
75	41	7' 2"	10.2	66	6.6	54
50	35	6' 7"	10.7	53	6.9	57
25	27	6' 1"	11.4	39	7.4	60
5	18	5' 4"	12.7	20	8.2	67
		GIRLS				
95	38	6' 3"	10.8	43	7.1	63
75	29	5' 7"	11.6	20	7.7	67
50	21	5' 0"	12.4	11	8.3	71
25	15	4' 7"	13.2	6	8.9	75
5	6	4' 1"	14.4	0	10.1	82

Age is for completed years.

(data from CAHPER Fitness Performance Test Manual for boys and girls 7 to 17 years of age, 1966).

TABLE 128

PERFORMANCE DATA FOR CANADIAN CHILDREN AGED 17 YEARS

Percen-tiles	Speed Sit up (No.)	Stand Long Jump (Ft. In.)	Shuttle Run (Sec.)	Flexed Arm Hang (Sec.)	50 Yard Run (Sec.)	300 Yard Run (Sec.)
		BOYS				
95	49	7' 10"	9.5	83	6.2	50
75	40	7' 4"	10.0	62	6.5	53
50	35	6' 11"	10.5	52	6.8	56
25	27	6' 3"	11.2	41	7.1	58
5	16	5' 6"	12.2	20	7.8	63
		GIRLS				
95	34	6' 0"	10.8	44	7.2	63
75	26	5' 4"	11.9	20	8.0	67
50	20	4' 11"	12.5	12	8.5	72
25	12	4' 5"	13.2	5	8.9	77
5	4	4' 2"	14.4	0	9.6	83

Age is for completed years.

(data from CAHPER Fitness Performance Test Manual for boys and girls 7 to 17 years of age, 1966).

TABLE 129

HAND-ARM REACTION TIME(sec) AND MOVEMENT TIME (sec)
FOR BOYS IN BERKELEY (CA)

		Age (years)		
		8	12	18
Movement B				
Reaction	Mean	.275	.214	.191
	S.D.	.042	.035	.029
Movement	Mean	.174	.097	.081
	S.D.	.043	.023	.021
Movement C				
Reaction	Mean	.295	.226	.202
	S.D.	.026	.033	.031
Movement	Mean	.762	.493	.399
	S.D.	.131	.108	.114

Movement B involves grasping a tennis ball hung by a string;
Movement C involves striking a suspended tennis ball, pushing
a button and striking another suspended tennis ball.

(data from Henry, 1961, Research Quarterly 32:353-366, by permission
of the American Alliance for Health, Physical Education, Recreation
and Dance, 1900 Association Drive, Reston VA 22091)

TABLE 130

REACTION TIME (units of $1/120$ sec) FOR CALIFORNIA CHILDREN

Age	Boys			Girls		
(years)	M	AV	RIV	M	AV	RIV
Group 1						
4½	47.00	18.59	.356	52.71	18.63	.312
5½	37.99	14.93	.325	43.08	16.84	.323
6½	29.70	9.31	.268	32.13	9.15	.285
7½	26.81	9.15	.275	26.54	8.11	.257
8½	24.56	10.80	.390	23.18	8.05	.277
9½	24.98	8.30	.337	23.13	8.02	.296
10½	19.96	7.07	.317	20.78	5.83	.246
11½	19.29	7.04	.342	19.11	5.94	.279
Group 2, right hand						
10	24.58	7.63	.185	23.16	6.05	.205
11	23.17	5.72	.195	24.81	6.89	.208
12	19.25	4.63	.186	19.56	5.59	.207
13	18.30	3.85	.173	18.46	3.96	.176
14	17.58	3.24	.147	17.89	3.63	.162
15	17.62	3.52	.170	17.51	3.78	.171
16	16.87	4.21	.208	19.30	4.30	.196
Group 2, left hand						
10	24.63	8.85	.201	24.62	5.81	.173
11	24.09	6.60	.214	25.86	7.40	.208
12	19.39	5.19	.208	19.79	5.53	.201
13	18.21	3.97	.180	18.67	4.33	.187
14	17.27	4.09	.164	18.00	3.66	.164
15	17.34	3.60	.171	17.65	3.59	.166
16	16.55	3.75	.181	19.10	4.23	.186

The time was measured from the sounding of a buzzer to the release of
pressure by fingers. M = mean; AV = average variability and RIV = relative
intra-individual variability.

(data from Eckert and Eichorn, 1977)

TABLE 131

REACTION TIMES (sec) IN ILLINOIS BOYS

Visual Reaction Time-Vertical Jump

Per-centiles	Age (years)							
	6	7	8	9	10	11	12	13
95	.249	.228	.218	.237	.194	.157	.247	.163
75	.366	.332	.320	.317	.290	.261	.311	.259
50	.512	.462	.447	.417	.410	.391	.391	.379
25	.659	.592	.575	.517	.530	.521	.471	.499
5	.776	.696	.677	.597	.626	.625	.535	.595

Auditory Reaction Time - Vertical Jump

Per-centiles	Age (years)							
	6	7	8	9	10	11	12	13
95	.268	.216	.217	.231	.218	.154	.204	.116
75	.362	.320	.313	.303	.294	.254	.273	.220
50	.479	.450	.433	.393	.389	.379	.358	.350
25	.597	.580	.553	.483	.484	.504	.443	.480
5	.691	.684	.649	.555	.560	.604	.511	.584

Combination Visual and Auditory Reaction Time - Vertical Jump

Per-centiles	Age (years)							
	6	7	8	9	10	11	12	13
95	.222	.225	.173	.213	.218	.115	.218	.105
75	.332	.325	.277	.293	.294	.227	.282	.217
50	.470	.450	.433	.393	.389	.367	.362	.357
25	.608	.575	.563	.493	.484	.507	.442	.497
5	.719	.675	.667	.573	.560	.619	.506	.609

(data from Cureton and Barry, 1964)

TABLE 132

SPEED FOR HAND REACTION TIME (RT), HAND MOVEMENT TIME (MT),
AND TOTAL BODY REACTION TIME
IN SASKATCHEWAN BOYS

AGE	HAND RT		HAND MT		TOTAL BODY RT	
(years)	Mean	S.D.	Mean	S.D.	Mean	S.D.
7	.364	.058	.111	.029	--	--
8	.340	.048	.095	.019	--	--
9	.328	.044	.076	.015	--	--
10	.303	.053	.080	.022	.577	.096
11	.275	.043	.083	.015	.483	.074
12	.282	.039	.080	.015	.456	.073
13	.271	.032	.067	.014	.424	.057

(data from "A longitudinal examination of speed of reaction and
speed of movement in young boys 7 to 13 years," Human Biology 45:
663-681, 1973, by Carron and Bailey, by permission of the Wayne
State University Press. Copyright 1973, Wayne State University
Press, Detroit, Michigan 48202)

TABLE 133

FLEXIBILITY FOR CHILDREN IN BALTIMORE (MD)

Touching Fingertips to Toes When Sitting on Floor

Age (years)	Boys Range of limitation	Mean	% can touch	Girls Range of limitation	Mean	% can touch
5	$\frac{1}{2}$" – 9"	$2\frac{3}{4}$"	86%	$3\frac{1}{2}$" – 4"	$3\frac{3}{4}$"	98%
6	1" – 10"	4"	74%	$\frac{1}{2}$" – 4"	3"	83%
7	$\frac{1}{2}$" – $10\frac{1}{2}$"	3"	56%	$\frac{1}{2}$" – $10\frac{1}{2}$"	$3\frac{1}{2}$"	63%
8	$\frac{1}{2}$" – $9\frac{1}{2}$"	$3\frac{1}{2}$"	52%	1" – $8\frac{1}{2}$"	4"	59%
9	$\frac{1}{2}$" – $10\frac{1}{2}$"	$4\frac{1}{2}$"	52%	1" – $13\frac{1}{2}$"	$4\frac{1}{2}$"	57%
10	1" – 10"	$4\frac{1}{2}$"	50%	$\frac{1}{2}$" – 8"	4"	59%
11	1" – $11\frac{1}{2}$"	$4\frac{1}{4}$"	41%	$\frac{1}{2}$" – 10"	$4\frac{1}{2}$"	49%
12	$\frac{1}{2}$" – $9\frac{1}{2}$"	4"	28%	$\frac{1}{2}$" – $11\frac{1}{2}$"	6"	43%
13	$1\frac{1}{2}$" – 13"	$4\frac{1}{2}$"	40%	$\frac{1}{2}$" – 10"	5"	30%
14	$\frac{1}{2}$" – 10"	$4\frac{1}{2}$"	50%	2" – 13"	$5\frac{1}{2}$"	37%
15	$\frac{1}{2}$" – $12\frac{1}{2}$"	$3\frac{1}{2}$"	60%	$\frac{1}{2}$" – 12"	5"	59%
16	$\frac{1}{2}$" – $12\frac{1}{2}$"	5"	64%	1" – 12"	5"	64%
17	1" – 12"	3"	87%	1" – 14"	5"	69%

Bending Forehead Towards Knees When Sitting

Age (years)	Boys Range of limitation	Mean	% can touch	Girls Range of limitation	Mean	% can touch
5	$\frac{1}{2}$" – 10"	5"	5%	$\frac{1}{2}$" – $7\frac{1}{2}$"	4"	16%
6	2" – $11\frac{1}{2}$"	7"	2%	$\frac{1}{2}$" – $10\frac{1}{2}$"	6"	5%
7	3" – 13"	$7\frac{1}{2}$"	2%	1" – $13\frac{1}{2}$"	7"	6%
8	$\frac{1}{2}$" – 11"	$6\frac{1}{2}$"	1%	1" – $11\frac{1}{2}$"	6"	5%
9	4" – 14"	9"	2%	1" – $12\frac{1}{2}$"	$7\frac{1}{2}$"	3%
10	1" – $12\frac{1}{2}$"	7"	0	1" – $10\frac{1}{2}$"	6"	2%
11	$1\frac{1}{2}$" – 15"	$7\frac{1}{2}$"	0	2" – $11\frac{1}{2}$"	$6\frac{3}{4}$"	4%
12	$3\frac{1}{2}$" – $13\frac{1}{2}$"	9"	1%	$\frac{1}{2}$" – $11\frac{1}{2}$"	6"	5%
13	1" – 18"	8"	1%	$1\frac{1}{2}$" – 20"	7"	4%
14	2" – 19"	10"	1%	$\frac{1}{2}$" – 12"	7"	6%
15	$1\frac{1}{2}$" – 19"	9"	1%	1" – $18\frac{1}{2}$"	8"	6%
16	$2\frac{1}{2}$" – $23\frac{1}{2}$"	11"	1%	1" – $18\frac{1}{2}$"	8"	0
17	$\frac{1}{2}$" – 18"	8"	1%	$1\frac{1}{2}$" – 20"	8"	1%

(data from Kendall and Kendall, 1948)

TABLE   134

MEAN DEXTRALITY INDICES IN CALIFORNIA   CHILDREN

Age (years)	Boys	Girls
12.5	106.1	112.2
13.5	106.0	110.0
14.5	106.7	111.9
15.5	106.9	110.9
16.5	106.2	110.2

(data from "Sex differences in physical abilities,"
Human Biology 19:12-25, 1947, by H. E. Jones,
by permission of the Wayne State University Press.
Copyright 1947, Wayne State University Press,
Detroit, Michigan 48202)

TABLE   135

STRUCTURAL PROPERTIES OF THE ANTERIOR CRUCIATE
LIGAMENT AT 16 TO 26 YEARS

Stiffness (kN/m)	Linear Force (kN)	Maximum Force (kN)	Energy to Failure (N-m)
182 ± 56	1.17 ± 0.75	1.73 ± 0.66	12.8 ± 5.5

(data from Noyes and Grood, 1976)

TABLE 136

RANGE OF HIP MOTION IN CALIFORNIA INFANTS

	6 weeks	3 months	6 months
Flexion Contracture			
Mean	19°	7°	7°
S.D.	6.9°	3.8°	4.2°
Internal Rotation			
Mean	24°	26°	21°
S.D.	5.0°	3.4°	4.3°
External Rotation			
Mean	48°	45°	46°
S.D.	11.0°	4.5°	4.8°

(data from Coon, Donato, Houser, et al., 1975, Clinical Orthopedics 110:256-260, by permission of Lippincott/Harper Company)

TABLE 137

RESULTS OF HIP FLEXIBILITY TEST IN WASHINGTON BOYS

Grade		Adapted Kraus-Weber	Wells Sit and Reach	Greater Trochanter to Floor
Third Grade	Mean	- .95	- .44	27.73
	S.D.	2.58	2.39	1.44
Fourth Grade	Mean	-2.00	- .92	29.74
	S.D.	2.62	2.61	1.54
Fifth Grade	Mean	-1.39	-1.01	30.37
	S.D.	2.69	2.52	2.14
Sixth Grade	Mean	- .97	1.18	32.29
	S.D.	2.85	3.01	1.87

The Kraus-Weber score is the distance (in) reached above or below toes. The Wells sit and reach was scored as the distance (in) reached. The distance from the greater trochanter to the floor was measured in inches.

(data from Mathews, Shaw, and Woods, 1959, Research Quarterly 3:297-302, by permission of the American Alliance for Health, Physical Education, Recreation and Dance, 1900 Association Drive, Reston VA 22091)

TABLE 138

TRUNK FLEXIBILITY (in) IN ILLINOIS BOYS

	Backward Extension						
				Age (years)			
Per-centiles	7	8	9	10	11	12	13
95	20.3	22.5	20.6	23.1	24.0	23.1	22.6
75	16.5	18.1	16.6	19.1	19.6	19.2	18.6
50	11.7	12.7	11.6	14.1	14.1	14.3	13.6
25	7.0	7.2	6.6	9.1	8.6	9.4	8.6
5	3.2	2.9	2.6	5.1	4.2	5.5	4.6

	Forward Flexion						
95	2.5	1.2	3.8	2.0	3.4	4.6	6.1
75	5.2	4.9	7.0	5.4	6.9	7.2	9.1
50	8.5	9.4	10.8	9.8	11.1	10.5	12.5
25	11.9	14.0	14.6	14.2	15.5	13.8	15.9
5	14.6	17.6	17.7	17.4	18.9	16.5	18.6

(data from Cureton and Barry, 1964)

TABLE 139

FLEXIBILITY AND MOTOR PERFORMANCE IN NEW MEXICO GIRLS

Flexibility Measures

Flexibility Measure	Mean Scores by Grades				Standard Deviations by Grades			
	7	8	9	All	7	8	9	All
Ankle	64.91	65.59	67.00	65.79	9.00	8.67	9.27	8.97
Knee	130.27	129.65	127.33	129.50	12.70	16.38	12.55	14.28
Hip-thigh	121.67	122.85	126.85	124.27	15.40	14.00	17.50	15.26
Wrist	146.82	152.32	149.00	149.34	13.95	20.80	16.59	17.92
Elbow	161.45	160.97	160.91	162.31	8.58	8.13	9.30	8.97
Shoulder	265.36	268.21	278.21	270.16	26.01	21.07	33.21	27.54

Long Jump, 50-Yard Dash, and Basketball Throw

Tests	Mean Scores			Standard Deviation		
	7	8	9	7	8	9
Long jump (in)	67.3	67.4	74.29	6.57	7.70	5.30
Basketball throw (ft)	31.40	31.12	36.21	4.97	6.70	9.76
Fifty-yard dash (secs)	7.43	7.85	7.44	.501	.594	.582

(data from Burley, Dobell and Farrell, 1961, Research Quarterly 32:443-448, by
permission of the American Alliance for Health, Physical Education, Recreation
and Dance, 1900 Association Drive, Reston VA 22091)

TABLE 140

FLEXIBILITY (degrees) IN OREGON GIRLS

Age (years)	Mean	S.D.	Mean	S.D.
	Hip flexion-extension		Trunk-hip flexion-extension	
6	121.30	16.70	163.10	18.40
9	126.50	19.90	176.60	17.15
12	139.10	18.15	185.20	18.25
15	126.90	17.75	175.10	15.95
18	128.60	11.10	179.40	19.30
	Side trunk flexion-extension		Head Rotation	
6	92.00	13.95	168.50	9.51
9	107.20	18.05	174.10	12.15
12	118.34	20.40	170.40	13.60
15	110.40	18.80	157.30	20.16
18	104.40	18.00	163.15	14.00
	Head flexion-extension		Elbow flexion-extension	
6	124.30	13.65	155.98	6.00
9	127.70	12.45	157.26	6.87
12	134.90	18.15	157.42	8.13
15	125.38	17.29	155.66	7.50
18	119.35	13.75	151.27	7.83
	Wrist flexion-extension		Shoulder flexion-extension	
6	147.40	11.15	228.40	12.90
9	152.30	12.85	219.70	11.04
12	155.10	13.15	215.46	12.04
15	152.60	17.65	213.02	11.91
18	152.10	10.71	212.75	12.00
	Ankle flexion-extension		Knee flexion	
6	75.06	9.18	130.62	9.64
9	76.34	9.00	127.68	9.90
12	76.94	9.24	125.88	8.46
15	74.00	11.30	121.48	6.63
18	74.38	8.52	115.95	8.85
	Leg abduction		Thigh flexion	
6	47.38	7.98	111.78	9.24
9	51.78	10.20	109.26	10.56
12	53.94	8.94	107.18	9.39
15	51.40	8.79	108.10	12.20
18	52.60	7.65	105.40	10.95

Measurements made with a Leighton flexometer

(data from Hupprich and Sigerseth, 1950, Research Quarterly 21:25-33, by permission of the American Alliance for Health, Physical Education, Recreation and Dance, 1900 Association Drive, Reston VA 22091)

TABLE 141

RANGES OF HIP MOTION IN NEWLY-BORN INFANTS
IN WASHINGTON (D.C.)

Motion	Mean	S.D.
Internal rotation	62.0°	+ 12.9°
External rotation	89.1°	+ 14.3°
Abduction	76.4°	+ 11.5°
Flexion contracture	27.9°	+ 8.2°

COMPARISON BY RACE AND SEX

Motion		Black	White		
Internal rotation	Mean	59.4°	64.7°	63.1°	60.8°
	Range	45–100°	45–90°	35–90°	45–100°
External rotation	Mean	90.3°	87.9°	90.6°	87.5°
	Range	45–110°	60–105°	60–110°	45–110°
Abduction	Mean	71.0°	81.8°	76.2°	76.7°
	Range	50–110°	60–90°	60–90°	50–90°
Flexion contracture	Mean	28.3°	27.6°	27.5°	28.0°
	Range	20–40°	10–75°	20–45°	20–75°

(data from Haas, Epps, and Adams, 1973, Clinical Orthopaedics 91:114-118, by permission of Lippincott/Harper Co., Philadelphia, PA)

TABLE 142

FLEXIBILITY (degrees) IN BOYS FROM THE STATE OF WASHINGTON

	Age (years)									
	10		12		14		16		18	
Test	Mean	S.D.	Mean	S.D.	Mean	S.D.	Mean	S.D.	Mean	S.D.
Neck										
Flexion-extension	126	18	138	13	131	11	123	12	127	15
Lateral flexion	97	12	97	9	92	12	88	10	98	17
Rotation	177	19	162	16	159	11	158	12	159	22
Shoulder										
Flexion-extension	241	16	247	12	247	14	258	10	236	19
Adduction-abduction	185	10	181	7	184	12	173	9	177	14
Rotation	181	12	179	14	175	10	170	15	171	17
Elbow										
Flexion-extension	150	9	145	8	147	11	142	7	149	11
Radial-ulnar										
Supination-pronation	194	16	188	12	181	16	162	11	168	18
Wrist										
Flexion-extension	168	15	145	15	137	19	131	14	126	14
Ulnar-radial flexion	92	12	87	11	78	8	76	10	84	13
Hip										
Extension-flexion	70	12	57	14	65	19	55	13	66	15
Adduction-abduction	52	10	58	8	59	11	63	14	53	9
Rotation	111	17	81	14	71	10	69	13	92	22
Knee										
Flexion-extension	145	13	149	6	138	9	136	13	132	10
Ankle										
Flexion-extension	81	11	61	9	57	7	63	10	64	10
Inversion-eversion	49	9	47	9	44	11	43	8	45	13
Trunk										
Lateral flexion	102	11	87	10	88	9	96	15	101	18
Rotation	145	8	148	9	149	13	129	10	126	17

Leighton Flexometer used.

(data from Leighton, 1956)

TABLE 143

DYNAMIC FLEXIBILITY (number of cycles in 20 sec)
FOR URBAN CHILDREN IN 7 AREAS OF THE U.S.

| BOYS | | | | GIRLS | | | |
| Age (years) | | | | Age (years) | | | |
14	15	16-18	Percentile	15	16	17	18
23	24	22	95th	20	18	--	16
--	19	--	75th	--	--	14	--
18	17	16	50th	14	13	--	12
--	14	14	25th	12	11	11	--
11	12	11	5th	9	9	9	8

Dynamic flexibility is the ability to make repeated flexion or stretching movements. One cycle included forward flexion of the vertebral column followed by lateral rotation.

(data from Fleishman, 1964)

TABLE   144

EXTENT FLEXIBILITY (in) FOR URBAN
CHILDREN IN 7 AREAS OF THE U.S.

| BOYS | | | | | | | GIRLS | | | | |
| Age (years) | | | | | | | Age (years) | | | | |
13	14	15	16	17	18	Percentile	14	15	16	17	18
21	23	31	28	31	33	95th	27	29	29	27	29
18	20	22	22	23	25	75th	--	23	21	20	21
14	15	17	18	18	18	50th	21	17	15	16	17
10	11	13	14	14	14	25th	14	12	--	11	13
4	6	7	8	7	7	5th	7	5	4	4	6

Extent flexibility is the ability to move or stretch the body. Lateral rotation of the vertebral column was measured in these tests.

(data from Fleishman, 1964)

TABLE 145

VISUAL ACUITY OF PAPAGO CHILDREN
IN SOUTHWESTERN ARIZONA

Visual acuity and sex	10-15 years	16-20 years
20/40 or less:		
Boys	8	7
Girls	30	13
20/35 to 20/25:		
Boys	17	5
Girls	14	10
20/22 to 20/18:		
Boys	24	5
Girls	23	9
20/17 or better:		
Boys	19	13
Girls	13	10

(data from Adams et al., 1970)

TABLE 146

MAXIMAL OXYGEN UPTAKE FOR U.S. AND CANADIAN CHILDREN

Reference	Age Group or Range (years)	Stature (cm) Mean	S.D.	Weight (kg) Mean	S.D.	Exercise Mode[a]	Maximal Oxygen Uptake l/min Mean	S.D.	ml/kg/min Mean	S.D.
MAXIMAL OXYGEN UPTAKE FOR U.S. BOYS										
Morse et al., (1949)	10-12	--	--	--	--	T	--	--	47.8	5.0
	13	--	--	--	--	--	--	--	44.8	--
Rodahl et al., (1961)	12	--	--	--	--	--	1.43	--	--	--
	14	--	--	--	--	--	1.85	--	--	--
	16	--	--	--	--	--	2.03	--	--	--
	18	--	--	--	--	--	2.07	--	--	--
Kramer & Lurie (1964)	≤11	143.0	0.8	35.0	2.1	CM	1.74	0.08[b]	49.8	1.3[b]
	11	149.9	3.8	42.2	3.9	--	1.97	0.15	45.8	1.8
	12	155.0	3.0	41.8	2.6	--	1.85	0.11	45.9	2.2
	13	162.1	2.6	52.9	3.3	--	2.51	0.10	48.8	3.6
	<13	153.4	2.0	44.1	2.0	--	2.08	0.08	47.6	1.4
	>13	171.1	2.7	58.3	2.0	--	2.76	0.10	46.7	2.2
Kramer & Lurie (1964)		173.9	2.4	64.3	3.1	--	3.95	0.21[b]	62.1	2.5[b]
Knuttgen (1967)	15-18	173.5	7.1	67.5	13.0	B	3.34	0.49	50.3	6.9
	15	--	--	--	--	--	3.1[c]	--	50.5[c]	--
	16	--	--	--	--	--	3.2	--	50.0	--
	17	--	--	--	--	--	3.4	--	49.5	--
	18	--	--	--	--	--	3.7	--	51.0	--
Metz & Alexander (1970)	12-13	160.3	--	49.0	--	T	2.55	0.56	50.9	--
	14-15	168.9	--	58.3	--	--	3.09	0.44	53.3	--
Maksud et al., (1971)	9	132.2	5.9	30.5	4.7	B	1.16[d]	0.19	38.1[d]	3.5
	10	138.2	8.5	33.2	6.1	--	1.25	0.26	37.5	4.1
	11	146.9	10.1	41.3	10.3	--	1.43	0.22	35.6	5.8
Daniels & Oldridge (1971)	13-14	157.4	10.9	44.8	9.8	T	2.73	0.74	60.6	5.6
		163.8	10.2	50.7	9.6	--	3.03	0.65	59.6	5.3
Boileau et al., (1977)	11-14	159.2	13.3	49.1	12.8	T	2.45	0.58	50.5	5.9
		--	--	--	--	B	2.27	0.49	47.0	6.3
Stewart & Gutin (1976)	10-12	--	--	--	--	T	--	--	49.2	--
Krahenbuhl et al., (1977)	8	133.5	4.4	28.9	4.1	T	--	--	47.6	7.1
Nagle et al., (1977)	14-17	169.3	8.9	59.1	11.1	T	3.16	0.48	54.0	5.9
	15	175.3	6.9	67.3	10.0	--	3.74	0.31	56.3	5.8
	16	177.2	5.5	68.9	8.7	--	3.71	0.42	54.0	3.5
	17	179.3	6.5	71.3	11.1	--	3.87	0.54	54.7	6.7

TABLE 146   (continued)

MAXIMAL OXYGEN UPTAKE FOR U.S. AND CANADIAN CHILDREN

Reference	Age Group or Range (years)	Stature (cm) Mean	S.D.	Weight (kg) Mean	S.D.	Exercise Mode	Maximal Oxygen Uptake l/min Mean	S.D.	ml/kg/min Mean	S.D.

MAXIMAL OXYGEN UPTAKE FOR U.S. GIRLS

Reference	Age	Stature Mean	S.D.	Weight Mean	S.D.	Mode	l/min Mean	S.D.	ml/kg/min Mean	S.D.
Rodahl et al., (1961)	12	--	--	--	--	B	1.35	--	--	--
	14	--	--	--	--	--	1.34	--	--	--
Wilmore & Sigerseth (1967)	7-9	134.0	7.0	30.3	3.9	B	1.59	0.27	53.5	6.5
	10-11	143.0	7.6	37.0	7.5	--	1.87	0.34	50.7	5.9
	12-13	159.0	5.7	49.0	5.1	--	2.40	0.42	48.7	8.7
Knuttgen (1967)	15-18	161.9	5.8	56.9	7.2	B	1.90	0.26	33.6	4.4
	15	--	--	--	--	--	1.90[c]	--	34.0[c]	--
	16	--	--	--	--	--	1.95	--	34.5	--
	17	--	--	--	--	--	1.90	--	33.5	--
	18	--	--	--	--	--	1.80	--	31.5	--
Maksud et al., (1971)	9	127.2	4.9	28.2	3.4	B	0.95[b]	0.16	33.7[d]	6.2
	10	136.7	8.1	31.5	7.9	--	1.07	0.22	34.3	4.8
	11	146.2	7.4	39.0	10.5	--	1.23	0.18	32.3	5.1
Flint et al., (1977)	12	155.1	45.9	--	--	T	--	--	45.6	--
		162.2	5.7	57.9	7.1	T	2.15	0.33	37.1	3.9
Krahenbuhl et al., (1977)	8	130.6	5.9	27.7	4.7	T	--	--	42.9	5.7
Nagle et al., (1977)	14-17	163.8	6.2	55.6	7.8	T	2.27	0.25	41.1	3.7
	15	165.0	5.8	56.2	8.7	--	2.30	0.28	41.2	3.7
	16	164.7	6.0	58.2	9.0	--	2.36	0.34	40.7	4.0
	17	163.6	8.8	58.4	6.8	--	2.34	0.26	40.2	4.7
Gutin et al., (1978)	11-12	153.8	7.0	50.4	13.67	T	1.81	0.33	37.0	5.9

MAXIMAL OXYGEN UPTAKE FOR U.S. CHILDREN (sexes combined)

Reference	Age	Stature Mean	S.D.	Weight Mean	S.D.	Mode	l/min Mean	S.D.	ml/kg/min Mean	S.D.
Gilliam et al., (1977)	6-8	--	--	--	--	B	1.28	0.23	49.1	9.4
	9-10	--	--	--	--	--	1.52	0.25	43.6	6.8
	11-13	--	--	--	--	--	1.84	0.33	44.7	6.5

MAXIMAL OXYGEN UPTAKE FOR CANADIAN BOYS

Reference	Age	Stature Mean	S.D.	Weight Mean	S.D.	Mode	l/min Mean	S.D.	ml/kg/min Mean	S.D.
Cumming & Friesen (1967)	11-15	154.6	--	44.7	--	B	--	--	53.8	--
Cumming (1967)	10	--	--	--	--	B	--	--	44	--
	12	--	--	--	--	--	--	--	47	--
	14	--	--	--	--	--	--	--	49	--
	16	--	--	--	--	--	--	--	52	--
	18	--	--	--	--	--	--	--	55	--
Shephard et al., (1969)	11	146.1	6.8	38.6	8.0	T	1.84	0.32	48.3	7.5
	12-13	148.2	8.3	40.5	8.8	--	1.78	0.34	45.8	6.9
	9-13	--	--	--	--	--	1.79	0.30	47.4	7.3

TABLE 146 (continued)

MAXIMAL OXYGEN UPTAKE FOR U.S. AND CANADIAN CHILDREN

Reference	Age Group or Range (years)	Stature (cm) Mean	S.D.	Weight (kg) Mean	S.D.	Exercise Mode	Maximal Oxygen Uptake l/min Mean	S.D.	ml/kg/min Mean	S.D.
MAXIMAL OXYGEN UPTAKE FOR CANADIAN BOYS										
Cunningham & Eynon (1973)	10-16	150.4	8.1	41.2	6.5	B	2.17	0.43	52.5	4.1
		164.8	10.1	52.1	13.4	--	2.65	0.52	52.9	9.5
		169.6	12.1	59.8	11.4	--	3.37	0.63	56.6	4.5
Larivière et al., (1974)	10	137.5	7.3	31.0	5.1	T	1.43	0.23	46.7	4.9
Bailey et al., (1974)	15-19	174.3	6.6	66.0	10.0	B	--	--	43.1[a]	8.2
		176.9	7.3	77.5	10.6	--	--	--	36.4[a]	8.0
Shephard et al.,(1974)	9-12	--	--	--	--	T	--	--	60.6[e]	10.9
	9-12	--	--	--	--	--	--	--	54.9	8.9
Seely et al., (1974)	13-18	160.9	8.1	52.0	10.7	B	--	--	55.3[f]	--
		164.6	10.6	57.6	14.2	--	--	--	57.0	--
		170.0	8.3	65.2	10.5	--	--	--	55.0	--
		171.6	6.5	67.9	12.3	--	--	--	55.8	--
Shephard (1974)	11-18	--	--	--	--	T	2.41	--	44.9	--
Cunningham et al., (1976)	10	140.5	5.6	35.5	5.4	B	2.00	0.30	56.6	7.7
Cunningham et al., (1977)	10	139.7	6.5	33.5	5.3	T	--	--	56.5	7.1
Bailey et al., (1978)	8-15	127.9	5.1	25.9	3.3	T	1.46	0.20	56.4	--
		133.4	5.1	28.9	3.7	--	1.72	0.22	59.5	--
		138.7	5.3	32.0	4.3	--	1.82	0.23	56.9	--
		143.8	5.6	34.8	4.8	--	1.96	0.25	56.3	--
		148.9	6.2	38.5	5.7	--	2.18	0.33	56.6	--
		155.3	7.6	43.4	7.7	--	2.39	0.39	55.1	--
		162.8	7.9	50.0	8.9	--	2.73	0.53	54.6	--
		169.9	7.1	55.9	8.5	--	2.94	0.50	52.6	--
MAXIMAL OXYGEN UPTAKE FOR CANADIAN GIRLS										
Cumming (1967)	6	--	--	--	--	B	--	--	52	--
	8	--	--	--	--	--	--	--	49	--
	10	--	--	--	--	--	--	--	40	--
	12	--	--	--	--	--	--	--	42	--
	14	--	--	--	--	--	--	--	38	--
	16	--	--	--	--	--	--	--	39	--
	18	--	--	--	--	--	--	--	44	--
Shephard et al., (1969)	9-10	140.5	6.0	33.7	4.7	T	1.23[e]	0.18	36.8[e]	5.2
	11	149.9	6.4	38.0	4.8	--	1.35	0.13	35.9	5.5
	12-13	156.6	3.8	42.9	8.3	--	1.68	0.30	38.3	7.6
	9-13	--	--	--	--	--	1.38	0.25	36.9	5.9

TABLE 146 (continued)

MAXIMAL OXYGEN UPTAKE FOR U.S. AND CANADIAN CHILDREN

Reference	Age Group or Range (years)	Stature (cm) Mean	Stature (cm) S.D.	Weight (kg) Mean	Weight (kg) S.D.	Exercise Mode	Maximal Oxygen Uptake l/min Mean	Maximal Oxygen Uptake l/min S.D.	Maximal Oxygen Uptake ml/kg/min Mean	Maximal Oxygen Uptake ml/kg/min S.D.
MAXIMAL OXYGEN UPTAKE FOR CANADIAN GIRLS										
Cunningham & Eynon (1973)	10-16	154.8	10.0	43.3	6.7	B	1.97	0.31	46.2	7.7
		160.0	4.5	52.1	8.1	--	2.24	0.34	43.4	6.5
		164.8	7.4	53.7	5.4	--	2.19	0.34	40.5	2.1
Larivière et al., (1974)	10	139.5	6.8	33.4	7.1	T	1.16	0.14	38.1	5.0
Bailey et al., (1974)	15-19	162.5	6.4	56.7	9.2	B	--	--	34.0[a]	9.0
		162.1	5.6	57.4	7.0	B	--	--	30.6[a]	7.2
Shephard et al., (1974)	9-12	--	--	--	--	T	--	--	49.4[e]	10.2
	9-12	--	--	--	--	--	--	--	42.4	7.5
Seely et al., (1974)	13-18	161.1	7.4	53.5	8.5	B	--	--	39.4[f]	--
		158.8	7.4	52.5	8.3	--	--	--	41.9	--
		159.7	7.2	56.9	8.9	--	--	--	46.5	--
		158.4	3.8	50.4	5.2	--	--	--	47.3	--
Shephard (1974)	13-18	--	--	--	--	T	1.86	--	33.5	--

[a] T = treadmill; B = Bicycle ergometer; CM = "cycle mill" - modified bicycle, placed on a treadmill of variable placed speed and incline.

[b] = standard error of the mean.

[c] = age-specific means estimated from a graph of age changes.

[d] = basic criterion for termination of exercise was a heart rate of 180 beats per minute.

[e] = predicted maximal oxygen intake based on submaximal treadmill exercise, using oxygen scale of Åstrand nomogram, and correcting to a maximal pulse rate of 212 in boys and 207 in girls.

[f] = predicted for a theoretical maximum heartrate of 247 (Cumming and Friesen [1967] modification of Åstrand-Rhyming technique).

TABLE 147

PHYSICAL WORKING CAPACITY IN NORMAL CALIFORNIA SCHOOL CHILDREN

Age (years)	Stature (cm)	Weight (kg)	Surface Area (m^2)	Blood Pressure (mm Hg)		Vital Capacity (cm^3)			Working Capacity (Kg/min)
				Syst.	Diast.	1-Sec.	3-Sec.	Total	
					BOYS				
6	121	24	0.90	101	59	990	1,210	1,290	331
7	127	29	1.02	102	58	1,190	1,510	1,540	368
8	131	30	1.06	104	61	1,260	1,640	1,720	438
9	140	35	1.20	110	58	1,670	2,040	2,130	472
10	145	40	1.31	106	60	1,720	2,130	2,230	551
11	152	46	1.41	114	65	2,030	2,530	2,640	650
12	155	48	1.45	114	67	1,940	2,850	2,905	703
13	160	51	1.51	118	68	2,140	3,070	3,070	739
14	170	59	1.68	117	65	2,290	3,600	3,600	964
					GIRLS				
6	120	24	0.92	97	57	960	1,220	1,300	265
7	124	25	0.94	91	59	1,010	1,300	1,340	287
8	132	30	1.06	98	57	1,309	1,540	1,610	343
9	133	32	1.06	104	59	1,280	1,570	1,660	337
10	144	38	1.25	99	56	1,470	1,960	2,020	406
11	148	44	1.36	106	59	1,760	2,270	2,370	488
12	158	46	1.43	112	63	1,730	2,600	2,540	483
13	163	55	1.59	116	70	2,060	2,970	3,010	564
14	165	60	1.63	115	73	2,130	3,020	3,050	542

(data from Adams, Linde, and Miyake, 1961, Pediatrics 28:55-64. Copyright American Academy of Pediatrics 1961)

TABLE 148

RELATIONSHIP OF WORKING CAPACITY TO SURFACE AREA
IN CALIFORNIA SCHOOL GIRLS

Surface Area (m^2)	Mean Working Capacity (kg/min)	Single 95% Confidence Interval (kg/min)	Mean 95% Confidence Interval (kg/min)
0.9	281	91-432	233-290
1.0	305	144-484	290-338
1.1	338	192-531	343-381
1.2	404	236-575	389-421
1.3	432	276-614	430-461
1.4	512	313-652	466-498
1.5	599	348-686	499-535
1.6	577	379-719	528-570
1.7	578	409-750	556-603

(data from Adams, Linde, and Miyake, 1961, Pediatrics 28:55-64.
Copyright American Academy of Pediatrics 1961)

TABLE   149

ANTHROPOMETRIC AND EXERCISE PHYSIOLOGY DATA
FROM 8-YEAR-OLD ARIZONA CHILDREN

Variables	Boys		Girls	
	Mean	S.D.	Mean	S.D.
Age (months)	103.8	4.3	103.4	3.3
Stature (cm)	133.5	4.4	130.6	5.9
Weight (kg)	28.9	4.1	37.7	4.7
Ve (1/min at VO_2 max)	54.6	12.6	47.5	8.3
VO_2 max (ml/kg min)	47.6	7.1	42.9	5.7
Max heart rate (beats/min)	200.6	10.8	204.0	11.4
549 meter run (min)	2.66	.32	2.77	.21
1207 meter run (min)	6.46	.54	7.23	.49
1609 meter run (min)	9.10	.86	10.29	.69

(data from Krahenbuhl, Pangrazi, Burkett, et al., 1977, Medicine and Science
in Sports 9:37-40)

TABLE 150

PHYSICAL WORK CAPACITY OF PHILADELPHIA CHILDREN

| Age (yr.) | Sex | Stature (cm) | | Weight (kg) | | Muscle Strength | | | | Manual Dexterity | | Pulse Response | | | | | |
| | | | | | | R. Biceps | | L. Biceps | | | | 300 Kpm/min | | 450 Kpm/min | | 600 Kpm/min | |
		Mean	S.D.	Mean	S.D.	Mean	S.D.	Mean	S.D.	Mean	S.D.	Mean	S.D.	Mean	S.D.	Mean	S.D.
8	Boys	130.6	6.5	29.1	5.6	9.2	2.0	8.8	2.0	12.4	2.7	171	12	187	14	--	--
	Girls	128.8	5.4	28.0	5.8	8.5	1.7	8.3	1.9	12.1	1.8	175	14	190	11	--	--
10	Boys	140.8	7.6	35.8	8.4	13.6	3.6	13.0	3.5	10.4	1.8	162	14	181	14	--	--
	Girls	140.3	7.0	35.3	7.4	11.1	2.7	10.5	2.3	10.1	1.5	170	13	189	9	--	--
12	Boys	152.3	8.5	44.8	9.0	16.2	4.7	14.8	3.7	9.0	1.0	147	16	167	14	--	--
	Girls	154.2	7.4	45.3	8.0	14.4	2.6	13.8	3.2	8.8	1.1	163	17	180	15	--	--
14	Boys	164.7	8.5	56.1	9.4	21.4	5.6	20.0	5.1	8.5	0.9	133	15	142	17	--	--
	Girls	158.7	6.2	49.3	--	16.2	3.3	15.7	3.0	8.1	0.8	154	18	172	15	--	--
16	Boys	172.4	6.8	63.8	9.4	30.8	6.8	30.0	6.3	8.6	0.9	122	14	--	--	148	14
	Girls	163.3	5.7	55.0	7.4	18.7	3.2	18.0	3.6	8.1	1.1	147	15	166	14	--	--
18	Boys	175.8	5.7	66.7	9.5	34.1	7.4	32.5	7.2	8.4	1.0	115	13	--	--	140	16
	Girls	162.0	6.6	43.3	--	17.9	3.7	17.2	3.6	8.0	1.3	145	14	164	14	--	--
20	Boys	177.2	6.4	73.2	10.8	31.9	5.4	31.2	5.3	8.2	1.1	121	14	--	--	145	14
	Girls	164.9	6.3	57.9	6.2	17.6	2.8	17.3	2.4	7.5	0.7	139	13	158	13	--	--
22	Boys	179.7	5.0	75.6	13.6	35.0	4.8	33.8	5.1	8.1	1.1	119	15	--	--	143	18
	Girls	163.3	5.4	69.0	8.0	19.0	3.2	18.7	3.5	7.5	0.7	140	14	155	13	--	--

Kpm = Kg per meter

(data from Rodahl et al., 1961)

TABLE 151

RESPONSE TO FIXED WORK LOADS ON BICYCLE ERGOMETER IN PHILADELPHIA CHILDREN

| Age (years) | Pulse Response (beats per minute) | | | | | | Oxygen Uptake (l/min) | | | | | |
| | 300 Kpm/min | | 450 Kpm/min | | 600 Kpm/min | | 300 Kpm/min | | 450 Kpm/min | | 600 Kpm/min | |
	Mean	S.D.	Mean	S.D.	Mean	S.D.	Mean	S.D.	Mean	S.D.	Mean	S.D.
BOYS												
8	171	12	187	14	--	--	--	--	--	--	--	--
10	162	14	181	14	--	--	0.85	0.09	1.11	0.11	--	--
12	147	16	167	14	--	--	0.97	0.13	1.20	0.09	1.53	0.16
14	133	15	142	17	166	17	1.04	0.06	1.20	0.12	1.53	0.16
16	122	14	--	--	148	14	--	--	1.14	0.07	--	--
18	115	13	--	--	140	16	--	--	--	--	1.47	0.18
GIRLS												
8	175	14	190	11	--	--	--	--	--	--	--	--
10	170	13	189	9	--	--	0.79	0.09	1.02	0.11	--	--
12	163	17	180	15	--	--	0.96	0.19	1.27	0.06	--	--
14	154	18	172	15	--	--	0.94	0.08	--	--	1.34	0.12
16	147	15	166	14	--	--	0.93	0.17	--	--	1.26	0.08
18	145	14	164	14	--	--	0.91	0.22	--	--	1.27	0.12

Age groups defined as 7.5 to 8.5, 9.5 to 10.5, 17.5 to 18.5. Kpm = kg/m

(data from Malina, unpublished b)

TABLE   152

FORCED VITAL CAPACITY (cm^3) FOR UNITED STATES CHILDREN

Age (years)	Mean	S.D.	Age (years)	Mean	S.D.
White boys			Negro boys		
6	1,326	263	6	1,153	275
7	1,524	282	7	1,331	262
8	1,732	296	8	1,528	265
9	1,947	358	9	1,651	309
10	2,122	364	10	1,812	347
11	2,406	419	11	2,021	366
White girls			Negro girls		
6	1,194	266	6	1,040	266
7	1,382	265	7	1,215	232
8	1,579	286	8	1,370	278
9	1,763	317	9	1,661	328
10	1,972	356	10	1,746	373
11	2,230	404	11	2,000	415

(data from Hamill et al., 1978)

TABLE 153

VITAL CAPACITY (ℓ) AND MAXIMUM BREATHING CAPACITY (ℓ/min) IN BOSTON BOYS

Age (years)	Vital Capacity		Maximum Breathing Capacity	
	Mean	S.D.	Mean	S.D.
5.0 - 5.9	1.29	0.23	42.0	6
6.0 - 6.9	1.65	0.19	45.0	10
7.0 - 7.9	1.93	0.24	65.0	6
8.0 - 8.9	2.16	0.41	69.0	10
9.0 - 9.9	2.17	0.24	73.0	19
10.0 - 10.9	2.23	0.24	79.0	18
11.0 - 11.9	2.54	0.37	75.0	7
12.0 - 12.9	3.75	0.27	109.0	17
13.0 - 13.9	3.81	0.64	117.0	25
14.0 - 14.9	4.29	0.76	117.0	25
15.0 - 15.9	4.47	0.62	129.0	34
16.0 - 16.9	4.51	0.42	134.0	13
17.0 - 17.9	4.49	0.41	155.0	11

(data from Ferris, Whittenberger, and Gallagher, 1952, Pediatrics 9:659-670. Copyright American Academy of Pediatrics 1952)

TABLE   154

VITAL CAPACITY (ℓ) AND MAXIMUM BREATHING CAPACITY (ℓ/min) IN BOSTON GIRLS

Age (years)	Vital Capacity		Maximum Breathing Capacity	
	Mean	S.D.	Mean	S.D.
5.0 - 5.9	1.08	0.17	41.2	10.9
6.0 - 6.9	1.45	0.19	52.6	5.2
7.0 - 7.9	1.51	0.16	52.8	10.1
8.0 - 8.9	1.87	0.32	60.3	12.1
9.0 - 9.9	2.04	0.33	67.0	10.5
10.0 - 10.9	2.29	0.24	71.9	11.2
11.0 - 11.9	2.57	0.43	79.3	14.9
12.0 - 12.9	3.03	0.43	95.8	24.3
13.0 - 13.9	3.49	0.68	103.9	18.7
14.0 - 14.9	3.40	0.49	99.4	30.2
15.0 - 15.9	3.66	0.39	105.0	23.2
16.0 - 16.9	3.63	0.36	92.0	18.2
17.0 - 17.9	4.00	0.42	108.3	24.6

(data from Ferris and Smith, 1953, Pediatrics 12:341-352. Copyright American Academy of Pediatrics 1953)

TABLE 155

VITAL CAPACITY (ℓ) AND MAXIMUM BREATHING CAPACITY (ℓ/min)
IN RELATION TO BODY SURFACE AREA (m^2) IN BOSTON (MA) BOYS

Surface Area Groups	Vital Capacity		Maximum Breathing Capacity	
	Mean	S.D.	Mean	S.D.
0.70 - 0.89	1.52	0.34	46.0	9.0
0.90 - 0.99	1.63	0.15	48.0	16.0
1.00 - 1.09	1.99	0.26	65.0	15.0
1.10 - 1.19	2.10	0.18	70.0	19.0
1.20 - 1.29	2.47	0.42	79.0	12.0
1.30 - 1.39	2.57	0.41	84.0	11.0
1.40 - 1.49	3.36	0.34	93.0	7.0
1.50 - 1.59	3.53	0.61	108.0	28.0
1.60 - 1.69	4.08	0.49	110.0	20.0
1.70 - 1.79	4.32	0.47	130.0	23.0
1.80 - 1.89	4.85	0.57	130.0	30.0
1.90 - 1.99	4.88	0.60	125.0	26.0

(data from Ferris, Whittenberger, and Gallagher, 1952, Pediatrics 9:659-670. Copyright
American Academy of Pediatrics 1952)

TABLE 156

VITAL CAPACITY (ℓ) AND MAXIMUM BREATHING CAPACITY (ℓ/min)
IN RELATION TO BODY SURFACE AREA (m^2) IN BOSTON GIRLS

Surface Area Groups (m^2)	Vital Capacity		Maximum Breathing Capacity	
	Mean	S.D.	Mean	S.D.
0.60 - 0.79	1.12	0.20	41.9	10.4
0.80 - 0.89	1.38	0.26	53.0	11.3
0.90 - 0.99	1.64	0.29	59.6	17.7
1.00 - 1.09	1.84	0.26	63.3	12.4
1.10 - 1.19	2.21	0.28	69.8	11.8
1.20 - 1.29	2.39	0.30	75.7	13.1
1.30 - 1.39	2.76	0.37	81.2	18.9
1.40 - 1.49	3.08	0.54	90.8	22.0
1.50 - 1.59	3.29	0.46	93.0	25.1
1.60 - 1.69	3.69	0.37	106.6	26.5
1.70 - 1.79	4.00	0.42	120.6	17.7

(data from Ferris and Smith, 1953, Pediatrics 12:341-352. Copyright American Academy
of Pediatrics 1953)

TABLE 157

VITAL CAPACITY (ℓ) AND MAXIMUM BREATHING CAPACITY (ℓ/min)
IN RELATION TO STATURE (cm) IN BOSTON BOYS

| | Vital Capacity | | Maximum Breathing Capacity | |
Stature Groups	Mean	S.D.	Mean	S.D.
110.0 - 119.9	1.46	0.32	44.0	7.0
120.0 - 124.9	1.64	0.16	45.0	12.0
125.0 - 129.9	1.79	0.18	63.0	9.0
130.0 - 134.9	2.05	0.21	69.0	18.0
135.0 - 139.9	2.27	0.45	68.0	19.0
140.0 - 144.9	2.41	0.41	79.0	8.0
145.0 - 149.9	2.38	0.31	82.0	12.0
150.0 - 154.9	2.69	0.45	86.0	19.0
155.0 - 159.9	3.52	0.44	102.0	15.0
160.0 - 164.9	3.72	0.47	115.0	27.0
165.0 - 169.9	4.07	0.45	113.0	24.0
170.0 - 174.9	4.47	0.55	127.0	26.0
175.0 - 179.9	4.82	0.54	129.0	11.0
180.0 - 184.9	5.24	0.49	148.0	19.0

(data from Ferris, Whittenberger, and Gallagher, 1952, Pediatrics 9:659-670. Copyright
American Academy of Pediatrics 1952)

TABLE 158

VITAL CAPACITY (ℓ) AND MAXIMUM BREATHING CAPACITY (ℓ/min) IN RELATION TO STATURE (cm) IN BOSTON GIRLS

Stature Groups (cm)	Vital Capacity		Maximum Breathing Capacity	
	Mean	S.D.	Mean	S.D.
100.0 - 109.9	--	--	41.5	13.6
110.0 - 114.9	1.12	0.17	45.2	8.6
115.0 - 119.9	1.37	0.19	49.0	11.0
120.0 - 124.9	1.70	0.35	61.7	16.3
125.0 - 129.9	1.70	0.22	57.9	9.8
130.0 - 134.9	1.97	0.33	62.1	10.6
135.0 - 139.9	2.19	0.27	72.0	11.3
140.0 - 144.9	2.32	0.32	73.7	13.2
145.0 - 149.9	2.53	0.33	77.9	18.4
150.0 - 154.9	2.94	0.30	89.8	20.9
155.0 - 159.9	3.28	0.43	94.7	21.8
160.0 - 164.9	3.57	0.41	102.7	26.4
165.0 - 169.9	3.85	0.44	107.3	28.4
170.0 - 174.9	3.91	0.26	124.3	14.6

(data from Ferris and Smith, 1953, Pediatrics 12:341-352. Copyright American Academy of Pediatrics 1953)

TABLE 159

VITAL CAPACITY (ℓ) AND MAXIMUM BREATHING CAPACITY (ℓ/min)
IN RELATION TO WEIGHT (kg) IN BOSTON BOYS

Weight Groups	Vital Capacity		Maximum Breathing Capacity	
	Mean	S.D.	Mean	S.D.
20.0 - 24.9	1.53	0.32	46.0	9.0
25.0 - 29.9	1.72	0.22	53.0	16.0
30.0 - 34.9	2.10	0.20	74.0	17.0
35.0 - 39.9	2.31	0.43	71.0	16.0
40.0 - 44.9	2.66	0.37	83.0	11.0
45.0 - 49.9	3.26	0.53	93.0	10.0
50.0 - 54.9	3.85	0.47	110.0	29.0
55.0 - 59.9	4.14	0.53	121.0	18.0
60.0 - 64.9	4.34	0.71	129.0	27.0
65.0 - 69.9	4.14	0.66	119.0	27.0
70.0 - 74.9	4.86	0.46	126.0	32.0
75.0 - 79.9	4.87	0.75	122.0	24.0

(data from Ferris, Whittenberger, and Gallagher, 1952, Pediatrics 9:659-670. Copyright
American Academy of Pediatrics 1952)

TABLE   160

VITAL CAPACITY (ℓ) AND MAXIMUM BREATHING CAPACITY (ℓ/min)
IN RELATION TO WEIGHT (kg) IN BOSTON GIRLS

Weight Groups (kg)	Vital Capacity		Maximum Breathing Capacity	
	Mean	S.D.	Mean	S.D.
15.0 - 19.9	1.04	0.22	36.0	5.8
20.0 - 24.9	1.40	0.25	53.0	10.0
25.0 - 29.9	1.75	0.26	60.7	10.0
30.0 - 34.9	1.98	0.32	65.0	13.5
35.0 - 39.9	2.37	0.37	77.6	13.4
40.0 - 44.9	2.63	0.48	77.3	17.2
45.0 - 49.9	3.02	0.42	87.6	19.7
50.0 - 54.9	3.24	0.51	97.6	28.9
55.0 - 59.9	3.53	0.58	100.0	27.7
60.0 - 64.9	3.65	0.36	105.9	23.3
65.0 - 69.9	3.55	0.99	97.9	23.6

(data from Ferris and Smith, 1953, Pediatrics 12:341-352. Copyright American Academy
of Pediatrics 1953)

TABLE 161

DIRECT MEASUREMENTS OF MAXIMAL OXYGEN INTAKE ON ESKIMO GROUPS

Population	Age (years)	Heart Rate (/min)		Gas Exchange Ratio		Ventilation (1/min BTPS)		Maximal Oxygen Intake			
								(1/min STPD)		(ml/kg min STPD)	
		Mean	S.D.	Mean	S.D.	Mean	S.D.	Mean	S.D.	Mean	S.D.
Igloolik:fair (step)	14 ± 2F	182	16	1.05	0.03	–	–	1.95	0.30	39.2	7.0
Point Hope (treadmill)	11 - 18M	186	–	–	–	84.8	–	2.41	–	44.9	–
	13 - 18F	189	–	–	–	69.3	–	1.86	–	33.5	–
Upernavik (bicycle ergometer)	14 - 19M	186	–	0.96	–	–	–	2.36	–	40.1	–

BTPS = body temperature and pressure saturated with water vapor; STPD = standard
temperature, pressure, dry (0°C; 760 mm Hg; dry).

(data from "Work physiology and activity patterns of circumpolar Eskimos and Ainu," Human
Biology 46:263-294, 1974, by R. J. Shephard, by permission of the Wayne State University
Press. Copyright 1974, Wayne State University Press, Detroit, Michigan 48202)

TABLE 162

VITAL CAPACITY (liters) IN BOSTON CHILDREN

Age (years)

	5.0 to 5.9	6.0 to 6.9	7.0 to 7.9	8.0 to 8.9	9.0 to 9.9	10.0 to 10.9	11.0 to 11.9	12.0 to 12.9	13.0 to 13.9	14.0 to 14.9	15.0 to 15.9	16.0 to 16.9	17.0 to 17.9
Mean	1.29	1.65	1.93	2.16	2.17	2.23	2.54	3.75	3.81	4.29	4.47	4.51	4.49
S.D.	0.23	0.19	0.24	0.41	0.24	0.24	0.37	0.27	0.64	0.76	0.62	0.42	0.41

Weight (kg)

	20.0 to 24.9	25.0 to 29.9	30.0 to 34.9	35.0 to 39.9	40.0 to 44.9	45.0 to 49.9	50.0 to 54.9	55.0 to 59.9	60.0 to 64.9	65.0 to 69.9	70.0 to 74.9	75.0 to 79.9	80.0 to 84.9
Mean	1.53	1.72	2.10	2.31	2.66	3.26	3.85	4.14	4.34	4.14	4.86	4.87	4.52
S.D.	0.32	0.22	0.20	0.43	0.37	0.53	0.47	0.53	0.71	0.66	0.46	0.75	0.28

Stature (cm)

	110.0 to 119.9	120.0 to 124.9	125.0 to 129.9	130.0 to 134.9	135.0 to 139.9	140.0 to 144.9	145.0 to 149.9	150.0 to 154.9	155.0 to 159.9	160.0 to 164.9	165.0 to 169.9	170.0 to 174.9	175.0 to 179.9	180.0 to 184.9
Mean	1.46	1.64	1.79	2.05	2.27	2.41	2.38	2.69	3.52	3.72	4.07	4.47	4.82	5.24
S.D.	0.32	0.16	0.18	0.21	0.45	0.41	0.31	0.45	0.44	0.47	0.45	0.55	0.54	0.49

Body Surface Area (m^2)

	0.70 to 0.89	0.90 to 0.99	1.00 to 1.09	1.10 to 1.19	1.20 to 1.29	1.30 to 1.39	1.40 to 1.49	to 1.59	1.60 to 1.69	1.70 to 1.79	1.80 to 1.89	1.90 to 1.99	2.00 to 2.09
Mean	1.52	1.63	1.99	2.10	2.47	2.57	3.36	3.53	4.08	4.32	4.85	4.88	5.39
S.D.	0.34	0.15	0.26	0.18	0.42	0.41	0.34	0.61	0.49	0.47	0.57	0.60	0.39

(data from Ferris, Whittenberger, and Gallagher, 1952, Pediatrics 9:659-670. Copyright American Academy of Pediatrics 1952)

TABLE   163

ESTIMATED VALUES FOR PULMONARY FUNCTION IN COLORADO BOYS

Stature (cm.)	FVC (L.)	FEV$_1$ (L.)	$\dfrac{FEV_1}{FVC}$ 100		VC (L.)	MVV (L min)	MEFR (L./sec)	PFR. (L./min)	Corr. PFR (L./min)
			Mean	95% limits					
100	0.69	0.68	99	--	0.70	14	1.44	94	115
102	0.73	0.73	100	--	0.76	15	1.44	96	117
104	0.78	0.77	99	99-100	0.81	17	1.44	99	121
106	0.83	0.82	99	88-100	0.87	18	1.46	103	126
108	0.89	0.87	98	94-100	0.93	20	1.47	106	129
110	0.94	0.92	98	90-100	0.99	21	1.50	111	135
112	1.00	0.97	97	87-100	1.06	23	1.53	115	140
114	1.07	1.03	96	96-100	1.12	25	1.56	120	147
116	1.13	1.09	96	91-100	1.20	26	1.60	125	152
118	1.20	1.15	96	95-100	1.27	28	1.64	131	160
120	1.28	1.21	95	89-100	1.35	30	1.69	136	166
122	1.35	1.28	95	94-100	1.42	32	1.75	143	175
124	1.43	1.35	94	83- 96	1.51	35	1.81	149	182
126	1.51	1.42	94	92-100	1.59	37	1.88	156	191
128	1.59	1.50	94	--	1.68	39	1.95	164	200
130	1.68	1.58	94	85- 98	1.77	42	2.03	171	209
132	1.77	1.66	94	93-100	1.86	44	2.11	179	218
134	1.86	1.74	93	84-100	1.95	47	2.20	188	230
136	1.96	1.82	93	80-100	2.05	50	2.29	196	240
138	2.06	1.91	93	81-100	2.15	52	2.39	206	252
140	2.16	2.00	93	80-100	2.26	55	2.50	215	263
142	2.27	2.10	92	86- 98	2.36	58	2.61	225	275
144	2.37	2.19	92	86-100	2.47	61	2.72	235	287
146	2.49	2.29	92	72- 97	2.58	65	2.84	246	300
148	2.60	2.39	92	90-100	2.70	68	2.97	257	314
150	2.72	2.49	91	90- 96	2.81	71	3.10	268	327
152	2.84	2.60	91	87- 98	2.93	75	3.24	279	341
154	2.96	2.71	91	81-100	3.05	78	3.38	291	355
156	3.09	2.82	91	89- 96	3.18	82	3.53	304	371
158	3.22	2.93	91	77- 98	3.31	85	3.69	316	386
160	3.35	3.05	91	88- 93	3.44	89	3.85	330	403
162	3.48	3.17	91	87- 98	3.57	93	4.01	343	420
164	3.62	3.29	91	82- 98	3.71	97	4.18	357	436
166	3.76	3.42	91	--	3.84	101	4.36	371	453
168	3.91	3.54	91	85- 92	3.99	105	4.54	385	470

TABLE 163 (continued)

ESTIMATED VALUES FOR PULMONARY FUNCTION IN COLORADO BOYS

Stature (cm.)	FVC (L.)	FEV₁ (L.)	$\frac{FEV_1}{FVC}100$ Mean	$\frac{FEV_1}{FVC}100$ 95% limits	VC (L.)	MVV (L./min)	MEFR (L./sec)	PFR. (L./min)	Corr. PFR (L./min)
170	4.05	3.67	91	85-100	4.13	109	4.72	400	488
172	4.20	3.80	91	--	4.28	113	4.91	415	506
174	4.36	3.94	90	--	4.43	118	5.11	431	526
176	4.51	4.07	90	--	4.58	122	5.31	447	545
178	4.67	4.21	90	--	4.73	127	5.52	463	565
180	4.83	4.36	90	--	4.89	131	5.74	480	586

S.D. (%) means standard deviation in % of measured values.
FVC = forced vital capacity; FEV₁ = one second forced expiratory volume; VC = slow
vital capacity; MVV = maximum voluntary ventilation; MEFR = maximum expiratory
flow rate; PFR = peak expiratory flow rate; Corr. PFR = PFR corrected for altitude.

(data from Kopetzky, Maselli, and Ellis, 1974, The Journal of Allergy and Clinical
Immunology 53:1-8)

TABLE 164

BASAL OXYGEN CONSUMPTION, OXYGEN CONSUMPTION PER KILOGRAM,
WEIGHT AND STATURE BY AGE AND SEX
IN OHIO CHILDREN

Sex	Age (years)	Stature (cm) Mean	Stature (cm) S.D.	Weight (kg) Mean	Weight (kg) S.D.	O₂ per 24 Hrs. Mean	O₂ per 24 Hrs. S.E.	O₂ per Kg Mean	O₂ per Kg S.E.
Boys	6- 7	124.4	4.8	24.3	3.3	1145	17.1	47.3	0.7
Girls	6- 7	125.3	7.3	27.9	3.9	1095	19.4	46.6	0.9
Boys	8- 9	136.0	4.5	30.8	4.4	1252	19.8	41.0	0.5
Girls	8- 9	135.4	5.2	28.8	4.7	1167	19.6	41.1	0.7
Boys	10-11	145.2	5.3	36.6	6.1	1322	23.8	36.7	0.9
Girls	10-11	145.7	6.8	35.8	6.3	1268	26.2	35.9	0.6
Boys	12-13	155.7	9.5	47.6	9.6	1546	35.6	33.5	0.8
Girls	12-13	156.4	7.5	44.3	6.8	1354	26.6	30.7	0.6
Boys	14-15	170.0	7.5	60.8	9.9	1779	36.2	29.7	0.7
Girls	14-15	160.2	7.7	52.4	7.9	1375	18.2	26.8	0.6
Boys	16-18	177.8	4.8	70.6	7.5	1841	42.1	26.3	0.3
Girls	16-18	162.9	6.6	54.7	8.1	1321	23.1	24.7	0.7

(data from Garn and Clark, 1953)

TABLE   165

ESTIMATED VALUES FOR PULMONARY FUNCTION IN COLORADO GIRLS

Stature (cm)	FVC (L.)	FEV$_1$ (L.)	$\dfrac{FEV_{1\%}}{FVC}$ Mean	95% limits	VC (L.)	MVV (L./min)	MEFR (L./sec)	PFR (L./min)	Corr. PFR (L./min)
100	0.76	0.74	97	--	0.69	10	1.43	53	65
102	0.79	0.77	98	--	0.74	12	1.45	62	76
104	0.83	0.80	96	--	0.79	13	1.48	70	85
106	0.86	0.84	98	--	0.86	15	1.52	79	96
108	0.91	0.88	97	92-100	0.90	17	1.55	88	108
110	0.95	0.92	97	--	0.96	18	1.59	97	118
112	1.00	0.97	97	89-100	1.03	20	1.64	107	131
114	1.05	1.02	98	--	1.09	22	1.68	116	141
116	1.11	1.07	97	85-100	1.16	23	1.73	125	152
118	1.17	1.13	97	92-100	1.23	25	1.78	134	163
120	1.23	1.18	96	95-100	1.30	27	1.84	144	176
122	1.29	1.24	96	95-100	1.37	29	1.90	153	187
124	1.36	1.31	96	--	1.45	31	1.96	153	199
126	1.43	1.37	96	93-100	1.53	33	2.03	173	211
126	1.51	1.44	95	94-100	1.61	35	2.10	182	222
130	1.59	1.51	95	--	1.69	37	2.17	192	234
132	1.67	1.58	95	80-100	1.78	39	2.24	201	245
134	1.76	1.66	94	91-100	1.86	41	2.32	211	257
136	1.84	1.74	94	85-100	1.96	43	2.41	221	270
138	1.94	1.82	94	90-100	2.05	45	2.49	231	282
140	2.03	1.91	94	89-100	2.14	47	2.58	241	294
142	2.13	1.99	93	85-100	2.24	49	2.67	251	306
144	2.23	2.08	93	85-100	2.34	52	2.77	261	318
146	2.34	2.18	93	81- 98	2.45	54	2.87	271	331
148	2.45	2.27	93	85-100	2.55	56	2.97	282	344
150	2.56	2.37	93	92-100	2.66	59	3.08	292	356
152	2.67	2.47	92	86-100	2.77	61	3.19	302	368
156	2.79	2.57	92	87-100	2.88	64	3.30	313	382
156	2.91	2.68	92	90-100	3.00	66	3.41	323	394
158	3.04	2.79	92	90-100	3.11	69	3.53	334	407
160	3.17	2.90	91	76-100	3.23	71	3.65	344	420
162	3.30	3.02	91	81-100	3.35	74	3.78	355	433
164	3.43	3.14	91	90-100	3.48	77	3.91	366	446
166	3.57	3.26	91	--	3.61	79	4.04	377	460
168	3.72	3.38	91	85- 96	3.74	82	4.18	388	474

TABLE 165 (continued)

ESTIMATED VALUES FOR PULMONARY FUNCTION IN COLORADO GIRLS

Stature (cm)	FVC (L.)	FEV_1 (L.)	$\dfrac{FEV_{1\%}}{FVC}$ Mean	95% limits	V (L.)	MVV (L.)	MEFR (L./sec)	PFR (L./min)	Corr. PFR (L./min)
170	3.86	3.50	91	--	3.87	85	4.32	398	485
172	4.01	3.63	91	--	4.00	87	4.46	409	499
174	4.16	3.76	90	--	4.14	90	4.60	420	512
176	4.32	3.90	90	--	4.28	93	4.75	432	526
178	4.48	4.04	90	--	4.42	96	4.90	443	540
180	4.64	4.18	90	--	4.56	99	5.06	454	554

FVC = forced vital capacity; FEV_1 = one second forced expiratory volume; VC = slow vital capacity; MVV = maximum voluntary ventilation; MEFR = maximum expiratory flow rate; PFR = peak expiratory flow rate; Corr. PFR = PFR corrected for altitude.

(data from Kopetzky, Maselli, and Ellis, 1974, The Journal of Allergy and Clinical Immunology 53:1-8)

TABLE 166

PERFORMANCE DATA FOR MICHIGAN CHILDREN

					GRADE					
	Second		Third		Fourth		Fifth		Sixth	
Sex	Fall	Spring	Fall	Spring	Fall	Spring	Fall	Spring	Fall	Spring

Vital Capacity - (in^3)

Boys

Mean	85.00	94.00	106.87	110.48	119.65	122.35	126.83	133.09	144.30	151.39
S.D.	17.78	13.98	17.18	16.46	16.42	17.92	18.89	19.03	22.65	25.63

Girls

Mean	80.48	89.24	101.16	104.20	112.52	115.60	120.80	130.40	142.32	155.00
S.D.	8.41	9.43	7.68	10.89	10.90	11.21	13.56	13.46	14.78	15.28

Ball Bounce (No. in 30 Sec)

Boys

Mean	50.65	60.11	65.00	72.78	79.09	88.85	85.93	99.65	101.11	109.70
S.D.	15.37	14.08	14.57	14.79	12.26	14.97	13.12	13.96	12.58	14.42

Girls

Mean	43.70	50.14	57.98	58.10	64.52	72.32	76.90	87.44	90.76	101.16
S.D.	15.59	11.65	11.84	14.08	11.94	13.50	12.77	10.24	11.89	10.73

(data from Govatos, 1966)

TABLE 167

OBSERVED LUNG FUNCTION VALUES FOR WHITE
CHILDREN IN CONNECTICUT AND SOUTH CAROLINA

		GIRLS Age (years)		BOYS Age (years)	
		7-9	10-14	7-9	10-14
Stature (cm)	Mean	131.1	150.9	130.8	151.8
	S.D.	7.6	10.6	7.4	12.1
Weight (kg)	Mean	29.0	43.7	28.7	43.4
	S.D.	6.6	13.0	5.6	12.1
FVC (liter)	Mean	1.69	2.55	1.84	2.80
	S.D.	0.34	0.61	0.38	0.74
$FEV_{1.0}$ (liter)	Mean	1.57	2.29	1.64	2.39
	S.D.	0.28	0.50	0.29	0.56
$FEV_{1.0}$/FVC(%)	Mean	93.4	90.2	89.9	86.2
	S.D.	5.5	6.3	7.4	7.0
PEF (liter/sec)	Mean	3.18	4.33	3.18	4.36
	S.D.	0.67	0.92	0.56	0.98
MEF50%	Mean	2.46	3.29	2.31	3.14
(liter/sec)	S.D.	0.52	0.78	0.51	0.82
MEF25%	Mean	1.33	1.81	1.21	1.63
(liter/sec)	S.D.	0.37	0.59	0.39	0.55

FVC = forced vital capacity; $FEV_{1.0}$ = forced expiratory volume in 1
second; PEF = peak expiratory flow; MEF50% = instantaneous maximum
expiratory flow at 50% of FVC; MEF25% = similar, at 25% of FVC
(=25% above point of maximum expiration).

(data from Schoenberg et al., 1978)

TABLE 168

OBSERVED LUNG FUNCTION VALUES FOR BLACK
CHILDREN IN CONNECTICUT AND SOUTH CAROLINA

		GIRLS Age (years)		BOYS Age (years)	
		7-9	10-14	7-9	10-14
Stature (cm)	Mean	132.2	154.6	132.2	151.8
	S.D.	6.9	9.1	7.5	11.8
Weight (kg)	Mean	28.9	47.0	28.8	42.8
	S.D.	6.4	12.3	5.7	11.8
FVC (liter)	Mean	1.43	2.33	1.50	2.37
	S.D.	0.30	0.57	0.28	0.64
$FEV_{1.0}$ (liter)	Mean	1.37	2.12	1.42	2.11
	S.D.	0.27	0.47	0.24	0.50
$FEV_{1.0}/FVC$ (%)	Mean	96.1	92.0	95.1	89.7
	S.D.	4.0	6.4	5.4	7.1
PEF (liter/sec)	Mean	3.03	4.35	3.10	4.27
	S.D.	0.67	1.04	0.60	1.05
MEF50% (liter/sec)	Mean	2.38	3.36	2.35	3.11
	S.D.	0.58	0.89	0.48	0.84
MEF25%	Mean	1.31	1.79	1.30	1.56
(liter/sec)	S.D.	0.37	0.62	0.33	0.53

FVC = forced vital capacity; $FEV_{1.0}$ = forced expiratory volume in 1
second; PEF = peak expiratory flow; MEF50% = instantaneous maximum
expiratory flow at 50% of FVC; MEF25% = similar, at 25% of FVC
(=25% above point of maximum expiration).

(data from Schoenberg et al., 1978)

TABLE   169

MAXIMUM BREATHING CAPACITY (liters/minute)
IN BOSTON CHILDREN

Age (years)

	5.0 to 5.9	6.0 to 6.9	7.0 to 7.9	8.0 to 8.9	9.0 to 9.9	10.0 to 10.9	11.0 to 11.9	12.0 to 12.9	13.0 to 13.9	14.0 to 14.9	15.0 to 15.9	16.0 to 16.9	17.0 to 17.9
Mean	42	45	65	69	73	79	75	109	117	117	129	134	155
S.D.	6	10	6	10	19	18	7	17	25	25	34	13	11

Weight (kg)

	20.0 to 24.0	25.0 to 29.9	30.0 to 34.9	35.0 to 39.9	40.0 to 44.9	45.0 to 49.9	50.0 to 54.9	55.0 to 59.9	60.0 to 64.9	65.0 to 69.9	70.0 to 74.9	75.0 to 79.9	80.0 to 84.9
Mean	46	53	74	71	83	93	110	121	129	119	126	122	136
S.D.	9	16	17	16	11	10	29	18	27	27	32	24	16

Stature (cm)

	110.0 to 119.9	120.0 to 124.9	125.0 to 129.9	130.0 to 134.9	135.0 to 139.9	140.0 to 144.9	145.0 to 149.9	150.0 to 154.9	155.0 to 159.9	160.0 to 164.9	165.0 to 169.9	170.0 to 174.9	175.0 to 179.9	180.0 to 184.9
Mean	44	45	63	69	68	79	82	86	102	115	113	127	129	148
S.D.	7	12	9	18	19	8	12	19	15	27	24	26	11	19

Body Surface Area (m^2)

	0.70 to 0.89	0.90 to 0.99	1.00 to 1.09	1.10 to 1.19	1.20 to 1.29	1.30 to 1.39	1.40 to 1.49	1.50 to 1.59	1.60 to 1.69	1.70 to 1.79	1.80 to 1.89	1.90 to 1.99	2.00 to 2.09
Mean	46	48	65	70	79	84	93	108	110	130	130	125	148
S.D.	9	16	15	19	12	11	7	28	20	23	30	26	12

(data from Ferris, Whittenberger, and Gallagher, 1952, Pediatrics 9:659-670. Copyright American Academy of Pediatrics 1952)

TABLE 170

DIFFERENCES (cub. in.) BETWEEN MEAN LUNG CAPACITIES
OF OREGON BOYS 10, 13 AND 16 YEARS OF AGE
CLASSIFIED INTO PUBESCENT DEVELOPMENT GROUPS

Age (years)	Groups					
	1	2	3	4	4+5	5
10	119.28	124.86	--	--	--	--
13	--	158.90	173.73	--	--	--
13	--	158.90	--	--	207.84	--
13	--	--	173.73	--	207.84	--
16	--	--	--	245.33	--	261.93

The pubertal stages are those of Greulich, et al., 1942.

(data from Clarke and Degutis, 1962, Research Quarterly 33:356-368, by permission of
The American Alliance for Health, Physical Education, Recreation and Dance, 1900
Association Drive, Reston VA 22091)

TABLE 171

PULMONARY FUNCTION IN UTAH CHILDREN AGES
5 THROUGH 18 YEARS SEPARATED BY STATURE

| | BOYS | | | | | | GIRLS | | | | | |
| Stature | FVC (ml) | | FEV$_1$ (ml) | | FEV$_1$/FVC (%) | | FVC (ml) | | FEV$_1$ (ml) | | FEV$_1$/FVC (%) | |
(in)	Mean	S.D.	Mean	S.D.	Mean	S.D.	Mean	S.D.	Mean	S.D.	Mean	S.D.
44-45	1,235	193	963	164	75.9	6.5	1,147	137	889	139	76.7	4.3
46-47	1,393	177	1,107	174	78.6	4.9	1,185	145	942	138	78.8	2.3
48-49	1,504	252	1,239	219	80.4	3.7	1,391	195	1,134	189	80.6	3.6
50-51	1,729	217	1,430	205	82.2	3.9	1,484	246	1,214	232	81.0	3.7
52-53	1,835	232	1,527	193	82.8	3.7	1,633	216	1,344	212	81.7	4.8
54-55	2,096	280	1,770	260	83.7	3.3	1,841	189	1,543	167	83.3	3.5
56-57	2,276	267	1,908	243	83.4	4.6	2,059	302	1,768	300	85.1	3.2
58-59	2,539	287	2,133	256	83.5	4.6	2,142	318	1,861	315	86.1	3.6
60-61	2,814	369	2,448	338	86.3	4.0	2,527	342	2,170	311	85.5	4.9
62-63	3,144	484	2,701	407	85.4	4.6	2,986	482	2,643	488	86.9	4.2
64-65	3,545	418	2,952	331	83.5	6.3	3,247	417	2,841	372	87.3	4.2
66-67	4,074	577	3,446	509	84.1	6.6	3,564	470	3,016	399	84.2	6.2
68-69	4,575	659	3,868	567	84.3	6.2	*3,746	506	3,346	570	88.1	5.9
70-71	5,031	607	4,266	473	84.2	6.5	--	--	--	--	--	--
72-78	5,532	686	4,630	515	83.2	5.0	--	--	--	--	--	--

*This group includes girls 68 inches to 73 inches tall.

FVC = forced vital capacity; FEV$_1$ = 1 second expiratory volume.

(data from Dickman, Schmidt, and Gardner, 1971, American Review of Respiratory Disease
104:680-687)

TABLE   172

PHYSICAL WORK CAPACITY MEASUREMENTS
OF BLACK CHILDREN IN AUGUSTA (GA)

Age (years)	Heart Rate At Maximum Exercise Boys	Heart Rate At Maximum Exercise Girls	PWC Boys	PWC Girls
7	186 ± 15	188 ± 22	294 ± 141	321 ± 118
8	186 ± 25	188 ± 23	359 ± 196	297 ± 126
9	193 ± 17	184 ± 22	419 ± 144	368 ± 182
10	193 ± 18	193 ± 15	486 ± 222	379 ± 193
11	192 ± 17	193 ± 22	538 ± 239	385 ± 144
12	194 ± 18	197 ± 18	511 ± 141	528 ± 177
13	187 ± 21	193 ± 13	614 ± 76	475 ± 156

PWC = physical work capacity of the heart at a heart rate of 170/min. Age
is at last birthday.

(data from Strong, Spencer, Miller, et al., 1978, American Journal of Diseases
of Children 132:244-248. Copyright 1978, American Medical Association)

TABLE   173

PHYSICAL WORK CAPACITY (PWC) AT VARIOUS BODY SURFACE AREAS
(BSA) FOR BLACK CHILDREN IN AUGUSTA (GA)

BSA	PWC Boys Mean	PWC Boys S.D.	PWC Girls Mean	PWC Girls S.D.	Total Work Load Boys Mean	Total Work Load Boys S.D.	Total Work Load Girls Mean	Total Work Load Girls S.D.
0.9	325	166	337	232	2,159	676	2,487	1,104
1.0	--	--	342	94	--	--	2,619	524
1.1	452	180	375	182	3,720	1,350	2,875	1,174
1.2	533	184	--	--	4,303	1,158	--	--
1.3	--	--	423	176	4,071	1,178	3,769	1,038
1.4	--	--	427	128	--	--	4,020	2,292

(data from Strong, Spencer, Miller, et al., 1978, American Journal of Diseases
of Children 132:244-248. Copyright 1978, American Medical Association)

TABLE 174

PHYSICAL WORK CAPACITY OF BLACK CHILDREN IN AUGUSTA (GA)

Age	PWC_{170}				PWC_{170} Index			
	Boys		Girls		Boys		Girls	
(years)	Mean	2S.D.	Mean	2S.D.	Mean	2S.D.	Mean	2S.D.
7	294	141	321	118	321	134	326	128
8	359	196	297	126	369	212	314	125
9	419	144	368	182	373	145	372	160
10	486	222	379	193	399	161	336	120
11	538	239	385	144	434	168	298	112
12	511	141	528	177	398	81	325	137
13	614	76	475	156	417	75	328	112
14	--	--	402	112	--	--	273	61

Age	Total Work Load				Total Work Load Index			
	Boys		Girls		Boys		Girls	
(years)	Mean	2S.D.	Mean	2S.D.	Mean	2S.D.	Mean	2S.D.
7	2,018	545	2,835	666	2,206	714	2,826	986
8	2,445	1,420	2,270	1,018	2,492	1,309	2,444	1,158
9	3,197	1,016	2,527	919	2,841	890	2,418	1,026
10	4,053	1,599	3,075	1,439	3,341	1,242	2,716	816
11	3,989	2,376	3,357	1,583	3,182	1,536	2,573	904
12	4,360	1,531	3,862	2,245	3,403	1,072	2,906	1,404
13	5,175	2,651	3,837	1,126	3,666	1,412	2,679	925
14	--	--	3,645	2,072	--	--	2,482	1,414

(data from Strong, Spencer, Miller, et al., 1978, American Journal of Diseases of Children 132:244-248. Copyright 1978, American Medical Association)

TABLE   175

MEAN PHYSICAL WORK **CAPACITY** IN WHITE
AND BLACK CHILDREN IN RELATION TO
BODY SURFACE AREA (BSA)

BSA (m^2)	Boys		Girls	
	White	Black	White	Black
0.8	258	353	236	253
0.9	334	325	281	337
1.0	385	378	305	342
1.1	450	452	338	375
1.2	504	533	404	380
1.3	589	553	432	423
1.4	655	548	512	427
1.5	763	578	599	438
1.6	901	640	577	520

Comparison of mean physical working capacity at a heart rate
of 170 between California children (white) and Augusta, GA,
children (black) using different bicycle ergometer protocols.

(data from Strong, Spencer, Miller, et al., 1978, American
Journal of Diseases of Children 132:244-248. Copyright 1978,
American Medical Association)

TABLE 176

PHYSICAL WORK CAPACITY IN CANADIAN CHILDREN

Percentiles	BOYS		GIRLS	
	PWC_{170}/kg	PWC_{170}	PWC_{170}	PWC_{170}/kg
		7-YEAR-OLDS		
95	19.19	485	336	14.18
75	14.19	356	283	11.17
50	11.61	287	229	9.49
25	9.62	231	193	7.83
5	6.40	170	117	5.51
		8-YEAR-OLDS		
95	17.44	506	440	17.08
75	15.08	409	334	12.38
50	13.01	354	271	10.09
25	10.59	280	214	8.16
5	7.57	203	161	5.97
		9-YEAR-OLDS		
95	17.18	518	440	14.17
75	14.65	442	366	12.22
50	12.89	392	302	10.42
25	10.62	315	246	8.67
5	8.21	241	165	6.00
		10-YEAR-OLDS		
95	16.81	602	509	14.32
75	14.34	467	394	11.96
50	12.81	419	325	10.10
25	11.11	361	273	8.28
5	9.12	288	177	6.13
		11-YEAR-OLDS		
95	18.12	703	573	14.62
75	14.79	563	414	11.44
50	13.06	492	343	9.75
25	11.40	390	283	8.53
5	8.02	309	218	5.56
		12-YEAR-OLDS		
95	18.83	850	607	14.93
75	15.31	626	491	11.51
50	13.06	541	405	9.75
25	11.28	444	339	8.06
5	8.50	355	239	5.70

TABLE   176   (continued)

PHYSICAL WORK CAPACITY IN CANADIAN CHILDREN

Percentiles	BOYS		GIRLS	
	PWC_{170}/kg	PWC_{170}	PWC_{170}	PWC_{170}/kg
13-YEAR-OLDS				
95	19.68	1059	673	13.94
75	16.16	730	522	10.72
50	13.99	629	446	9.14
25	11.19	519	362	7.29
5	8.46	356	236	5.12
14-YEAR-OLDS				
95	18.73	1039	681	13.42
75	15.72	830	511	10.19
50	13.59	725	414	8.28
25	11.34	593	349	6.80
5	9.19	452	239	4.56
15-YEAR-OLDS				
95	17.18	1091	762	13.10
75	14.73	862	517	9.75
50	12.71	729	418	7.83
25	11.08	615	354	6.74
5	8.47	441	253	4.49
16-YEAR-OLDS				
95	18.49	1238	720	12.44
75	15.69	987	514	9.62
50	13.42	839	444	8.20
25	11.11	723	359	6.95
5	8.96	569	281	4.96
17-YEAR-OLDS				
95	18.48	1294	681	12.01
75	15.39	1008	558	10.12
50	13.30	869	470	8.50
25	10.83	709	377	6.87
5	9.41	528	280	4.84

Age is that at last completed year.
PWC_{170} = physical work capacity (KPM/min) at a heart rate of 170/min.

(data from The Canadian Association for Health, Physical Education and Recreation, 1968)

TABLE   177

PHYSICAL WORK CAPACITY IN CANADIAN YOUTHS
AGED 18 AND 19 YEARS

Percentiles	PWC_{170}	PWC_{170}/kg
BOYS		
95	1381	19.18
75	1101	16.13
50	1005	14.52
25	870	12.43
5	598	9.05
GIRLS		
95	819	14.06
75	687	11.71
50	541	10.06
25	466	8.52
5	316	5.90

Age is that at last completed year.
PWC_{170} = physical work capacity (KPM/min)
at a heart rate of 170/min.

(data from The Canadian Association for
Health, Physical Education and Recreation,
n.d.)

TABLE 178

PWC_{170} AND PWC_{170}/kg BODY WEIGHT

FOR CANADIAN CHILDREN

Age (years)	BOYS				GIRLS			
	PWC_{170}/kg		PWC_{170}		PWC_{170}		PWC_{170}/kg	
	Mean	S.D.	Mean	S.D.	Mean	S.D.	Mean	S.D.
7	11.97	3.75	306.7	105.3	236.1	66.8	9.61	2.59
8	12.71	2.94	351.0	92.0	285.0	100.2	10.67	3.75
9	12.73	2.75	384.7	90.1	306.4	83.1	10.27	2.50
10	12.79	2.23	427.1	95.1	337.3	101.6	10.13	2.59
11	13.21	3.11	493.6	129.2	361.1	115.2	10.08	2.78
12	13.33	3.15	553.6	150.3	417.2	106.3	9.92	2.60
13	13.98	3.85	655.4	235.9	450.5	144.6	9.17	2.75
14	13.86	3.24	728.4	183.2	436.8	127.0	8.51	2.65
15	12.84	2.79	740.0	178.4	443.8	135.3	8.27	2.37
16	13.45	2.96	852.6	191.3	459.3	146.8	8.48	2.63
17	13.34	2.79	873.8	224.5	476.7	147.3	8.51	2.23

Age is that at last completed year.
PWC = physical work capacity (KPM/min) at a heart rate of 170/min.

(data from The Canadian Association for Health, Physical Education and Recreation, 1968)

TABLE   179

PHYSICAL WORK CAPACITY (Kgm/min) AT A HEART
RATE OF 170/mm IN CANADIAN CHILDREN

Age (years)	Mean	S.D.
	BOYS	
10	317.0	112.4
11	569.9	115.5
14	797.6	225.9
15	948.2	180.4
	GIRLS	
10	287.4	96.4
11	422.6	72.2
14	487.1	100.3
15	592.7	136.4

Kgm = force necessary to move 1 kg one meter.

(data from Alderman, 1969, Research Quarterly 40:
1-5, by permission of the American Alliance for Health,
Physical Education, Recreation and Dance, 1900 Association
Drive, Reston VA 22091)

TABLE   180

OXYGEN CONSUMPTION AND CIRCULATORY RESPONSES OF
TORONTO BOYS AND GIRLS TO THREE WORK LOADS

Work Load (kg-m/min)	250		425
	Boys	Girls	Girls
Vo_2, liters/min STPD	0.81 +0.04	0.75 +0.11	1.04 +0.14
Q, liters/min	6.1 +1.3	6.2 +1.3	7.9 +1.3
$(a-v)O_2$, ml/liter	131 +30	121 +24	132 +25
SV, ml	50 +8	43 +6	46 +5
HR	123 +10	144 +15	173 +11

Values are means +S.D. at each work load.
Vo_2 = maximal oxygen uptake
Q = cardiac output
$(a-v)O_2$ = arteriovenous oxygen difference
SV = stroke volume
HR = heart rate

(data from Bar-Or, Shephard, and Allen, 1971, Journal
of Applied Physiology 30:219-223)

TABLE 181

PHYSICAL WORK CAPACITY
IN GERMAN CHILDREN

Age (years)	PWC_{130} (kgm - s^{-1})		$VO_{2\,130}$ (1 - min^{-1})	
	Mean	S.D.	Mean	S.D.
8	2.2	0.9	0.6	0.1
9	2.6	0.8	0.6	0.1
10	3.6	1.6	0.7	0.2
11	3.6	1.3	0.7	0.1
12	4.2	1.6	0.8	0.1
13	5.4	1.8	0.9	0.2
14	6.7	2.1	1.0	0.2
15	7.4	2.9	1.1	0.3
16	8.3	2.7	1.2	0.2
17	10.3	3.3	1.4	0.3
18	9.6	3.8	1.4	0.4

PWC_{130} = physical work capacity at a heart rate of 130/min;
$VO_{2\,130}$ = oxygen consumption at a heart rate of 130/min.

(data from Bouchard et al., 1977)

TABLE   182

SIZE AND EXERCISE PHYSIOLOGY DATA FOR
BOYS STUDIED LONGITUDINALLY OVER 8 YEARS
(Saskatoon Child Growth and Development Study)

Age at test occasion (yr)		Stature (cm)		Weight (kg)		Treadmill (run, sec)		VO$_2$ max (1/min^{-1})		Terminal heart rate (beats/min)	
Mean	S.D.	Mean	S.D.	Mean	S.D.	Mean	S.D.	Mean	S.D.	Mean	S.D.
8.07	0.26	127.9	5.1	25.9	3.3	448	45	1.46	0.20	193.4	10.2
9.06	0.29	133.4	5.1	28.9	3.7	457	72	1.72	0.22	196.2	10.8
10.05	0.30	138.7	5.3	32.0	4.3	471	56	1.82	0.23	195.5	99.0
11.05	0.30	143.8	5.6	34.8	4.8	489	58	1.96	0.24	197.6	8.9
12.04	0.29	148.9	6.2	38.5	5.7	501	70	2.18	0.33	195.3	8.9
13.03	0.29	155.3	7.6	43.4	7.7	536	67	2.39	0.39	198.3	7.9
14.05	0.29	162.8	7.9	50.0	8.9	563	74	2.73	0.53	197.3	9.8
15.03	0.29	169.9	7.1	55.9	8.5	575	81	2.94	0.50	197.3	10.2

(data from Bailey et al., 1978)

TABLE 183

INCISIVE BITING FORCE (1b) IN INDIANA CHILDREN

Age (years)	Males		Females	
	Mean	S.D.	Mean	S.D.
10	27.34	10.60	24.13	6.56
11	23.72	8.13	29.98	13.92
12	35.10	11.27	33.37	10.53
13	25.09	16.06	34.64	16.27
14	29.31	7.09	34.43	11.56
15	39.10	19.48	25.95	8.05
16	28.43	11.94	30.97	9.97
17	36.74	17.59	34.72	12.67
18	40.12	10.35	16.57	8.99

(data from Garner and Kotwal, 1973)

TABLE 184

BITE STRENGTH (1b pressure) IN CHILDREN
OF PHILADELPHIA (PA) AND BERKELEY-OAKLAND (CA)

Age (years)	Philadelphia				Berkeley-Oakland			
	MN	FN	MW	FW	MN	FN	MW	FW
3	19.5	23.3	16.2	14.1	24.0	19.3	19.1	17.8
4	28.3	28.9	23.7	21.4	32.7	24.1	29.1	26.1
5	33.6	35.5	32.6	33.3	49.4	39.1	39.0	37.4
6	48.7	48.2	46.0	35.8	58.4	47.5	54.0	48.6

MN = male Negro; FN = female Negro; MW = male white; FW = female white.

(data from Krogman, 1971)

TABLE   185

TOTAL MAXIMUM MEAN VOLTAGE AMPLITUDE ($\mu\nu$) DURING ALL CHEWING CYCLES
IN SWEDISH CHILDREN

Occlusion	Anterior temporal		Posterior temporal		Masseter	
	Mean	S.D.	Mean	S.D.	Mean	S.D.
Normal						
	6.13	2.73	3.52	1.53	2.02	1.45

DURATION OF EMG ACTIVITY (mSec) IN THE INDIVIDUAL CHEWING CYCLES

Occlusion	Anterior temporal		Posterior temporal		Masseter	
	Mean	S.D.	Mean	S.D.	Mean	S.D.
Normal	366.73		408.33		321.40	
		66.08	25.20	94.30	20.14	75.36

(data from Ahlgren, Ingervall, and Thilander, 1973, American Journal of Orthodontics
64:445-456)

FUNCTION

TABLE 186

STATURE, EYE HEIGHT, SHOULDER HEIGHT,
ELBOW HEIGHT FROM FLOOR (WITH SHOES; in)
IN MICHIGAN CHILDREN

Age (years)	Sex	Stature		Eye height from floor		Shoulder height from floor		Elbow height from floor	
		Mean	S.D.	Mean	S.D.	Mean	S.D.	Mean	S.D.
5	Boys	45.4	1.47	40.8	1.47	35.6	1.79	27.4	1.49
	Girls	43.9	2.26	39.2	2.26	34.0	2.13	26.1	1.55
6	Boys	46.0	2.04	41.3	2.04	36.0	1.91	27.7	1.50
	Girls	45.8	1.88	41.3	1.88	35.9	1.51	27.5	1.33
7	Boys	48.9	2.14	44.2	2.14	38.8	2.08	29.8	1.50
	Girls	48.2	2.33	43.6	2.33	38.2	2.18	29.3	1.75
8	Boys	51.0	2.40	46.4	2.40	40.6	2.24	31.3	1.66
	Girls	51.0	2.37	46.4	2.37	40.4	2.05	31.0	1.59
9	Boys	53.6	2.22	49.0	2.22	43.1	1.97	33.1	1.82
	Girls	52.7	2.23	48.2	2.23	42.1	2.10	32.4	1.66
10	Boys	55.2	2.79	50.6	2.79	44.5	2.67	34.4	2.14
	Girls	55.5	2.63	50.9	2.63	44.6	2.44	34.4	2.39
11	Boys	57.6	2.82	52.9	2.82	46.7	2.59	36.1	2.99
	Girls	58.4	3.18	53.8	3.18	47.3	2.87	36.4	2.20
12	Boys	58.7	3.26	54.2	3.26	47.9	3.09	36.9	2.38
	Girls	60.4	2.51	55.8	2.51	49.1	2.11	37.7	1.80
13	Boys	61.6	3.80	57.0	3.80	50.4	3.35	38.8	2.58
	Girls	62.0	2.59	57.4	2.59	50.4	2.42	38.6	1.95
14	Boys	64.5	3.72	59.8	3.72	52.9	3.34	40.6	2.53
	Girls	63.6	2.33	59.0	2.33	51.8	2.24	39.6	1.87
15	Boys	67.5	3.35	62.8	3.35	55.5	2.94	42.5	2.24
	Girls	64.2	2.26	59.6	2.26	52.4	2.20	40.0	1.80
16	Boys	68.8	2.72	64.1	2.72	56.5	2.44	43.2	1.93
	Girls	64.5	2.60	59.9	2.60	52.5	2.39	40.2	1.97

TABLE   186   (continued)

STATURE, EYE HEIGHT, SHOULDER HEIGHT,
ELBOW HEIGHT FROM FLOOR (WITH SHOES; in)
IN MICHIGAN CHILDREN

Age (years)	Sex	Stature		Eye height from floor		Shoulder height from floor		Elbow height from floor	
		Mean	S.D.	Mean	S.D.	Mean	S.D.	Mean	S.D.
17	Boys	69.9	2.91	65.3	2.91	57.4	2.76	43.8	2.23
	Girls	64.7	2.34	60.2	2.34	52.7	2.35	40.4	1.71
18	Boys	70.2	2.54	65.7	2.54	57.7	2.42	44.3	1.98
	Girls	65.0	2.45	60.2	2.45	53.0	2.02	40.6	1.65

(data from Martin, 1955)

TABLE   187

AURICULAR HEIGHT (mm) IN PHILADELPHIA CHILDREN

Age (Years)	White Boys		White Girls		Negro Boys		Negro Girls	
	Mean	S.D.	Mean	S.D.	Mean	S.D.	Mean	S.D.
7	122.0	3.6	119.3	3.6	121.5	5.1	118.1	5.5
8	122.4	3.9	119.7	4.2	121.0	4.9	119.4	4.8
9	123.7	4.3	120.8	4.8	122.6	5.5	119.2	5.3
10	123.8	4.2	120.6	6.6	124.1	5.6	121.0	5.2
11	124.7	4.8	121.2	4.6	125.9	5.9	121.6	5.2
12	125.6	4.9	122.7	5.1	124.7	5.1	120.9	4.7
13	126.7	4.7	123.1	4.7	126.0	5.6	123.1	5.9
14	127.8	4.5	124.2	4.4	127.9	5.7	124.1	6.6
15	130.5	5.5	125.7	3.2	126.5	4.6	125.0	4.1
16	131.6	5.2	125.2	4.2	--	--	--	--
17	130.6	4.2	126.7	4.5	--	--	--	--

(data from Krogman, 1970)

TABLE 188

WRIST HEIGHT, FIST CARRYING HEIGHT, FINGER TIP
HEIGHT, AND TROCHANTERIC HEIGHT FROM FLOOR (WEARING SHOES)
AND KNEE JOINTS TO SOLE OF FOOT (in)
FOR MICHIGAN CHILDREN

Age (years)	Sex	Wrist height from floor		Fist (carrying) height		Finger tip height		Trochanteric height		Knee joint to sole of foot	
		Mean	S.D.	Mean	S.D.	Mean	S.D.	Mean	S.D.	Mean	S.D.
5	Boys	20.8	1.05	18.1	1.05	15.3	1.05	21.4	1.05	12.9	0.72
	Girls	19.9	1.26	17.4	1.26	14.7	1.26	20.9	1.26	12.5	0.72
6	Boys	21.0	1.24	18.3	1.24	15.5	1.24	21.7	1.24	13.0	0.87
	Girls	20.9	1.20	18.3	1.20	15.6	1.20	22.0	1.20	13.0	0.69
7	Boys	22.7	1.37	19.9	1.37	16.9	1.37	23.5	1.37	14.0	0.93
	Girls	22.4	1.50	19.6	1.50	16.8	1.50	23.5	1.50	13.8	1.01
8	Boys	23.8	1.42	20.8	1.42	17.7	1.42	24.8	1.42	14.7	1.00
	Girls	23.6	1.61	20.8	1.61	17.7	1.61	24.9	1.61	14.7	0.91
9	Boys	25.2	1.37	22.1	1.37	18.9	1.37	26.4	1.37	15.5	0.98
	Girls	24.8	1.47	21.9	1.47	18.7	1.47	26.0	1.47	15.3	0.94
10	Boys	26.2	1.70	23.0	1.79	19.7	1.79	27.5	1.79	16.1	1.20
	Girls	26.2	1.33	23.2	1.33	19.8	1.33	27.7	1.33	16.1	0.93
11	Boys	27.5	1.71	24.2	1.71	20.8	1.71	29.1	1.71	16.9	1.14
	Girls	27.7	1.84	24.4	1.84	21.0	1.84	29.4	1.84	17.0	1.10
12	Boys	28.0	2.09	24.7	2.09	21.1	2.09	29.7	2.09	17.3	1.77
	Girls	28.6	1.55	25.3	1.55	21.6	1.55	30.1	1.55	17.5	1.02
13	Boys	29.5	2.11	26.0	2.11	22.3	2.11	31.2	2.11	18.0	1.30
	Girls	29.3	1.67	24.8	1.67	22.1	1.67	30.6	1.67	17.5	1.05
14	Boys	30.8	2.04	27.1	2.04	23.1	2.64	32.6	2.04	18.8	1.30
	Girls	29.9	1.41	26.4	1.41	22.6	1.41	31.2	1.41	17.7	1.14
15	Boys	32.1	1.88	28.2	1.88	24.1	1.88	33.9	1.88	19.7	1.28
	Girls	30.3	1.61	26.7	1.61	22.9	1.61	31.3	1.61	17.9	1.22
16	Boys	32.6	1.81	28.7	1.81	24.5	1.81	34.3	1.81	19.9	1.10
	Girls	30.5	2.14	26.9	2.14	23.1	2.14	31.3	2.14	17.9	1.14

TABLE 188 (continued)

WRIST HEIGHT, FIST CARRYING HEIGHT, FINGER TIP
HEIGHT, AND TROCHANTERIC HEIGHT FROM FLOOR (WEARING SHOES)
AND KNEE JOINTS TO SOLE OF FOOT (in)
FOR MICHIGAN CHILDREN

Age (years)	Sex	Wrist height from floor		Fist (carring) height		Finger tip height		Trochanteric height		Knee joint to sole of foot	
		Mean	S.D.	Mean	S.D.	Mean	S.D.	Mean	S.D.	Mean	S.D.
17	Boys	33.0	2.01	29.0	2.01	24.8	2.01	34.5	2.01	20.1	1.08
	Girls	30.6	1.66	27.0	1.66	23.2	1.66	31.1	1.66	18.0	1.13
18	Boys	33.6	1.89	29.6	1.89	25.5	1.89	34.4	1.89	20.1	1.49
	Girls	30.8	1.63	27.2	1.63	23.4	1.63	31.4	1.63	18.0	1.10

(data from Martin, 1955)

TABLE 189

FUNCTIONAL ARM MEASUREMENTS (in)
FOR MICHIGAN CHILDREN

Age (years)	Sex	Maximum Upward Reach		Total arm length (arc)		Upper arm length (arc)		Shoulder to wrist (arc)		Wrist to finger tip (arc)	
		Mean	S.D.	Mean	S.D.	Mean	S.D.	Mean	S.D.	Mean	S.D.
5	Boys	55.2	2.22	20.3	0.95	8.2	0.4	14.9	0.93	5.5	0.26
	Girls	52.6	3.10	19.3	1.28	7.9	0.68	14.1	0.9	5.2	0.28
6	Boys	55.8	2.74	20.5	1.05	8.3	0.64	15.0	0.94	5.5	0.29
	Girls	55.8	2.39	20.3	1.19	8.3	0.73	15.0	0.95	5.4	0.30
7	Boys	59.7	2.78	21.9	1.27	8.9	0.85	16.1	0.98	5.8	0.28
	Girls	58.8	3.08	21.4	1.33	8.9	0.62	15.8	0.98	5.6	0.31
8	Boys	62.6	3.22	22.9	1.32	9.4	0.65	16.9	1.05	6.0	0.30
	Girls	62.7	3.14	22.8	1.26	9.4	0.69	16.8	1.06	5.9	0.30
9	Boys	66.2	3.03	24.1	1.11	10.0	0.60	17.9	1.10	6.3	0.32
	Girls	64.8	3.45	23.4	1.32	9.6	0.70	17.3	1.10	6.1	0.32
10	Boys	68.5	3.89	24.8	1.58	10.1	0.86	18.3	1.16	6.5	0.33
	Girls	68.4	3.24	24.8	1.72	10.2	1.39	18.4	1.16	6.4	0.37

TABLE 189   (continued)

FUCTIONAL ARM MEASUREMENTS (in)
FOR MICHIGAN CHILDREN

Age (years)	Sex	Maximum Upward Reach		Total arm length (arc)		Upper arm length (arc)		Shoulder to wrist (arc)		Wrist to finger tip (arc)	
		Mean	S.D.	Mean	S.D.	Mean	S.D.	Mean	S.D.	Mean	S.D.
11	Boys	71.8	3.78	25.9	2.41	10.7	0.84	19.4	1.20	6.7	0.36
	Girls	72.4	4.16	26.3	1.63	10.9	0.91	19.6	1.26	6.7	0.40
12	Boys	73.6	4.62	26.8	1.83	11.0	1.01	19.9	1.30	6.9	0.42
	Girls	75.0	3.32	27.5	1.49	11.4	0.78	20.4	1.27	7.1	0.36
13	Boys	77.6	4.95	28.1	2.01	11.6	1.04	20.9	1.40	7.2	0.47
	Girls	77.2	3.71	28.3	1.59	11.8	0.72	21.1	1.17	7.2	0.33
14	Boys	81.7	5.02	29.8	1.99	12.3	0.91	21.4	1.45	7.7	0.48
	Girls	79.1	3.28	29.2	1.39	12.2	0.66	21.9	1.12	7.4	0.33
15	Boys	85.5	4.38	31.4	1.78	13.1	0.99	23.4	1.35	8.0	0.46
	Girls	79.7	3.45	29.5	1.53	12.3	0.74	22.1	1.08	7.4	0.36
16	Boys	86.9	4.02	32.1	1.75	13.3	0.82	23.9	1.24	8.1	0.38
	Girls	80.0	3.33	29.4	1.35	12.2	0.71	22.0	1.08	7.4	0.36
17	Boys	88.3	4.17	32.6	1.56	13.6	0.82	24.4	1.20	8.2	0.34
	Girls	80.4	3.20	29.5	1.51	12.3	0.72	22.1	1.11	7.4	0.28
18	Boys	88.2	3.69	32.2	1.47	13.4	0.79	24.1	1.20	8.1	0.34
	Girls	80.4	3.46	29.5	1.35	12.3	0.69	22.2	1.11	7.4	0.34

(data from Martin, 1955)

TABLE 190

HEIGHTS FROM SEAT IN ERECT SITTING
POSITION AT MAXIMUM SEAT HEIGHT (in)
IN MICHIGAN CHILDREN

Age (years)	Sex	Sitting height		Eye height		Shoulder height*		Apex of lumbar curve		Elbow height	
		Mean	S.D.	Mean	S.D.	Mean	S.D.	Mean	S.D.	Mean	S.D.
5	Boys	24.7	0.57	20.1	.70	14.9	1.02	8.4	.62	6.7	.74
	Girls	23.7	1.08	19.1	1.06	13.8	.99	8.3	.59	6.0	.47
6	Boys	25.0	1.10	20.3	1.07	15.0	1.04	8.3	.65	6.7	.76
	Girls	24.6	0.93	20.0	1.00	14.7	.86	8.5	.60	6.3	.68
7	Boys	26.1	0.96	21.4	1.02	16.0	1.02	8.8	.73	7.0	.86
	Girls	25.6	0.99	21.0	1.06	15.6	1.06	8.9	1.05	6.7	.81
8	Boys	27.0	1.20	22.4	1.25	16.6	1.11	8.9	.93	7.2	.73
	Girls	26.7	1.24	22.2	1.27	16.2	.97	9.1	.88	6.8	.72
9	Boys	27.9	1.09	23.3	1.25	17.4	.94	9.3	1.12	7.4	.88
	Girls	27.4	1.08	22.8	1.13	16.7	.99	9.4	.74	7.1	.78
10	Boys	28.6	1.38	24.0	1.24	17.9	1.31	9.4	1.10	7.8	.95
	Girls	28.6	1.14	24.0	1.15	17.8	1.03	9.5	.37	7.4	.92
11	Boys	29.5	1.44	24.8	1.34	18.5	1.25	9.7	.96	7.8	.75
	Girls	29.9	1.55	25.3	1.55	18.8	1.39	9.9	.97	7.9	.99
12	Boys	29.9	1.37	25.4	1.28	19.1	1.21	9.8	1.33	8.0	1.04
	Girls	30.9	1.18	26.4	1.27	19.6	1.03	10.4	.93	8.2	.87
13	Boys	31.0	1.95	26.4	1.88	19.8	1.52	9.8	1.20	8.3	1.10
	Girls	31.7	1.38	27.1	1.40	20.1	1.12	10.3	1.04	8.3	1.09
14	Boys	32.4	1.94	27.7	1.82	20.9	1.56	10.1	1.10	8.5	1.16
	Girls	32.6	1.37	28.0	1.35	20.8	1.20	10.6	1.12	8.6	1.22
15	Boys	33.7	1.76	29.0	1.78	21.8	1.37	10.6	1.26	8.7	1.25
	Girls	32.9	1.19	28.3	1.17	21.1	1.03	10.2	.72	8.7	1.12
16	Boys	34.7	1.53	30.0	1.50	22.4	1.24	11.0	1.25	9.1	1.15
	Girls	33.3	1.22	28.7	1.41	21.2	1.17	10.9	1.15	9.0	1.09

TABLE 190 (continued)

HEIGHTS FROM SEAT IN ERECT SITTING
POSITION AT MAXIMUM SEAT HEIGHT (in)
IN MICHIGAN CHILDREN

Age (years)	Sex	Sitting height		Eye height		Shoulder height*		Apex of lumbar curve		Elbow height	
		Mean	S.D.	Mean	S.D.	Mean	S.D.	Mean	S.D.	Mean	S.D.
17	Boys	35.4	1.43	30.7	1.52	22.9	1.17	11.1	1.26	9.3	1.13
	Girls	33.5	1.13	28.9	1.27	21.4	1.08	11.0	1.31	9.1	1.03
18	Boys	35.7	1.07	31.1	1.34	23.1	1.15	11.2	.84	9.8	1.20
	Girls	33.6	1.26	28.8	2.79	21.5	.87	11.1	1.36	9.2	.96

* "point of arm rotation"

(data from Martin, 1955)

TABLE 191

HIP DEPTH (cm)
FOR CHILDREN IN 8 STATES

Age	(months)	BOYS		GIRLS	
		Mean	S.D.	Mean	S.D.
0	- 2	7.3	1.0	7.3	1.3
3	- 4	8.8	1.3	8.6	0.9
5	- 6	8.9	1.0	9.5	1.2

(data from Snyder et al., 1975)

TABLE 192

HEIGHTS FROM SEAT IN ERECT SITTING POSITION (in)
FOR MICHIGAN CHILDREN

Age (years)	Sex	Top of thigh		Top of knee	
		Mean	S.D.	Mean	S.D.
5	Boys	4.3	.44	2.8	.22
	Girls	3.8	.60	2.5	.16
6	Boys	4.0	.65	2.8	.28
	Girls	4.0	.66	2.6	.11
7	Boys	4.3	.69	3.1	.19
	Girls	4.2	.71	2.8	.19
8	Boys	4.5	.67	3.1	.23
	Girls	4.5	.81	3.0	.14
9	Boys	4.8	.66	3.2	.18
	Girls	4.9	.75	3.1	.23
10	Boys	5.1	.87	3.5	.28
	Girls	5.0	.88	3.3	.16
11	Boys	5.3	.81	3.5	.27
	Girls	5.3	.80	3.5	.10
12	Boys	5.4	.85	3.6	.24
	Girls	5.3	.80	3.7	.23
13	Boys	5.9	.81	3.3	.23
	Girls	5.9	.75	3.4	.16
14	Boys	6.3	.84	3.3	.25
	Girls	6.2	.77	3.4	.16
15	Boys	6.6	.76	3.6	.20
	Girls	6.1	.75	3.8	.25
16	Boys	6.6	.77	3.8	.10
	Girls	6.0	.73	3.9	.19

TABLE 192 (continued)

HEIGHTS FROM SEAT IN ERECT SITTING POSITION (in)
FOR MICHIGAN CHILDREN

Age (years)	Sex	Top of thigh		Top of knee	
		Mean	S.D.	Mean	S.D.
17	Boys	6.7	.73	3.8	.05
	Girls	6.0	.70	4.0	.11
18	Boys	6.8	.77	4.0	.44
	Girls	6.0	.76	4.4	.05

(data from Martin, 1955)

TABLE 193

CROTCH HEIGHT (cm)
FOR CHILDREN IN 8 STATES

Age (months)	Boys		Girls	
	Mean	S.D.	Mean	S.D.
25 - 30	32.9	1.2	35.6	3.7
31 - 36	37.5	2.2	37.8	2.7
37 - 42	40.1	2.3	40.3	2.1
43 - 48	41.7	2.4	42.6	2.7
49 - 54	44.5	3.0	45.0	2.8
55 - 60	46.9	2.9	47.1	2.9
61 - 66	48.6	3.4	49.2	3.1
67 - 72	50.5	3.0	50.7	3.1
73 - 78	51.8	2.9	52.7	3.2
79 - 84	54.9	3.1	54.7	3.1
85 - 96	57.8	3.7	57.4	3.4
97 - 108	60.6	3.5	61.8	3.6
109 - 120	63.9	3.9	64.3	3.5
121 - 132	68.3	3.2	68.3	3.9
133 - 144	70.6	4.1	70.9	4.2
145 - 156	69.9	3.9	75.9	3.8

(data from Snyder et al., 1975)

TABLE   194

SOME VERTICAL LENGTHS IN ERECT SITTING POSITION (in)
IN MICHIGAN CHILDREN

Age (Years)	Sex	Top of knee to sole of foot		Back length*	
		Mean	S.D.	Mean	S.D.
5	Boys	14.3	0.72	3.3	0.35
	Girls	13.8	0.72	2.9	0.47
6	Boys	14.4	0.87	3.4	0.50
	Girls	14.3	0.69	3.1	0.54
7	Boys	15.5	0.93	3.6	0.51
	Girls	15.2	1.01	3.3	0.46
8	Boys	16.2	1.00	3.8	0.58
	Girls	16.2	0.91	3.6	0.55
9	Boys	17.1	0.98	3.9	0.55
	Girls	16.8	0.94	3.7	0.45
10	Boys	17.9	1.20	4.2	0.64
	Girls	17.8	0.93	4.1	0.61
11	Boys	18.6	1.14	4.4	0.77
	Girls	18.8	1.10	4.4	0.73
12	Boys	19.1	1.77	4.7	0.95
	Girls	19.3	1.02	4.8	0.86
13	Boys	19.7	1.30	5.1	1.00
	Girls	19.2	1.05	5.1	0.96
14	Boys	20.5	1.30	5.6	0.99
	Girls	19.4	1.14	5.4	0.93
15	Boys	21.4	1.28	6.0	0.98
	Girls	19.8	1.22	5.4	0.93
16	Boys	21.8	1.10	6.0	0.74
	Girls	19.9	1.14	5.4	0.74

TABLE 194 (continued)

SOME VERTICAL LENGTHS IN ERECT SITTING POSITION (in)
IN MICHIGAN CHILDREN

Age (Years)	Sex	Top of knee to sole of foot		Back length*	
		Mean	S.D.	Mean	S.D.
17	Boys	22.0	1.08	6.2	0.75
	Girls	20.0	1.13	5.4	0.62
18	Boys	22.1	1.49	6.2	0.72
	Girls	20.2	1.10	5.4	0.55

* lumbar curve to inferior angle of scapula.

(data from Martin, 1955)

TABLE 195

PERCENTILES FOR SEAT BREADTH (cm) IN U.S. CHILDREN

Sex and age	5	10	25	50	75	90	95
Boys							
6 years	18.1	18.6	19.5	20.5	21.5	22.6	23.5
7 years	19.1	19.4	20.3	21.3	22.4	23.6	24.5
8 years	19.6	20.2	21.2	22.3	23.5	24.9	26.3
9 years	20.3	21.0	22.1	23.3	24.7	26.8	28.8
10 years	21.1	21.7	22.7	24.1	25.6	27.5	28.9
11 years	22.1	22.7	23.9	25.5	27.3	29.3	30.6
Girls							
6 years	18.1	18.5	19.4	20.5	21.7	22.8	23.7
7 years	18.7	19.4	20.4	21.6	22.9	24.6	25.7
8 years	19.7	20.3	21.4	22.8	24.4	25.9	26.9
9 years	20.6	21.3	22.4	23.6	25.7	28.0	29.2
10 years	21.3	22.1	23.4	25.2	27.3	29.5	31.2
11 years	22.3	23.2	24.9	26.6	28.8	31.6	33.8

(data from Malina et al., 1973)

TABLE 196

KNEE HEIGHT (cm) IN UNITED STATES CHILDREN

BOYS

Age (years)	5	10	25	50	75	90	95
6	32.9	33.5	34.6	35.9	37.4	38.8	39.7
7	34.8	35.5	36.7	38.2	39.6	41.3	42.2
8	36.3	37.3	38.6	40.2	41.7	42.9	43.8
9	38.1	39.1	40.7	42.4	43.8	45.6	46.7
10	39.7	40.7	42.4	44.3	45.9	47.5	48.6
11	41.7	42.8	44.4	46.3	48.2	49.8	50.9

GIRLS

Age (years)	5	10	25	50	75	90	95
6	32.4	33.1	34.5	35.9	37.3	38.7	39.7
7	34.3	35.2	36.5	37.8	39.5	40.7	41.6
8	36.3	37.2	38.5	40.1	41.8	43.3	44.3
9	38.2	39.1	40.5	42.3	44.4	46.1	47.3
10	39.6	40.7	42.4	44.4	46.4	47.8	49.3
11	42.1	43.0	44.8	46.6	48.3	50.3	51.2

(data from Malina et al., 1973)

TABLE 197

POPLITEAL HEIGHT (cm) IN UNITED STATES CHILDREN

BOYS

Age (years)	5	10	25	50	75	90	95
6	26.3	26.9	28.0	29.3	30.5	31.6	32.6
7	28.1	28.6	29.7	31.1	32.4	33.7	34.6
8	29.2	30.1	31.3	32.7	33.9	35.2	35.8
9	30.8	31.5	32.9	34.3	35.7	37.2	38.0
10	32.2	33.0	34.4	35.9	37.4	39.0	39.7
11	33.7	34.5	35.7	37.3	39.1	40.4	41.3

GIRLS

Age (years)	5	10	25	50	75	90	95
6	26.0	26.5	27.7	29.0	30.2	31.4	32.1
7	27.4	28.2	29.3	30.6	32.0	33.3	34.0
8	29.1	29.6	31.1	32.5	33.7	34.9	35.8
9	30.3	31.3	32.6	34.2	35.7	38.7	38.8
10	31.8	32.6	34.1	35.6	37.4	39.1	39.8
11	33.3	34.2	35.7	37.5	39.3	40.7	41.7

(data from Malina et al., 1973)

TABLE 198

KNEE HEIGHT (mm) IN CLEVELAND CHILDREN

	BOYS		GIRLS	
Age	Mean	S.D.	Mean	S.D.
3 mos.	129.79	8.15	126.32	10.38
6 "	145.92	10.38	143.40	11.08
9 "	160.72	11.21	155.79	10.98
12 "	171.88	9.80	168.40	10.81
18 "	193.83	10.98	191.01	13.51
2 yrs.	211.27	11.80	209.73	11.62
2½ yrs.	224.57	11.93	223.78	12.86
3 "	238.97	13.56	237.70	13.58
3½ "	251.40	13.19	250.88	13.25
4 "	264.47	13.82	263.29	14.06
4½ "	275.50	15.60	276.06	16.32
5 "	286.25	14.78	286.65	16.41
6 "	310.28	17.52	312.12	18.34
7 "	331.36	18.40	334.00	19.83
8 "	351.18	18.16	352.58	19.85
9 "	368.93	22.56	370.37	22.44
10 "	387.64	21.35	390.69	22.63
11 "	407.07	22.65	410.88	23.16
12 "	423.90	24.74	428.82	21.93
13 "	442.56	26.89	442.01	20.40
14 "	463.67	25.55	448.24	22.77
15 "	478.94	21.82	448.73	20.36
16 "	483.72	20.78	448.36	21.15
17 "	488.75	22.71	449.11	18.25

(data from Simmons, 1944)

TABLE 199

KNEE HEIGHT (in) IN OKLAHOMA CHILDREN

Age (years)	BOYS		GIRLS	
	Mean	S.D.	Mean	S.D.
5	12.84	0.56	12.83	0.39
6	13.62	0.92	13.55	0.73
7	14.70	0.70	14.38	0.84
8	15.54	0.80	15.16	0.72
9	16.11	0.91	16.20	0.80
10	17.16	0.97	16.96	1.00
11	17.66	1.03	18.09	1.13
12	18.85	1.36	18.72	0.92
13	19.50	1.26	19.17	0.92
14	20.30	1.10	19.57	0.86
15	21.23	0.98	19.65	0.80
16	21.30	1.11	19.73	1.03
17	21.29	0.81	20.18	0.74
18	21.79	0.79	20.19	0.59

(data from Swearingen and Young, 1965)

TABLE 200

FUNCTIONAL MEASUREMENTS (in)
FOR MICHIGAN CHILDREN

Age (years)	Sex	Maximum arm span		Shoulder to opposite finger-tip		Back to finger-tip		Manubrium to finger-tip		Buttocks to abdomen	
		Mean	S.D.	Mean	S.D.	Mean	S.D.	Mean	S.D.	Mean	S.D.
5	Boys	43.9	2.15	23.6	1.67	22.4	0.96	17.3	1.00	5.8	.58
	Girls	42.0	2.52	22.7	1.97	21.0	1.35	16.3	1.30	6.3	.78
6	Boys	44.5	2.08	24.0	1.76	22.6	1.14	17.6	1.07	6.0	.79
	Girls	44.3	2.15	23.9	1.82	22.3	0.91	17.5	1.11	6.5	.61
7	Boys	47.6	2.28	25.7	2.01	23.9	1.43	18.7	1.14	6.3	.67
	Girls	46.4	2.74	25.0	2.41	23.3	1.25	18.4	1.17	6.5	.79
8	Boys	50.0	2.40	27.1	2.31	25.1	1.34	19.9	1.17	6.6	.72
	Girls	49.6	2.29	26.9	2.21	24.8	1.25	19.7	1.20	7.1	.72
9	Boys	52.3	2.46	28.2	2.16	26.2	1.27	20.8	1.15	7.0	.69
	Girls	51.2	2.38	27.9	2.10	25.6	1.45	20.5	1.17	7.4	.79
10	Boys	54.1	3.10	29.3	2.66	27.2	1.65	21.6	1.57	7.2	1.32
	Girls	54.5	2.94	29.5	2.52	26.8	1.33	21.4	1.09	7.8	.48
11	Boys	56.6	3.14	30.7	2.94	28.4	1.57	22.8	1.45	7.6	.89
	Girls	57.2	3.29	30.9	3.08	28.4	1.61	22.7	1.36	8.0	.91
12	Boys	58.2	3.63	31.4	2.93	29.0	1.86	23.3	1.57	7.6	1.03
	Girls	59.5	2.99	32.0	2.41	29.4	1.51	23.6	1.31	8.0	.83
13	Boys	61.2	4.05	33.1	3.07	30.4	1.85	24.4	1.48	7.7	.96
	Girls	61.6	3.22	33.3	2.45	30.4	1.75	24.2	1.41	8.2	.88
14	Boys	64.6	4.26	34.7	3.26	32.1	2.14	25.8	1.68	7.8	1.01
	Girls	63.3	3.06	34.1	2.34	31.1	1.57	24.9	1.24	8.4	.99
15	Boys	68.0	3.76	36.5	3.07	33.8	1.83	27.1	1.42	8.1	1.12
	Girls	64.1	3.13	34.6	2.55	31.4	1.47	24.9	1.45	8.4	.88
16	Boys	69.7	3.54	37.6	3.18	34.5	1.73	27.4	1.57	8.1	1.33
	Girls	64.2	2.81	34.8	2.52	31.5	1.51	25.0	1.44	8.5	.71

TABLE 200 (continued)

FUNCTIONAL MEASUREMENTS (in)
FOR MICHIGAN CHILDREN

Age (years)	Sex	Maximum arm span		Shoulder to opposite finger-tip		Back to finger-tip		Manubrium to finger-tip		Buttocks to abodomen	
		Mean	S.D.	Mean	S.D.	Mean	S.D.	Mean	S.D.	Mean	S.D.
17	Boys	70.7	3.35	38.2	3.15	35.1	1.68	28.0	1.56	8.0	1.24
	Girls	64.7	2.79	35.2	2.53	31.8	1.50	25.2	1.29	8.4	.82
18	Boys	70.6	2.84	38.4	2.71	35.0	1.63	27.7	1.74	8.1	1.20
	Girls	64.9	3.13	35.4	2.97	31.6	1.53	25.0	1.45	8.4	.86

(data from Martin, 1955)

TABLE 201

MINIMUM HAND CLEARANCE (cm)
FOR CHILDREN IN 8 STATES

Age (months)	Boys		Girls	
	Mean	S.D.	Mean	S.D.
0 - 3	3.33	0.30	3.21	0.29
4 - 6	3.73	0.26	3.55	0.28
7 - 9	4.03	0.25	3.84	0.24
10 - 12	4.14	0.29	3.86	0.25
13 - 18	4.18	0.24	3.90	0.29
19 - 24	4.24	0.32	4.10	0.27
25 - 30	4.29	0.22	4.26	0.27
31 - 36	4.51	0.24	4.30	0.32
37 - 42	4.57	0.28	4.39	0.28
43 - 48	4.59	0.28	4.50	0.27
49 - 54	4.73	0.29	4.58	0.26
55 - 60	4.82	0.32	4.66	0.30
61 - 66	4.90	0.29	4.73	0.31
67 - 72	4.99	0.30	4.79	0.28
73 - 78	5.01	0.27	4.88	0.33
79 - 84	5.16	0.30	5.01	0.31
85 - 96	5.28	0.38	5.08	0.34
97 - 108	5.42	0.35	5.22	0.33
109 - 120	5.56	0.33	5.42	0.33
121 - 132	5.85	0.38	5.60	0.40
133 - 144	6.03	0.37	5.82	0.34
145 - 156	6.06	0.40	6.16	0.37

(data from Snyder et al., 1975)

TABLE 202

ACROMIAL HEIGHT (mm) IN CLEVELAND CHILDREN

	BOYS		GIRLS	
Age	Mean	S.D.	Mean	S.D.
3 mos.	458.40	19.53	443.66	20.47
6 "	509.97	19.00	496.22	22.36
9 "	549.68	22.30	536.86	23.84
12 "	585.90	26.37	574.08	26.15
18 "	625.95	26.02	616.10	35.22
2 yrs.	662.66	28.68	654.53	26.24
2½ "	699.82	27.82	694.55	30.48
3 "	737.32	29.74	730.82	30.97
3½ "	767.49	30.10	764.57	32.13
4 "	801.04	31.50	797.57	32.85
4½ "	832.04	34.47	831.38	37.09
5 "	860.82	34.72	862.00	38.27
6 "	922.24	38.26	925.17	43.28
7 "	978.28	41.30	981.40	46.14
8 "	1035.90	43.06	1035.50	49.61
9 "	1084.47	46.14	1085.26	49.44
10 "	1134.47	51.02	1138.14	51.60
11 "	1182.31	55.44	1196.39	56.77
12 "	1229.92	59.52	1255.54	62.73
13 "	1282.00	69.93	1306.02	54.45
14 "	1348.52	72.05	1337.30	49.25
15 "	1409.54	64.67	1349.64	48.13
16 "	1433.88	54.49	1350.36	49.64
17 "	1446.83	53.24	1349.14	43.71

(data from Simmons, 1944)

TABLE 203

ARM REACH (in) IN MICHIGAN CHILDREN

Age (years)	Sex	Maximum upward reach, arm at 45°		Height of hand, arm straight forward		Height of downward reach, arm at 45°		Maximum space for forward reach (back to finger tips)	
		Mean	S.D.	Mean	S.D.	Mean	S.D.	Mean	S.D.
5	Boys	50.1	2.65	35.6	1.79	21.2	1.09	22.4	0.96
	Girls	46.7	3.15	34.0	2.13	20.3	1.29	21.0	1.35
6	Boys	50.6	2.95	36.0	1.91	21.4	1.50	22.6	1.14
	Girls	50.3	2.33	35.9	1.51	21.5	1.18	22.3	0.91
7	Boys	54.3	2.96	38.8	2.08	23.2	1.48	23.9	1.43
	Girls	53.4	3.10	38.2	2.18	23.0	1.56	23.3	1.25
8	Boys	56.9	3.34	40.6	2.24	24.4	1.63	25.1	1.34
	Girls	56.6	3.06	40.4	2.05	24.3	1.49	24.8	1.25
9	Boys	60.2	3.15	43.1	1.97	25.9	1.39	26.2	1.27
	Girls	58.7	3.35	42.1	2.10	25.5	1.49	25.6	1.45
10	Boys	63.1	3.83	44.5	2.67	26.9	1.67	27.2	1.65
	Girls	62.3	3.49	44.6	2.44	27.0	1.53	26.8	1.33
11	Boys	65.0	3.85	46.7	2.59	28.3	1.63	28.4	1.57
	Girls	65.9	4.27	47.3	2.87	28.6	1.81	28.4	1.61
12	Boys	66.9	5.06	47.9	3.09	28.9	2.15	29.0	1.86
	Girls	68.6	3.46	9.1	2.11	29.6	1.47	29.4	1.51
13	Boys	70.3	4.98	50.4	3.35	30.4	2.20	30.4	1.85
	Girls	70.4	3.60	50.4	2.42	30.3	1.58	30.4	1.75
14	Boys	74.1	5.31	52.9	3.34	31.8	2.38	32.1	2.14
	Girls	72.5	3.55	51.8	2.24	31.0	1.60	31.1	1.57
15	Boys	77.9	5 1	55.5	2.94	33.2	2.40	33.8	1.83
	Girls	73.3	4.13	52.4	2.20	31.5	1.80	31.4	1.47
16	Boys	79.3	5.15	56.5	2.44	33.8	2.54	34.5	1.73
	Girls	73.3	5.05	52.5	2.39	31.6	2.48	31.5	1.51

TABLE 203   (continued)

ARM REACH (in) IN MICHIGAN CHILDREN

Age (years)	Sex	Maximum upward reach, arm at 45°		Height of hand, arm straight forward		Height of downward reach, arm at 45°		Maximum space for forward reach (back to finger tips)	
		Mean	S.D.	Mean	S.D.	Mean	S.D.	Mean	S.D.
17	Boys	80.5	5.85	57.4	2.76	34.3	2.74	35.1	1.68
	Girls	73.6	4.99	52.7	2.35	31.8	2.34	31.8	1.50
18	Boys	80.6	5.88	57.7	2.42	34.8	2.56	35.0	1.63
	Girls	73.9	4.90	53.0	2.02	32.0	2.14	31.6	1.53

(data from Martin, 1955)

TABLE 204

SITTING - MID-SHOULDER HEIGHT (cm)

Age (months)	BOYS		GIRLS	
	Mean	S.D.	Mean	S.D.
0 - 3	0.0	0.0	31.9	2.2
4 - 6	0.0	0.0	32.4	2.1
7 - 9	0.0	0.0	34.1	1.9
10 - 12	0.0	0.0	34.5	2.1
13 - 18	0.0	0.0	35.5	2.2
19 - 24	0.0	0.0	36.7	2.4
25 - 30	31.3	1.7	37.8	2.3
31 - 36	32.7	1.6	38.5	2.3
37 - 42	34.0	2.0	39.4	2.4
43 - 48	35.2	2.2	40.4	2.4
49 - 54	36.2	2.1	41.8	2.4
55 - 60	37.1	2.2	44.0	2.3
61 - 66	38.4	2.3	45.6	2.6
67 - 72	39.2	2.3	47.7	2.6
73 - 78	40.0	2.6	49.5	2.8
79 - 84	41.6	2.3	53.6	2.6
85 - 96	42.8	2.3	--	--
97 - 108	44.3	2.4	--	--
109 - 120	45.7	2.6	--	--
121 - 132	47.8	2.2	--	--
133 - 144	49.4	2.6	--	--
145 - 156	50.2	2.6	--	--

(data from Snyder et al., 1975)

TABLE 205

INSIDE GRIP DIAMETER (cm)
FOR CHILDREN IN 8 STATES

Age (months)	Boys		Girls	
	Mean	S.D.	Mean	S.D.
0 – 3	1.73	0.19	1.68	0.18
4 – 6	1.97	0.19	1.91	0.19
7 – 9	2.06	0.20	2.12	0.21
10 – 12	2.20	0.23	2.17	0.18
13 – 18	2.41	0.22	2.35	0.24
19 – 24	2.59	0.17	2.57	0.20
25 – 30	2.76	0.25	2.72	0.24
31 – 36	2.91	0.26	2.87	0.23
37 – 42	3.00	0.23	2.96	0.19
43 – 48	3.04	0.24	3.02	0.22
49 – 54	3.19	0.23	3.13	0.23
55 – 60	3.27	0.23	3.23	0.25
61 – 66	3.35	0.27	3.29	0.26
67 – 72	3.46	0.26	3.40	0.26
73 – 78	3.57	0.25	3.52	0.25
79 – 84	3.65	0.25	3.58	0.26
85 – 96	3.81	0.30	3.77	0.27
97 – 108	3.93	0.30	4.00	0.30
109 – 120	4.12	0.32	4.06	0.37
121 – 132	4.32	0.33	4.34	0.38
133 – 144	4.42	0.36	4.51	0.38
145 – 156	4.54	0.36	4.87	0.36

Measured as the maximum diameter of a cone able to be gripped with
the right hand so that the tip of the middle finger touches the
tip of the thumb.

(data from Snyder et al., 1975)

TABLE 206

BUTTOCKS DEPTH (cm)
FOR CHILDREN IN 8 STATES

Age (months)	BOYS		GIRLS	
	Mean	S.D.	Mean	S.D.
0 - 2	7.6	1.2	7.8	1.4
3 - 4	9.4	1.1	9.3	1.0
5 - 6	9.6	1.4	10.1	1.3

(data from Snyder et al., 1975)

TABLE 207

BUTTOCKS DEPTH (in)
IN OKLAHOMA CHILDREN

Age (years)	BOYS		GIRLS	
	Mean	S.D.	Mean	S.D.
5	5.11	0.44	5.10	0.45
6	5.38	0.61	5.36	0.51
7	5.66	0.55	5.61	0.55
8	6.04	0.81	5.83	0.63
9	6.30	0.83	6.29	0.87
10	6.79	0.97	6.46	0.76
11	6.75	0.90	7.14	1.06
12	7.26	1.12	7.55	1.05
13	7.97	0.89	7.97	0.82
14	8.35	0.81	8.13	0.96
15	8.56	0.56	8.26	0.77
16	9.07	0.78	8.30	0.84
17	9.18	0.83	8.26	0.71
18	9.42	0.69	8.62	0.77

(data from Swearingen and Young, 1965)

TABLE 208

DEPTHS IN ERECT SITTING POSITION
AT MAXIMUM SEAT HEIGHT (in)
FOR MICHIGAN CHILDREN

Age (years)	Sex	Buttocks to seat front		Buttocks to front of knee		Buttocks to sole of foot		Knee joint to sole of foot	
		Mean	S.D.	Mean	S.D.	Mean	S.D.	Mean	S.D.
5	Boys	11.6	.69	15.0	0.71	23.0	.93	12.9	0.72
	Girls	11.8	.94	14.5	0.96	23.1	1.11	12.5	0.72
6	Boys	11.9	.77	14.9	1.05	23.5	1.40	13.0	0.87
	Girls	12.2	.75	15.3	0.80	24.0	1.36	13.0	0.69
7	Boys	12.8	.80	16.0	0.92	25.3	1.26	14.0	0.93
	Girls	13.1	.79	16.0	1.10	25.5	1.38	13.8	1.01
8	Boys	13.4	.71	16.9	0.96	26.6	1.33	14.7	1.00
	Girls	13.8	.85	17.1	0.96	27.2	1.51	14.7	0.91
9	Boys	14.3	.76	18.0	1.01	28.2	1.56	15.5	0.98
	Girls	14.4	.89	17.8	1.15	28.1	1.68	15.3	0.94
10	Boys	14.8	.93	18.6	1.53	29.2	2.09	16.1	1.20
	Girls	15.3	.99	19.0	0.55	29.8	1.55	16.1	0.93
11	Boys	15.6	1.01	19.7	1.26	30.8	1.77	16.9	1.14
	Girls	16.2	1.05	20.0	1.30	31.4	1.91	17.0	1.10
12	Boys	16.1	1.17	20.1	1.45	31.6	2.18	17.3	1.77
	Girls	16.7	.97	20.8	1.16	32.4	1.62	17.5	1.02
13	Boys	16.8	1.19	21.0	1.37	33.1	2.22	18.0	1.30
	Girls	17.1	1.02	21.2	1.25	32.9	1.76	17.5	1.05
14	Boys	17.4	1.09	21.9	1.37	34.6	2.46	18.8	1.30
	Girls	17.7	.98	21.7	1.34	33.7	1.70	17.7	1.14
15	Boys	17.9	.99	22.9	1.34	35.8	2.31	19.7	1.28
	Girls	17.8	.91	22.0	1.04	33.7	1.97	17.9	1.22
16	Boys	18.3	.89	23.3	1.33	36.3	2.05	19.9	1.10
	Girls	18.2	.87	22.2	1.14	34.1	2.43	17.9	1.14

TABLE 208 (continued)

DEPTHS IN ERECT SITTING POSITION
AT MAXIMUM SEAT HEIGHT (in)
FOR MICHIGAN CHILDREN

Age (years)	Sex	Buttocks to seat front		Buttocks to front of knee		Buttocks to sole of foot		Knee joint to sole of foot	
		Mean	S.D.	Mean	S.D.	Mean	S.D.	Mean	S.D.
17	Boys	18.7	.92	23.4	1.24	36.9	2.59	20.1	1.08
	Girls	18.1	.91	22.2	1.17	34.1	2.13	18.0	1.13
18	Boys	18.7	.93	23.5	1.20	36.8	2.44	20.1	1.49
	Girls	18.2	.85	22.3	1.12	33.9	2.10	18.0	1.10

Age (years)	Sex	Lumbar support to front of knee		Abdomen to front of knee	
		Mean	S.D.	Mean	S.D.
5	Boys	15.1	0.71	9.2	0.76
	Girls	14.8	1.03	8.3	0.72
6	Boys	15.0	0.75	8.9	0.77
	Girls	15.6	0.70	8.8	0.73
7	Boys	16.0	0.78	9.7	0.67
	Girls	16.0	0.97	9.5	0.87
8	Boys	16.9	0.87	10.3	0.76
	Girls	17.2	1.07	10.1	0.68
9	Boys	18.0	0.96	11.0	0.75
	Girls	17.8	1.03	10.5	0.77
10	Boys	18.7	0.95	11.5	1.00
	Girls	18.8	1.07	11.3	0.91
11	Boys	19.8	0.93	12.1	0.88
	Girls	20.1	1.39	12.0	1.14
12	Boys	20.1	1.20	12.5	1.05
	Girls	20.7	0.98	12.8	0.92

TABLE 208   (continued)

DEPTHS IN ERECT SITTING POSITION
AT MAXIMUM SEAT HEIGHT (in)
FOR MICHIGAN CHILDREN

Age (years)	Sex	Lumbar support to front of knee		Abdomen to front of knee	
		Mean	S.D.	Mean	S.D.
13	Boys	20.9	1.33	13.3	1.10
	Girls	20.9	0.99	13.0	1.09
14	Boys	21.8	1.19	14.1	1.32
	Girls	21.5	1.14	13.3	1.02
15	Boys	23.0	1.64	14.8	1.17
	Girls	21.6	0.98	13.6	1.05
16	Boys	23.1	0.95	15.2	1.16
	Girls	21.6	0.85	13.7	1.07
17	Boys	23.1	0.87	15.5	1.07
	Girls	21.7	0.87	13.9	1.10
18	Boys	23.3	0.77	15.4	1.10
	Girls	21.6	0.79	13.8	1.04

Age (years)	Sex	Ischium to seat front		Ischium to front of knee		Ischium to lumbar back support		Seat front to front of knee	
		Mean	S.D.	Mean	S.D.	Mean	S.D.	Mean	S.D.
5	Boys	9.2	0.51	12.6	0.51	2.6	.39	3.4	0.47
	Girls	9.0	0.76	11.7	0.76	3.0	.53	2.7	0.73
6	Boys	9.3	0.62	12.3	0.62	2.7	.41	3.0	0.69
	Girls	9.5	0.62	12.6	0.62	3.0	.39	3.1	0.68
7	Boys	10.0	0.78	13.2	0.78	2.9	.33	3.2	0.67
	Girls	10.2	0.85	13.1	0.85	2.9	.33	3.0	0.73
8	Boys	10.4	0.84	13.8	0.84	3.1	.33	3.5	0.73
	Girls	10.8	0.72	14.0	0.72	3.3	.33	3.2	0.71

TABLE 208   (continued)

DEPTHS IN ERECT SITTING POSITION
AT MAXIMUM SEAT HEIGHT (in)
FOR MICHIGAN CHILDREN

Age (years)	Sex	Ischium to seat front		Ischium to front of knee		Ischium to lumbar back support		Seat front to front of knee	
		Mean	S.D.	Mean	S.D.	Mean	S.D.	Mean	S.D.
9	Boys	11.2	0.80	14.9	0.80	3.1	.42	3.7	0.73
	Girls	11.3	0.78	14.7	0.78	3.1	.50	3.4	0.79
10	Boys	11.6	0.82	15.5	0.82	3.2	.45	3.9	0.71
	Girls	11.9	0.90	15.5	0.90	3.3	.49	3.6	0.73
11	Boys	12.3	1.02	16.4	1.02	3.4	.58	4.1	0.85
	Girls	12.6	1.16	16.4	1.16	3.7	.60	3.8	0.98
12	Boys	12.7	1.15	16.8	1.15	3.4	.56	4.0	0.76
	Girls	13.1	1.14	17.1	1.14	3.5	.46	4.1	0.86
13	Boys	13.4	1.27	17.6	1.27	3.3	.55	4.2	0.93
	Girls	13.6	1.00	17.6	1.00	3.3	.47	4.0	0.80
14	Boys	13.9	1.07	18.4	1.07	3.5	.52	4.5	0.92
	Girls	13.8	0.94	17.9	0.94	3.6	.46	4.1	0.99
15	Boys	14.5	0.98	19.4	0.98	3.6	.65	5.0	0.95
	Girls	13.9	0.97	18.2	0.97	3.4	.59	4.2	0.85
16	Boys	14.8	0.92	19.7	0.92	3.4	.46	5.0	0.99
	Girls	14.2	0.96	18.3	0.96	3.3	.44	4.1	0.87
17	Boys	15.0	0.95	19.8	0.95	3.3	.44	4.8	0.91
	Girls	14.3	0.76	18.4	0.76	3.2	.44	4.1	1.02
18	Boys	15.0	0.87	19.8	0.87	3.4	.43	4.8	0.95
	Girls	14.0	0.93	18.5	0.93	3.1	.39	4.1	0.94

(data from Martin, 1955)

TABLE 209

BUTTOCKS-KNEE LENGTH (cm) IN UNITED STATES CHILDREN

BOYS

Age (years)	5	10	25	50	75	90	95
6	31.5	33.6	35.7	37.4	39.1	40.8	41.6
7	33.7	36.1	38.1	39.9	41.6	43.4	44.6
8	35.7	37.6	40.2	41.8	43.8	45.4	46.5
9	37.7	39.7	41.9	44.2	46.2	47.9	49.5
10	39.8	41.5	44.2	46.3	48.2	50.1	51.0
11	42.2	44.1	46.2	48.3	50.5	52.5	53.7

GIRLS

Age (years)	5	10	25	50	75	90	95
6	32.2	33.5	36.1	37.9	39.6	41.2	41.9
7	34.2	35.7	38.2	40.1	41.9	43.5	44.4
8	37.1	38.6	40.5	42.5	44.5	46.4	47.6
9	38.6	40.4	42.6	44.7	47.3	49.4	50.5
10	40.5	42.3	44.7	47.3	49.5	51.4	52.7
11	43.7	45.2	47.3	49.5	52.1	54.8	55.9

(data from Malina et al., 1973)

TABLE   210

BUTTOCKS-KNEE LENGTH (cm) IN CHILDREN OF 8 STATES

Age (months)	BOYS		GIRLS	
	Mean	S.D.	Mean	S.D.
0 - 3	13.7	1.3	13.8	1.3
4 - 6	16.5	1.3	16.2	1.3
7 - 9	18.4	1.2	17.6	0.9
10 - 12	19.2	1.3	19.3	1.4
13 - 18	20.7	2.0	21.2	1.7
19 - 24	23.9	1.8	23.8	2.3
25 - 30	25.4	1.8	25.5	1.8
31 - 36	27.2	1.9	27.7	2.1
37 - 42	28.7	1.6	28.9	1.9
43 - 48	29.8	1.7	30.1	1.8
49 - 54	31.1	2.2	31.5	1.7
55 - 60	32.5	1.8	32.5	2.0
61 - 66	33.8	2.1	34.1	2.0
67 - 72	35.1	2.1	35.3	2.1
73 - 78	35.9	1.9	36.5	2.4
79 - 84	37.8	2.2	37.7	2.0
85 - 96	39.6	2.6	39.6	2.4
97 - 108	41.5	2.4	42.5	2.6
109 - 120	43.5	2.8	44.1	2.7
121 - 132	46.4	2.4	47.1	3.2
133 - 144	48.5	2.4	48.7	2.9
145 - 156	50.0	3.1	52.0	3.1

(data from Snyder et al., 1975)

TABLE 211

BUTTOCKS-KNEE LENGTH (in) IN OKLAHOMA CHILDREN

Age (years)	BOYS		GIRLS	
	Mean	S.D.	Mean	S.D.
5	13.59	0.75	14.01	0.57
6	14.47	0.99	14.67	0.77
7	15.61	0.83	15.62	0.92
8	16.42	0.85	16.50	0.82
9	17.05	0.95	17.58	0.97
10	18.56	1.04	18.45	1.08
11	19.00	1.18	19.78	1.38
12	20.36	1.76	21.03	1.22
13	21.34	1.34	21.85	1.16
14	22.16	1.37	22.12	1.20
15	23.08	1.16	22.10	0.96
16	23.44	1.17	22.44	1.12
17	23.36	1.08	22.71	0.87
18	23.85	1.08	22.98	0.98

(data from Swearingen and Young, 1965)

TABLE   212

BUTTOCKS-POPLITEAL LENGTH (cm) IN U.S. CHILDREN

Age	Percentiles						
(years)	5	10	25	50	75	90	95

BOYS

6	28.6	29.3	30.4	31.9	33.7	35.7	37.4
7	30.4	31.2	32.4	33.8	35.7	38.0	38.9
8	32.3	33.1	34.3	35.8	37.8	40.1	42.2
9	34.1	34.7	36.3	38.2	39.9	42.7	45.0
10	35.3	36.2	37.8	39.7	41.9	44.3	46.5
11	36.9	38.2	39.7	41.7	43.7	46.4	48.3

GIRLS

6	28.8	29.7	31.1	32.6	34.4	37.0	38.6
7	30.6	31.6	32.8	34.6	36.5	38.5	40.3
8	32.7	33.5	35.1	36.6	38.6	41.1	43.1
9	34.3	35.4	37.2	38.9	41.2	43.8	45.2
10	35.8	37.0	39.1	41.2	43.6	45.8	47.7
11	38.1	39.2	40.9	43.1	45.7	48.7	50.5

(data from Malina et al., 1973)

TABLE   213

THIGH CLEARANCE (cm) IN U.S. CHILDREN

Age	Percentiles						
(years)	5	10	25	50	75	90	95

BOYS

6	7.4	7.7	8.3	9.1	9.9	10.7	11.0
7	7.9	8.2	8.8	9.6	10.5	11.4	11.7
8	8.3	8.8	9.4	10.3	11.2	11.9	12.6
9	8.4	9.1	9.8	10.7	11.7	12.9	13.9
10	9.0	9.3	10.1	11.1	11.9	13.1	13.7
11	9.3	9.8	10.6	11.6	12.8	13.9	14.7

GIRLS

6	7.4	7.8	8.4	9.2	10.0	10.8	11.5
7	8.0	8.2	8.8	9.6	10.5	11.5	12.2
8	8.2	8.7	9.4	10.3	11.3	12.4	12.9
9	8.5	9.1	9.8	10.7	11.8	13.3	13.8
10	9.0	9.4	10.3	11.4	12.6	13.6	14.3
11	9.4	10.1	10.7	11.9	13.1	14.3	14.9

(data from Malina et al., 1973)

TABLE 214

THIGH-CLEARANCE HEIGHT (in)
IN OKLAHOMA CHILDREN

Age (years)	BOYS		GIRLS	
	Mean	S.D.	Mean	S.D.
5	3.51	0.33	3.64	0.40
6	3.71	0.43	3.72	0.33
7	3.94	0.39	3.88	0.39
8	4.15	0.64	4.04	0.43
9	4.22	0.57	4.38	0.47
10	4.43	0.56	4.40	0.49
11	4.50	0.48	4.73	0.58
12	4.63	0.78	5.10	0.58
13	5.17	0.73	5.42	0.57
14	5.37	0.59	5.55	0.68
15	5.69	0.53	5.58	0.57
16	5.88	0.68	5.43	0.49
17	5.97	0.71	5.43	0.52
18	6.13	0.48	5.52	0.49

(data from Swearingen and Young, 1965)

TABLE 215

THIGH-CLEARANCE AND SITTING-HEIGHT MEASUREMENTS (in)
IN OKLAHOMA CHILDREN

Age (years)	Sex	Thigh Clearance		Sitting Height	
		Mean	S.D.	Mean	S.D.
5	F	3.6	0.38	23.4	0.83
	M	3.5	0.33	23.5	0.92
6	F	3.8	0.33	24.4	1.2
	M	3.7	0.43	24.6	1.2
7	F	3.9	0.39	25.3	1.2
	M	3.9	0.39	25.8	1.1
8	F	4.0	0.43	26.3	1.0
	M	4.2	0.64	26.9	1.2
9	F	4.4	0.47	27.7	1.0
	M	4.2	0.57	27.5	1.3
10	F	4.4	0.49	28.2	1.2
	M	4.4	0.56	28.8	1.2
11	F	4.7	0.58	29.9	1.7
	M	4.5	0.48	29.2	1.5
12	F	5.1	0.58	31.1	1.8
	M	4.6	0.78	30.4	1.7
13	F	5.4	0.57	32.3	1.3
	M	5.2	0.73	31.8	1.9
14	F	5.6	0.68	32.5	1.5
	M	5.4	0.59	33.3	1.8
15	F	5.6	0.57	33.2	1.2
	M	5.7	0.53	34.6	1.6
16	F	5.4	0.49	33.6	1.3
	M	5.9	0.68	35.3	1.8
17	F	5.4	0.52	34.1	1.2
	M	6.0	0.71	35.7	1.3
18	F	5.5	0.50	33.7	1.2
	M	6.1	0.48	36.6	1.1

(data from Swearingen and Young, 1965)

TABLE 216

CENTER OF GRAVITY (height from floor, in) IN THE STANDING POSITION
FOR OKLAHOMA CHILDREN

| Age (years) | Sex | Height Above Floor | | % of Stature | |
		Mean	S.D.	Mean	S.D.
5	F	24.8	0.7	57.5	0.73
	M	24.8	1.2	57.5	1.4
6	F	25.8	1.1	57.2	1.0
	M	26.1	1.3	57.5	0.75
7	F	27.0	1.2	57.0	0.83
	M	27.6	1.2	57.0	0.89
8	F	28.3	1.1	56.8	0.71
	M	28.8	1.1	56.9	0.84
9	F	29.9	1.2	56.7	1.0
	M	29.5	1.2	56.7	0.84
10	F	31.0	1.5	56.8	0.85
	M	31.2	1.4	56.6	0.74
11	F	32.5	1.7	56.5	0.88
	M	31.8	1.5	56.5	0.93
12	F	33.7	1.6	56.4	0.84
	M	33.6	2.1	56.8	0.74
13	F	34.6	1.5	56.0	1.1
	M	34.9	1.9	56.6	0.73
14	F	35.3	1.3	56.4	0.74
	M	36.3	2.0	56.6	0.69
15	F	35.4	1.3	56.2	0.66
	M	38.2	1.9	56.8	1.1
16	F	35.6	1.5	56.0	0.76
	M	38.5	1.7	56.7	0.69
17	F	36.5	1.2	56.3	0.62
	M	38.9	1.7	57.0	0.81
18	F	36.3	1.2	56.3	0.85
	M	40.0	1.2	57.1	0.70

(data from Swearingen and Young, 1965)

TABLE 217

POSITION OF CENTER OF GRAVITY (in) IN OKLAHOMA CHILDREN
IN SITTING AND STANDING POSITIONS (horizontal from seat back)

Age (years)	Sex	Sitting				Standing			
		Vertical		Horizontal		Vertical		Horizontal	
		Mean	S.D.	Mean	S.D.	Mean	S.D.	Mean	S.D.
5	F	8.0	0.37	5.4	0.63	5.1	0.41	2.8	0.38
	M	8.0	0.43	5.2	0.60	5.3	0.71	2.8	0.55
6	F	8.0	0.33	5.3	0.66	5.1	0.44	2.9	0.49
	M	8.2	0.36	5.4	0.66	5.3	0.46	2.9	0.39
7	F	8.1	0.42	5.7	0.63	4.9	0.55	2.9	0.38
	M	8.2	0.33	6.0	0.83	5.0	0.54	3.0	0.41
8	F	8.1	0.39	5.7	0.90	4.8	0.55	2.9	0.45
	M	8.4	0.42	6.1	0.91	5.1	0.52	3.2	0.45
9	F	8.2	0.38	6.1	0.68	4.9	0.61	3.2	0.41
	M	8.3	0.39	6.2	0.77	5.0	0.60	3.2	0.55
10	F	8.2	0.39	6.2	0.75	4.7	0.63	3.2	0.39
	M	8.5	0.37	6.6	0.77	4.9	0.67	3.2	0.40
11	F	8.5	0.50	6.9	0.80	4.9	0.68	3.4	0.33
	M	8.5	0.43	6.8	0.79	4.7	0.72	3.3	0.41
12	F	8.6	0.48	7.6	0.77	5.0	0.60	3.7	0.52
	M	8.4	0.51	7.6	0.99	4.8	0.83	3.5	0.58
13	F	8.9	0.53	7.9	0.65	5.2	0.72	3.7	0.52
	M	8.5	0.67	8.0	0.72	5.0	0.81	3.6	0.53
14	F	8.8	0.58	8.0	0.73	5.3	0.73	3.3	0.44
	M	8.6	0.60	8.3	0.52	5.5	0.73	3.6	0.41
15	F	8.9	0.41	7.9	0.81	5.6	0.68	3.4	0.46
	M	8.8	0.67	8.3	0.81	5.5	0.82	3.6	0.52
16	F	8.9	0.52	7.8	0.52	5.6	0.76	3.4	0.50
	M	9.1	0.69	8.5	0.58	5.9	0.83	3.7	0.74
17	F	8.9	0.50	8.0	0.50	5.7	0.66	3.4	0.40
	M	9.5	0.55	8.8	0.70	6.5	0.92	3.7	0.46
18	F	9.0	0.49	8.3	0.59	5.5	0.66	3.6	0.45
	M	9.5	0.52	9.0	0.57	6.6	1.01	3.8	0.39

(data from Swearingen and Young, 1965)

TABLE 218

STANDING CENTER OF GRAVITY AS PERCENT OF STATURE
FOR CHILDREN IN 8 STATES

Age (months)	Boys		Girls	
	Mean	S.D.	Mean	S.D.
0 - 3	58.5	2.1	59.4	1.9
4 - 6	59.1	2.3	58.1	2.5
7 - 9	57.7	2.7	58.3	2.2
10 - 12	58.5	2.4	58.1	1.8
13 - 18	59.4	2.1	57.4	0.8
19 - 24	57.5	1.1	57.5	1.0
25 - 30	59.1	2.5	60.5	2.1
31 - 36	58.9	1.0	59.3	2.5
37 - 42	59.2	2.0	58.9	2.0
43 - 48	59.7	1.7	58.8	2.0
49 - 54	59.2	2.0	59.2	2.1
55 - 69	58.9	1.7	59.3	2.0
61 - 66	59.1	1.1	59.2	1.7
67 - 72	59.1	1.3	59.3	1.5
73 - 78	58.9	1.2	58.8	1.2
79 - 84	58.7	1.3	58.6	1.1
85 - 96	58.6	1.1	58.0	1.7
97 - 108	57.9	1.1	58.0	1.4
109 - 120	58.0	1.1	57.5	0.9
121 - 132	57.7	1.0	57.4	0.7
133 - 144	57.8	1.0	57.4	1.1
145 - 156	58.0	1.5	57.4	1.3

(data from Snyder et al., 1975)

TABLE 219

SEATED CENTER OF GRAVITY AS PERCENT OF SITTING HEIGHT
FOR CHILDREN IN 8 STATES

Age	(Months)	Boys		Girls	
		Mean	S.D.	Mean	S.D.
0 –	3	48.0	4.4	50.2	3.3
4 –	6	46.6	3.7	47.1	2.8
7 –	9	--	--	--	--
10 –	12	--	--	--	--
13 –	18	--	--	--	--
19 –	24	--	--	--	--
25 –	30	--	--	41.9	2.7
31 –	36	--	--	39.1	1.9
37 –	42	38.2	1.7	39.5	2.8
43 –	48	37.2	2.2	37.9	3.3
49 –	54	36.1	1.7	36.7	2.6
55 –	60	36.6	2.6	35.3	2.6
61 –	66	34.9	2.2	36.8	5.0
67 –	72	35.0	1.9	34.0	2.3
73 –	78	33.8	2.7	33.6	2.3
79 –	84	33.1	2.1	33.3	1.8
85 –	96	32.3	2.2	32.2	2.4
97 –	108	32.1	1.8	30.7	1.6
109 –	120	31.1	1.9	30.2	2.0
121 –	132	30.0	1.6	29.5	1.6
133 –	144	30.1	2.1	29.4	1.4
145 –	156	29.7	1.5	29.2	1.3

(data from Snyder et al., 1975)

TABLE 220

MEANS AND VARIABILITY, FOR INTERVALS OF STATURE, OF HEIGHT
OF TRANSVERSE PLANE OF GRAVITY FROM SOLES (cm)
FOR CHILDREN IN MINNESOTA AND ILLINOIS

Interval of Stature	Boys		Girls	
	Mean	S.D.	Mean	S.D.
90- 95	53.47	1.20	53.93	1.25
95-100	56.82	0.92	56.33	1.03
100-105	59.62	1.09	58.97	1.22
105-110	62.31	1.16	62.20	1.08
110-115	64.87	1.18	65.13	0.66
115-120	67.45	0.99	67.02	1.25
120-125	69.93	0.95	69.67	0.94
125-130	72.30	1.39	72.46	0.99
130-135	75.37	1.10	74.94	1.35
135-140	77.84	1.04	77.67	1.75
140-145	80.68	1.26	80.37	1.31
145-150	83.54	1.27	82.78	1.58
150-155	85.97	1.18	85.28	1.76
155-160	89.24	1.41	88.37	1.22
160-165	92.16	1.05	90.47	1.30
165-170	94.14	1.30	93.83	1.36
170-175	97.25	1.51	--	--
175-180	99.67	1.29	--	--

(data from Palmer, 1944)

TABLE 221

MEANS AND VARIABILITY FOR INTERVALS OF STATURE OF DISTANCE
OF FRONTAL PLANE OF GRAVITY FROM BACK (cm)
FOR CHILDREN IN MINNESOTA AND ILLINOIS

Interval of Stature	Boys		Girls	
	Mean	S.D.	Mean	S.D.
90-100	6.86	0.90	6.58	0.69
100-110	7.00	0.53	6.91	0.71
110-120	7.20	0.43	7.18	0.52
120-130	7.50	0.49	7.31	0.54
130-140	7.76	0.60	7.63	0.57
140-150	8.15	0.49	8.15	0.69
150-160	8.64	0.57	8.59	0.58
160-170	9.52	0.61	9.00	0.63
170-180	9.82	0.54	9.47	0.72

(data from Palmer, 1944)

BODY COMPOSITION

TABLE   222

CHEMICAL COMPOSITION OF MALE REFERENCE INFANT AT BIRTH

| | Data from Chemical Analyses of Stillborn Infants | | | | Male Reference Infant | |
| | Uncorrected* | | Corrected† | | | |
Component	gm	% of Body Weight	gm	% of Body Weight	gm	% of Body Weight
Body Weight	3,197	--	3,477	--	3,500	--
Water	2,331	73.0	2,608	75.1	2,628	75.1
Lipid	382	12.0	382	11.0	385	11.0
Protein	392	12.4	395	11.4	399	11.4
"Other"	85	2.6	85	2.5	88	2.5
Fat-free body mass	2,815	88.0	3,092	89.0	3,115	89.0

*Assuming no loss of water from body between time of death and chemical analysis.
†Assuming 277 gm loss of water between time of death and chemical analysis.

(data from Fomon, 1967, Pediatrics 40:863-870. Copyright American Academy of Pediatrics 1967)

TABLE 223

BODY COMPOSITION OF MALE REFERENCE INFANT AT VARIOUS AGES

| | | Percentage Composition (gm/100 gm) | | | | | |
| Age (months) | Weight (kg) | Whole Body | | | | Fat-Free Body Mass | |
		Water	Protein	Lipid	Other*	Water	Protein
Birth	3.50	75.1	11.4	11.0	2.5	84.3	12.8
2	5.45	63.7	11.4	22.4	2.5	82.0	14.7
4	7.00	60.2	11.4	25.9	2.5	81.0	15.4
6	8.28	59.9	12.3	25.3	2.5	80.0	16.5
8	9.08	59.6	13.1	24.8	2.5	79.2	17.4
10	9.82	59.3	13.7	24.5	2.5	78.5	18.1
12	10.50	59.0	14.6	23.9	2.5	77.5	19.4

*Includes minerals, carbohydrate, and non-protein nitrogenous compounds.

(data from Fomon, 1967, Pediatrics 40:863-870. Copyright American Academy of Pediatrics 1967)

TABLE   224

COLLATED DATA FOR BODY COMPOSITION AT DIFFERENT STAGES OF GROWTH

Subject and age (years)	Stature (cm)	Body Weight (kg)	Σ organ weight[a]	Muscle mass[b]	Body fat[c]	ECF volume[d]
Low birth weight	--	1.1	21	<10	3	50
Newborn	50	3.5	18	20	12	40
Child 0.25	60	5.5	15	22	11	32
Child 1.5	80	11.0	14	23	20	26
Child 5	110	19.0	10	35	15	24
Child 10	140	31.0	8.4	37	15	25
Boys 14	160	50	5.7	42	12	21
Girls 13	160	45	4.8	39	18	19

[a]Sum of brain, liver, heart, and kidney (Holliday, 1971)
[b]Derived from creatinine excretion (Holliday, 1971)
[c]Interpolated values from literature (Widdowson and Dickerson, 1964; Friis-Hansen, 1971).
[d]Average of bromide space (Gamble, 1946) and thiosulfate space (Cheek, 1961; Friis-Hansen, 1957)

(data from Holliday, 1978, pp. 117-139 in F. Falkner and J. M. Tanner (eds), Human Growth, Vol. II. Plenum Publishing Corp., New York)

TABLE   225

GAIN IN BODY COMPONENTS OF REFERENCE BOY
AT VARIOUS INTERVALS BETWEEN BIRTH AND THREE YEARS

Age Interval (months)	Gain in Weight (kg)	Composition of Gain (gm/100 gm)				Gain in FFBM*	
		Water	Protein	Lipid	Other	(kg)	(gm/day)
0-4	3.5	45.3	11.4	41.6	1.7	2.04	17.0
4-12	3.5	56.6	21.0	19.1	3.3	2.83	11.6
12-24	2.5	69.4	20.3	6.8	3.5	2.33	6.4
24-36	2.0	68.5	20.9	3.4	7.2	1.93	5.5

*Fat-free body mass

(data from Fomon, 1974)

TABLE 226

ANTHROPOMETRIC, BODY COMPOSITIONAL, RESPIRATORY,
CARDIOVACULAR AND WORK CAPACITY VALUES FOR
CALIFORNIA BOYS AGED 8 TO 12 YEARS

Variable	Mean	S.D.
Body composition		
Stature (cm)	142.3	8.1
Weight (kg)	35.2	7.1
Lean body weight (kg)	28.2	4.6
Fat weight (kg)	6.6	3.3
Relative Fat * (%)	18.7	5.5
Density (gm./c.c.)	1.056	0.013
Lung volumes		
Vital capacity (liters)	2.48	0.43
FEV_1 (liters)	2.04	0.35
$FEV1$ (%)	82.5	6.17
FEV_3 (liters)	2.44	0.43
FEV_3 (%)	98.4	2.19
RV (liters)	0.57	0.14
TLC (liters)	3.05	0.51
RV/TLC (%)	18.7	3.7
Blood lipid analyses		
Cholesterol (mg/100 ml)	178.9	27.0
Triglycerides (mg/100 ml)	60.6	38.1
Blood pressure		
Systolic (mm.Hg)	99.7	14.7
Diastolic (mm.Hg)	62.5	9.9
Work capacity		
VO_2, max (ml/Kg/min)	53.3	6.3
V_E max (liters/min.) (BTPS)	62.1	12.4
Heart Rate max (beats/min.)	196.8	7.7

FEV = forced expiratory volume; RV = residual volume; TLC = total
lung capacity; BTPS = body temperature and pressure saturated with
water vapor.

*Determined by the equation of Siri (1956): Fat per cent = 495/
density - 450.

(data from Wilmore and McNamara, 1974, The Journal of Pediatrics
84:527-533)

TABLE 227

BODY COMPOSITION MEASUREMENTS IN CALIFORNIA BOYS, AGE 17 YEARS

Variable	Mean	S.D.	Range
Stature (cm)	178.6	7.50	159.3-195.9
Weight (kg)	72.11	10.42	53.8-94.6
Underwater weight (kg)	3.436	0.917	0.630-5.150
Body density (g/ml)	1.047	0.018	1.021-1.104
Lean body weight (kg)	63.34	7.64	48.7-79.1
Percent fat	11.6	7.3	1.0-33.4
Pulmonary function (liters, BTPS)			
Functional residual capacity	2.810	0.533	1.804-3.776
Expiratory reserve capacity	1.172	0.408	0.591-2.891
Residual volume	1.638	0.497	0.441-2.909
Vital capacity	4.810	0.754	3.065-6.286
Circumferences (cm)			
Shoulder	103.3	4.8	93.7-114.8
Chest	92.0	7.3	63.1-103.7
Buttocks	92.7	6.5	83.0-108.4
Abdomen 1,2	77.7	7.2	64.6-94.7
Thigh	55.6	6.6	44.9-84.6
Knee	36.3	2.4	27.3-40.2
Calf	36.7	2.8	31.5-44.1
Ankle	22.3	1.3	18.9-25.0
Biceps flexed	31.3	2.8	25.7-38.2
Forearm	26.4	1.7	22.6-30.9
Wrist	16.8	0.8	14.9-18.6
Skinfold (mm)			
Triceps	13.5	5.0	32.2-5.6
Chest	10.7	4.3	26.3-5.6
Iliac	19.0	9.8	47.3-5.5
Abdomen	18.0	10.7	52.6-5.3
Scapula	12.3	5.6	34.7-5.7
Thigh	14.4	5.8	35.9-6.2

Abdomen 1 is the minimum circumference; abdomen 2 is at the level of the iliac crests.

(data from Michael and Katch, 1968, Journal of Applied Physiology 25: 747-750)

TABLE 228

DATA RELEVANT TO BODY COMPOSITION IN CHILDREN
AGED 4 TO 8 YEARS IN COLUMBIA (MO)

	Girls			Boys		
	White	Negro	All	White	Negro	All
Stature (cm)	107.8	109.0	108.5	111.0	109.0	110.0
Weight (kg)	18.2	18.7	18.5	19.3	18.2	18.8
Triceps skinfold (mm)	10.8	9.4	10.1	9.7	7.8	8.8
Waist circumference (cm)	50.3	50.0	50.1	52.4	50.7	51.5
Upper arm circumference (cm)	17.3	16.3	16.8	17.4	16.1	16.7
Bicristal diameter (cm)	17.4	18.3	17.8	18.2	17.8	17.9
Potassium (gm)	39.7	40.4	40.1	41.6	41.5	41.5
Potassium (mg/cm)	366.6	370.5	370.5	374.4	378.3	374.4
Lean Body Mass (kg)	14.9	15.2	15.1	15.6	15.6	15.6
Fat (kg)	3.3	3.5	3.4	3.7	2.6	3.2
Fat (%)	18.1	18.7	18.4	19.2	14.3	17.0

(data from Flynn et al., 1970)

TABLE 229

BODY COMPOSITION DATA FOR ILLINOIS BOYS
AGED 7 TO 12 YEARS

Variable	Mean	S.D.
Stature (cm)	141.8	12.20
Weight (kg)	35.16	9.0
% Fat (^{40}K)	22.0	8.1
Lean Body Mass (^{40}K)	26.8	6.9
Triceps Skinfold (mm)	11.6	4.2
Subscapular Skinfold (mm)	7.0	4.9
Suprailiac Skinfold (mm)	6.4	4.2
Calf Skinfold (mm)	6.7	4.3
Biceps Circumference (cm)	21.0	2.7
Calf Circumference (cm)	28.0	3.0
Elbow Width (cm)	5.6	.6
Knee Width (cm)	8.2	.7
Endomorphic Component (Sheldon)	3.8	1.1
Mesomorphic Component (Sheldon)	4.2	.91
Ectomorphic Component (Sheldon)	3.5	1.1
First Component (Heath and Carter)	2.6	1.3
Second Component (Heath and Carter)	4.1	.98
Third Component (Heath and Carter)	3.2	1.2

(data from Slaughter and Lohman, 1977)

TABLE   230

BODY COMPOSITION AND OTHER MEASUREMENTS FOR ILLINOIS CHILDREN

Age	Stature (cm)		Weight (kg)		K (g)		LBM (kg)	
(years)	Mean	S.D.	Mean	S.D.	Mean	S.D.	Mean	S.D.
7.0- 7.9	126.5	6.1	25.7	4.0	53.5	8.0	20.1	3.0
8.0- 8.9	131.1	6.2	28.1	4.7	58.3	8.5	21.9	3.2
9.0- 9.9	136.2	4.6	31.6	5.0	64.9	8.2	24.4	3.1
10.0-10.9	144.1	7.3	37.6	7.3	74.5	11.2	28.0	4.2
11.0-11.9	148.2	4.8	41.1	8.7	81.9	12.8	30.8	4.8

Age	% fat		Midaxillary Skinfold (mm)		Paraumbilical Skinfold (mm)		Forearm Circumference (cm)		Chest (deflated) Circumference (cm)	
(years)	Mean	S.D.	Mean	S.D.	Mean	S.D.	Mean	S.D.	Mean	S.D.
7.0- 7.9	20.5	5.6	5.2	2.0	6.7	4.4	18.9	1.1	59.3	4.1
8.0- 8.9	21.7	6.2	6.0	3.2	7.8	5.1	19.5	1.4	60.1	4.2
9.0- 9.9	22.3	7.5	6.4	3.8	8.6	5.9	20.4	1.5	63.8	3.9
10.0-10.9	24.4	9.6	9.5	6.5	13.6	8.8	21.6	1.7	66.7	5.2
11.0-11.9	23.8	8.3	9.0	5.5	14.1	8.7	22.1	1.8	69.1	6.5

K = potassium; LBM = lean body mass.

(data from Slaughter et al., 1978, Annals of Human Biology 5:469-482)

TABLE   231

ANTHROPOMETRIC DATA FOR ILLINOIS CHILDREN

Age (years)	Stature (cm)		Weight (kg)		LBM (kg)		% BF		Triceps Skinfold (mm)		Subscapular Skinfold (mm)	
	Mean	S.D.	Mean	S.D.	Mean	S.D.	Mean	S.D.	Mean	S.D.	Mean	S.D.
6.3- 8.5	129.0	7.0	28.4	6.5	22.2	3.9	20.4	10.1	9.0	5.1	6.3	4.5
8.6- 9.5	136.1	7.0	30.9	5.9	25.1	3.6	17.7	8.3	9.6	3.7	6.3	4.1
9.6-10.5	142.6	4.9	35.3	5.7	27.3	3.4	21.6	8.5	11.7	4.8	7.4	4.2
10.6-11.5	146.3	5.5	38.3	8.1	30.0	4.6	22.0	7.7	11.9	5.3	7.9	5.1
11.6-12.9	152.3	8.1	42.7	8.3	34.7	6.8	18.9	8.2	10.5	4.1	6.5	2.7

LBM = lean body mass from ^{40}K.
%BF = total body fat weight as a percentage of total body weight.

(data from "Prediction of lean body mass in young boys from skinfold thickness and body weight", Human Biology 47:245-262, 1975, by Lohman, Boileau, and Massey by permission of the Wayne State University Press. Copyright, Wayne State University Press, Detroit, Michigan 48202)

TABLE 232

DATA RELATIVE TO BODY COMPOSITION IN MISSOURI INFANTS
STUDIED SERIALLY. THE BREAST-FED INFANTS RECEIVED
BREAST MILK FOR AT LEAST 4 MONTHS

		Birth		2-4 weeks		3 - 5 months		7 - 9 months		10-14 months	
		M	F	M	F	M	F	M	F	M	F
Breast-fed											
Length (cm)	Mean	51.4	50.7	54.8	53.7	64.8	62.8	72.0	69.2	77.0	75.1
	S.E.	0.6	1.5	0.6	0.5	0.5	0.4	0.6	0.5	0.6	0.5
Weight (kg)	Mean	3.6	3.6	4.4	4.2	7.4	6.6	9.6	8.4	11.0	10.0
	S.E.	0.1	0.1	0.1	0.1	0.1	0.1	0.2	0.2	0.2	0.2
TBK (g)	Mean	--	--	7.7	6.2	10.8	10.2	14.2	13.7	19.7	17.8
	S.E.	--	--	0.5	0.4	0.4	0.3	0.7	0.6	1.4	1.1
TBK (g/cm)	Mean	--	--	0.14	0.12	0.17	0.16	0.20	0.20	0.26	0.23
	S.E.	--	--	0.01	0.01	0.01	0.01	0.01	0.01	0.02	0.01
TBK (g/kg)	Mean	--	--	1.7	1.5	1.5	1.5	1.5	1.6	1.8	1.8
	S.E.	--	--	0.1	0.1	0.1	0.1	0.1	0.1	0.1	0.1
Bottle-fed											
Length (cm)	Mean	52.5	51.0	54.7	53.2	64.0	62.6	71.7	70.1	76.6	76.0
	S.E.	0.6	0.6	0.5	0.5	0.6	0.6	0.6	0.6	0.9	0.8
Weight (kg)	Mean	3.6	3.4	4.4	4.2	7.3	6.7	9.6	8.6	10.3	10.0
	S.E.	0.1	0.1	0.1	0.1	0.2	0.2	0.2	0.2	0.4	0.3
TBK (g)	Mean	--	--	7.2	7.1	11.3	10.7	14.7	12.8	18.7	16.7
	S.E.	--	--	0.3	0.3	0.4	0.4	0.4	0.4	0.8	0.8
TBK (g/cm)	Mean	--	--	0.13	0.13	0.18	0.17	0.22	0.19	0.24	0.22
	S.E.	--	--	0.01	0.01	0.01	0.01	0.01	0.01	0.01	0.01
TBK (g/kg)	Mean	--	--	1.6	1.7	1.6	1.6	1.5	1.5	1.8	1.7
	S.E.	--	--	0.0	0.0	0.1	0.1	0.1	0.1	0.1	0.1

TBK = total body potassium

(data from Rutledge et al., 1976)

TABLE 233

BODY COMPOSITION DATA FOR CHILDREN IN ROCHESTER (NY)

Age (years)	Stature (cm) 50*	Weight (kg) 50	LBM (kg) 25	LBM (kg) 50	LBM (kg) 75	Fat (kg) 50	Fat (kg) %	LBM (kg) Mean	LBM (kg) 95% limits	Fat Mean (kg)	Fat %
					BOYS						
9	134.7	29.9	24.0	25.5	26.2	4.7	15.7	25.4	24.6-26.2	5.21	17.0
10	140.6	32.0	24.9	26.6	28.9	5.7	17.8	27.1	25.7-28.5	5.83	17.7
11	143.8	37.1	27.2	29.8	32.1	5.4	14.6	30.3	28.4-32.2	7.26	19.4
12	150.3	40.0	30.1	33.7	37.4	7.1	17.8	33.6	32.4-34.9	8.37	19.9
13	153.4	45.9	33.0	36.9	41.5	8.0	17.5	38.6	36.3-40.9	9.38	19.6
14	165.8	56.6	43.0	46.4	50.3	8.4	14.8	47.1	44.8-49.5	9.69	20.6
15	171.7	62.3	46.0	54.0	57.3	9.5	15.2	53.0	50.2-55.9	10.1	16.0
16	172.3	63.0	51.4	56.4	58.6	7.4	11.7	56.3	52.8-59.7	7.24	11.4
17	175.0	64.3	54.4	59.0	65.8	5.5	8.6	60.1	57.4-62.8	6.82	10.2
18	180.9	70.3	57.0	63.4	68.0	7.6	10.8	63.3	59.2-67.4	8.87	12.3
					GIRLS						
8	127.0	26.6	19.9	23.0	24.0	4.0	14.8	22.4	20.9-23.9	6.12	21.5
9	133.8	28.6	21.5	23.1	25.4	5.5	19.2	23.8	22.6-25.0	6.32	21.0
10	139.2	33.7	23.3	26.1	27.2	7.9	23.4	26.0	24.5-27.5	8.09	23.8
11	144.2	37.1	25.4	27.8	34.1	9.2	24.9	29.5	27.1-31.9	9.23	23.8
12	151.0	43.0	30.3	33.0	37.0	9.4	21.7	33.3	31.8-34.8	10.1	23.3
13	156.0	46.0	30.8	36.1	37.9	11.0	24.0	35.0	32.8-37.2	12.2	25.8
14	158.2	50.0	36.0	39.1	41.0	11.1	22.2	38.8	36.2-41.4	13.5	25.8
15	162.7	51.8	37.2	40.8	44.3	12.0	23.2	40.7	39.0-42.4	12.6	23.7
16	161.7	54.8	38.0	40.6	44.4	12.9	23.6	40.9	39.4-42.5	13.7	25.1
17	161.7	52.8	34.8	38.9	41.4	14.6	28.1	38.9	36.0-41.9	15.3	28.2
18	163.2	56.0	36.9	42.9	45.8	12.7	22.7	42.7	39.9-45.5	13.9	24.6

*Percentile

(data from Forbes, 1972)

TABLE 234

BODY SIZE AND BODY COMPOSITION IN BOYS
OF THREE ETHNIC GROUPS
IN BOSTON (MA)

Measurements	10			50			90		
	I*	J*	N*	I*	J*	N*	I*	J*	N*

10 years

Skinfolds

	I*	J*	N*	I*	J*	N*	I*	J*	N*
Abdomen	3.6	3.5	3.5	5.5	6.0	4.7	21.8	25.5	6.9
Chest	2.6	2.9	2.2	4.2	4.6	3.1	11.6	11.8	4.3
Lateral arm	5.8	6.1	4.5	9.6	11.0	7.4	21.8	18.7	9.9
Posterior arm	6.4	6.5	5.4	9.8	10.0	7.5	17.6	18.5	9.9
Scapula	4.1	4.2	4.2	5.3	5.7	5.1	9.9	10.5	6.2

Other

	I*	J*	N*	I*	J*	N*	I*	J*	N*
Stature	127.4	132.6	129.6	136.6	139.6	137.0	143.9	144.8	144.4
Weight	60.5	57.2	61.0	75.5	74.1	69.6	98.5	90.8	82.5
Chest circ.	58.8	56.8	58.3	62.8	61.8	62.0	71.8	72.7	64.8
Upper arm circ.	18.2	18.0	18.4	21.6	21.4	19.8	26.3	26.8	21.9
Thigh circ.	37.1	36.8	37.1	43.0	43.2	40.7	50.8	51.4	45.9
Biiliac diameter	19.7	19.8	18.1	21.3	21.4	19.9	23.5	23.4	21.6

11 years

Skinfolds

	I*	J*	N*	I*	J*	N*	I*	J*	N*
Abdomen	4.0	4.2	3.5	7.3	8.3	4.9	25.8	33.4	10.2
Chest	3.0	2.8	2.2	5.0	5.8	3.1	13.2	15.8	5.2
Lateral arm	5.8	8.0	4.5	10.7	13.0	7.4	20.6	26.4	11.8
Posterior arm	7.0	8.3	5.9	10.5	12.5	7.5	18.8	19.8	10.7
Scapula	4.3	4.4	4.2	5.8	6.0	5.2	11.2	13.4	7.6

Other

	I*	J*	N*	I*	J*	N*	I*	J*	N*
Stature	135.8	138.0	136.9	143.7	144.3	143.0	153.3	153.8	152.1
Weight	66.3	65.5	65.8	87.3	87.7	78.8	112.5	122.2	90.1
Chest circ.	62.2	61.2	59.2	67.0	65.5	63.5	74.8	76.8	67.6
Upper arm circ.	19.4	19.7	19.1	22.0	23.5	21.0	28.3	29.2	23.2
Thigh circ.	40.3	39.1	38.8	44.8	48.0	42.1	55.5	54.9	48.2
Biiliac diameter	20.8	21.1	19.6	22.3	22.7	21.0	23.9	24.8	22.1

12 years

Skinfolds

	I*	J*	N*	I*	J*	N*	I*	J*	N*
Abdomen	4.0	4.3	3.6	5.6	10.0	5.2	21.4	31.8	13.9
Chest	2.8	2.7	2.3	4.3	5.6	3.6	9.9	13.8	7.0
Lateral arm	6.2	6.4	4.6	9.4	13.0	8.2	20.4	20.1	15.9
Posterior arm	8.5	6.4	4.8	9.9	12.3	7.7	17.9	18.6	14.8
Scapula	4.2	4.3	4.1	5.4	6.7	5.7	9.3	12.9	9.2

TABLE 234 (continued)

BODY SIZE AND BODY COMPOSITION IN BOYS
OF THREE ETHNIC GROUPS
IN BOSTON (MA)

Measurements	10			50			90		
	I*	J*	N*	I*	J*	N*	I*	J*	N*

12 years

Other

	I*	J*	N*	I*	J*	N*	I*	J*	N*
Stature	137.7	143.2	141.1	146.9	149.5	147.7	158.9	157.7	155.2
Weight	69.6	72.5	70.8	89.1	99.5	87.7	125.8	128.8	111.5
Chest circ.	62.0	63.0	61.3	67.8	68.6	65.0	76.9	76.9	70.4
Upper arm circ.	19.4	20.2	18.7	22.9	24.7	21.6	29.9	27.9	24.5
Thigh circ.	40.0	41.2	40.0	45.7	48.4	44.6	53.9	56.2	52.9
Biiliac diameter	21.1	21.2	20.0	22.3	23.7	21.3	25.5	25.7	23.3

I* = Italian; J* = Jewish and N* = Negro.

(data from Piscopo, 1962, Research Quarterly 33:255-264, by permission of the American Alliance for Health, Physical Education, Recreation and Dance, 1900 Association Drive, Reston VA 22091)

TABLE 235

MEAN BODY DENSITY (gm/cc) FOR
MEXICAN-AMERICAN BOYS IN AUSTIN (TX)

Age (years)	Mean	S.D.
9	1.0726	.014
10	1.0681	.009
11	1.0608	.013
12	1.0664	.010
13	1.0634	.011
14	1.0702	.013

(data from Zavaleta, 1976)

TABLE 236

DATA RELATIVE TO BODY COMPOSITION OF
CZECH BOYS EXAMINED SERIALLY

Measurements	Age (years)							
	10	11	12	13	14	15	16	17
Stature (cm)								
Mean	144.1	149.7	155.1	162.4	169.3	174.9	177.7	179.0
S.D.	5.0	4.8	6.0	7.4	7.0	5.9	5.6	5.6
Weight (kg)								
Mean	36.0	39.4	44.0	50.0	56.8	62.8	66.6	69.2
S.D.	4.09	4.50	5.72	7.44	8.27	7.56	6.40	6.28
Bone age (years)								
Mean	10.8	11.8	12.8	13.6	14.6	15.8	17.1	18.1
S.D.	0.76	0.88	0.74	0.76	0.88	1.34	1.24	1.04
Body density								
Mean	1.057	1.059	1.062	1.070	1.066	1.065	1.071	1.076
S.D.	0.014	0.018	0.015	0.014	1.012	0.013	0.011	1.012
Lean body mass (kg)								
Mean	30.2	33.4	37.8	44.4	49.6	54.6	59.4	60.0
S.D.	4.7	3.8	4.8	6.4	7.9	6.7	6.0	5.8
Fat (% body wt.)								
Mean	16.1	15.3	14.2	11.3	12.7	13.1	10.9	9.1
S.D.	5.6	7.1	6.0	5.6	4.8	5.2	4.4	4.8

(data from Pařízková, 1976)

TABLE 237

BODY COMPOSITION OF CANADIAN ESKIMOS AGED 15 TO 19 YEARS

	Stature (cm; 5.5)	Weight (kg; 8.1)	Average skinfold (mm; 2.9)	Total body water (ℓ; 4.7)	(%; 5.3)
Male	163.1	61.6	5.5	39.3	63.9
Female	155.0	54.0	9.6	31.8	59.0

	Body solids (kg; 5.4)	(%; 5.3)	Body fat (kg; 5.2)	(%; 7.3)	Fat-free solids (kg ± 1.7)	(% ± 2.0)
Male	22.3	36.1	7.9	12.7	14.4	23.4
Female	22.4	41.0	10.7	19.4	11.6	21.6

S.D. is in parentheses

(data from Shephard et al., 1973)

TABLE 238

ESTIMATED UPPER ARM MUSCLE CIRCUMFERENCE (cm)
IN U.S. CHILDREN

Age			Percentiles						
(years)	Mean	S.D.	5	10	25	50	75	90	95

BOYS

6	15.0	1.23	13.1	13.3	14.0	14.9	15.8	16.7	17.3
7	15.6	1.28	13.4	14.0	14.6	15.5	16.4	17.3	17.8
8	16.1	1.46	14.0	14.3	15.1	16.0	16.9	17.9	18.7
9	16.7	1.57	14.2	14.8	15.6	16.6	17.6	18.7	19.3
10	17.2	1.52	15.0	15.3	16.1	17.1	18.0	19.0	19.8
11	18.0	1.71	15.4	16.1	16.8	17.9	19.0	20.3	21.0
12	19.6	1.92	16.7	17.3	18.2	19.4	20.9	22.2	22.9
13	21.0	2.36	17.3	18.2	19.3	20.8	22.4	24.3	25.2
14	22.8	2.66	18.6	19.6	20.9	22.7	24.3	26.1	27.1
15	24.1	2.36	20.5	21.2	22.4	24.1	25.7	27.2	28.0
16	25.0	2.40	21.1	22.3	23.6	25.0	26.4	28.0	24.3
17	25.8	2.22	22.4	23.2	24.1	25.7	27.3	28.9	29.8

GIRLS

6	14.5	1.20	12.5	13.0	13.6	14.5	15.4	16.2	16.7
7	15.0	1.45	13.0	13.3	14.1	14.9	15.8	16.8	17.4
8	15.6	1.54	13.3	13.8	14.5	15.5	16.5	17.6	18.4
9	16.2	1.61	13.8	14.2	15.1	16.1	17.1	18.3	19.1
10	16.9	1.91	14.1	14.6	15.6	16.7	18.0	19.3	20.1
11	17.7	2.06	14.8	15.3	16.2	17.5	18.9	20.4	21.6
12	19.4	1.94	16.3	16.9	18.1	19.3	20.6	21.9	22.6
13	20.0	1.96	16.9	17.6	18.6	19.8	21.2	22.5	23.3
14	20.4	2.02	17.6	18.1	19.1	20.2	21.5	23.0	23.8
15	20.9	2.06	17.8	18.4	19.4	20.7	22.1	23.6	24.5
16	21.0	2.15	18.0	18.7	19.4	20.6	22.2	23.8	25.0
17	20.9	2.26	17.9	18.6	19.5	20.6	22.1	23.5	24.4

(data from Cycle II and Cycle III of the Health Examination Survey 1963-1965 and
1966-1970 by the National Center for Health Statistics).

TABLE 239

ESTIMATED UPPER ARM MUSCLE CIRCUMFERENCE
(cm) IN U.S. WHITE CHILDREN

Age (years)	Mean	S.D.	Percentiles						
			5	10	25	50	75	90	95
BOYS									
6	15.0	1.21	13.0	13.3	14.0	14.8	15.8	16.7	12.2
7	15.6	1.30	13.4	13.9	14.6	15.5	16.4	17.3	17.8
8	16.1	1.49	14.0	14.3	15.1	16.0	16.9	17.9	18.7
9	16.6	1.57	14.1	14.7	15.6	16.6	17.6	18.6	19.3
10	17.1	1.54	15.0	15.3	16.1	17.0	18.0	18.9	19.8
11	18.0	1.72	15.4	16.1	16.8	17.8	19.0	20.3	21.1
12	19.6	1.94	16.6	17.3	18.2	19.4	20.9	22.3	23.0
13	20.9	2.32	17.3	18.2	19.2	20.8	22.4	24.0	25.0
14	22.8	2.68	18.6	19.6	20.9	22.7	24.3	26.0	27.1
15	24.2	2.35	20.6	21.4	22.6	24.2	25.8	27.3	28.1
16	25.0	2.39	21.2	22.2	23.5	25.0	26.3	27.9	29.3
17	25.8	2.23	22.4	23.1	24.1	25.6	27.3	28.9	29.8
GIRLS									
6	14.5	1.20	12.5	13.0	13.6	14.5	15.3	16.2	16.8
7	14.9	1.38	13.0	13.3	14.1	14.8	15.7	16.7	17.3
8	15.6	1.50	13.4	13.8	14.5	15.4	16.5	17.5	18.2
9	16.2	1.62	13.9	14.2	15.1	16.0	17.1	18.2	19.1
10	16.8	1.85	14.1	14.6	15.6	16.6	17.9	19.1	19.8
11	17.7	2.01	14.8	15.2	16.2	17.4	18.8	20.4	21.5
12	19.3	1.90	16.3	16.9	18.0	19.2	20.4	21.9	22.4
13	20.0	1.93	16.9	17.6	18.7	19.9	21.2	22.5	23.2
14	20.4	1.98	17.6	18.1	19.1	20.1	21.3	23.0	23.8
15	20.9	2.09	17.9	18.4	19.4	20.8	22.1	23.6	24.6
16	20.9	2.12	18.0	18.7	19.4	20.6	22.1	23.8	25.1
17	20.8	2.21	17.9	18.5	19.4	20.6	21.9	23.3	24.1

(data from Cycle II and Cycle III of the Health Examination Survey 1963-1965
and 1966-1970 by the National Center for Health Statistics).

TABLE   240

ESTIMATED UPPER ARM MUSCLE CIRCUMFERENCE
(cm) IN U.S. NEGRO CHILDREN

| Age | | | Percentiles | | | | | | |
(years)	Mean	S.D.	5	10	25	50	75	90	95
				BOYS					
6	15.2	1.34	13.1	13.4	14.1	15.2	16.2	17.1	17.7
7	15.7	1.18	13.5	14.1	14.9	15.6	16.5	17.1	17.8
8	16.2	1.19	14.3	14.7	15.4	16.2	16.9	17.9	18.5
9	16.8	1.58	14.6	15.1	15.6	16.6	18.1	18.8	19.7
10	17.4	1.43	15.0	15.6	16.5	17.4	18.3	19.3	19.8
11	18.1	1.60	15.3	15.9	17.2	18.1	19.1	20.4	20.9
12	19.6	1.76	17.3	17.6	18.3	19.5	20.7	21.6	22.3
13	21.8	2.55	18.0	19.0	19.7	21.5	23.6	25.2	26.0
14	22.8	2.56	18.6	19.4	21.2	22.7	24.5	26.2	26.8
15	23.4	2.30	19.3	20.4	22.0	23.4	24.8	26.5	27.1
16	25.2	2.39	21.4	22.4	23.8	25.0	26.5	28.5	29.4
17	26.2	2.16	22.7	23.5	24.5	26.4	27.6	28.8	29.8
				GIRLS					
6	14.6	1.25	12.5	12.8	13.6	14.7	15.6	16.3	16.7
7	15.3	1.73	13.0	13.3	14.2	15.2	16.2	17.2	17.7
8	15.8	1.74	13.2	13.5	14.4	15.6	16.8	18.2	18.8
9	16.3	1.61	14.0	14.3	15.2	16.3	17.2	18.6	19.3
10	17.4	2.16	14.2	15.0	15.8	17.0	18.6	20.3	21.5
11	18.1	2.34	15.0	15.6	16.6	17.7	19.2	20.8	23.1
12	19.8	2.12	16.0	16.8	18.6	19.7	21.1	22.6	24.1
13	20.0	2.13	17.1	17.7	18.4	19.8	21.2	22.7	23.6
14	20.7	2.22	17.3	18.3	19.4	20.6	22.1	23.5	24.6
15	20.8	1.79	18.2	18.5	19.7	20.7	21.9	23.3	24.2
16	21.5	2.25	18.5	19.2	20.2	21.1	22.7	23.8	25.0
17	21.6	2.43	18.2	18.9	20.0	21.4	22.8	24.4	25.5

(data from Cycle II and Cycle III of the Health Examination Survey 1963-1965
and 1966-1970 by the National Center for Health Statistics).

TABLE 241

LEAN UPPER LIMB DIAMETERS (cm)
FOR BLACK BOYS IN TEXAS

| | Upper Arm | | | |
| Age | Lower Income | | Middle Income | |
(years)	Mean	S.D.	Mean	S.D.
10.1 - 11	5.9	0.24	6.2	0.76
11.1 - 12	6.2	0.32	6.2	0.54
12.1 - 13	6.4	0.58	6.7	0.68
13.1 - 14	6.8	0.66	6.8	0.81
14.1 - 15	7.3	0.12	7.4	0.78
15.1 - 16	8.2	1.56	8.3	0.77
16.1 - 17	8.4	0.54	8.9	1.06
17.1 - 18	8.4	0.56	8.2	0.63

| | Forearm | | | |
| Age | Lower Income | | Middle Income | |
(years)	Mean	S.D.	Mean	S.D.
10.1 - 11	6.1	0.38	6.4	0.87
11.1 - 12	6.3	0.35	6.2	0.44
12.1 - 13	6.4	0.47	6.7	0.58
13.1 - 14	6.9	0.54	6.9	0.64
14.1 - 15	7.1	0.71	7.4	0.53
15.1 - 16	7.6	0.72	7.9	0.46
16.1 - 17	8.0	0.51	8.2	0.60
17.1 - 18	8.0	0.41	8.0	0.42

(data from Schutte, 1979)

TABLE 242

MUSCLE AREA IN RADIOGRAPHS OF THE CALF IN BOSTON CHILDREN

Age	Boys Percentiles			Girls Percentiles		
(years)	10	50	90	10	50	90
6	45.2	52.9	65.4	43.4	55.0	66.2
7	50.8	59.2	69.8	49.2	62.2	74.4
8	55.1	66.0	79.1	51.6	66.9	80.8
9	61.2	71.8	86.3	56.7	70.0	84.4
10	61.9	74.4	89.2	56.0	72.5	89.3

(data from Stuart and Dwinell, 1942)

TABLE 243

TISSUE BREADTHS IN CALF RELATIVE TO
TOTAL CALF BREADTH (%)
IN OHIO CHILDREN

Age-Level	Fat	Muscle	Bone
Boys			
Birth	20	50	30
1 month	24	50	26
1 year	30	46	24
7.5 years	14	53	33
12.5 years	14	54	32
15.5 years	9	59	32
Girls			
Birth	23	56	21
1 month	26	48	26
1 year	32	45	23
7.5 years	17	52	31
12.5 years	16	54	30
15.5 years	17	54	29

(data from Reynolds and Grote, 1948)

TABLE 244

TISSUE BREADTHS (mm) OF THE UPPER ARM IN PHILADELPHIA CHILDREN

Age (years)	Total Arm		Total Muscle (Abs.)		Muscle (Rel.)		Total Fat (Abs.)		Total Fat (Rel.)	
	Mean	S.D.	Mean	S.D.	Mean	S.D.	Mean	S.D.	Mean	S.D.
Boys										
6	67.5	6.5	39.5	3.6	58.6	5.2	14.5	5.5	21.2	9.1
7	67.2	7.7	39.5	3.6	58.8	3.2	13.6	3.3	20.1	3.4
8	70.7	8.6	40.1	3.2	57.0	4.8	15.5	6.6	21.4	5.9
9	80.8	15.7	44.4	6.0	55.5	4.3	20.8	9.8	24.1	6.7
10	81.4	5.3	44.5	4.3	55.0	5.3	19.5	7.2	23.4	6.2
11	80.1	8.7	44.5	4.7	55.7	4.3	17.9	5.5	22.0	4.9
12	84.4	10.9	47.2	5.9	56.0	3.9	18.9	5.4	22.1	4.5
13	88.5	7.8	54.1	7.0	61.1	6.0	14.2	6.7	15.9	6.8
14	85.3	8.4	54.1	7.0	63.4	3.7	11.4	3.5	13.4	3.9
15-16	95.6	9.9	61.7	8.2	64.4	3.4	10.9	2.4	11.4	2.9
Girls										
6	65.2	4.9	36.9	5.5	56.3	5.1	14.8	3.1	22.8	5.0
7	68.0	5.8	38.4	4.1	55.5	4.4	15.7	4.2	22.9	5.1
8	69.1	7.8	37.9	5.1	55.0	4.6	16.6	4.4	23.8	4.6
9	71.5	9.4	37.8	4.0	53.2	4.7	19.1	6.6	26.2	5.7
10	76.6	8.7	41.9	4.6	54.8	3.5	18.4	5.0	23.7	4.4
11	77.7	11.7	41.8	4.4	54.2	4.8	19.1	8.1	23.9	6.0
12	79.2	7.4	43.1	3.5	54.6	4.4	18.9	5.4	23.6	5.3
13	84.5	6.6	47.3	3.6	56.1	4.1	19.1	4.9	22.4	4.7
14	85.5	6.4	48.3	3.6	56.5	3.2	18.7	4.3	21.6	3.7
15-16	86.9	10.0	47.7	4.6	55.2	5.0	20.7	7.5	23.3	6.1

TABLE 244   (continued)

TISSUE BREADTHS (mm) OF THE UPPER ARM IN PHILADLPHIA CHILDREN

Age (years)	Total Bone (Abs.)		Total Bone (Rel.)		Marrow Space		Cortex	
	Mean	S.D.	Mean	S.D.	Mean	S.D.	Mean	S.D.
Boys								
6	13.4	1.2	20.0	2.5	7.5	1.3	5.8	0.7
7	13.9	1.1	20.9	2.8	7.8	1.0	6.0	0.7
8	15.1	1.3	21.5	2.7	8.3	1.0	6.8	0.8
9	15.9	1.1	20.2	2.9	8.9	0.7	7.0	0.9
10	17.4	1.1	21.5	1.9	9.7	0.9	7.6	0.8
11	17.6	1.4	22.2	2.3	10.3	1.5	7.3	0.7
12	18.2	2.3	21.7	2.6	10.2	2.0	7.9	0.9
13	20.0	1.9	22.7	2.0	10.7	1.3	9.2	1.4
14	19.5	1.5	22.9	1.8	10.9	1.4	8.5	1.1
15-16	23.0	2.0	24.1	2.3	12.6	2.0	10.3	1.7
Girls								
6	13.4	0.8	20.6	1.4	8.1	0.5	5.2	0.5
7	13.8	0.8	20.4	1.6	7.6	0.7	6.1	0.6
8	14.5	1.2	21.1	2.1	7.9	1.0	6.5	0.5
9	14.5	1.5	20.5	2.9	8.1	1.3	6.4	0.8
10	16.2	1.4	21.3	1.8	9.2	1.2	6.9	1.0
11	16.7	1.4	21.7	2.3	9.4	1.4	7.2	0.8
12	17.2	1.9	21.7	2.1	9.2	1.4	8.0	1.2
13	18.0	1.1	21.4	1.7	9.6	1.2	8.4	0.9
14	18.4	0.8	21.6	1.6	9.6	1.2	8.8	1.0
15-16	18.3	1.2	21.2	2.8	9.8	1.2	8.5	0.9

(data from "Age changes in the composition of the upper arm in Philadelphia children,"
Human Biology 38:1-21, 1966, by Johnston and Malina, by permission of the Wayne
State University Press. Copyright 1966, Wayne State University Press, Detroit,
Michigan 48202)

TABLE 245

ESTIMATED ARM MUSCLE CIRCUMFERENCE (cm)
AT EXAMINATIONS ONE YEAR APART
IN PHILADELPHIA CHILDREN

Ages (years)	Examination 1 Mean	S.D.	Examination 2 Mean	S.D.	Increase Mean	S.D.
			Negro Boys			
6- 7	14.1	1.0	14.7	1.2	0.6	0.5
7- 8	14.9	1.1	15.5	1.0	0.6	0.5
8- 9	15.6	1.1	16.3	1.2	0.7	0.5
9-10	16.1	1.1	17.0	1.1	0.9	0.6
10-11	16.7	1.3	17.7	1.3	1.0	0.5
11-12	17.5	1.4	18.5	1.9	1.0	0.8
12-13	17.6	1.2	18.7	1.6	1.1	0.8
			White Boys			
6- 7	14.5	1.3	15.1	1.3	0.6	0.6
7- 8	14.8	0.8	15.6	1.0	0.8	0.7
8- 9	15.1	1.0	16.0	1.1	0.9	0.6
9-10	15.8	1.2	16.4	1.1	0.6	0.5
10-11	16.7	1.5	17.7	1.8	1.0	0.8
11-12	17.1	1.0	18.2	1.5	1.1	0.8
			Negro Girls			
6- 7	13.8	0.8	14.5	1.1	0.7	0.6
7- 8	14.8	1.2	15.7	1.3	0.9	0.6
8- 9	15.1	0.9	16.0	0.9	0.9	0.5
9-10	15.7	1.3	16.6	1.4	0.9	0.6
10-11	16.7	1.5	17.7	1.7	1.0	0.6
11-12	16.9	1.5	18.0	3.2	1.1	0.7
12-13	17.8	1.5	19.1	3.2	1.3	0.8
			White Girls			
6- 7	13.8	0.8	14.5	0.9	0.7	0.1
7- 8	14.3	0.9	15.0	0.9	0.7	0.5
8- 9	14.7	1.3	15.0	1.2	0.3	0.6
9-10	15.5	1.3	16.2	1.6	0.7	0.7
10-11	15.7	1.2	16.2	1.2	0.5	0.6
11-12	16.6	1.3	16.9	1.1	0.3	0.6

(data from Malina, 1972)

TABLE   246

LIMB CIRCUMFERENCES AND LIMB ESTIMATED MUSCLE CIRCUMFERENCES
FOR MEXICAN-AMERICAN BOYS IN TEXAS

	Age (years)	Austin Mean	Austin S.D.	Brownsville Mean	Brownsville S.D.
Arm Circumference					
(cm)	9	16.8	1.4	18.9	2.2
	10	20.1	2.1	19.4	1.7
	11	21.8	2.0	20.3	2.1
	12	21.8	1.6	21.2	2.0
	13	23.6	2.5	22.6	2.3
	14	23.7	2.4	23.2	2.4
Arm Estimated Muscle					
Circumference (cm)	9	16.8	1.4	15.8	1.3
	10	17.6	1.5	16.4	1.2
	11	19.1	1.4	17.1	1.4
	12	19.2	1.3	17.9	1.5
	13	20.6	2.0	19.6	1.8
	14	21.3	1.8	20.4	1.9
Calf Circumference					
(cm)	9	25.2	1.6	--	--
	10	26.1	1.7	--	--
	11	27.9	2.3	--	--
	12	28.4	1.6	--	--
	13	30.1	3.1	--	--
	14	30.9	2.6	--	--
Calf Estimated Muscle					
Circumference (cm)	9	22.8	1.3	--	--
	10	24.1	1.2	--	--
	11	25.4	1.8	--	--
	12	26.1	1.4	--	--
	13	27.9	2.8	--	--
	14	29.1	2.2	--	--

(data from Zavaleta, 1976)

TABLE 247

LEAN LOWER LIMB DIAMETERS (cm)
FOR BLACK BOYS IN TEXAS

Age	Thigh			
	Lower Income		Middle Income	
(years)	Mean	S.D.	Mean	S.D.
10.1 - 11	11.8	0.68	11.6	0.66
11.1 - 12	12.1	0.81	12.2	1.04
12.1 - 13	12.6	1.37	13.2	1.39
13.1 - 14	13.3	1.29	13.4	1.45
14.1 - 15	13.3	1.14	14.2	0.98
15.1 - 16	14.2	1.23	15.0	0.84
16.1 - 17	15.3	0.69	15.8	0.88
17.1 - 18	15.7	0.79	15.1	1.02

Age	Calf			
	Lower Income		Middle Income	
(years)	Mean	S.D.	Mean	S.D.
10.1 - 11	8.4	0.37	8.4	0.56
11.1 - 12	8.8	0.43	9.1	1.51
12.1 - 13	8.9	0.80	9.3	0.85
13.1 - 14	9.7	0.93	9.4	0.77
14.1 - 15	9.9	1.65	9.9	0.75
15.1 - 16	10.3	1.28	10.5	0.64
16.1 - 17	10.6	0.53	10.9	0.69
17.1 - 18	10.8	0.61	10.7	0.55

(data from Schutte, 1979)

TABLE 248

MUSCLE WIDTHS (mm) MEASURED ON RADIOGRAPHS OF COLORADO BOYS

Age	Max. Forearm Percentiles			Mid-thigh (Lateral half) Percentiles			Max. Calf Percentiles		
(yr.-mo.)	10	50	90	10	50	90	10	50	90
0 - 2	16.42	19.52	23.47	11.87	14.22	17.19	19.48	21.91	25.15
0 - 4	17.82	20.64	23.54	12.61	15.55	17.83	19.77	23.59	28.60
0 - 6	17.23	20.16	24.58	13.06	16.58	19.40	20.27	24.58	28.67
1 - 0	19.95	23.06	25.63	15.01	18.43	21.59	23.73	28.24	32.84
1 - 6	19.58	22.72	26.12	15.85	18.26	22.21	27.64	30.60	35.31
2 - 0	20.22	23.80	27.24	15.93	18.93	22.29	28.71	32.34	36.79
2 - 6	20.56	24.23	27.34	17.34	20.10	24.10	29.12	33.57	38.82
3 - 0	21.18	24.96	28.75	19.02	21.93	25.49	31.59	35.35	40.86
3 - 6	21.23	24.90	29.55	19.82	22.81	26.26	31.90	36.50	42.20
4 - 0	21.96	25.54	29.88	20.28	23.82	27.19	33.27	37.78	43.76
4 - 6	22.90	26.69	29.98	21.24	24.67	28.83	34.72	38.38	45.26
5 - 0	23.75	27.30	31.06	23.56	26.43	29.62	34.40	39.82	45.44
5 - 6	24.56	28.19	32.37	23.09	26.32	30.55	35.86	40.99	47.72
6 - 0	25.46	28.87	33.28	23.51	27.12	32.27	36.14	41.28	48.03
6 - 6	26.99	30.81	35.17	23.89	28.83	31.40	37.41	42.51	49.68
7 - 0	27.24	31.13	35.85	24.50	28.70	32.84	38.06	44.59	50.64
7 - 6	28.18	32.75	36.55	25.84	29.80	35.08	38.79	44.42	50.89
8 - 0	29.32	33.24	37.35	26.62	30.58	34.55	40.75	46.14	51.86
8 - 6	29.94	33.76	37.79	27.06	31.55	37.00	41.02	47.65	53.50
9 - 0	31.64	34.80	40.67	27.43	32.25	38.43	42.58	48.14	54.59
9 - 6	32.17	35.88	39.51	28.04	33.33	38.31	42.02	49.49	56.47
10 - 0	32.87	36.84	40.32	30.49	35.03	39.90	43.77	50.62	55.38
10 - 6	32.90	37.38	42.30	30.11	35.86	40.93	43.65	50.56	58.12
11 - 0	33.92	38.79	43.05	31.20	37.10	41.36	42.77	52.06	57.37
11 - 6	34.21	38.68	45.43	31.62	36.81	41.90	44.11	51.14	58.68
12 - 0	35.55	40.60	45.55	34.44	38.92	44.73	45.37	53.08	60.24
12 - 6	36.04	41.46	47.36	34.64	39.82	45.28	46.20	52.55	62.51
13 - 0	36.26	42.20	48.93	35.00	40.95	47.54	46.54	55.35	64.18
13 - 6	38.00	44.30	51.05	35.44	41.05	49.04	45.34	55.36	66.84
14 - 0	38.39	45.33	52.02	36.32	42.33	49.56	47.27	57.25	68.02
14 - 6	40.16	46.79	54.97	37.54	43.09	50.26	48.51	56.36	69.22
15 - 0	43.37	49.72	57.52	38.11	44.05	50.65	50.30	57.90	71.20
15 - 6	44.78	51.35	58.51	37.69	43.49	49.93	51.06	58.36	70.31
16 - 0	47.00	52.35	59.57	38.66	45.65	53.94	52.91	62.11	70.40
16 - 6	49.26	51.48	61.49	41.15	46.88	51.82	55.54	62.07	71.51
17 - 0	49.82	52.32	59.64	42.38	47.05	51.29	54.22	63.06	71.48
18 - 0	52.08	56.44	62.37	39.07	47.35	51.53	56.03	66.08	72.84

(data from McCammon, Human Growth and Development, 1970. Courtesy of Charles C Thomas, Publisher, Springfield, Illinois)

TABLE 249

MUSCLE WIDTHS (mm) MEASURED ON RADIOGRAPHS OF COLORADO GIRLS

Age (yr.-mo.)	Max. Forearm Percentiles			Mid-thigh (Lateral half) Percentiles			Max. Calf Percentiles		
	10	50	90	10	50	90	10	50	90
0 - 2	16.72	19.27	21.86	10.84	13.52	16.30	17.55	21.05	25.04
0 - 4	16.00	19.22	21.63	11.80	15.20	18.11	19.14	22.03	26.35
0 - 6	17.16	20.47	22.84	13.60	16.09	18.87	20.85	23.60	27.28
1 - 0	19.74	22.16	24.85	15.03	17.85	21.12	25.08	28.31	31.60
1 - 6	19.83	22.85	25.12	15.72	18.06	21.59	26.60	30.58	33.89
2 - 0	20.67	23.41	27.04	15.92	19.13	24.00	28.98	31.98	35.54
2 - 6	20.45	23.15	26.05	17.39	20.96	24.28	30.10	33.95	36.53
3 - 0	21.41	23.86	26.86	18.97	22.26	25.75	31.58	34.86	38.97
3 - 6	21.86	24.50	27.64	19.38	23.43	27.25	33.21	37.08	42.00
4 - 0	23.25	25.71	28.10	21.42	24.11	27.58	34.25	38.93	42.74
4 - 6	23.87	26.53	29.29	21.28	25.12	28.80	36.22	40.97	45.21
5 - 0	23.95	27.08	31.04	23.02	27.10	30.16	36.57	41.69	45.76
5 - 6	23.77	27.54	31.74	23.71	27.35	31.23	37.48	41.96	46.40
6 - 0	24.88	28.66	32.14	22.90	28.30	31.80	38.69	43.60	47.62
6 - 6	24.85	29.48	34.25	24.95	29.72	32.81	38.43	43.94	49.04
7 - 0	25.80	30.89	34.94	25.66	30.55	34.16	40.39	44.28	50.74
7 - 6	27.16	30.59	36.09	25.01	30.75	35.08	39.90	46.53	54.03
8 - 0	28.56	32.44	37.88	27.05	31.48	36.36	41.22	47.48	51.74
8 - 6	29.41	32.86	37.99	27.78	32.99	37.87	41.62	48.06	54.16
9 - 0	30.00	33.30	38.24	29.22	33.23	37.68	40.95	48.20	57.24
9 - 6	29.79	34.03	38.55	29.43	33.72	39.38	43.32	47.82	56.69
10 - 0	30.13	34.99	39.56	30.46	35.10	40.68	42.28	49.50	57.05
10 - 6	31.18	35.56	40.45	31.92	35.11	41.72	43.24	49.59	58.95
11 - 0	32.58	37.27	40.90	32.28	36.69	43.81	43.44	50.51	61.20
11 - 6	31.70	37.30	41.76	32.75	37.43	42.78	44.45	52.06	61.44
12 - 0	33.53	37.75	41.77	33.44	39.38	43.99	44.73	53.13	61.80
12 - 6	35.09	39.01	43.29	35.76	39.70	45.10	46.40	54.05	64.83
13 - 0	35.72	39.91	45.96	36.35	41.27	45.70	47.04	56.08	65.32
13 - 6	35.19	40.47	46.81	33.84	41.10	46.11	47.02	55.81	70.12
14 - 0	36.94	41.01	45.93	36.44	42.15	48.17	52.30	57.31	70.07
14 - 6	36.21	40.36	46.26	37.62	42.13	45.75	50.44	57.61	67.14
15 - 0	35.83	42.08	47.27	37.65	43.52	48.04	52.79	61.40	71.31
16 - 0	39.21	43.21	47.66	36.98	44.60	50.46	56.23	62.60	73.40
17 - 0	41.86	42.89	49.58	35.65	42.10	52.85	61.42	65.22	74.59

(data from McCammon, Human Growth and Development, 1970. Courtesy of Charles C Thomas, Publisher, Springfield, Illinois)

TABLE   250

MEAN MEASURED CALF CIRCUMFERENCES AND CIRCUMFERENCES
CALCULATED FROM RADIOGRAPHIC BREADTHS AT THE SAME LEVEL (cm)

Age Level	Measured	Calculated
Birth	10.6	11.3
1 month	12.0	12.6
1 year	18.9	21.0
7.5 years	25.2	25.4
12.5 years	30.4	30.8
15.5 years	33.3	35.2

(data from Reynolds, 1948)

TABLE   251

MUSCLE ELECTROLYTE LEVELS IN CHILDREN IN BALTIMORE (MD)

		Age (years)	% Fat in Fresh Muscle	H_2O_t (ml)	Cl sp (mls)	Cl (mEq)	Na (mEq)	K (mEq)	Mg (mEq)	Zn (μg/gm)
Male infants	Mean	0.58	0.67	371.9	133.8	16.44	21.28	40.96	7.22	329
	S.D.	0.45	0.37	19.5	12.5	3.08	3.41	2.33	0.52	5
Boys	Mean	9.38	2.2	355.0	95.49	11.30	14.84	37.90	6.94	211
	S.D.	3.40	1.0	6.1	11.48	1.51	2.34	1.89	0.53	30
Girls	Mean	9.59	2.6	349.3	92.49	10.97	15.30	39.65	7.46	194
	S.D.	3.80	1.4	8.3	12.18	1.57	1.92	2.23	0.46	26

		Age (years)	% Fat in Fresh Muscle	ml per 100 gm Fat-Free Fresh Muscle			mEq per 100 gm Fat-Free Fresh Muscle				Zn (μg/gm)	$(K)_i$
				H_2O_t	Cl sp	H_2O_i	Cl	Na	K	Mg		
Boys	Mean	9.38	2.2	780.18	209.76	570.42	24.82	32.59	83.31	15.26	46.49	146.2
	S.D.	3.40	1.0	2.95	24.04	23.12	3.17	4.89	4.45	1.16	6.64	9.8
Girls	Mean	9.59	2.6	777.34	206.07	570.99	24.31	34.03	88.26	16.60	43.06	154.3
	S.D.	3.80	1.4	4.04	26.19	8.27	3.37	3.92	4.73	1.07	5.39	6.2

Data expressed as ml or mEq per 100 gm fat-free dried muscle.
H_2O_t = total water content; H_2O_i = intracellular water content; Cl sp = chloride space;
$(K)_i$ = mEq of K per liter of cell water.

(data from Elliott and Cheek, pp. 260-273 in D. B. Cheek (ed.), Body Composition, Cell
Growth, Energy, and Intelligence. Lea & Febiger, Philadelphia, PA, 1968)

TABLE 252

BREADTH OF CALF BONE PLUS MUSCLE (cm) IN BOSTON (MA) CHILDREN

			Percentiles				
Age	10	25	50	75	90	Mean	S.D.
			BOYS				
			Three - foot tube distance				
3 months	3.3	3.4	3.6	3.8	3.9	3.59	0.24
6 months	3.6	3.6	3.9	4.2	4.3	3.91	0.28
9 months	3.9	4.0	4.2	4.4	4.6	4.24	0.29
1 year	4.2	4.4	4.6	4.8	5.1	4.61	0.39
1½ year	4.5	4.8	5.0	5.2	5.5	5.03	0.35
2 year	4.8	5.0	5.2	5.5	5.7	5.27	0.36
2½ year	4.9	5.4	5.6	5.9	6.1	5.60	0.45
3 year	5.3	5.6	5.8	6.1	6.4	5.83	0.46
3½ year	5.6	5.6	6.2	6.4	6.6	6.10	0.47
4 year	5.7	5.8	6.2	6.6	6.8	6.25	0.49
4½ year	5.7	6.1	6.5	6.8	7.1	6.49	0.52
5 year	5.8	6.2	6.5	6.9	7.2	6.54	0.53
5½ year	6.1	6.4	6.8	7.2	7.4	6.81	0.54
6 year	6.3	6.6	7.0	7.3	7.6	6.98	0.48
			Six - foot tube distance				
6 year	6.2	6.4	6.8	7.2	7.4	6.81	0.55
7 year	6.4	6.6	7.0	7.5	7.7	7.04	0.54
8 year	6.6	7.1	7.4	7.8	8.2	7.41	0.61
9 year	7.0	7.2	7.7	8.0	8.5	7.68	0.63
10 year	7.0	7.6	8.0	8.6	8.9	8.06	0.72
11 year	7.5	7.8	8.3	8.8	9.4	8.32	0.68
12 year	7.7	8.3	8.6	9.0	9.5	8.57	0.70
13 year	7.9	8.4	8.9	9.4	10.3	9.07	0.90

TABLE 252   (continued)

BREADTH OF CALF BONE PLUS MUSCLE (cm) IN BOSTON (MA) CHILDREN

Age	Percentiles 10	25	50	75	90	Mean	S.D.
			BOYS				
14 years	8.4	8.9	9.5	10.3	10.9	9.59	0.95
15 years	9.3	9.6	10.2	10.8	11.3	10.19	1.02
16 years	9.0	9.8	10.3	11.2	11.4	10.40	1.07
			Three - foot tube distance				
			GIRLS				
3 months	3.1	3.2	3.4	3.6	3.7	3.40	0.26
6 months	3.4	3.6	3.8	4.0	4.2	3.79	0.31
9 months	3.7	3.8	4.1	4.3	4.6	4.11	0.41
1 year	4.0	4.2	4.5	4.8	5.0	4.46	0.44
1½ years	4.4	4.6	5.0	5.2	5.4	4.90	0.40
2 years	4.6	5.0	5.2	5.4	5.6	5.18	0.37
2½ years	4.9	5.4	5.6	5.8	6.0	5.54	0.42
3 years	5.2	5.5	5.8	6.2	6.4	5.80	0.45
3½ years	5.5	5.8	6.2	6.4	6.8	6.13	0.49
4 years	5.6	6.0	6.3	6.6	7.0	6.30	0.52
4½ years	5.8	6.1	6.5	6.8	7.2	6.49	0.54
5 years	5.9	6.2	6.5	7.0	7.2	6.57	0.55
5½ years	6.0	6.4	6.8	7.3	7.5	6.79	0.58
6 years	6.2	6.4	6.8	7.2	7.6	6.84	0.51
			Six - foot tube distance				
6 years	6.0	6.4	6.9	7.2	7.5	6.84	0.55
7 years	6.4	6.8	7.1	7.5	7.9	7.13	0.57
8 years	6.4	6.8	7.4	7.8	8.3	7.35	0.65
9 years	6.8	7.1	7.4	8.0	8.5	7.56	0.67

TABLE 252 (continued)

BREADTH OF CALF BONE PLUS MUSCLE (cm) IN BOSTON (MA) CHILDREN

Age	Percentiles					Mean	S.D.
	10	25	50	75	90		
			GIRLS				
10 years	6.8	7.1	7.6	8.2	8.8	7.68	0.75
11 years	7.2	7.4	7.9	8.9	9.2	8.11	0.84
12 years	7.3	7.8	8.7	9.2	10.1	8.64	0.95
13 years	7.6	8.1	9.1	9.9	10.6	9.06	1.09
14 years	8.4	8.6	9.2	10.1	10.8	9.40	0.82
15 years	8.7	9.1	9.6	10.1	11.3	9.72	0.84
16 years	8.3	9.0	9.3	10.1	10.8	9.57	0.89

(data from Lombard, 1950)

TABLE 253

TISSUE THICKNESSES (mm) IN THE CALF
FOR PHILADELPHIA CHILDREN

Age (years)	Total Leg		Total Muscle		Total Bone		Medial Fat	
	Mean	S.D.	Mean	S.D.	Mean	S.D.	Mean	S.D.
Boys								
6	83.8	9.3	45.5	5.6	26.6	2.6	6.0	2.5
7	85.9	7.0	47.7	6.2	27.4	1.7	5.9	1.5
8	90.6	9.1	49.8	6.4	29.0	2.7	6.5	2.1
9	100.8	16.1	54.1	10.5	31.0	3.1	9.1	5.3
10	102.6	11.3	53.2	7.1	33.6	3.7	9.0	3.3
11	101.5	8.2	52.5	4.9	34.3	2.1	8.4	2.8
12	107.7	11.2	55.9	7.0	35.9	4.1	9.4	3.3
13	113.3	11.9	60.3	7.7	40.6	3.6	7.5	3.3
14	106.5	10.1	58.0	8.4	38.4	3.5	5.7	2.4
15.5	121.8	8.9	65.6	7.6	44.2	5.0	6.8	3.5
Girls								
6	79.5	5.6	41.4	4.3	25.7	2.1	6.2	1.3
7	86.3	7.7	47.3	5.7	26.3	2.1	7.1	1.9
8	88.9	8.3	48.5	6.2	27.3	2.1	7.0	2.1
9	93.8	9.7	50.9	4.7	28.1	2.1	8.2	2.9
10	100.0	12.6	52.9	8.3	31.5	3.2	8.8	2.4
11	103.1	10.9	54.3	6.1	31.4	2.7	9.6	4.1
12	110.1	14.1	58.3	7.1	32.5	3.1	11.4	5.7
13	115.3	7.7	62.1	7.2	34.4	2.9	10.7	3.7
14	117.7	10.0	63.2	8.3	35.5	3.7	11.3	3.7
15.5	120.7	12.3	64.6	9.2	34.2	2.0	12.6	5.4

Age (years)	Lateral Fat		Total Tibia		Tibial Cortex		Tibial Medullary Cavity	
	Mean	S.D.	Mean	S.D.	Mean	S.D.	Mean	S.D.
Boys								
6	5.7	2.1	19.1	2.3	6.6	1.0	12.5	2.0
7	5.0	1.3	19.6	1.6	6.4	1.0	13.1	1.8
8	5.4	1.7	20.7	2.1	7.5	1.2	13.1	2.0
9	6.5	2.7	22.3	2.5	8.3	1.0	14.0	1.8
10	6.8	1.6	24.3	2.5	8.4	1.1	15.9	2.1
11	6.4	2.0	25.0	1.8	8.7	1.4	16.3	1.8
12	6.5	1.7	26.4	3.3	8.5	1.1	17.9	3.1
13	5.0	1.2	30.0	3.2	8.4	1.1	21.6	3.1
14	4.3	1.0	27.9	2.6	8.8	1.3	19.2	2.1
15.5	5.1	1.8	32.7	4.0	9.3	1.6	23.4	3.9

TABLE 253   (continued)

TISSUE THICKNESSES (mm) IN THE CALF
FOR PHILADELPHIA CHILDREN

Age (years)	Lateral Fat		Total Tibia		Tibial Cortex		Tibial Medullary Cavity	
	Mean	S.D.	Mean	S.D.	Mean	S.D.	Mean	S.D.
Girls								
6	6.1	1.1	18.3	1.8	6.0	0.8	12.3	1.9
7	5.7	1.5	19.1	1.5	6.3	0.7	12.7	1.6
8	6.0	1.4	19.6	1.7	6.5	0.9	13.1	1.6
9	6.7	2.3	20.4	1.6	7.2	1.0	13.2	1.7
10	6.8	1.8	23.1	2.4	7.5	1.2	15.6	2.1
11	7.8	2.2	22.8	2.0	7.5	1.0	15.3	2.3
12	7.9	3.6	24.0	2.5	8.4	0.9	15.6	2.2
13	7.7	1.9	25.2	2.3	8.5	1.7	16.8	2.7
14	7.8	1.7	25.7	2.9	7.9	1.3	17.8	3.7
15.5	9.2	2.9	24.9	1.6	8.9	1.1	16.0	1.6

(data from "Relations between bone, muscle and fat widths in the upper arms and calves
of boys and girls studied cross-sectionally at ages 6 to 16 years," Human Biology 39:211-
223, 1967, by Malina and Johnston, by permission of the Wayne State University Press.
Copyright 1967, Wayne State University Press, Detroit, Michigan 48202)

TABLE 254

WIDTH OF MUSCLE (cm);
TOTAL WIDTH, INCLUDING BONE SHADOWS AT GREATEST
WIDTH OF CALF
IN BOSTON CHILDREN

Age (years)	Boys Percentiles			Girls Percentiles		
	10	50	90	10	50	90
6	6.06	6.60	7.68	6.09	6.94	7.48
7	6.35	6.92	8.02	6.42	7.13	7.91
8	6.60	7.26	8.20	6.49	7.24	8.28
9	6.88	7.64	8.67	6.64	7.38	8.46
10	7.10	7.72	8.85	6.69	7.50	8.72

(data from Stuart and Dwinell, 1942)

TABLE 255

BODY COMPOSITION DATA FOR
BLACK ADOLESCENT BOYS IN TEXAS

Age	Low Income		Middle Income	
(years)	Mean	S.D.	Mean	S.D.
Lean Body Mass (kg)				
10.1-11	29.8	3.62	28.2	3.16
11.1-12	30.3	2.09	32.4	4.42
12.1-13	33.5	5.53	38.4	6.95
13.1-14	40.3	6.37	40.8	8.88
14.1-15	40.1	7.80	47.0	8.71
16.1-17	55.1	4.57	--	--
Total Body Fat (kg)				
10.1-11	5.5	3.16	5.5	1.30
11.1-12	6.5	2.42	8.6	3.31
12.1-13	9.3	7.06	7.9	5.04
13.1-14	6.0	3.02	8.5	4.25
14.1-15	7.9	2.76	7.8	2.56
16.1-17	9.2	4.79	--	--
Total Body Fat as a Percent of Weight				
10.1-11	14.9	5.37	16.4	3.66
11.1-12	17.3	4.94	20.5	6.22
12.1-13	20.2	7.44	16.4	7.01
13.1-14	12.0	6.64	17.3	7.78
14.1-15	16.2	2.83	14.5	5.24
16.1-17	13.8	5.53	--	--

(data from "Growth differences between lower and middle
income black male adolescents," Human Biology 52:193-
204, 1980, by J. E. Schutte, by permission of the Wayne
State University Press. Copyright 1980, Wayne State
University Press, Detroit, Michigan 48202)

TABLE   256

LEAN BODY MASS (LBM) IN POUNDS
FOR CHILDREN IN BALTIMORE (MD)

Age*	Percentiles						
(years)	3	10	25	50	75	90	97

BOYS

5	33	35	37	40	43	45	49
6	35	38	40	42	45	49	52
7	39	41	44	46	50	53	59
8	43	45	48	52	55	62	69
9	45	47	51	54	60	67	77
10	47	52	55	60	68	77	88
11	51	54	59	68	79	88	103
12	57	63	70	79	91	102	118
13	58	71	81	94	107	121	140
14	70	84	95	109	121	132	148
15	76	92	107	119	133	146	158
16	91	106	116	127	138	152	167
17	--	--	--	127	--	--	--

GIRLS

5	30	32	33	35	40	43	50
6	32	34	36	40	44	47	52
7	34	38	41	45	49	54	60
8	39	42	46	50	56	62	70
9	41	45	50	54	61	68	76
10	46	50	55	61	69	77	88
11	52	57	64	71	78	87	93
12	58	63	70	77	85	92	100
13	62	71	78	85	92	100	110
14	71	78	83	89	96	104	115
15	75	80	85	91	98	106	116
16	77	82	87	93	99	106	116
17	--	--	--	93	--	--	--

Age is defined at subject's last birthday

(data from Cheek, 1968, Human Growth: Body Composition, Cell Growth, Energy
and Intelligence, by permission of Lea and Febiger, Publishers, Philadelphia)

TABLE   257

BODY COMPOSITION AND OTHER DATA FOR CHILDREN IN ROCHESTER (MN)

Age	BOYS				GIRLS			
	Stature (cm)		Weight (kg)		Stature (cm)		Weight (kg)	
(yr and mo)	Mean	S.D.	Mean	S.D.	Mean	S.D.	Mean	S.D.
11.6 - 12.5	150.3 + 4.94		41.5 + 3.66		154.8 + 7.79		46.2 + 7.79	
12.6 - 13.5	157.6 + 8.57		50.2 + 11.92		158.5 + 5.92		49.6 + 7.37	
13.6 - 14.5	165.0 + 7.24		54.8 + 9.02		161.0 + 6.41		55.4 + 9.30	
14.6 - 15.5	167.8 + 7.75		57.9 + 7.24		164.9 + 4.99		56.2 + 9.40	

Total Body Potassium

Age	BOYS				GIRLS			
	mEq		Gm		mEq		Gm	
(yr and mo)	Mean	S.D.	Mean	S.D.	Mean	S.D.	Mean	S.D.
11.6 - 12.5	2,390 + 241.9		93.5 + 9.46		2,508 + 343.1		98.0 + 13.42	
12.6 - 13.5	2,805 + 602.7		109.7 + 23.57		2,556 + 281.0		99.9 + 10.99	
13.6 - 14.5	3,232 + 522.3		126.4 + 20.42		2,774 + 260.7		108.5 + 10.20	
14.6 - 15.5	3,420 + 388.7		133.7 + 15.20		2,840 + 412.0		111.0 + 16.11	

Total Body Potassium/kg body weight

Age	BOYS				GIRLS			
	mEq/kg		Gm/kg		mEq/kg		Gm/kg	
(yr and mo)	Mean	S.D.	Mean	S.D.	Mean	S.D.	Mean	S.D.
11.6 - 12.5	57.8 + 5.36		2.26 + 0.21		54.8 + 4.98		2.14 + 0.21	
12.6 - 13.5	56.5 + 5.65		2.21 + 0.22		52.0 + 5.05		2.03 + 0.22	
13.6 - 14.5	59.1 + 4.38		2.31 + 0.17		50.7 + 5.35		1.98 + 0.17	
14.6 - 15.5	59.3 + 5.40		2.32 + 0.21		50.8 + 3.90		1.99 + 0.21	

Corrected Limb Diameters

Age	BOYS				GIRLS			
	Upper Arm		Calf		Upper Arm		Calf	
(yr and mo)	Mean	S.D.	Mean	S.D.	Mean	S.D.	Mean	S.D.
11.6 - 12.5	6.1 + 0.28		8.9 + 0.81		6.1 + 0.53		8.8 + 0.64	
12.6 - 13.5	6.6 + 0.68		9.2 + 0.79		6.2 + 0.51		8.9 + 0.54	
13.6 - 14.5	7.2 + 0.67		9.6 + 0.74		6.5 + 0.49		8.9 + 0.69	
14.6 - 15.5	7.4 + 0.47		9.8 + 0.93		6.5 + 0.56		8.9 + 0.67	

TABLE   257   (continued)

BODY COMPOSITION AND OTHER DATA FOR CHILDREN IN ROCHESTER (MN)

Corrected Limb Volumes

Age	BOYS				GIRLS			
	Upper Arm		Lower Leg		Upper Arm		Lower Leg	
(yr and mo)	Mean	S.D.	Mean	S.D.	Mean	S.D.	Mean	S.D.
11.6 - 12.5	904 + 87.6		2,404 + 511.6		944 + 189.1		2,377 + 391.6	
12.6 - 13.5	1,151 + 305.2		2,714 + 604.8		996 + 193.2		2,472 + 335.8	
13.6 - 14.5	1,419 + 312.5		3,109 + 576.0		1,138 + 208.6		2,657 + 491.4	
14.6 - 15.5	1,489 + 210.0		3,264 + 601.6		1,140 + 197.5		2,583 + 441.9	

The corrected limb diameters are values corrected for subcutaneous fat thickness.

(data from Novak, Tauxe, and Orvis, 1973, Medicine and Science in Sports, 5: 147-155, by permission of the American College of Sports Medicine, Madison WI)

TABLE   258

TOTAL BODY POTASSIUM (meq) IN GERMAN BOYS

Age (years)		Percentiles						
		3	10	25	50	75	90	97
8-9	W	21.8	23.6	24.5	26.8	28.1	29.3	35.2
	S	119.4	121.9	124.5	129.5	132.1	134.6	137.2
	K	1128.9	1249.36	1325.58	1401.79	1515.86	1592.33	1753.71
9-10	W	22.7	24.0	26.4	28.6	31.8	34.5	38.1
	S	121.9	127.0	129.5	134.6	137.2	139.7	144.8
	K	1164.45	1285.68	1437.08	1564.45	1712.28	1864.71	2041.94
10-11	W	24.5	27.1	29.5	31.8	35.0	40.9	49.5
	S	121.9	129.5	134.6	137.2	142.2	144.8	149.9
	K	1283.63	1439.64	1571.36	1728.64	1882.10	2021.74	2125.32
11-12	W	26.3	29.1	32.2	35.5	41.0	44.9	51.6
	S	132.1	134.6	139.7	142.2	147.3	149.9	154.9
	K	1449.10	1619.95	1765.47	1906.14	2086.19	2236.83	2463.94
12-13	W	30.5	32.7	35.4	39.0	44.6	51.3	59.9
	S	134.6	139.7	142.2	147.3	152.4	157.5	162.6
	K	1625.58	1766.50	1892.07	2099.49	2303.84	2582.35	2911.51

TABLE 258 (continued)

TOTAL BODY POTASSIUM (meq) IN GERMAN BOYS

Age (years)		3	10	25	Percentiles 50	75	90	97
13-14	W	32.9	35.8	39.5	44.9	51.6	57.3	62.3
	S	142.2	144.8	149.9	154.9	160.0	165.1	170.2
	K	1766.24	1921.74	2118.93	2367.52	2710.23	3131.20	3397.19
14-15	W	36.3	39.5	45.0	50.3	56.7	64.6	73.5
	S	144.8	147.3	154.9	160.0	167.6	170.2	177.8
	K	1921.23	2080.05	2385.68	2803.58	3138.11	3501.53	3858.57
15-16	W	42.4	48.2	52.2	57.2	63.8	71.3	78.8
	S	154.9	157.5	162.6	167.6	172.7	175.3	180.3
	K	2198.47	2585.42	2977.75	3261.38	3600.51	3954.99	4102.30
16-17	W	45.6	51.3	54.7	61.7	66.2	73.5	79.0
	S	157.5	162.6	165.1	170.2	175.3	180.3	182.9
	K	2547.06	2882.86	3249.36	3560.10	3894.12	4175.19	4444.50
17-18	W	52.7	54.9	58.5	63.7	70.5	76.2	88.6
	S	162.9	165.1	170.2	172.7	175.3	180.3	185.4
	K	2906.39	3176.21	3347.06	3707.42	4073.40	4462.66	4761.13
18-19	W	51.3	58.1	62.01	66.4	71.7	78.0	83.6
	S	157.5	165.1	167.6	172.7	177.8	182.9	185.4
	K	3002.56	3146.04	3415.35	3809.21	4162.66	4403.58	4778.77

W = weight (kg); S = stature (cm) and K = potassium (meq).

(data from Oberhausen, Burmeister, and Huycke, 1966, Annales Paediatrici 205:381-400, by permission of S. Karger AG, Basel)

TABLE 259

TOTAL BODY POTASSIUM (meq) IN GERMAN GIRLS

Age (years)					Percentiles			
		3	10	25	50	75	90	97
6-8	W	18.3	19.4	22.2	24.5	35.9	29.5	32.9
	S	106.7	114.3	119.4	121.9	127.0	129.5	134.6
	K	928.39	1015.35	1095.40	1223.79	1318.67	1418.67	1503.32
8-9	W	22.3	23.7	24.6	28.1	31.9	37.7	40.4
	S	121.9	121.9	127.0	129.5	134.6	139.7	142.2
	K	1117.65	1190.28	1253.20	1377.75	1516.62	1690.28	1826.85
10-11	W	24.0	26.8	29.0	32.3	36.7	41.4	46.8
	S	124.5	129.5	134.6	137.2	142.2	147.3	149.9
	K	1217.65	1325.06	1454.48	1593.86	1756.78	1894.37	2038.87
11-12	W	27.2	29.5	32.3	38.1	43.0	48.2	56.6
	S	132.1	137.2	139.7	144.8	149.9	154.9	157.5
	K	1399.74	1470.84	1596.93	1802.05	2017.14	2189.51	2422.25
12-13	W	30.5	34.0	37.2	42.6	49.0	56.2	62.8
	S	137.2	139.7	144.8	149.9	154.9	157.5	162.6
	K	1536.06	1656.27	1815.09	2033.50	2266.75	2460.10	2686.70
13-14	W	34.5	38.1	43.1	48.5	54.0	60.5	72.4
	S	142.2	147.3	152.4	154.9	160.0	162.6	167.6
	K	1644.25	1863.17	2063.17	2261.64	2470.84	2662.40	2839.39
14	W	39.0	41.4	46.1	50.8	55.8	62.3	69.7
	S	147.3	152.4	154.9	157.5	162.6	165.1	167.6
	K	1871.61	2011.76	2130.43	2320.20	2514.58	2687.72	2965.73
15-17	W	42.2	46.3	50.9	55.3	59.9	66.4	73.4
	S	149.9	154.9	157.5	162.6	165.1	170.2	172.7
	K	1822.76	2002.56	2223.53	2468.80	2691.30	2867.26	3053.45
17-19	W	45.4	48.8	52.7	57.6	63.0	71.2	78.5
	S	149.9	152.4	157.5	162.6	167.6	170.2	175.3
	K	2052.69	2186.45	2306.91	2491.56	2720.97	2919.44	3225.83

W = weight (kg); S = stature (cm) and K = total body potassium (meq).

(data from Oberhausen, Burmeister, and Huycke, 1966, Annales Paediatrici 205:381-400, by permission of S. Karger AG, Basel)

TABLE   260

BODY COMPOSITION VALUES FOR GERMAN
BOYS FROM TOTAL BODY POTASSIUM

Age		Percentiles		
(years)		10	50	90
8-9	W	23,6	26,8	29,3
	S	121,9	129,5	131,6
	CM	13,5	15,1	17,2
	F	1,16	3,84	5,60
9-10	W	24,0	28,6	34,5
	S	127,0	134,6	139,7
	CM	13,9	16,9	20,1
	F	1,46	3,69	7,72
10-11	W	27,1	31,8	40,9
	S	129,5	137,2	144,8
	CM	15,6	18,7	21,8
	F	2,00	4,72	10,07
11-12	W	29,1	35,5	44,9
	S	134,6	142,2	149,9
	CM	17,5	20,6	24,2
	F	2,34	5,26	12,23
12-13	W	32,7	39,0	51,3
	S	139,7	147,3	157,5
	CM	19,1	22,7	27,9
	F	2,81	6,22	14,62
13-14	W	35,8	44,9	57,3
	S	144,8	154,9	165,1
	CM	20,6	25,6	33,8
	F	3,63	7,12	13,59
14-15	W	39,5	50,3	64,6
	S	147,3	160,0	170,2
	CM	22,5	30,4	37,8
	F	3,69	7,50	15,53
15-16	W	48,2	57,2	71,3
	S	157,5	167,6	175,3
	CM	28,0	35,2	42,7
	F	2,98	8,97	17,50

TABLE 260 (continued)

BODY COMPOSITION VALUES FOR GERMAN
BOYS FROM TOTAL BODY POTASSIUM

Age		Percentiles		
(years)		10		90
16-17	W	51,3	61,7	73,5
	S	162,6	170,2	180,3
	CM	31,2	38,5	45,1
	F	2,69	7,46	15,00
17-18	W	54,9	63,7	76,2
	S	165,1	172,7	180,3
	CM	34,2	40,1	48,3
	F	4,07	8,77	16,66
18-19	W	58,1	66,4	78,0
	S	165,1	172,7	182,9
	CM	34,0	41,2	47,6
	F	4,75	9,85	18,47

W = weight (kg); S = stature (cm); CM = lean body mass
(kg) and F = total body fat (kg).

(data from Burmeister and Bingert, 1967)

TABLE 261

BODY COMPOSITION VALUES FOR GERMAN
GIRLS FROM TOTAL BODY POTASSIUM

Age (years)		Percentiles		
		10	50	90
8-10	W	23,7	28,1	37,7
	S	121,9	129,5	139,7
	CM	12,9	14,9	18,3
	F	2,35	5,94	11,50
10-11	W	26,8	32,3	41,4
	S	129,5	137,2	147,3
	CM	14,3	17,2	20,5
	F	2,93	6,91	13,85
11-12	W	29,5	38,1	48,2
	S	137,2	144,8	154,9
	CM	15,9	19,5	23,7
	F	2,58	7,78	16,68
12-13	W	34,0	42,6	56,2
	S	139,7	149,9	157,5
	CM	17,9	21,9	26,6
	F	5,28	10,31	18,63
13-14	W	38,1	48,5	60,5
	S	147,3	154,9	162,6
	CM	20,1	24,4	28,8
	F	6,76	12,15	21,54
14-15	W	41,4	50,8	62,3
	S	152,4	157,5	165,1
	CM	21,8	25,1	29,1
	F	8,06	13,85	22,50
15-17	W	46,3	55,3	66,4
	S	154,9	162,6	170,2
	CM	21,6	26,8	31,0
	F	11,20	16,06	23,35

W = weight (kg); S = stature (cm); CM = lean body mass (kg)
and F = total body fat (kg).

(data from Burmeister and Bingert, 1967)

TABLE   262

MEAN TOTAL BODY WATER (ℓ) IN NORMAL MASSACHUSETTS INFANTS
AND CHILDREN

				Total Body Water	
Age	Weight (kg)	Surface area (m^2)	Liters	Per cent body weight	L. per m^2
Infants, under 1 month of age					
14 Days	3.16	0.198	2.42	76.7	12.1
1 to 12 months of age					
5.2 Months	6.94	0.328	4.27	62.6	12.9
Children, 1 to 9 years of age					
4.5 Years	16.6	0.65	9.77	58.9	14.2
Adolescents, 10 to 16 years of age					
Males					
13.0 Years	46.8	1.42	27.9	59.0	19.2
Females					
12.8 Years	44.6	1.42	25.0	56.2	17.6

(data from Edelman, Haley, Schloerb, et al., 1952. By permission of
SURGERY, GYNECOLOGY & OBSTETRICS)

TABLE   263

TOTAL BODY WATER (ℓ), TOTAL BODY WATER (%), TOTAL BODY SOLIDS (kg), AND
TOTAL BODY SOLIDS (%) OF MINNESOTA CHILDREN
AGES 6 TO 7 YEARS

Compartment	Boys		Girls	
	Mean	S.D.	Mean	S.D.
Total body water (ℓ)	15.16	1.92	13.53	1.94
Total body water (%)	65.42	3.54	61.67	3.82
Total body solids (kg)	8.04	1.42	8.48	1.78
Total body solids (%)	34.58	3.47	38.33	3.81

(data from Novak, 1966, Pediatrics 38:483-489. Copyright American Academy of Pediatrics
1966)

TABLE   264

TOTAL BODY WATER FOR BLACK BOYS IN TEXAS

| Age (years) | Total body water (liters) | | | | Total body water as a percent of weight | | | |
| | Lower Income | | Middle Income | | Lower Income | | Middle Income | |
	Mean	S.D.	Mean	S.D.	Mean	S.D.	Mean	S.D.
10.1 - 11	21.8	2.65	20.6	2.31	62.3	3.93	61.2	2.67
11.1 - 12	22.2	1.52	23.7	3.25	60.5	3.62	58.2	4.55
12.1 - 13	24.5	4.05	28.1	5.09	58.4	5.45	61.2	5.13
13.1 - 14	29.5	4.67	29.9	6.50	63.9	4.28	60.5	5.63
14.1 - 15	29.4	5.71	34.4	6.37	61.4	2.07	62.6	3.84
15.1 - 16	36.4	6.83	37.3	3.21	--	--	--	--
16.1 - 17	40.4	3.25	40.6	3.89	63.1	4.33	--	--

(data from Schutte, 1979)

TABLE   265

VOLUME OF BODY WATER AND PERCENTAGE BODY FAT OF
IGLOOLIK ESKIMOS AS ESTIMATED BY A DEUTERIUM DILUTION METHOD

Age (years)	Sex	Volume of Body Water	Body Fat (%)
15-19	Boys	39.3	12.7
	Girls	31.8	19.5

(data from "Work physiology and activity patterns of circumpolar
Eskimos and Ainu," Human Biology 46:263-294, 1974, by R. J.
Shephard, by permission of the Wayne State University Press.
Copyright 1974, Wayne State University Press, Detroit, Michigan
48202)

TABLE 266

PUBERTY AND BODY MASS IN CALIFORNIA CHILDREN

Stage	Age (years)	Body Weight (kg) Mean	S.D.	TBW Absolute (liters) Mean	S.D.	% of Body Weight Mean	S.D.	BF Absolute (kg) Mean	S.D.	% of Body Weight Mean	S.D.	Gonadotropin Excretion (IU/24 hr) FSH Mean	S.D.	LH Mean	S.D.
						Girls									
I	9-10	32.3	4.9	19.7	3.1	61.1	3.5	4.9	2.0	15.2	4.9	2.6	0.5	3.0	0.5
II	9-10	32.3	4.8	18.7	1.9	58.5	4.2	6.3	2.9	17.2	4.1	4.7	0.9	5.6	1.2
						Boys									
I	9-10	33.1	3.9	19.8	2.1	61.2	3.1	5.0	1.5	15.1	4.2	2.8	0.7	3.8	1.2

TBW = total body water; BF = body fat; FSH = follicle stimulating hormone; and LH = luteinizing hormone

(data from Penny, Goldstein and Frasier, 1978, Pediatrics 61:294-299. Copyright American Academy of Pediatrics 1978)

TABLE 267

BODY COMPOSITION AND GONADROTROPIN EXCRETION BY STAGE
OF SEXUAL DEVELOPMENT IN CALIFORNIA CHILDREN

Stage	TBW Absolute (liters) Mean	S.D.	% of Body Weight Mean	S.D.	BF Absolute (kg) Mean	S.D.	% of Body Weight Mean	S.D.	Gonadotropin Excretion (IU/24 hr) FSH Mean	S.D.	LH Mean	S.D.
					Normal Girls							
I	13.8	1.7	60.5	2.6	4.0	2.3	15.7	4.2	2.5	0.7	2.7	1.1
II	21.6	3.5	58.6	4.9	7.4	3.9	18.9	6.7	5.2	1.3	6.2	1.7
III	24.3	3.4	56.6	5.5	9.8	4.7	21.6	7.5	8.2	2.0	11.2	3.3
IV	28.0	2.4	52.7	2.7	14.4	3.0	26.7	3.8	8.4	2.1	12.0	5.5
					Normal Boys							
I	15.5	2.0	61.8	2.0	3.8	1.0	14.3	2.7	2.3	1.0	2.9	1.1
II	25.0	4.0	63.8	2.4	4.5	1.7	11.2	3.5	4.4	1.9	7.3	2.9
III	29.4	3.5	64.2	2.1	5.1	1.9	10.8	2.9	6.1	2.2	11.3	4.5
IV	36.0	3.8	64.0	2.1	6.3	1.8	11.2	2.9	7.4	2.0	14.5	4.6
V	39.7	3.3	63.9	2.6	7.3	2.9	11.3	3.5	9.5	2.9	19.4	7.3

TBW = total body water; BF = body fat; FSH = follicle stimulating hormone; and LH = luteinizing hormone

(data from Penny, Goldstein and Frasier, 1978, Pediatrics 61:294-299. Copyright American Academy of Pediatrics 1978)

TABLE   268

BODY COMPOSITION AND GONADOTROPIN EXCRETION BY AGE
IN CALIFORNIA CHILDREN

| Age (years) | TBW | | | | BF | | | | Gonadotropin Excretion (IU/24 hr) | | | |
| | Absolute (Liters) | | % of Body Weight | | Absolute (kg) | | % of Body Weight | | FSH | | LH | |
	Mean	S.D.	Mean	S.D.	Mean	S.D.	Mean	S.D.	Mean	S.D.	Mean	S.D.
					Normal Girls							
3-4	9.0	0.9	59.6	0.8	2.5	0.4	17.2	0.9	2.0	0.5	2.0	0.4
5-6	13.1	2.0	61.6	3.1	3.3	1.3	14.6	4.5	2.5	0.8	2.6	1.6
7-8	16.1	1.7	62.2	2.7	3.6	1.1	13.6	3.7	2.9	1.0	3.0	1.0
9-10	19.4	2.5	59.6	3.8	5.7	1.3	17.1	5.3	4.0	1.9	4.8	2.8
11-12	23.8	3.2	56.0	5.0	10.0	4.6	21.8	7.3	7.0	2.7	8.8	3.9
13-14	26.0	2.9	54.5	4.3	12.4	4.6	24.6	6.0	7.6	2.2	11.9	5.5
15-16	29.2	2.8	52.7	2.9	15.5	4.1	26.8	4.0	8.1	1.8	11.5	3.8
					Normal Boys							
3-4	10.7	1.5	62.6	1.6	2.3	0.6	13.1	2.3	1.8	1.1	2.1	1.2
5-6	13.9	1.9	61.3	1.8	3.5	1.1	14.9	2.3	2.0	0.8	2.5	1.0
7-8	16.7	2.7	61.8	1.4	3.9	0.9	14.2	2.0	2.4	1.0	3.2	1.3
9-10	20.7	2.3	61.8	2.8	4.7	1.6	14.1	4.0	3.2	1.0	4.4	1.6
11-12	26.1	3.3	63.7	2.2	4.8	1.5	11.6	3.1	4.4	2.2	7.4	4.7
13-14	32.2	5.5	64.0	2.4	5.9	2.4	11.0	3.4	6.4	2.4	12.0	4.4
15-16	37.2	4.4	64.0	2.4	6.7	2.4	11.2	3.3	8.7	2.9	19.1	9.7

TBW = total body water; BF = body fat; FSH = follicle stimulating hormone and
LH = luteinizing hormone.

(data from Penny, Goldstein and Frasier, 1978, Pediatrics 61:294-299. Copyright
American Academy of Pediatrics 1978)

TABLE 269

ESTIMATED SKELETAL WEIGHT (gm) FOR BOYS IN
THE TEN-STATE NUTRITION SURVEY

Age (years)	White Percentiles			Black Percentiles		
	15	50	85	15	50	85
1	170.99	252.17	373.34	179.36	254.60	325.99
2	255.23	378.36	487.84	310.68	411.01	539.72
3	339.43	465.58	620.75	400.12	502.68	627.85
4	438.23	571.97	720.67	534.73	643.24	861.27
5	517.19	672.27	830.91	653.23	802.64	997.41
6	620.84	789.42	1005.47	755.57	950.74	1181.17
7	714.04	923.35	1179.64	860.84	1052.15	1371.13
8	862.42	1066.70	1408.97	991.69	1230.30	1583.21
9	1005.59	1254.75	1581.74	1098.60	1393.30	1692.04
10	1139.46	1396.60	1724.13	1241.95	1553.85	2017.61
11	1248.06	1570.60	2002.99	1404.58	1786.30	2329.38
12	1440.14	1837.60	2412.48	1593.04	1940.40	2406.48
13	1769.48	2238.40	3081.17	1759.02	2344.30	3008.79
14	1916.37	2592.20	3681.48	2079.50	2698.60	3525.65
15	2514.78	3108.30	3822.86	2261.71	3091.50	4090.10
16	2616.07	3495.90	4142.77	3543.06	3690.60	4475.46
17	2988.03	3437.60	4204.18	3139.75	3854.90	4669.26

(data from Garn et al., 1976)

TABLE 270

ESTIMATED SKELETAL WEIGHT (gm) IN GIRLS
IN THE TEN-STATE NUTRITION SURVEY

Age (years)	White Percentiles			Black Percentiles		
	15	50	85	15	50	85
1	158.53	232.61	360.32	129.72	202.36	360.04
2	272.31	348.47	454.80	289.56	388.14	468.11
3	340.08	431.53	541.64	432.25	544.20	661.32
4	434.02	531.74	736.54	496.28	671.51	879.75
5	465.74	658.96	831.30	601.87	750.53	936.47
6	608.80	770.56	933.25	698.07	890.52	1171.28
7	702.13	896.36	1140.97	789.70	1019.36	1291.24
8	816.58	1044.40	1317.62	996.06	1226.30	1596.46
9	938.00	1191.45	1482.02	1088.99	1413.60	1866.83
10	1057.19	1433.20	1737.52	1214.45	1584.10	2117.31
11	1294.57	1623.55	2087.05	1487.15	1964.30	2563.30
12	1464.89	1950.10	2419.90	1753.74	2241.96	2912.27
13	1704.88	2206.90	2773.84	2139.30	2446.30	3084.40
14	2003.56	2441.60	3046.40	2333.83	2714.70	3324.90
15	2112.43	2581.40	3006.19	2375.94	2877.40	3448.93
16	2243.02	2610.35	3118.53	2342.02	2839.40	3589.96
17	2201.33	2605.30	3243.59	2504.34	2942.40	3521.58

(data from Garn et al., 1976)

TABLE 271

ESTIMATED SKELETAL WEIGHT (gm) FOR BOYS
WITH A POVERTY INCOME RATIO OF 0.0 TO 1.49
IN THE   TEN-STATE NUTRITION SURVEY

Age (years)	White Mean	White S.D.	Black Mean	Black S.D.	Mexican-American Mean	Mexican-American S.D.
1	--	--	259.9	108.4	--	--
2	359.6	93.0	--	--	--	--
3	503.0	143.3	500.4	121.0	--	--
4	479.1	163.1	673.3	164.9	--	--
5	680.9	153.1	803.5	154.5	--	--
6	781.2	162.2	972.8	199.3	886.7	219.6
7	957.2	236.2	1120.1	289.2	1002.2	191.1
8	1075.5	265.2	1287.1	305.5	1072.1	271.1
9	1229.0	243.6	1378.6	298.7	--	--
10	1393.8	292.9	1547.7	336.0	1263.9	216.4
11	1654.0	366.8	1794.9	437.1	1665.8	229.6
12	1923.0	481.8	1978.3	440.0	1811.4	353.8
13	2325.6	633.9	2432.6	646.7	2025.9	594.4
14	2611.9	707.9	2756.0	696.8	2474.4	681.8
15	3183.2	610.6	3022.4	817.1	--	--
16	3532.1	876.0	3572.9	793.1	--	--
17	3752.4	668.2	3861.4	781.1	--	--

Poverty income ratio corresponding to 00 to $750 per capita per annum.

(data from Garn et al., 1976)

TABLE 272

ESTIMATED SKELETAL WEIGHT (gm) FOR BOYS WITH A
POVERTY INDEX RATIO OF 1.5 AND OVER

Age (years)	White		Black	
	Mean	S.D.	Mean	S.D.
2	382.9	94.8	--	--
3	433.4	137.4	--	--
4	575.2	119.1	--	--
5	653.8	141.4	837.7	151.2
6	826.8	186.0	877.2	88.8
7	930.2	202.6	1136.6	226.1
8	1155.8	270.9	1267.7	272.9
9	1305.6	273.9	1394.2	231.6
10	1430.5	295.9	1779.9	326.8
11	1613.9	392.0	2073.9	530.9
12	1896.4	466.7	2015.3	309.2
13	2451.3	703.4	2385.5	739.6
14	2631.7	873.6	2669.0	553.7
15	3131.4	768.5	3643.2	1138.3
16	3450.1	653.6	4064.1	754.6
17	3577.8	608.8	--	--

Poverty index ratio corresponding to $1170 per capita and over per
annum; children from the Ten-State Nutrition survey.

(data from Garn et al., 1976)

TABLE   273

ESTIMATED SKELETAL WEIGHT (gm) FOR GIRLS WITH A
POVERTY INCOME RATIO OF 0.0 TO 1.49
IN THE TEN-STATE NUTRITION SURVEY

Age (years)	White Mean	White S.D.	Black Mean	Black S.D.	Mexican-American Mean	Mexican-American S.D.
1	--	--	228.0	127.6	--	--
2	--	--	389.6	97.3	--	--
3	423.5	107.7	562.9	119.8	--	--
4	552.7	150.1	642.0	159.3	--	--
5	633.3	180.6	787.8	211.6	678.2	138.6
6	764.3	145.7	902.9	198.2	839.3	232.6
7	910.2	218.6	1020.1	237.7	895.6	231.8
8	1071.9	298.7	1222.9	248.3	1038.5	292.9
9	1201.1	294.7	1414.6	377.0	1120.1	343.2
10	1348.0	314.3	1653.2	412.9	1326.5	214.6
11	1625.3	452.0	2012.1	526.8	1496.9	392.5
12	1898.5	458.0	2294.2	563.3	2014.1	508.1
13	2188.7	473.0	2474.6	504.8	2701.2	629.4
14	2530.7	540.2	2780.5	427.6	2569.8	456.8
15	2604.3	458.0	2894.2	458.1	2464.5	527.3
16	2583.7	413.2	2858.8	517.2	--	--
17	2593.1	469.6	2991.6	535.0	--	--

Poverty income ratio corresponding to 00 to $750 per capita per annum.

(data from Garn et al., 1976)

TABLE 274

ESTIMATED SKELETAL WEIGHT (gm) IN GIRLS WITH A
POVERTY INCOME RATIO OF 1.5 AND OVER

Age (years)	White		Black	
	Mean	S.D.	Mean	S.D.
2	361.5	92.6	--	--
3	436.8	96.1	--	--
4	556.7	136.5	786.6	185.1
5	669.0	165.4	--	--
6	790.1	181.9	1013.8	266.9
7	960.2	218.1	1126.4	255.6
8	1097.0	241.1	1353.8	315.2
9	1191.6	243.5	1419.4	252.1
10	1432.9	311.6	1811.6	452.4
11	1768.2	398.8	1932.9	532.1
12	1948.5	449.6	2299.3	453.7
13	2356.4	478.4	2580.0	523.0
14	2471.9	566.5	2883.9	719.5
15	2611.1	533.8	--	--
16	2679.9	465.3	--	--
17	2715.5	483.5	--	--

Poverty income ratio corresponding to $1170 per capita per annum;
children from the Ten-State Nutrition survey.

(data from Garn et al., 1976)

TABLE 275

COMPOSITION OF CORTICAL BONE IN NEW YORK
CHILDREN AGED 2 TO 19 YEARS

Parameter	Mean	S.D.
Molar Ratio Ca/P	1.62	± 0.075
Specific Gravity	1.84	± 0.058
Calcium as % of Ash	37.9	± 1.7

Constituent	G/100 G	G/100 cc.
Calcium	20.8 ± 0.83	38.2 ± 2.8
Phosphorus	10.2 ± 0.47	18.7 ± 1.5
Ash	54.7 ± 2.1	99.9 ± 6.8
Water	15.4 ± 2.2	28.8 ± 2.6
Protein	25.7 ± 0.77	46.5 ± 2.6
Weight not accounted for	4.0	7.3

(data from Woodard, 1964, Clinical Orthopedics 37:
187-193, by permission of Lippincott/Harper Co.)

TABLE   276

BONE MINERAL MEASUREMENTS OF WHITE SCHOOL CHILDREN IN WISCONSIN

Age	Mineral		Radius Width		M/W	
(years)	F	M	F	M	F	M
Means						
6	436	466	9.1	9.5	47.5	48.6
7	457	510	9.1	10.0	49.9	50.9
8	490	557	9.4	10.2	52.2	54.3
9	542	584	9.7	10.5	55.6	55.4
10	565	633	9.9	11.1	56.6	57.3
11	645	691	10.8	11.3	59.6	61.0
12	716	763	11.3	12.0	63.1	62.9
13	742	781	11.5	12.6	64.0	61.6
Coefficients of variation						
6	18	11	12	9	10	8
7	12	14	10	8	8	8
8	12	15	9	10	9	9
9	13	12	10	10	6	7
10	17	13	11	10	9	13
11	18	16	11	12	12	8
12	12	16	10	11	8	9
13	13	17	10	12	8	8
Humerus						
Means						
6	943	1018	13.7	14.4	69.2	70.9
7	1001	1109	14.3	14.7	69.8	75.7
8	1106	1249	14.9	15.6	74.2	80.2
9	1216	1276	15.7	15.9	77.3	79.9
10	1226	1442	15.8	16.8	77.4	85.7
11	1354	1478	16.5	17.2	81.7	85.9
12	1533	1662	17.3	19.0	88.1	87.1
13	1627	1624	18.2	18.0	89.4	90.6
Coefficients of variation						
6	11	12	13	12	10	8
7	11	15	8	14	7	7
8	14	12	9	11	8	7
9	11	13	9	8	7	9
10	12	14	9	10	8	9
11	18	10	11	8	12	8
12	16	13	12	9	9	8
13	12	14	9	11	9	13

TABLE 276   (continued)

BONE MINERAL MEASUREMENTS OF WHITE SCHOOL CHILDREN IN WISCONSIN

Age	Mineral		Ulna Width		M/W	
(years)	F	M	F	M	F	M
Means						
6	354	343	7.6	8.1	46.6	42.1
7	374	411	8.2	8.6	45.9	48.2
8	413	480	8.8	9.1	47.2	52.7
9	481	475	8.4	9.2	57.5	51.6
10	473	554	8.6	9.5	54.6	58.6
11	540	567	9.6	9.8	55.0	58.0
12	--	595	--	10.4	--	57.0
Coefficients of variation						
6	16	10	7	7	15	8
7	14	12	9	13	12	9
8	17	12	9	11	17	11
9	9	15	13	10	7	12
10	16	14	12	7	9	12
11	26	10	12	14	16	9
12	--	10	--	7	--	8

Mineral in mg; bone widths in mm; ratios in mg/mm.

(data from Mazess and Cameron, 1972)

TABLE   277

AGE-ADJUSTED MEAN DENSITIES OF SELECTED BONES
ACCORDING TO RACE-SEX GROUPS

Age Group	Bone	White Male	White Female	Negro Male	Negro Female
Young[1] (Adjusted to geometric mean age of 5.0 years)	Humerus	0.77	0.76	0.84	0.82
	Radius	0.91	0.87	0.95	0.97
	Femur	0.68	0.64	0.74	0.73
	Tibia	0.67	0.65	0.72	0.71

[1]For the young series entries are geometric means since analyses of densities were made in logarithms. Adapted from Trotter, M., 1971, Growth, 35: 221-231.

(data from Trotter and Hixon, 1974)

TABLE 278

BONE AREA IN RADIOGRAPHS OF THE CALF IN BOSTON CHILDREN

Age	Boys Percentiles			Girls Percentiles		
(years)	10	50	90	10	50	90
6	51.0	57.6	67.4	49.9	58.0	67.2
7	57.1	65.4	76.6	54.8	63.8	76.8
8	64.8	73.4	86.0	60.3	71.6	84.8
9	70.5	81.3	96.0	65.4	80.0	93.3
10	73.0	87.2	103.0	68.5	86.2	99.3

(data from Stuart and Dwinell, 1942)

TABLE 279

BONE MINERAL MEASUREMENTS BY PHOTON ABSORPTIOMETRY OF THE
DISTAL RADIUS IN CANADIAN ESKIMO CHILDREN

Age	Mineral (mg/cm)		Width ($m \times 10^{-5}$)		M/W	
(years)	Mean	S.D.	Mean	S.D.	Mean	S.D.
MALES						
5-7	480	142	1516	296	.316	.057
8-9	472	74	1454	278	.330	.043
10-11	589	102	1727	241	.342	.045
12-14	749	152	1994	299	.376	.049
15-16	988	138	2384	293	.416	.052
17-19	1234	190	2538	258	.485	.047
FEMALES						
5-7	407	112	1359	223	.298	.051
8-9	466	118	1454	253	.317	.038
10-11	523	72	1715	247	.307	.032
12-14	702	189	1923	336	.363	.051
15-16	912	107	2112	168	.434	.045
17-19	973	147	2128	287	.461	.069

(data from "Bone mineral content in Canadian Eskimos," Human Biology 47:45-
63, 1975, by Mazess and Mather, by permission of the Wayne State University
Press. Copyright 1975, Wayne State University Press, Detroit, Michigan
48202)

TABLE   280

BONE MINERAL MEASUREMENTS BY PHOTON ABSORPTIOMETRY OF THE
MIDSHAFT OF THE RADIUS IN CANADIAN ESKIMO CHILDREN

Age (years)	Mineral (mg/cm)		Width (m x 10^{-5})		M/W	
	Mean	S.D.	Mean	S.D.	Mean	S.D.
MALES						
5-7	438	44	929	57	.471	.031
8-9	477	42	948	83	.505	.033
10-11	574	58	1066	90	.538	.035
12-14	710	120	1170	88	.604	.063
15-16	881	129	1293	116	.678	.053
17-19	1055	155	1409	128	.746	.048
FEMALES						
5-7	382	67	862	70	.441	.059
8-9	432	44	909	75	.475	.021
10-11	493	47	949	65	.519	.032
12-14	673	96	1094	145	.615	.042
15-16	814	57	1199	65	.679	.041
17-19	823	81	1186	96	.694	.036

TABLE 281

BONE MINERAL MEASUREMENTS BY PHOTON ABSORPTIOMETRY OF THE
MIDSHAFT OF THE ULNA IN CANADIAN ESKIMO CHILDREN

| Age | Mineral (mg/cm) | | Width (m x 10^{-5}) | | M/W | |
(years)	Mean	S.D.	Mean	S.D.	Mean	S.D.
MALES						
5-7	372	48	851	64	.437	.043
8-9	427	40	916	87	.469	.048
10-11	493	46	992	80	.498	.040
12-14	612	91	1049	92	.584	.075
15-16	743	139	1094	54	.680	.130
17-19	920	143	1204	74	.762	.095
FEMALES						
5-7	333	65	799	89	.414	.055
8-9	385	67	892	148	.433	.033
10-11	431	36	905	71	.477	.033
12-14	597	87	966	75	.617	.069
15-16	696	59	957	60	.728	.054
17-19	737	85	981	76	.749	.050

(data from "Bone mineral content in Canadian Eskimos," Human Biology 47:45-63, 1975, by Mazess and Mather, by permission of the Wayne State University Press. Copyright 1975, Wayne State University Press, Detroit, Michigan 48202)

TABLE   282

BONE MINERAL MEASUREMENTS BY PHOTON ABSORPTIOMETRY OF THE
DISTAL ULNA IN CANADIAN ESKIMO CHILDREN

Age	Mineral (mg/cm)		Width (m x 10^{-5})		M/W	
(years)	Mean	S.D.	Mean	S.D.	Mean	S.D.
MALES						
5-7	266	46	856	104	.311	.048
8-9	293	51	879	154	.337	.045
10-11	321	41	953	119	.339	.044
12-14	414	72	1083	147	.386	.067
15-16	518	102	1216	273	.437	.100
17-19	633	121	1217	137	.519	.070
FEMALES						
5-7	228	47	799	139	.287	.046
8-9	254	52	856	184	.298	.030
10-11	284	39	963	132	.298	.044
12-14	570	83	1009	194	.369	.061
15-16	470	50	1047	112	.451	.046
17-19	515	74	1003	117	.514	.062

(data from "Bone mineral content in Canadian Eskimos," Human Biology 47:45-
63, 1975, by Mazess and Mather, by permission of the Wayne State University
Press. Copyright 1975, Wayne State University Press, Detroit, Michigan
48202)

TABLE 283

BONE MINERAL CONTENT (BMC) OF THE OS CALCIS IN GERMAN CHILDREN

BOYS

Age (years)	BMC		C-WIDTH	C-THICKNESS	STATURE	WEIGHT
	M± SD (mg/cm^3)	M±SD (g/cm)	M ± SD (mm)	M ± SD (mm)	M ± SD (cm)	M±SD (kg)
3- 4	145 ± 24	0.82 ± 0.23	32.0 ±1.8	24.8± 1.9	108 ± 4.0	16.8 ± 2.1
5- 6	176 ± 36	1.07 ± 0.21	35.5 ±1.6	25.6± 4.0	117 ± 8.2	21.3 ± 2.5
7- 8	188 ± 20	1.48 ± 0.26	39.3 ± 2.8	29.2 ± 2.5	132 ± 5.9	27.0 ± 4.2
9-10	195 ± 28	1.83 ± 0.39	43.6 ± 5.3	32.0 ± 3.3	139 ± 9.2	34.4 ± 7.1
11-12	206 ± 26	2.23 ± 0.38	47.3 ± 4.3	33.9 ± 2.6	152 ± 5.6	42.3 ± 5.9
13-16	222 ± 31	2.68 ± 0.40	49.8 ± 2.6	36.3 ± 2.7	161 ± 9.6	51.0 ± 9.1

GIRLS

Age (years)	BMC		C-WIDTH	C-THICKNESS	STATURE	WEIGHT
	M± SD (mg/cm^3)	M±SD (g/cm)	M ± SD (mm)	M ± SD (mm)	M ± SD (cm)	M ± SD (kg)
3- 4	165 ± 34	0.90 ± 0.18	33.0 ± 4.1	24.0 ± 3.7	108 ± 5.4	17.7 ± 1.0
5- 6	172 ± 37	1.05 ± 0.25	34.3 ± 4.0	26.2 ± 3.3	118 ± 7.0	21.4 ± 3.7
7- 8	186 ± 26	1.33 ± 0.28	37.1 ± 3.6	27.9 ± 2.6	129 ± 6.6	26.9 ± 4.1
9-10	195 ± 24	1.74 ± 0.32	43.1 ± 3.1	30.8 ± 1.9	142 ± 6.5	32.6 ± 4.1
11-12	221 ± 36	2.29 ± 0.42	45.7 ± 4.1	33.5 ± 1.8	149 ±12.6	37.5 ± 6.0
13-16	236 ± 34	2.88 ± 0.50	49.3 ± 1.2	36.3 ± 2.7	165 ± 8.3	50.3 ± 8.0
17-20	239 ± 35	2.80 ± 0.38	48.7 ± 3.4	36.7 ± 1.8	165 ± 7.5	53.2 ± 6.7

c = cortical. BMC measurements made by photon aborptiometry.

(data from Klemm, Bamzer and Schneider, 1976, American Journal of Roentgenology 126:1283-1284. Copyright 1976, American Roentgen Ray Society)

TABLE   284

METACARPAL II SIZE (mm) IN WHITE BOYS
IN THE TEN-STATE NUTRITION SURVEY

Age (years)	Length		Diameters					
			Total		Medullary		Cortical	
	Mean	S.D.	Mean	S.D.	Mean	S.D.	Mean	S.D.
1	24.62	2.30	4.59	.52	2.85	.67	1.74	.47
2	30.19	2.71	5.09	.52	3.20	.73	1.88	.47
3	34.22	2.49	5.33	.58	3.33	.70	2.00	.48
4	37.02	2.70	5.41	.45	3.10	.63	2.31	.47
5	39.99	2.89	5.57	.51	3.12	.61	2.54	.43
6	42.72	3.02	5.83	.54	3.23	.63	2.61	.43
7	45.59	3.14	6.02	.60	3.20	.69	2.82	.52
8	47.84	3.43	6.34	.68	3.34	.74	3.00	.49
9	50.37	3.25	6.60	.61	3.42	.70	3.18	.50
10	52.06	3.84	6.83	.58	3.47	.68	3.35	.52
11	54.43	3.89	7.11	.67	3.63	.78	3.47	.57
12	57.19	4.40	7.52	.73	3.83	.82	3.69	.65
13	61.33	4.98	7.87	.78	3.77	.77	4.10	.69
14	63.96	5.19	8.21	.90	3.90	.75	4.31	.84
15	66.86	4.58	8.56	.78	3.81	.82	4.75	.77
16	68.36	4.35	8.84	.78	3.92	.81	4.92	.77
17	68.36	3.96	8.90	.75	3.69	1.04	5.21	.75

(data from Garn et al., 1976)

TABLE 285

METACARPAL II SIZE (mm) IN BLACK BOYS
IN THE TEN-STATE NUTRITION SURVEY

Age (years)	Length		Diameters					
			Total		Medullary		Cortical	
	Mean	S.D.	Mean	S.D.	Mean	S.D.	Mean	S.D.
1	26.79	2.99	4.83	.41	3.35	.49	1.48	.44
2	32.41	3.30	5.21	.52	3.35	.62	1.86	.37
3	35.64	2.96	5.37	.53	3.21	.79	2.16	.48
4	39.60	3.35	5.84	.57	3.52	.71	2.32	.46
5	43.03	3.34	5.97	.56	3.45	.70	2.52	.42
6	45.40	3.12	6.31	.57	3.62	.68	2.68	.41
7	47.99	3.45	6.46	.61	3.55	.68	2.90	.50
8	50.60	3.77	6.71	.66	3.65	.70	3.05	.44
9	52.46	3.76	6.91	.64	3.77	.72	3.14	.46
10	54.76	3.76	7.14	.67	3.81	.69	3.33	.43
11	57.38	4.32	7.49	.72	4.03	.70	3.46	.51
12	59.63	4.42	7.72	.61	4.18	.73	3.54	.54
13	62.90	4.93	8.08	.82	4.18	.83	3.89	.69
14	66.16	5.17	8.44	.80	4.31	.83	4.13	.72
15	68.33	5.08	8.68	.91	4.30	.87	4.39	.75
16	71.51	4.12	9.03	.85	4.22	.85	4.81	.72
17	71.78	4.64	9.20	.82	4.16	.94	5.04	.73

(data from Garn et al., 1976)

TABLE   286

METACARPAL II SIZE (mm) IN MEXICAN-AMERICAN BOYS
IN THE TEN-STATE NUTRITION SURVEY

| Age (years) | Length | | Diameters | | | | | |
| | | | Total | | Medullary | | Cortical | |
	Mean	S.D.	Mean	S.D.	Mean	S.D.	Mean	S.D.
4	38.21	2.73	5.63	.55	3.15	.64	2.47	.28
5	40.28	2.96	5.50	.60	2.97	.62	2.53	.33
6	43.59	2.70	5.90	.61	3.12	.59	2.78	.35
7	45.79	2.88	6.20	.56	3.32	.71	2.88	.35
8	47.91	3.65	6.33	.49	3.36	.61	2.96	.41
9	48.61	3.04	6.45	.61	3.46	.61	2.99	.34
10	51.77	3.11	6.84	.69	3.57	.75	3.28	.49
11	53.48	3.41	6.94	.57	3.40	.74	3.54	.46
12	55.78	4.64	7.37	.75	3.60	.76	3.77	.48
13	59.00	4.25	7.62	.98	3.64	1.03	3.98	.57
14	62.41	4.98	7.83	.70	3.54	.79	4.30	.72
15	66.46	4.41	8.46	.63	3.92	.69	4.55	.48
17	64.60	3.07	8.45	.81	3.06	.43	5.39	.65

(data from Garn et al., 1976)

TABLE   287

METACARPAL II SIZE (mm) IN WHITE GIRLS
IN THE TEN-STATE NUTRITION SURVEY

Age	Length		Diameters					
(years)			Total		Medullary		Cortical	
	Mean	S.D.	Mean	S.D.	Mean	S.D.	Mean	S.D.
1	25.96	3.52	4.34	.57	2.72	.48	1.62	.31
2	31.77	2.23	4.76	.42	2.91	.61	1.85	.41
3	34.68	2.60	4.99	.49	2.97	.69	2.01	.45
4	37.63	3.55	5.21	.59	2.91	.68	2.30	.45
5	40.66	3.02	5.38	.56	2.92	.60	2.46	.47
6	43.13	3.10	5.54	.50	2.86	.61	2.68	.44
7	45.81	2.91	5.83	.58	2.95	.64	2.88	.46
8	48.22	3.27	6.07	.67	2.96	.67	3.11	.54
9	50.38	3.87	6.26	.60	3.00	.68	3.27	.55
10	53.36	3.89	6.57	.62	3.13	.67	3.43	.51
11	56.20	4.08	6.89	.62	3.17	.70	3.72	.64
12	58.81	4.13	7.19	.66	3.21	.74	3.98	.65
13	61.20	4.55	7.43	.65	3.07	.73	4.36	.66
14	62.82	3.83	7.64	.75	2.92	.88	4.72	.73
15	63.13	3.34	7.74	.64	2.88	.74	4.86	.64
16	63.78	3.58	7.85	.60	3.00	.83	4.85	.64
17	63.92	3.31	7.83	.70	2.88	.97	4.95	.66

(data from Garn et al., 1976)

TABLE 288

METACARPAL II SIZE (mm) IN BLACK GIRLS
IN THE TEN-STATE NUTRITION SURVEY

| Age (years) | Length | | Diameters | | | | | |
| | | | Total | | Medullary | | Cortical | |
	Mean	S.D.	Mean	S.D.	Mean	S.D.	Mean	S.D.
1	25.39	4.69	4.11	.50	2.57	.29	1.54	.51
2	32.83	2.09	4.88	.60	2.99	.73	1.90	.38
3	36.91	2.72	5.38	.51	3.17	.66	2.21	.49
4	40.63	4.46	5.56	.64	3.08	.69	2.49	.60
5	42.83	3.49	5.68	.61	3.08	.78	2.60	.46
6	45.66	3.72	5.93	.62	3.12	.72	2.81	.48
7	48.12	3.48	6.09	.60	3.13	.67	2.96	.48
8	50.81	3.31	6.42	.58	3.24	.62	3.18	.47
9	53.45	4.12	6.78	.73	3.47	.77	3.31	.55
10	56.46	4.22	6.97	.70	3.48	.73	3.50	.59
11	59.93	4.53	7.36	.70	3.50	.79	3.86	.71
12	62.32	4.49	7.71	.74	3.62	.83	4.09	.65
13	64.65	4.57	7.84	.66	3.48	.83	4.36	.69
14	65.52	4.04	8.16	.67	3.53	.93	4.63	.70
15	65.95	3.99	8.24	.69	3.48	.82	4.76	.63
16	66.40	3.76	8.14	.78	3.33	.90	4.79	.58
17	66.74	3.68	8.21	.71	3.29	.90	4.92	.57

(data from Garn et al., 1976)

TABLE 289

METACARPAL II SIZE (mm) IN MEXICAN-AMERICAN GIRLS
IN THE TEN-STATE NUTRITION SURVEY

| Age (years) | Length | | Diameters | | | | | |
| | | | Total | | Medullary | | Cortical | |
	Mean	S.D.	Mean	S.D.	Mean	S.D.	Mean	S.D.
3	34.33	2.38	4.99	.49	2.86	.53	2.13	.33
4	37.51	3.00	5.01	.47	2.73	.44	2.28	.33
5	40.39	2.91	5.47	.45	2.92	.41	2.55	.33
6	44.14	3.61	5.56	.53	2.79	.54	2.76	.47
7	45.90	3.83	5.92	.61	2.93	.57	2.99	.47
8	47.61	3.29	6.12	.66	3.06	.66	3.06	.52
9	48.82	4.05	6.09	.66	2.83	.72	3.26	.53
10	53.20	3.94	6.45	.53	3.11	.61	3.34	.40
11	55.37	3.62	6.74	.70	2.92	.69	3.82	.60
12	59.11	3.63	7.19	.68	2.91	.92	4.28	.78
13	61.60	4.30	7.54	.79	2.95	.75	4.59	.49
14	62.56	3.09	7.68	.58	3.08	.66	4.61	.55
15	61.16	3.86	7.60	.59	2.94	.65	4.66	.70
16	60.93	2.99	7.51	.59	2.63	1.00	4.88	.77

(data from Garn et al., 1976)

TABLE 290

CORTICAL AREA (mm^2) OF METACARPAL II
IN WHITE BOYS IN THE TEN-STATE NUTRITION SURVEY

Age (years)	Total Area		Medullary Area		Cortical Area		Percent Cortical Area	
	Mean	S.D.	Mean	S.D.	Mean	S.D.	Mean	S.D.
1	16.72	3.74	6.70	3.31	10.02	2.43	60.88	12.39
2	20.53	4.10	8.47	3.55	12.06	2.72	59.83	12.81
3	22.59	4.89	9.08	3.77	13.51	3.38	60.88	12.81
4	23.14	3.90	7.85	3.15	15.30	2.91	66.64	10.25
5	24.54	4.48	7.92	3.03	16.62	3.10	68.24	8.93
6	26.95	4.99	8.48	3.22	18.47	3.34	69.09	8.42
7	28.70	5.76	8.41	3.54	20.29	4.00	71.24	8.41
8	31.97	7.02	9.19	4.16	22.78	4.69	72.01	8.72
9	34.53	6.40	9.58	3.96	24.95	4.43	72.77	7.97
10	36.88	6.27	9.84	3.85	27.04	4.66	73.71	7.83
11	40.02	7.53	10.87	4.45	29.15	5.45	73.34	8.69
12	44.81	8.78	12.02	4.99	32.79	6.65	73.60	8.33
13	49.16	9.79	11.64	4.65	37.55	7.89	76.54	7.44
14	53.59	11.74	12.40	4.65	41.22	10.18	76.73	7.47
15	58.08	10.70	11.94	5.04	46.13	8.88	79.57	7.23
16	61.86	10.84	12.56	5.17	49.30	9.11	79.84	6.99
17	62.70	10.60	11.54	6.17	51.15	7.61	82.22	7.61

(data from Garn et al., 1976)

TABLE 291

CORTICAL AREA (mm^2) OF METACARPAL II
IN BLACK BOYS IN THE TEN-STATE NUTRITION SURVEY

Age (years)	Total Area		Medullary Area		Cortical Area		Percent Cortical Area	
	Mean	S.D.	Mean	S.D.	Mean	S.D.	Mean	S.D.
1	18.44	3.18	9.01	2.62	9.43	2.82	51.12	10.96
2	21.52	4.27	9.10	3.27	12.43	2.66	58.47	9.86
3	22.84	4.60	8.56	4.00	14.28	2.60	63.81	11.71
4	27.06	5.35	10.13	4.21	16.93	3.40	63.35	10.20
5	28.28	5.36	9.73	4.02	18.49	3.07	66.35	9.15
6	31.49	5.68	10.66	3.96	20.83	3.41	66.82	8.25
7	33.01	6.31	10.27	3.91	22.74	4.59	69.33	8.56
8	35.69	6.93	10.87	4.08	24.82	4.51	70.17	7.47
9	37.81	7.06	11.55	4.28	26.26	4.56	70.07	7.75
10	40.37	7.56	11.75	4.22	28.62	4.89	71.42	6.79
11	44.46	8.64	13.14	4.50	31.32	6.14	70.79	6.89
12	47.14	7.48	14.15	4.99	32.99	5.28	70.36	7.46
13	51.74	10.62	14.28	5.72	37.46	8.16	72.71	8.08
14	56.42	10.78	15.11	5.68	41.31	8.70	73.46	7.76
15	59.86	12.45	15.09	6.07	44.76	9.66	75.03	7.68
16	64.67	12.07	14.57	5.87	50.09	9.46	77.75	6.96
17	67.02	11.92	14.29	6.23	52.73	8.81	79.08	6.90

(data from Garn et al., 1976)

TABLE 292

CORTICAL AREA (mm^2) OF METACARPAL II
IN MEXICAN-AMERICAN BOYS IN THE TEN-STATE NUTRITION SURVEY

Age (years)	Total Area		Medullary Area		Cortical Area		Percent Cortical Area	
	Mean	S.D.	Mean	S.D.	Mean	S.D.	Mean	S.D.
4	25.09	4.77	8.11	3.02	16.97	2.43	68.61	7.68
5	23.99	5.11	7.19	2.90	16.79	3.05	70.85	7.57
6	27.66	5.76	7.93	2.86	19.73	3.71	71.88	6.97
7	30.43	5.54	9.05	3.69	21.38	2.94	71.20	8.05
8	31.61	4.88	9.17	3.34	22.45	3.31	71.49	7.84
9	32.90	6.46	9.68	3.37	23.22	4.07	71.12	6.42
10	37.15	7.51	10.42	4.35	26.73	5.04	72.63	7.89
11	38.09	6.06	9.49	4.03	28.60	3.88	75.73	7.88
12	43.08	8.35	10.64	4.36	32.45	5.67	75.95	6.89
13	46.30	12.05	11.21	6.48	35.09	7.51	77.03	8.23
14	48.56	8.69	10.30	4.18	38.26	7.49	78.93	8.07
15	56.54	8.63	12.40	4.29	44.15	6.14	78.38	5.75
17	56.54	10.71	7.49	2.12	49.05	9.61	86.70	3.05

(data from Garn et al., 1976)

TABLE 293

CORTICAL AREA (mm^2) OF METACARPAL II
IN WHITE GIRLS IN THE TEN-STATE NUTRITION SURVEY

Age (years)	Total Area		Medullary Area		Cortical Area		Percent Cortical Area	
	Mean	S.D.	Mean	S.D.	Mean	S.D.	Mean	S.D.
1	15.04	3.96	5.98	2.14	9.05	2.45	60.50	7.64
2	17.95	3.13	6.94	2.75	11.00	2.11	62.16	11.59
3	19.71	3.92	7.32	3.37	12.39	2.48	63.95	11.68
4	21.59	4.86	7.02	3.06	14.57	3.27	68.31	10.30
5	23.01	4.65	6.99	2.85	16.02	3.47	70.08	9.04
6	24.31	4.38	6.71	2.75	17.60	3.06	72.91	8.61
7	26.98	5.41	7.16	3.05	19.83	3.91	74.01	8.32
8	29.29	6.69	7.22	3.28	22.07	5.08	75.82	7.88
9	31.10	6.07	7.41	3.37	23.69	4.64	76.60	8.21
10	34.24	6.39	8.07	3.39	26.17	4.78	76.88	7.41
11	37.60	6.77	8.28	3.54	29.31	5.68	78.19	7.73
12	40.95	7.43	8.51	3.80	32.44	6.12	79.50	7.61
13	43.68	7.64	7.81	3.66	35.87	6.37	82.36	6.98
14	46.28	9.19	7.30	4.34	38.98	7.30	84.70	7.33
15	47.40	7.79	6.94	3.37	40.46	6.53	85.65	5.97
16	48.73	7.43	7.63	4.06	41.10	5.74	84.79	6.77
17	48.49	8.71	7.25	4.64	41.25	6.24	85.78	7.54

(data from Garn et al., 1976)

TABLE 294

CORTICAL AREA (mm^2) OF METACARPAL II
IN U.S. BLACK GIRLS

Age (years)	Total Area Mean	Total Area S.D.	Medullary Area Mean	Medullary Area S.D.	Cortical Area Mean	Cortical Area S.D.	Percent Cortical Area Mean	Percent Cortical Area S.D.
1	13.48	3.33	5.25	1.24	8.22	3.20	59.35	12.24
2	19.01	4.69	7.41	3.42	11.60	2.47	62.47	11.51
3	22.95	4.51	8.22	3.25	14.74	3.33	64.77	10.96
4	24.60	5.99	7.79	3.33	16.81	4.96	68.63	10.83
5	25.65	5.61	7.94	4.09	17.71	3.27	70.29	10.21
6	27.96	5.82	8.06	3.61	19.90	3.82	71.93	8.69
7	29.41	5.84	8.04	3.54	21.37	4.00	73.25	8.24
8	32.59	5.87	8.52	3.19	24.07	4.23	74.23	7.04
9	36.51	7.84	9.90	4.29	26.61	5.48	73.49	8.03
10	38.58	7.57	9.92	4.08	28.66	5.79	74.69	7.96
11	42.95	8.12	10.12	4.39	32.84	6.73	76.69	8.27
12	47.05	9.05	10.82	4.99	36.23	6.66	77.47	7.62
13	48.65	8.12	10.08	4.45	38.57	6.56	79.64	7.77
14	52.60	8.62	10.45	5.53	42.15	6.25	80.73	7.91
15	53.69	9.09	10.04	4.98	43.65	6.55	81.74	6.72
16	52.47	10.33	9.35	5.02	42.92	6.98	82.81	6.76
17	53.28	9.23	9.13	4.73	44.15	6.40	83.52	6.79

(data from Garn et al., 1976)

TABLE   295

CORTICAL AREA   (mm^2) OF METACARPAL
II IN MEXICAN-AMERICAN GIRLS

Age (years)	Total Area		Medullary Area		Cortical Area		Percent Cortical Area	
	Mean	S.D.	Mean	S.D.	Mean	S.D.	Mean	S.D.
3	19.73	3.83	6.62	2.38	13.11	2.41	66.98	7.95
4	19.85	3.64	6.00	2.00	13.84	2.63	70.07	6.65
5	23.67	3.82	6.81	1.93	16.86	2.79	71.35	5.70
6	24.45	4.67	6.34	2.51	18.11	3.68	74.25	7.63
7	27.84	5.75	6.99	2.69	20.85	4.37	75.16	7.34
8	29.78	6.52	7.70	3.21	22.08	4.77	74.54	7.77
9	29.45	6.38	6.67	3.46	22.78	4.59	77.96	8.02
10	32.87	5.32	7.87	3.09	25.00	3.46	76.57	6.49
11	36.06	7.72	7.06	3.20	29.00	6.08	80.75	6.90
12	40.97	7.73	7.28	4.10	33.69	6.55	82.63	8.95
13	45.14	9.23	7.28	2.87	37.86	7.24	84.32	5.10
14	46.61	7.26	7.77	3.22	38.84	5.59	83.58	5.37
15	45.62	6.84	7.11	3.05	38.51	6.47	84.44	6.22
16	44.53	7.16	6.13	4.53	38.40	5.49	86.83	8.57

(data from Garn et al., 1976)

TABLE 296

DEVELOPMENT OF TOTAL SUBPERIOSTEAL AREA, MEDULLARY CAVITY AREA,
CORTICAL AREA AND PERCENT CORTICAL AREA OF METACARPAL II IN
WHITE MALES AND FEMALES IN THE TEN-STATE NUTRITION SURVEY

Age (years)	Total Area (mm^2)		Medullary Area (mm^2)		Cortical Area (mm^2)		Percent Cortical Area	
	Mean	S.D.	Mean	S.D.	Mean	S.D.	Mean	S.D.
MALES								
8	34.1	5.6	8.0	3.4	26.1	3.9	77.0	7.5
10	40.6	6.6	8.8	3.5	31.8	5.0	78.7	6.5
12	47.3	7.9	9.7	3.9	37.7	6.4	79.9	6.7
14	57.6	10.3	10.8	4.2	46.8	8.5	81.5	5.9
16	65.7	10.7	11.9	4.4	53.8	7.5	82.3	5.1
18	68.6	9.8	10.8	4.9	57.8	7.6	84.3	6.0
FEMALES								
8	31.1	5.8	7.6	3.1	23.5	4.0	76.0	6.3
10	36.6	6.9	8.7	3.5	27.9	4.9	76.6	6.8
12	43.5	8.1	8.8	3.9	34.7	6.0	80.3	6.6
14	47.8	7.8	7.2	3.4	40.6	6.3	85.2	5.7
16	48.0	7.7	6.2	3.2	41.8	6.3	87.3	5.4
18	49.5	8.3	6.4	3.0	43.2	7.2	87.4	5.1

(data from "Subperiosteal and endosteal bone apposition during adolescence," Human
Biology 42:639-664, 1970, by Frisancho, Garn, and Ascoli, by permission of the Wayne
State University Press. Copyright 1970, Wayne State University Press, Detroit, Michigan
48202)

TABLE   297

RELATIVE BRACHIAL TISSUE AREAS IN BOSTON CHILDREN
EXPRESSED AS PERCENTAGES OF TOTAL BRACHIAL MIDSECTIONS

Sample	Mean Age (years)	Percentages of Total Brachial Area				Percentage of Marrow in Humerus
		Marrow	Compact Bone	Muscle	Fat	
Boys	8.0	1.4	4.4	56.6	37.6	24.1
	9.8	1.4	4.2	61.5	32.9	24.9
	11.9	1.6	4.5	63.4	30.4	26.0
	13.9	1.2	4.8	67.1	26.8	20.5
Girls	6.0	1.1	3.5	49.3	46.1	23.4
	8.1	1.0	3.5	53.2	42.2	22.0
	10.0	1.1	3.5	52.2	43.2	23.9
	12.2	1.3	4.1	55.2	39.4	23.8
	13.9	1.2	4.1	55.7	39.0	22.6

(data from Baker et al., 1958)

TABLE   298

MEAN AREAS OF CROSS SECTION (mm^2) OF BRACHIAL TISSUES
IN BOSTON CHILDREN

Sample	Mean Age (years)	Areas of Cross Section						
		Marrow	Compact Bone	Total Humerus	Muscle	Fat	Lean Brachium	Total Brachium
Boys	8.0	53	167	220	2145	1426	2365	3791
	9.8	63	190	253	2776	1487	3029	4516
	11.9	78	222	300	3102	1488	3402	4890
	13.9	69	267	336	3715	1486	4051	5537
Girls	6.0	40	131	171	1855	1733	2026	3759
	8.1	41	145	186	2184	1732	2370	4102
	10.0	49	156	205	2298	1903	2503	4406
	12.2	69	221	290	2954	2110	3244	5354
	13.8	69	236	305	3187	2231	3492	5723

(data from Baker et al., 1958)

TABLE 299

TISSUE AREAS AS PERCENTS OF TOTAL
AREA IN RADIOGRAPH OF CALF
IN BOSTON CHILDREN

Age	Boys Percentiles			Girls Percentiles		
(years)	10	50	90	10	50	90

Bone Area

6	41.3	43.3	47.2	37.6	40.6	44.4
7	40.2	44.3	48.0	37.7	41.3	45.2
8	40.1	44.4	48.1	37.8	41.6	45.7
9	40.9	45.1	48.7	38.6	41.8	46.3
10	40.7	45.5	48.8	37.4	42.3	47.0

Muscle Area

6	36.2	39.6	43.1	33.9	38.6	41.9
7	35.1	40.6	42.6	33.7	38.6	42.4
8	35.3	39.2	43.3	34.0	37.9	41.9
9	34.6	38.2	43.3	34.3	37.5	41.6
10	34.8	38.4	43.5	31.0	36.6	42.0

Skin + Subcutaneous Area

6	13.6	17.2	21.3	16.6	20.8	24.7
7	12.7	16.6	20.9	16.2	20.4	24.9
8	12.4	16.6	20.9	16.5	20.5	24.7
9	11.9	16.1	21.1	16.5	19.9	26.3
10	12.2	16.2	20.5	16.2	20.8	24.8

(data from Stuart and Dwinell, 1942)

TABLE   300

BONE WIDTHS (mm) MEASURED ON RADIOGRAPHS OF COLORADO BOYS

Age (yr.-mo.)	Max. Forearm Percentiles			Mid-thigh Percentiles			Max. Calf Percentiles		
	10	50	90	10	50	90	10	50	90
0 - 2	9.34	10.35	11.54	6.50	7.36	8.14	9.88	12.11	13.96
0 - 4	10.67	11.95	13.56	7.40	8.25	9.16	10.41	13.26	15.17
0 - 6	11.52	13.07	14.50	8.14	8.95	9.89	12.26	14.65	16.06
1 - 0	13.41	15.56	17.04	9.58	10.56	11.74	14.03	16.12	18.18
1 - 6	14.32	16.91	18.75	11.01	11.85	12.96	15.86	17.56	20.02
2 - 0	15.44	17.54	19.52	11.83	12.75	13.88	16.64	18.46	20.60
2 - 6	15.90	18.10	20.60	12.32	13.41	14.67	17.49	19.40	21.40
3 - 0	16.69	18.87	21.04	12.84	14.10	15.32	18.42	20.32	22.01
3 - 6	16.82	19.40	21.97	13.36	14.37	15.94	19.26	20.74	22.77
4 - 0	17.29	19.69	22.14	13.82	14.96	16.20	19.46	21.29	23.23
4 - 6	18.22	20.04	22.48	14.14	15.30	16.90	20.17	21.95	23.53
5 - 0	18.21	20.70	23.38	14.54	15.75	17.12	20.69	22.72	24.32
5 - 6	18.77	21.33	23.66	15.03	16.36	17.71	21.14	23.37	24.76
6 - 0	19.06	21.58	24.47	15.37	16.64	18.28	21.64	24.22	25.66
6 - 6	18.63	21.82	24.22	15.67	17.22	18.32	22.41	24.54	26.20
7 - 0	18.02	22.16	25.47	16.49	17.64	19.35	22.78	25.30	27.43
7 - 6	18.99	22.70	26.66	16.89	18.15	19.78	23.42	25.86	28.02
8 - 0	18.52	22.33	26.89	17.42	18.74	20.46	24.12	26.56	28.89
8 - 6	19.96	23.18	26.89	17.52	19.15	21.42	24.46	26.98	29.48
9 - 0	19.30	23.85	27.62	17.98	19.90	21.76	25.13	27.53	29.91
0 - 6	20.09	23.97	27.75	18.29	20.10	22.49	25.40	28.51	30.94
10 - 0	19.67	24.35	28.62	18.64	20.82	23.03	26.48	29.16	31.60
10 - 6	19.62	25.05	28.10	19.16	21.34	22.81	26.68	29.86	32.51
11 - 0	19.36	25.07	29.12	19.51	21.72	23.64	27.70	30.70	33.48
11 - 6	20.63	25.32	29.67	19.43	22.42	24.21	27.60	31.34	34.04
12 - 0	20.18	25.52	29.47	20.05	22.78	24.64	28.09	31.80	35.07
12 - 6	22.42	26.17	30.80	20.67	23.45	25.39	28.95	32.72	36.33
13 - 0	20.02	26.48	31.81	21.22	24.10	26.13	29.53	33.40	36.97
13 - 6	22.05	27.63	31.56	21.20	24.47	27.03	29.58	34.28	37.28
14 - 0	24.85	29.73	33.39	22.17	24.90	27.22	30.90	34.95	38.40
14 - 6	20.97	29.93	33.66	22.32	25.58	27.93	30.44	35.74	39.04
15 - 0	24.79	29.86	34.14	22.99	26.19	28.27	31.76	36.10	40.18
15 - 6	24.00	29.24	33.52	23.59	26.53	29.92	32.42	35.81	41.33
16 - 0	22.75	28.60	34.13	24.62	27.36	30.86	34.07	37.35	41.59
16 - 6	23.83	30.11	34.67	24.53	26.94	29.34	33.98	36.75	40.05
17 - 0	24.20	31.98	35.74	24.53	27.86	29.71	35.29	37.03	41.54
18 - 0	23.34	30.65	33.88	24.72	28.72	29.67	34.61	36.60	40.26

(data from McCammon, Human Growth and Development, 1970. Courtesy of Charles C Thomas, Publisher, Springfield, Illinois)

TABLE 301

BONE WIDTHS (mm) MEASURED ON RADIOGRAPHS OF COLORADO GIRLS

Age (yr. - mo.)	Max. Forearm Percentiles			Mid-thigh Percentiles			Max. Calf Percentiles		
	10	50	90	10	50	90	10	50	90
0 - 2	9.17	10.06	11.01	6.39	7.09	7.78	9.15	11.72	12.93
0 - 4	10.07	11.45	12.44	7.18	7.84	8.81	10.94	12.91	14.24
0 - 6	10.84	12.24	13.10	7.65	8.53	9.54	11.53	13.56	15.19
1 - 0	12.72	14.66	16.11	8.92	10.40	11.24	13.27	15.85	17.87
1 - 6	13.67	15.83	17.50	10.42	11.71	12.74	14.54	17.13	19.05
2 - 0	13.41	16.33	18.66	11.12	12.62	13.79	15.80	18.30	20.36
2 - 6	15.25	17.16	19.79	11.86	13.28	14.60	16.84	19.04	21.02
3 - 0	15.69	17.48	19.98	12.54	13.71	15.19	17.49	19.63	21.74
3 - 6	15.79	18.03	20.24	12.93	14.41	15.70	18.14	20.24	22.41
4 - 0	16.37	18.16	20.19	13.50	14.79	16.20	18.43	20.65	22.76
4 - 6	16.40	18.57	20.98	13.93	15.21	16.75	19.16	21.32	23.20
5 - 0	17.03	19.14	21.60	14.46	15.68	17.27	19.72	21.96	23.81
5 - 6	17.11	19.30	21.92	14.75	16.12	17.70	20.39	22.55	24.41
6 - 0	18.00	19.74	22.11	15.04	16.48	17.98	21.12	23.21	25.06
6 - 6	16.70	19.95	22.78	15.54	16.83	18.42	21.13	23.55	25.63
7 - 0	17.21	19.99	22.70	15.88	17.17	19.06	21.79	24.08	26.33
7 - 6	16.86	20.32	23.32	16.33	17.53	19.34	22.31	24.72	26.74
8 - 0	17.25	20.70	23.89	16.71	18.06	19.88	22.74	25.00	27.15
8 - 6	15.95	20.85	23.99	17.21	18.31	20.22	23.03	25.63	27.75
9 - 0	16.90	20.75	24.21	17.36	18.82	20.94	23.24	26.09	28.66
9 - 6	16.78	21.74	24.80	17.61	19.22	21.51	24.00	26.37	29.15
10 - 0	16.59	21.42	25.94	18.03	19.68	21.99	24.17	27.05	29.74
10 - 6	17.48	21.96	25.25	18.69	20.09	22.71	24.85	27.38	30.26
11 - 0	16.83	22.09	25.07	19.24	20.71	23.27	25.20	28.15	30.74
11 - 6	18.73	22.16	26.05	19.63	21.26	23.92	26.19	28.97	31.41
12 - 0	16.94	23.20	27.81	20.41	21.88	24.12	26.60	29.41	32.57
12 - 6	16.77	24.32	28.19	20.60	22.44	25.04	27.25	30.25	32.53
13 - 0	17.04	23.33	28.37	21.41	23.08	24.99	27.74	30.62	34.14
13 - 6	16.71	24.01	28.59	21.88	23.51	25.52	28.23	31.00	34.06
14 - 0	17.91	23.43	28.69	22.33	24.05	25.56	28.63	31.33	34.40
14 - 6	19.14	25.66	28.25	23.44	24.38	25.64	29.59	32.14	34.81
15 - 0	16.28	23.96	30.04	22.57	24.54	26.05	29.09	31.33	34.69
16 - 0	19.18	24.04	28.68	23.16	24.72	26.43	29.67	31.48	35.39
17 - 0	18.74	22.31	30.13	23.94	25.42	26.53	29.25	32.72	34.45

(data from McCammon, Human Growth and Development, 1970. Courtesy of Charles C Thomas, Publisher, Springfield, Illinois)

TABLE   302

BONE WIDTHS (mm) AT MID-LENGTH IN RADIOGRAPHS OF COLORADO CHILDREN

Age	BOYS						GIRLS					
(yr.-mo.)	Humerus Percentiles			Radius Percentiles			Humerus Percentiles			Radius Percentiles		
	10	50	90	10	50	90	10	50	90	10	50	90
0 - 2	6.05	6.88	7.96	3.87	4.58	5.37	6.01	6.78	7.64	4.10	4.57	5.06
0 - 4	7.80	8.38	9.51	4.85	5.58	6.31	7.09	7.99	9.18	4.63	5.30	5.92
0 - 6	8.27	9.17	10.20	5.58	6.30	6.80	7.93	8.93	9.72	5.14	5.70	6.51
1 - 0	9.92	11.08	12.01	6.63	7.22	8.00	9.32	10.74	11.68	5.98	6.68	7.66
1 - 6	10.73	11.79	12.96	7.09	8.01	8.70	10.19	11.56	12.70	6.69	7.57	8.25
2 - 0	10.98	12.41	13.43	7.46	8.15	8.84	10.41	12.30	13.05	6.78	7.72	8.72
2 - 6	11.58	12.66	14.07	7.39	8.32	9.27	10.65	12.25	13.33	7.05	7.99	8.89
3 - 0	11.63	13.02	14.07	7.82	8.59	9.52	11.16	12.57	13.66	7.40	8.19	9.08
3 - 6	11.80	13.16	14.22	8.07	8.86	9.96	11.24	12.79	13.68	7.55	8.71	9.70
4 - 0	12.32	13.29	14.40	8.34	9.19	10.54	11.48	12.98	14.44	7.86	8.93	9.90
4 - 6	12.72	13.56	15.05	8.57	9.54	10.52	11.67	13.38	14.64	8.01	9.15	10.14
5 - 0	12.79	13.91	15.22	8.77	9.88	10.82	11.89	13.52	14.92	8.37	9. 7	10.42
5 - 6	12.91	14.37	15.66	9.00	10.25	11.23	12.17	13.80	15.02	8.45	9.63	10.80
6 - 0	13.26	14.56	15.98	9.30	10.52	11.60	12.50	14.09	15.25	8.74	9.81	11.21
6 - 6	13.21	14.77	16.39	9.60	10.63	11.70	12.61	14.41	15.52	9.01	10.02	11.21
7 - 0	13.72	14.92	16.51	9.85	11.04	11.99	12.90	14.61	15.94	9.18	10.41	11.42
7 - 6	13.89	15.39	17.10	9.92	11.04	12.55	12.72	14.86	16.50	9.10	10.62	11.78
9 - 0	14.05	15.50	17.62	10.21	11.10	12.71	13.16	15.19	16.71	9.44	10.70	11.82
8 - 6	14.56	15.95	17.95	10.14	11.58	13.17	13.46	15.34	16.89	9.31	10.87	11.99
9 - 0	14.67	16.09	18.13	10.50	11.88	13.51	13.51	15.61	17.25	9.75	10.88	12.49
9 - 6	14.52	16.48	18.50	10.74	11.88	13.47	13.72	15.89	17.24	9.55	11.23	12.72
10 - 0	15.10	16.79	19.22	10.90	12.35	13.82	13.90	16.03	17.70	9.91	11.38	12.79
10 - 6	15.36	16.98	18.89	10.86	12.54	14.23	14.48	16.15	17.95	9.97	11.52	13.00
11 - 0	15.64	17.56	19.31	10.75	12.98	14.46	14.57	16.40	18.38	10.28	11.77	13.32
11 - 6	15.68	17.92	19.72	11.22	12.92	14.75	15.03	16.86	19.16	10.66	12.07	13.46
12 - 0	16.16	18.04	20.10	11.40	13.22	14.70	15.53	17.32	19.03	10.40	12.06	13.52
12 - 6	16.73	18.78	20.88	11.45	13.54	15.09	16.09	17.69	19.73	10.74	12.11	13.58
13 - 0	16.70	19.14	21.42	11.82	13.67	15.38	16.31	18.20	20.22	10.92	12.37	13.87
13 - 6	17.04	19.39	22.00	11.91	13.95	15.74	16.52	18.17	20.48	11.12	12.64	14.13
14 - 0	17.46	20.16	22.69	12.45	14.68	16.27	16.75	18.94	20.38	11.13	12.52	14.15
14 - 6	17.50	20.57	23.08	12.27	14.43	16.50	17.44	19.22	20.47	11.34	12.74	14.26
15 - 0	18.70	20.88	23.49	13.48	15.18	17.29	17.29	19.26	20.41	11.03	12.61	14.23
15 - 6	18.18	21.04	24.12	12.88	15.36	17.38	17.21	19.76	21.03	11.42	12.94	--
16 - 0	19.82	22.40	24.84	13.62	15.30	17.52	17.74	19.22	20.58	11.46	12.89	14.58
16 - 6	19.94	21.60	23.70	13.79	15.12	17.60	--	--	--	--	--	--
17 - 0	20.35	22.02	24.84	14.12	16.04	17.35	--	--	--	--	--	13.71
18 - 0	20.52	21.92	24.88	14.61	15.62	17.34	--	--	--	--	--	--

Note - no data for 16.6, 17.0, 18.0 in girls

(data from McCammon, Human Growth and Development, 1970. Courtesy of Charles C Thomas, Publisher, Springfield, Illinois)

TABLE 303

MEANS AND COEFFICIENTS OF VARIATION FOR RADIOGRAPHIC
MORPHOMETRY OF THE RADIUS SHAFT IN WISCONSIN CHILDREN

Age (years)	Thickness (mm)					
	Total		Medullary		Compact	
	Girls	Boys	Girls	Boys	Girls	Boys
Means						
6	9.4	10.0	4.4	4.6	5.0	5.4
7	9.7	10.3	4.4	4.8	5.2	5.5
8	9.6	10.6	4.2	4.9	5.4	5.7
9	9.9	10.8	4.4	5.2	5.5	5.6
10	10.3	11.2	4.6	5.0	5.7	6.2
11	11.2	11.4	4.8	5.2	6.3	6.2
12	11.5	12.0	5.2	5.7	6.3	6.3
13	11.6	12.2	4.9	5.5	6.7	6.6
14	11.7	12.3	4.8	5.5	7.0	6.8
Coefficients of Variation						
6	12	10	24	20	15	13
7	12	8	20	17	10	10
8	8	10	18	20	12	13
9	12	9	28	16	9	11
10	12	11	21	20	12	13
11	10	9	19	21	15	10
12	9	9	28	16	11	11
13	10	9	22	21	12	14
14	11	13	18	23	7	14

Age (years)	Area (mm^2)			
	Total		Compact	
	Girls	Boys	Girls	Boys
6	70	79	54	62
7	75	84	58	65
8	73	89	59	69
9	78	93	62	71
10	84	101	67	80
11	99	103	80	81
12	105	113	82	87
13	107	117	87	92
14	109	120	91	95
Coefficient of Variation				
6	25	20	23	20
7	23	15	20	14
8	16	20	16	19
9	23	18	17	17

TABLE   303   (continued)

MEANS AND COEFFICIENTS OF VARIATION FOR RADIOGRAPHIC
MORPHOMETRY OF THE RADIUS SHAFT IN WISCONSIN CHILDREN

| Age | Area (mm^2) | | | |
| (years) | Total | | Compact | |
	Girls	Boys	Girls	Boys
Coefficient of				
Variation				
10	24	22	22	21
11	20	19	22	16
12	18	17	11	17
13	19	18	18	18
14	22	26	19	25

| Age | Compact Total (%) | | | |
| (years) | Thickness | | Area | |
	Girls	Boys	Girls	Boys
Means				
6	54	54	78	79
7	54	54	79	78
8	56	54	81	78
9	56	52	80	77
10	56	55	80	80
11	56	55	81	79
12	55	53	79	77
13	58	55	82	79
14	60	56	84	80
Coefficient of				
Variation				
6	15	12	10	7
7	9	10	6	7
8	11	12	7	8
9	14	10	9	6
10	11	11	7	7
11	12	12	7	7
12	17	10	11	6
13	12	14	7	9
14	6	13	3	8

(data from Mazess and Cameron, 1972)

TABLE 304

MEAN DENSITY OF BONES OF MALE NEGROS

Bones	Age range (years)	Shaft	Epiphysis Proximal	Distal	Sum of weights Sum of volumes
Humerus	3.1-11	.918	.483	.623	.836
Radius	6. -14	1.003	.780	.617	.964
Femur	0.3-14	.708	.524	.389	.661
Tibia	0.3-17	.730	.395	.490	.669

(data from Trotter, 1971)

TABLE 305

MEAN DENSITIES OF BONES a, WITH NO
EPIPHYSIS FUSED, AND b, WITH ALL EPIPHYSES FUSED

Group	Age range (years)	Bones Humerus	Radius	Femur	Tibia
WM a	Birth-12.7	0.702	0.818	0.609	0.627
WF a	Birth-6.7	0.701	0.794	0.603	0.625
NM a	Birth-11.0	0.766	0.861	0.686	0.691
b	17.0-21.0	0.913	1.108	0.808	0.756
NF a	Birth-11.0	0.785	0.910	0.711	0.697
b	17.0-22.4	0.869	1.107	0.759	0.737

Density determined from mean weight divided by mean volume.
WM = white male; WF = white female; NM = negro male and NF = negro female.

(data from Trotter, 1971)

TABLE 306

ABSOLUTE TISSUE BREADTHS (mm) IN THE CALF
IN OHIO CHILDREN

Age Level	Fat		Muscle		Bone		Total Breadth	
	Mean	S.D.	Mean	S.D.	Mean	S.D.	Mean	S.D.
Boys								
Birth	7.4	1.7	18.0	2.2	10.6	1.1	36.0	3.3
1 month	9.9	1.6	20.2	2.6	10.8	.7	40.9	3.3
1 year	20.2	3.6	30.5	3.5	16.2	1.4	66.9	3.9
7.5 years	11.3	2.8	44.1	4.4	27.0	2.0	82.4	5.8
12.5 years	14.1	3.4	55.3	5.3	32.8	2.3	102.1	8.8
15.5 years	11.0	2.2	68.7	8.7	37.3	3.3	117.1	8.5
Girls								
Birth	8.4	1.9	20.2	2.0	7.7	1.1	36.4	3.1
1 month	10.5	2.0	19.2	2.0	10.7	.6	40.4	3.3
1 year	21.5	2.9	29.9	3.8	15.5	2.1	66.9	4.6
7.5 years	13.4	3.4	42.4	4.6	25.2	2.3	81.0	6.4
12.5 years	15.7	2.9	52.8	6.8	30.0	2.2	98.5	8.7
15.5 years	19.5	4.9	60.3	9.1	32.5	3.3	112.3	14.4

(data from Reynolds and Grote, 1948)

TABLE 307

MEANS FOR ABSOLUTE AND RELATIVE TISSUE BREADTHS IN CALF AT
LEVEL OF MAXIMUM CALF BREADTH
IN OHIO CHILDREN

Absolute Breadth (mm)

Age	Fat	Muscle	Bone	Total
Birth	8.4	20.2	7.8	36.4
1 month	10.5	19.2	10.7	40.4
1 year	21.5	29.9	15.5	66.9
7.5 years	13.4	42.4	25.2	81.0
12.5 years	15.7	52.8	30.0	98.5
15.5 years	19.5	60.3	32.5	112.3

Relative Breadth (%) In Relation To Total Breadth

Age	Fat	Muscle	Bone
Birth	23.1	55.6	21.3
1 month	26.0	47.5	26.5
1 year	32.1	44.6	23.3
7.5 years	16.5	52.3	31.2
12.5 years	16.0	53.6	30.4
15.5 years	17.1	53.6	29.2

(data from Reynolds, 1948)

TABLE   308

RATIO OF MEAN TISSUE BREADTHS AT LEVEL OF MAXIMUM CALF BREADTH
TO ADULT MEANS (%)
IN OHIO CHILDREN

Age	Fat	Muscle	Bone	Total Breadth
Birth	44	33	24	32
1 month	56	31	33	36
1 year	114	48	48	59
7.5 years	71	68	77	71
12.5 years	83	85	92	87
15.5 years	103	97	100	99

(data from Reynolds, 1948)

TABLE   309

FAT WIDTH/BONE WIDTH AT MAXIMUM DIAMETER
OF THE CALF
IN OHIO CHILDREN

Age (years)	Boys		Girls	
	Mean	S.D.	Mean	S.D.
7.5	44.2	11.8	51.2	15.0
10.5	41.9	12.9	51.7	13.5
13.5	37.9	11.3	50.8	15.3
16.5	30.4	13.2	58.6	20.4

(data from "The fat/bone index as a sex-differentiating character
in man," Human Biology 21:199-204, 1949, by E. L. Reynolds, by
permission of the Wayne State University Press. Copyright 1949,
Wayne State University Press, Detroit, Michigan 48202)

TABLE 310

BODY COMPOSITION DATA IN ILLINOIS BOYS

Age (years)	Stature (cm)		Weight (kg)		Density (gm/cc)		Potassium (gm)		Sum 10 Skinfolds (mm)	
	Mean	S.D.	Mean	S.D.	Mean	S.D.	Mean	S.D.	Mean	S.D.
9	140.8	7.4	33.5	5.3	1.051	.015	71.7	7.7	87.6	46.1
10	144.2	6.0	38.1	10.8	1.054	.017	79.5	15.3	116.6	71.0
11	146.5	5.8	38.2	6.8	1.052	.009	81.9	12.8	80.4	25.4

(data from "A comparison of densitometric, potassium-40 and skinfold estimates of body composition in prepubescent boys," Human Biology 47:321-336, 1975, by Cureton, Boileau, and Lohman, by permission of the Wayne State University Press. Copyright 1975, Wayne State University Press, Detroit, Michigan 48202)

TABLE 311

BODY DENSITY (gm/cc) OF MINNESOTA CHILDREN

Age (years)	Body Density	
	Mean	S.D.
Boys		
$12\frac{1}{2}$-$14\frac{1}{2}$	1.0654	0.020
$14\frac{1}{2}$-$16\frac{1}{2}$	1.0743	0.020
$16\frac{1}{2}$-$18\frac{1}{2}$	1.0743	0.030
Girls		
$12\frac{1}{2}$-$14\frac{1}{2}$	1.0643	0.022
$14\frac{1}{2}$-$16\frac{1}{2}$	1.0556	0.014
$16\frac{1}{2}$-$18\frac{1}{2}$	1.0409	0.011

(data from Novak, 1963, Annals of the New York Academy of Science 110:545-577. By permission of the New York Academy of Science)

TABLE 312

DATA RELATIVE TO SIZE AND BODY COMPOSITION FOR BOYS
IN ROCHESTER (NY)

							Skinfold Thickness					
Age (years)	Stature (cm)		Weight (kg)		Fat (kg)		Triceps (mm)		Subscapular (mm)		Ave. 6 Sites (mm)	
	Mean	S.D.	Mean	S.D.	Mean	S.D.	Mean	S.D.	Mean	S.D.	Mean	S.D.
8.5-9	134	4.6	29.7	2.7	4.4	1.9	13	3.8	6.5	2.4	8.6	2.9
9 - 9.5	135	4.4	31.4	4.2	5.8	3.2	11	4.4	7.6	3.8	8.6	4.1
10 -10.5	139	6.0	32.7	5.6	5.2	3.0	13	3.0	7.3	3.7	8.5	3.2
10.5-11	142	5.8	34.4	5.2	6.4	3.9	14	5.3	9.8	5.5	10.3	5.4
11 -11.5	146	7.1	40.9	9.9	8.2	6.8	15	6.4	11	7.8	12	6.8
11.5-12	149	7.2	40.2	7.6	7.8	4.5	14	5.6	9.2	5.0	12	6.4
12 -12.5	151	7.5	43.6	7.7	8.9	4.8	13	5.6	10	5.6	12.1	6.0
12.5-13	154	8.4	47.4	12.1	9.5	6.1	14	5.8	11	6.7	12.4	7.1
13 -13.5	158	8.2	49.4	10.1	9.0	5.6	11	5.4	11	6.4	11.4	6.2
13.5-14	164	7.7	55.5	9.8	10.6	5.9	12	6.0	10	5.4	11.5	5.9
14 -14.5	168	8.0	58.3	9.8	8.6	4.5	11	3.3	9.7	4.6	9.4	3.5
14.5-15	169	5.6	60.0	8.9	9.7	5.3	12	6.2	11	4.2	12	6.2
15 -15.5	174	7.3	64.3	8.9	7.9	5.3	9.6	2.0	9.9	3.6	9.2	2.5
16 -16.5	175	5.0	63.8	4.0	5.5	2.5	9.5	3.0	8.5	1.9	8.5	1.2
16.5-17	174	7.1	67.9	10.2	7.5	4.9	9.7	4.3	11	4.5	11.2	5.9
17 -17.5	176	6.8	66.2	10.6	6.3	4.3	8.6	4.2	10	4.2	10.8	4.3
17.5-18	181	6.1	71.6	10.9	8.8	4.1	12	4.0	10	4.0	10.9	3.7

(data from "Skinfold thickness and body fat in children," Human Biology 42:401-418, 1970, by Forbes and Amirhakimi, by permission of the Wayne State University Press. Copyright 1970, Wayne State University Press, Detroit, Michigan 48202)

TABLE 313

DATA RELATING TO SIZE AND BODY COMPOSITION FOR GIRLS
IN ROCHESTER (NY)

Age (years)	Stature (cm)		Weight (kg)		Fat (kg)		Skinfold Thickness					
							Triceps (mm)		Subscapular (mm)		Ave. 6 Sites (mm)	
	Mean	S.D.	Mean	S.D.	Mean	S.D.	Mean	S.D.	Mean	S.D.	Mean	S.D.
7.5- 8.5	128	6.8	28.5	5.9	6.1	4.0	17	8.4	10	5.4	11	5.7
9 - 9.5	135	5.1	32.0	6.5	7.6	4.2	17	6.3	12	8.3	13	7.2
9.5-10	139	7.2	35.0	8.1	8.9	4.8	18	9.0	15	8.0	16	7.5
10 -10.5	141	6.5	33.6	4.0	7.4	2.4	16	4.5	10	2.9	12	3.3
11 -11.5	147	5.7	40.7	7.6	9.3	6.1	18	5.7	12	5.6	15	6.3
12 -12.5	152	5.0	44.1	6.8	10.6	4.9	16	5.9	13	6.4	14	5.9
12.5-13	153	7.6	42.9	8.3	10.0	5.4	16	6.9	11	6.7	12	7.2
15 -15.5	163	5.8	55.1	8.3	13.2	5.6	15	4.5	14	3.3	12	3.5
15.5-16	163	6.2	56.0	5.7	13.8	4.6	17	5.4	15	5.5	14	4.3
16.5-18	163	7.7	56.0	8.3	13.9	4.5	17	7.7	15	5.5	15	4.3

(data from "Skinfold thickness and body fat in children," Human Biology 42:401-418, 1970,
by Forbes and Amirhakimi, by permission of the Wayne State University Press. Copyright
1970, Wayne State University Press, Detroit, Michigan 48202)

TABLE 314

NON-LEAN BODY MASS PERCENTAGE (FAT PERCENTAGE)
FOR CHILDREN IN BALTIMORE (MD)

Age* (years)	Percentiles						
	3	10	25	50	75	90	97
				BOYS			
5	7.3	8.8	10.0	12.5	14.2	16.0	18.8
6	5.5	9.0	10.8	12.7	15.0	16.6	21.4
7	8.8	10.4	12.0	14.0	16.3	18.3	21.9
8	9.8	11.7	13.7	16.1	18.5	21.4	25.3
9	9.6	12.0	14.2	16.9	19.7	23.0	25.8
10	9.0	11.6	14.7	17.6	20.2	22.8	24.6
11	5.6	10.4	12.9	16.9	19.7	22.2	24.6
12	4.9	8.5	11.4	14.9	17.8	20.7	24.0
13	4.9	6.9	9.8	12.6	15.9	20.0	23.7
14	3.9	5.8	7.9	11.4	14.4	17.9	21.5
15	4.2	6.8	8.5	11.4	15.0	18.1	22.7
16	3.6	5.8	8.5	11.8	15.3	18.3	20.9
17	--	--	--	12.3	--	--	--
				GIRLS			
5	5.5	8.3	12.1	15.3	20.0	24.4	35.1
6	5.8	8.5	11.1	14.0	16.6	20.4	25.5
7	7.6	9.5	11.3	14.0	16.6	20.3	24.6
8	7.5	9.6	11.7	14.7	17.7	21.6	25.8
9	8.0	10.0	12.0	15.4	19.1	23.6	26.6
10	5.6	10.0	12.8	16.0	20.0	23.8	28.4
11	8.8	11.7	14.9	18.8	22.4	25.8	28.7
12	9.8	13.3	16.3	19.5	23.5	25.9	29.3
13	12.5	15.6	18.8	22.0	25.0	27.5	30.4
14	15.2	17.5	20.0	23.1	25.7	28.6	32.0
15	16.4	18.3	20.7	23.3	26.1	28.9	32.9
16	16.9	19.6	21.8	23.8	26.1	28.7	32.0
17	--	--	--	24.6	--	--	--

*Age is defined at subject's last birthday.

Fat percentage = $\dfrac{\text{fat mass}}{\text{weight}}$ X 100

(data from Cheek, 1968, Human Growth: Body Composition, Cell Growth, Energy
and Intelligence, by permission of Lea and Febiger, Publishers, Philadelphia)

TABLE   315

NON-LEAN BODY MASS (FAT MASS) IN POUNDS
FOR CHILDREN IN BALTIMORE (MD)

Age*	Percentiles						
(years)	3	10	25	50	75	90	97

BOYS

5	3	4	4	6	7	8	10
6	2	4	5	6	8	10	14
7	4	5	6	8	9	12	17
8	5	6	8	10	13	16	22
9	5	7	9	11	14	19	24
10	6	7	10	13	15	20	25
11	5	8	10	13	17	22	28
12	4	8	10	13	17	22	29
13	5	6	10	13	18	26	37
14	4	7	9	13	19	26	36
15	5	8	11	15	21	30	40
16	4	7	12	17	25	32	43
17	--	--	--	18	--	--	--

GIRLS

5	2	3	5	7	9	13	19
6	2	4	5	7	8	11	15
7	3	4	5	7	9	12	18
8	4	5	6	9	11	16	24
9	4	5	7	10	14	21	26
10	3	6	8	12	16	24	35
11	5	8	11	16	22	29	36
12	7	10	14	19	26	31	41
13	10	14	18	24	30	38	47
14	13	17	21	27	33	40	51
15	15	18	23	28	34	42	52
16	16	20	24	29	35	42	53
17	--	--	--	31	--	--	--

*Age is defined at subject's last birthday.

(data from Cheek, 1968, Human Growth: Body Composition, Cell Growth, Energy
and Intelligence, by permission of Lea and Febiger, Publishers, Philadelphia)

TABLE 316

TOTAL BODY FAT FOR BLACK BOYS IN TEXAS

Age (years)	Total body fat (kg)				Total body fat as a percent of weight			
	Lower Income		Middle Income		Lower Income		Middle Income	
	Mean	S.D.	Mean	S.D.	Mean	S.D.	Mean	S.D.
10.1 - 11	5.5	3.16	5.5	1.30	14.9	5.37	16.4	3.66
11.1 - 12	6.5	2.42	8.6	3.31	17.3	4.94	20.5	6.22
12.1 - 13	9.3	7.06	7.9	5.04	20.2	7.44	16.4	7.01
13.1 - 14	6.0	3.02	8.5	4.25	12.0	6.64	17.3	7.78
14.1 - 15	7.9	2.76	7.8	2.56	16.2	2.83	14.5	5.24
16.1 - 17	9.2	4.79	--	--	13.8	5.53	--	--

(data from Schutte, 1979)

TABLE 317

COMPARISON OF PER CENT BODY FAT ESTIMATED FROM BODY DENSITY,
BODY POTASSIUM AND SKINFOLD THICKNESS
IN ILLINOIS CHILDREN

Age Group (years)	% Body Fat Estimated From					
	Body Density		Body Potassium		Skinfolds	
	Mean	S.D.	Mean	S.D.	Mean	S.D.
9	20.7	6.1	19.1	9.4	21.2	5.4
10	19.4	7.1	21.1	6.5	24.8	6.3
11	20.4	3.6	19.1	6.2	20.9	4.0

(data from "A comparison of densitometric, potassium-40 and skinfold estimates
of body composition in prepubescent boys," Human Biology 47:321-336, 1975,
by Cureton, Boileau, and Lohman, by permission of the Wayne State University
Press. Copyright 1975, Wayne State University Press, Detroit, Michigan
48202)

TABLE 318

SUBSCAPULAR SKINFOLD THICKNESS (mm) IN U.S. CHILDREN

ALL BOYS

Age (years)	Mean	Median	S.E. of Mean
1	6.4	6.4	0.25
2	5.5	5.5	0.28
3	5.5	5.5	0.29
4	5.2	5.3	0.22
5	5.0	5.0	0.18
6	5.1	4.8	0.27
7	5.4	4.8	0.38
8	5.0	4.9	0.26
9	7.7	5.8	0.72
10	7.2	5.9	0.62
11	7.1	6.1	0.51
12	7.7	6.0	0.53
13	8.9	6.8	1.05
14	8.8	7.0	1.16
15	10.4	7.7	0.79
16	10.4	8.4	0.94
17	10.0	8.3	0.66

WHITE BOYS

Age (years)	Mean	Median	S.E. of Mean
1	6.5	6.5	0.31
2	5.3	5.4	0.27
3	5.6	5.5	0.32
4	5.2	5.3	0.25
5	5.0	5.0	0.20
6	5.2	4.8	0.32

TABLE 318 (continued)

SUBSCAPULAR SKINFOLD THICKNESS (mm) IN U.S. CHILDREN

WHITE BOYS

Age (years)	Mean	Median	S.E. of Mean
7	5.5	5.0	0.43
8	5.0	4.8	0.26
9	7.7	6.0	0.71
10	7.2	6.0	0.70
11	7.3	6.2	0.62
12	7.8	6.1	0.63
13	9.1	7.2	1.30
14	9.4	7.3	1.39
15	10.1	7.5	0.78
16	10.7	8.4	1.10
17	10.0	8.3	0.72

NEGRO BOYS

Age (years)	Mean	Median	S.E. of Mean
1	5.6	5.8	0.28
2	6.6	5.8	0.96
3	5.2	5.4	0.30
4	5.0	5.3	0.23
5	5.3	4.8	0.53
6	4.8	4.8	0.47
7	4.1	4.2	0.36
8	5.5	5.3	0.75
9	7.1	5.5	1.27
10 ⎤			
11 ⎦	6.1	5.7	0.55
12	7.5	5.6	1.33
13			

TABLE 318 (continued)

SUBSCAPULAR SKINFOLD THICKNESS (mm) IN U.S. CHILDREN

NEGRO BOYS

Age (years)	Mean	Median	S.E. of Mean
14			
	9.2	8.2	1.53
15			
16	8.4	8.2	0.68
17	10.1	8.7	1.00

ALL GIRLS

1	6.2	6.3	0.35
2	6.1	6.1	0.21
3	5.9	5.9	0.25
4	5.6	5.6	0.16
5	6.3	5.5	0.31
6	6.2	6.2	0.53
7	6.0	5.5	0.45
8	8.2	5.9	0.63
9	8.3	7.4	0.66
10	9.2	6.8	0.68
11	10.0	8.1	0.94
12	11.0	8.9	0.98
13	12.7	10.5	0.91
14	13.8	10.8	0.98
15	12.6	10.5	1.12
16	13.4	10.6	1.46
17	15.9	12.7	1.48

TABLE 318 (continued)

SUBSCAPULAR SKINFOLD THICKNESS (mm) IN U.S. CHILDREN

WHITE GIRLS

Age (years)	Mean	Median	S.E. of Mean
1	6.2	6.2	0.37
2	6.0	6.1	0.25
3	5.9	6.0	0.29
4	5.7	5.7	0.17
5	6.3	5.5	0.34
6	6.3	6.2	0.66
7	5.9	5.4	0.54
8	7.9	5.9	0.71
9	8.6	7.6	0.76
10	9.3	6.9	0.74
11	9.9	7.8	1.06
12	11.1	8.9	1.24
13	12.5	10.2	1.02
14	14.1	11.3	1.19
15	13.4	10.8	1.42
16	13.4	10.4	1.59
17	15.7	12.7	1.67

NEGRO GIRLS

Age (years)	Mean	Median	S.E. of Mean
1	6.6	6.6	0.60
2	6.3	5.9	0.66
3	5.6	5.5	0.27
4	5.1	5.2	0.17
5	6.2	5.5	0.78
6	5.8	5.7	0.49

TABLE 318 (continued)

SUBSCAPULAR SKINFOLD THICKNESS (mm) IN U.S. CHILDREN

NEGRO GIRLS

Age (years)	Mean	Median	S.E. of Mean
7			
	8.5	10.1	1.05
8			
9	5.7	5.2	0.62
10	8.2	6.5	1.31
11	10.3	10.4	1.24
12	10.5	7.3	1.60
13	13.5	12.0	1.79
14	13.2	10.1	1.31
15	9.3	7.6	1.08
16	14.2	11.8	2.48
17	15.9	12.0	3.17

(data from Abraham et al., 1975)

TABLE   319

SUBSCAPULAR SKINFOLD THICKNESS (mm) IN U.S. CHILDREN

Race, sex and age	Percentiles						
	5	10	25	50	75	90	95
WHITE							
Boys							
6 years......	3.0	3.5	4.0	4.0	5.0	6.5	7.0
7 years......	3.0	4.0	4.0	4.5	6.0	7.0	8.0
8 years......	3.5	4.0	4.0	5.0	6.0	8.0	12.0
9 years......	3.5	4.0	4.0	5.0	6.0	10.0	15.0
10 years......	3.5	4.0	4.0	5.0	7.0	10.5	15.0
11 years......	4.0	4.0	4.5	6.0	8.0	14.0	17.5
Girls							
6 years......	3.5	4.0	4.0	5.0	6.0	8.0	10.5
7 years......	4.0	4.0	4.0	5.0	7.0	9.5	12.0
8 years......	4.0	4.0	4.5	6.0	8.0	12.0	16.0
9 years......	4.0	4.0	5.0	6.0	9.0	15.0	19.0
10 years......	4.0	4.0	5.0	6.0	9.5	16.0	20.0
11 years......	4.0	4.0	6.0	7.0	10.0	16.0	20.0
NEGRO							
Boys							
6 years......	3.0	4.0	4.0	4.0	5.0	6.0	6.5
7 years......	3.0	4.0	4.0	4.0	5.0	6.0	7.0
8 years......	3.0	4.0	4.0	5.0	6.0	6.5	7.0
9 years......	3.0	4.0	4.0	5.0	6.0	6.5	8.0
10 years......	3.5	4.0	4.0	5.0	6.0	8.0	12.0
11 years......	3.5	4.0	4.0	5.0	6.0	8.0	14.0
Girls							
6 years......	4.0	4.0	4.0	5.0	5.5	6.0	7.0
7 years......	3.0	4.0	4.0	5.0	6.0	7.0	10.0
8 years......	4.0	4.0	4.0	5.0	7.0	10.0	13.0
9 years......	4.0	4.0	5.0	6.0	7.0	10.0	14.0
10 years......	4.0	5.0	5.0	6.0	8.0	11.0	12.0
11 years......	4.0	4.0	5.5	7.0	9.0	14.0	20.0

(data from Johnston et al., 1973)

TABLE 320

SUBSCAPULAR SKINFOLD THICKNESS (mm) IN U.S. CHILDREN

Race, sex and age	Percentiles						
	5	10	25	50	75	90	95
WHITE							
Boys							
12 years......	3.6	4.1	4.7	5.7	7.8	14.0	21.1
13 years......	4.0	4.3	5.1	6.2	8.7	16.5	22.3
14 years......	4.2	4.6	5.4	6.5	8.7	14.6	20.6
15 years......	4.5	5.1	5.8	6.9	9.2	14.8	21.3
16 years......	5.1	5.3	6.2	7.2	9.3	15.0	20.5
17 years......	5.1	5.6	6.5	7.8	10.7	17.1	22.0
Girls							
12 years......	4.4	5.0	5.8	7.6	11.6	18.6	23.7
13 years......	4.9	5.3	6.5	8.5	12.8	21.3	25.5
14 years......	5.5	6.2	7.4	9.7	13.8	20.4	26.3
15 years......	5.8	6.4	7.7	10.3	15.3	24.4	30.0
16 years......	6.2	6.8	8.1	10.8	15.5	24.0	29.8
17 years......	6.2	6.6	8.2	11.2	17.0	23.4	31.0
NEGRO							
Boys							
12 years......	4.1	4.2	4.7	5.5	6.5	9.4	16.5
13 years......	4.2	4.4	5.1	6.2	7.6	11.6	17.2
14 years......	4.1	4.4	5.2	6.3	7.6	10.5	15.3
15 years......	4.4	4.8	5.4	6.4	7.5	9.4	11.7
16 years......	4.3	4.7	5.6	7.3	8.8	11.2	12.3
17 years......	5.2	5.5	6.6	7.6	9.6	11.9	20.7
Girls							
12 years......	4.6	5.3	6.6	8.4	11.8	20.5	24.6
13 years......	5.2	5.7	7.0	8.6	11.6	18.7	30.0
14 years......	5.7	6.4	7.5	9.5	13.6	18.1	25.1
15 years......	6.2	6.7	8.4	10.4	13.4	21.5	26.0
16 years......	6.1	7.1	8.3	10.5	16.4	26.6	32.8
17 years......	6.9	7.3	8.3	12.3	16.0	22.1	30.5

(data from Johnston et al., 1974)

TABLE 321

SUBSCAPULAR SKINFOLD THICKNESS (mm)
IN DENVER CHILDREN

Age (Years)	Boys			Girls		
	Median	Mean	S.D.	Median	Mean	S.D.
4.00	4.45	4.65	1.15	5.00	4.89	0.79
4.50	4.43	4.77	1.36	5.00	5.06	1.09
5.00	4.52	4.58	1.31	4.89	4.90	1.04
5.50	4.15	4.44	0.81	4.36	5.07	1.24
6.00	4.23	4.51	1.05	4.77	5.14	1.40
6.50	4.11	4.34	0.83	4.94	4.98	1.34
7.00	4.11	4.56	1.01	4.99	5.38	1.86
7.50	4.49	4.63	1.05	5.00	6.09	2.41
8.00	4.50	4.54	1.02	5.73	6.52	2.98
8.50	4.94	4.69	0.93	5.04	6.54	3.13
9.00	5.00	4.96	1.33	5.08	6.16	2.54
9.50	5.00	5.47	1.76	5.35	7.16	3.84
10.00	5.19	5.96	2.23	5.64	6.94	2.98
10.50	5.19	6.01	2.96	5.89	7.30	4.15
11.00	5.51	6.55	2.95	6.49	8.22	4.18
11.50	5.04	6.66	3.86	5.79	7.54	4.00
12.00	5.21	6.85	3.74	6.69	8.36	4.21
12.50	5.63	7.20	4.08	6.80	8.56	4.72
13.00	5.66	7.15	3.61	7.00	8.35	3.81
13.50	5.80	7.50	4.88	7.36	8.65	3.73
14.00	6.14	7.74	4.60	7.24	8.73	3.76
14.50	6.02	7.23	3.87	7.13	8.53	2.99
15.00	6.61	7.63	4.22	8.70	10.21	4.41

TABLE 321 (continued)

SUBSCAPULAR SKINFOLD THICKNESS (mm)
IN DENVER CHILDREN

Age (Years)	Boys			Girls		
	Median	Mean	S.D.	Median	Mean	S.D.
15.50	6.43	8.35	4.48	7.77	9.75	4.60
16.00	7.46	8.98	4.89	8.99	11.04	5.07
16.50	7.02	9.94	6.68	--	--	--
17.00	6.76	8.48	4.00	9.00	10.12	4.32
17.50	7.01	9.08	3.93	--	--	--
18.00	7.89	9.32	3.24	9.93	11.47	5.36

(data from McCammon, unpublished)

TABLE   322

SKINFOLD THICKNESSES (mm) ON THE MEDIAL ASPECT OF THE RIGHT
SCAPULA (mm) IN NEW JERSEY CHILDREN

Age	Boys			Girls		
	Mean	Median	S.D.	Mean	Median	S.D.
Newborn	3.8	3.7	0.8	4.1	4.1	0.7
3 months	5.7	5.6	1.1	6.1	6.0	1.5
6 months	5.9	5.6	1.2	6.4	6.0	1.5
9 months	6.3	6.0	1.3	6.6	6.8	1.6
12 months	6.4	6.3	1.3	6.8	6.8	1.6
18 months	7.2	7.3	1.3	7.1	6.7	1.5
2 years	6.0	5.7	1.5	6.3	5.9	1.9
2½ years	5.9	5.8	1.4	6.4	5.7	2.0
3 years	5.8	5.7	1.5	6.4	6.2	1.6
3½ years	5.4	5.3	1.2	5.8	5.7	1.3
4 years	5.4	5.3	1.3	5.9	5.3	1.8
4½ years	5.1	4.9	1.3	6.2	5.2	2.8
5 years	5.3	5.0	1.5	6.2	5.7	1.9
5½ years	5.7	5.1	1.9	6.1	5.7	1.7
6 years	5.5	5.0	2.3	5.9	5.3	2.0
6½ years	6.1	5.4	2.4	6.1	5.5	2.1
7 years	5.5	5.0	1.4	6.3	5.7	2.3
7½ years	5.1	4.9	1.4	6.6	5.4	2.9
8 years	5.5	5.1	1.4	6.3	5.3	2.5
8½ years	5.5	5.1	1.8	6.9	6.5	1.9
9 years	5.6	5.2	1.9	7.1	6.5	3.4
9½ years	5.7	5.4	1.8	7.7	7.0	2.8
10 years	5.7	5.2	1.7	6.6	6.2	1.9
10½ years	6.0	5.1	3.0	7.2	6.4	2.5
11 years	5.8	5.7	1.7	7.2	6.5	2.8
11½ years	6.1	5.6	2.0	7.2	7.5	1.3
12 years	7.0	6.2	2.3	7.6	7.3	2.4
12½ years	7.0	6.2	2.8	7.6	7.7	1.1
12 years	7.5	6.3	3.3	8.2	8.0	2.0
13½ years	6.5	6.1	1.8	8.5	8.2	1.5
14 years	7.1	5.4	2.7	--	--	--
14½ years	7.1	6.7	--	--	--	--
15 years	6.9	6.2	--	--	--	--

(data from "Typical and atypical changes in the soft tissue distribution during
childhood," Human Biology 29:62-82, 1957, by W. Kornfeld, by permission of the Wayne
State University Press. Copyright 1957, Wayne State University Press, Detroit,
Michigan 48202)

TABLE   323

SUBSCAPULAR SKINFOLD THICKNESS (mm)
IN OHIO CHILDREN

Age (Years)	Boys			Girls		
	Median	Mean	S.D.	Median	Mean	S.D.
3.00	5.59	5.77	1.05	5.83	6.56	1.34
3.50	5.40	5.70	1.36	6.60	6.67	1.94
4.00	5.40	5.65	1.22	6.16	6.44	1.48
4.50	5.07	5.68	1.55	6.31	6.25	1.46
5.00	5.10	5.52	1.83	6.20	6.71	1.91
5.50	5.37	5.73	1.27	6.01	6.53	1.72
6.00	5.10	5.23	1.16	6.57	7.22	2.87
6.50	5.10	5.62	1.22	5.62	7.29	4.15
7.00	5.00	5.21	1.07	5.61	6.89	3.16
7.50	5.31	6.15	2.73	5.74	7.18	3.29
8.00	5.21	5.95	2.53	5.65	7.51	3.50
8.50	5.08	6.05	1.87	6.97	8.52	4.55
9.00	5.40	6.29	2.17	6.41	9.11	5.45
9.50	--	--	--	6.92	9.20	5.07
10.00	5.39	6.13	1.60	8.01	9.88	5.68
10.50	5.41	6.52	2.22	7.17	9.57	5.51
11.00	5.61	6.66	2.41	8.07	8.87	3.51
11.50	5.51	7.83	6.31	7.44	8.86	4.02
12.00	5.90	7.13	3.62	8.20	10.54	5.34
12.50	6.01	7.72	4.44	8.41	10.63	5.70
13.00	5.92	8.21	5.46	7.77	9.84	4.71
13.50	6.30	8.74	6.60	8.66	10.69	5.86
14.00	5.93	7.27	3.71	9.12	9.51	3.04

TABLE   323   (continued)

SUBSCAPULAR SKINFOLD THICKNESS (mm)
IN OHIO CHILDREN

Age (Years)	Boys			Girls		
	Median	Mean	S.D.	Median	Mean	S.D.
14.50	6.68	8.38	5.25	8.60	10.29	4.08
15.00	6.89	8.28	3.97	9.51	10.57	3.74
15.50	7.01	8.50	3.79	8.77	10.03	3.16
16.00	7.63	8.40	2.93	9.12	10.99	4.57
16.50	7.82	8.29	2.59	8.92	11.59	5.39
17.00	8.49	8.85	2.51	9.79	11.03	3.68
17.50	8.38	8.80	2.38	10.10	11.46	4.28
18.00	7.89	8.66	2.35	10.01	11.37	3.85

(data from Roche, unpublished)

TABLE 324

SUBSCAPULAR SKINFOLD THICKNESS
(mm) FOR PHILADELPHIA CHILDREN

Age (years)	WHITE BOYS Mean	WHITE BOYS S.D.	BLACK BOYS Mean	BLACK BOYS S.D.
6	6.12	3.68	5.11	3.81
7	6.81	4.94	5.59	3.53
8	7.22	4.04	5.70	3.30
9	8.02	5.96	6.95	4.81
10	10.32	7.93	7.54	6.23
11	9.72	6.81	8.27	7.30
12	10.56	7.38	7.07	5.51
13	--	--	7.77	6.17

Age (years)	WHITE GIRLS Mean	WHITE GIRLS S.D.	BLACK GIRLS Mean	BLACK GIRLS S.D.
6	6.71	2.70	5.81	3.74
7	7.97	4.62	7.27	4.90
8	8.90	5.79	7.97	5.10
9	10.55	6.33	8.76	6.87
10	12.37	6.86	10.37	7.67
11	11.49	6.00	10.25	7.51
12	--	--	10.23	7.54
13	--	--	11.39	8.54

Age is for completed years.

(data from Malina, unpublished a)

TABLE   325

SUBSCAPULAR SKINFOLD THICKNESS (mm) FOR BOYS IN TECUMSEH (MI)

Per centile	5	6	7	8	9	Age (years) 10	11	12	13	14	15	16	17	18
95	8	11	10	14	11	17	19	25	20	18	23	26	29	23
90	7	7	7	9	9	13	15	19	15	15	17	21	21	18
75	6	6	6	7	7	10	9	11	11	11	12	14	14	13
50	5	5	5	5	5	6	6	7	7	8	9	10	10	11
25	4	4	4	4	5	5	5	6	6	6	7	8	8	9
10	4	4	4	4	4	4	4	5	5	4	6	6	7	8
5	3	4	3	4	4	4	3	4	4	4	5	6	6	7
Mean	5.1	5.5	5.7	6.4	6.3	7.7	8.3	9.7	9.5	9.4	10.6	12.2	12.3	12.2
S.D.	1.5	2.4	3.8	4.7	4.5	5.2	6.0	6.8	7.5	5.8	5.9	6.5	7.0	5.0

(data from Montoye, 1978, 1970. Copyright 1978, 1970 by Allyn & Bacon, Inc., Boston.
Originally published by Phi Epsilon Kappa Fraternity, Indianapolis)

TABLE   326

SUBSCAPULAR SKINFOLD THICKNESS (mm) FOR GIRLS IN TECUMSEH (MI)

Per- centile	5	6	7	8	9	Age (years) 10	11	12	13	14	15	16	17	18
95	12	12	13	15	16	23	29	29	28	27	38	31	30	41
90	9	8	10	12	13	18	24	23	23	22	30	25	23	30
75	7	7	8	9	8	12	15	13	15	15	20	17	16	19
50	6	6	6	6	6	7	10	9	10	11	13	12	11	13
25	5	5	5	5	5	6	7	7	7	8	10	9	9	11
10	4	4	4	4	5	5	5	6	6	7	8	8	8	10
5	4	4	4	4	4	4	5	5	5	7	7	7	7	8
Mean	6.2	6.2	6.6	7.4	7.9	10.0	10.2	11.4	12.2	13.3	16.6	14.4	13.9	17.4
S.D.	2.5	2.4	3.3	4.2	5.1	6.2	6.3	7.2	7.8	7.3	10.3	7.5	7.7	11.9

(data from Montoye, 1978, 1970. Copyright 1978, 1970 by Allyn & Bacon, Inc., Boston.
Originally published by Phi Epsilon Kappa Fraternity, Indianapolis)

TABLE 327

SUBSCAPULAR SKINFOLD THICKNESS (log units) IN
JAPANESE-AMERICAN CHILDREN IN LOS ANGELES

Age (years)	Boys		Girls	
	Mean	S.D.	Mean	S.D.
6	150.1	26.8	163.6	18.6
7	160.2	28.5	166.3	22.9
8	158.5	30.7	164.4	37.6
9	156.7	27.8	173.7	32.3
10	169.7	30.0	181.3	25.5
11	167.8	35.0	178.4	30.3
12	175.0	34.7	189.8	23.9
13	169.7	28.7	196.1	25.3
14	166.1	31.4	201.8	19.8
15	178.7	27.2	203.0	20.5
16	175.1	18.7	--	--
18	--	--	218.9	22.1

(data from Kondo and Eto, 1975, pp. 13-45 in Comparative
Studies on Human Adaptability of Japanese, Caucasians and
Japanese Americans; S. Horvath et al. (eds), University
of Tokyo Press)

TABLE   328

SUBSCAPULAR SKINFOLD THICKNESS (mm) IN RELATION
TO INCOME LEVEL IN U.S. CHILDREN

Age (years)	Mean Age	Income Below Poverty Level			Income Above Poverty Level		
		Mean	Median	S.E. of Mean	Mean	Median	S.E. of Mean
				BOYS			
1	1.55	6.5	6.3	0.77	6.4	6.4	0.24
2	2.45	5.5	5.4	0.62	5.5	5.5	0.31
3	3.46	5.6	5.6	0.33	5.6	5.5	0.33
4	4.49	5.2	5.6	0.24	5.2	5.2	0.26
5	5.53	5.4	5.5	0.61	4.9	5.0	0.16
6	6.46	4.7	4.8	0.28	5.4	4.8	0.36
7	7.47	5.3	4.6	0.63	5.2	5.0	0.47
8	8.46	5.0	5.4	0.44	5.0	4.8	0.31
9	9.46	7.6	4.7	1.53	8.0	6.6	0.78
10	10.52	5.7	5.5	0.61	7.5	6.1	0.75
11	11.48	6.7	5.7	1.43	7.3	6.3	0.60
12	12.56	7.4	5.7	1.23	7.9	6.1	0.59
13	13.48	7.9	6.2	1.93	9.0	7.0	1.21
14	14.46	8.5	6.7	1.98	9.0	7.2	1.28
15	15.50	8.6	8.3	0.80	10.5	7.6	0.79
16	16.52	9.8	7.9	1.86	11.1	8.7	1.14
17	17.51				10.0	8.4	0.63
				GIRLS			
1	1.49	6.6	6.7	0.56	6.2	6.2	0.38
2	2.45	5.9	5.6	0.44	6.2	6.2	0.23
3	3.50	5.8	5.6	0.83	5.9	6.1	0.27
4	4.54	5.3	5.2	0.33	5.7	5.6	0.18
5	5.55	5.8	5.3	0.86	6.3	5.5	0.35
6	6.47	5.7	6.3	0.76	6.4	6.1	0.65
7	7.51	6.2	5.8	0.44	6.0	5.4	0.63
8	8.47	7.3	5.5	1.49	8.5	6.6	0.86
9	9.52	8.1	7.2	1.44	8.4	7.6	0.75
10	10.47	6.4	6.4	0.81	9.6	7.5	0.74
11	11.55	9.4	7.5	1.91	10.1	8.1	1.26
12	12.49	16.7	19.0	2.72	9.5	8.5	0.67
13	13.52	12.0	10.8	1.90	12.9	10.5	1.04
14	14.50	13.0	10.5	2.13	14.2	11.4	1.27
15	15.48	10.9	9.5	0.92	12.9	10.6	1.57
16	16.55	13.7	9.5	2.38	13.9	11.3	1.68
17	17.46	17.4	17.2	3.26	15.4	12.4	1.60

(data from Abraham et al., 1975)

TABLE 329

PECTORAL SKINFOLD THICKNESS (mm) IN DENVER (CO) CHILDREN

Age (years)	Boys			Girls		
	Median	Mean	S.D.	Median	Mean	S.D.
4.00	4.02	4.34	1.46	4.00	3.97	1.05
4.50	4.10	4.24	1.35	4.02	4.43	1.55
5.00	4.03	4.03	1.21	4.23	4.51	1.75
5.50	3.99	3.99	1.22	4.29	4.66	1.49
6.00	4.01	3.91	1.03	4.04	4.46	1.75
6.50	3.30	3.64	1.15	4.51	4.73	1.71
7.00	3.20	3.60	1.35	4.04	4.86	2.42
7.50	3.48	3.92	1.34	5.07	5.97	3.41
8.00	3.21	3.87	1.45	5.10	6.35	3.76
8.50	3.92	4.28	1.51	5.00	6.71	4.32
9.00	4.01	4.68	2.05	4.88	6.42	4.08
9.50	4.18	5.62	3.47	6.33	7.14	3.50
10.00	4.00	5.80	3.92	6.08	7.00	3.87
10.50	5.01	6.79	5.27	7.02	7.69	4.41
11.00	6.81	7.19	4.07	5.99	8.55	5.49
11.50	6.18	7.66	5.25	6.17	7.49	4.28
12.00	5.98	8.33	6.07	7.43	8.09	3.75
12.50	5.77	8.35	6.09	6.36	7.79	4.38
13.00	6.18	8.13	6.28	6.42	7.47	3.68
13.50	5.67	7.44	6.21	5.96	7.63	3.95
14.00	5.47	6.89	5.91	5.92	7.69	4.28
14.50	5.05	6.98	6.18	5.83	7.43	3.96
15.00	4.79	6.49	5.74	6.32	8.02	4.40
15.50	4.38	7.04	6.24	5.93	7.45	3.74
16.00	5.43	7.70	6.72	7.11	8.27	4.52
16.50	6.39	8.16	8.31	--	--	--
17.00	4.13	6.09	5.18	7.07	8.03	4.20
17.50	4.42	6.49	6.45	--	--	--
18.00	3.98	5.91	4.64	6.18	8.03	4.41

Measurements over the left axillary fold

(data from McCammon, unpublished)

TABLE 330

SKINFOLD THICKNESSES OF THE ANTERIOR
PART OF THE CHEST (mm) IN NEW JERSEY CHILDREN

Age	Boys			Girls		
	Mean	Median	S.D.	Mean	Median	S.D.
Newborn	3.9	3.7	1.0	4.0	3.9	0.8
3 months	6.7	6.4	2.0	7.0	6.3	2.3
6 months	6.3	5.1	1.6	7.4	7.1	2.5
9 months	6.0	5.5	1.5	6.2	5.9	1.8
12 months	5.5	5.2	1.4	6.0	5.7	1.7
18 months	5.9	5.8	2.0	6.2	6.0	1.6
2 years	6.2	6.0	1.7	6.6	6.5	1.9
2½ years	6.2	6.0	2.2	7.0	6.7	2.2
3 years	6.3	6.2	1.8	7.7	7.5	2.3
3½ years	6.0	5.9	1.5	7.5	7.1	2.2
4 years	5.9	5.8	1.8	7.2	6.6	2.6
4½ years	5.4	4.9	1.5	7.2	6.2	3.9
5 years	5.8	5.1	2.4	7.3	6.7	2.9
5½ years	6.5	5.2	3.5	6.9	6.5	2.8
6 years	6.3	5.1	3.5	7.7	6.5	3.0
6½ years	6.5	5.2	3.9	8.0	6.7	3.7
7 years	5.9	5.1	3.2	8.1	7.2	3.4
7½ years	6.1	4.9	2.3	8.9	7.8	4.2
8 years	6.2	5.1	2.9	8.8	7.7	3.9
8½ years	7.1	5.3	4.5	9.3	8.5	3.6
9 years	7.0	5.4	4.0	10.4	8.9	5.1
9½ years	7.6	6.2	4.3	10.9	11.0	4.7
10 years	7.5	6.0	4.4	10.2	9.0	5.0
10½ years	8.1	6.0	5.7	11.0	8.6	5.4
11 years	7.8	6.0	4.4	10.6	8.4	4.9
11½ years	9.7	7.0	6.5	9.5	9.0	3.6
12 years	9.9	8.0	5.9	10.2	8.8	3.9
12½ years	10.0	8.5	5.5	9.4	9.0	3.7
13 years	9.1	6.8	5.6	11.2	10.5	4.0
13½ years	7.6	6.7	3.8	10.6	10.5	2.9
14 years	8.3	5.3	5.2	--	--	--
14½ years	6.9	5.7	--	--	--	--
15 years	7.5	5.5	--	--	--	--

These measurements were made on the chest about one inch below the middle of the
right clavicle.

(data from "Typical and atypical changes in the soft tissue distribution during
childhood," Human Biology 29:62-82, 1957, by W. Kornfeld, by permission of the Wayne
State University Press. Copyright 1957, Wayne State University Press, Detroit,
Michigan 48202)

TABLE 331

MIDAXILLARY SKINFOLD THICKNESS (mm) IN U.S. CHILDREN

Race, sex and age	Percentiles						
	5	10	25	50	75	90	95
WHITE							
Boys							
6 years......	2.5	3.0	3.0	4.0	4.0	5.0	7.0
7 years......	3.0	3.0	3.0	4.0	5.0	6.0	8.0
8 years......	3.0	3.0	3.0	4.0	5.0	7.0	10.0
9 years......	3.0	3.0	3.5	4.0	6.0	10.0	15.0
10 years......	3.0	3.0	3.5	4.0	6.0	10.0	15.0
11 years......	3.0	3.0	4.0	5.0	7.0	13.0	18.0
Girls							
6 years......	3.0	3.0	3.5	4.0	5.0	8.0	10.0
7 years......	3.0	3.0	4.0	4.5	6.0	9.0	11.0
8 years......	3.0	3.0	4.0	5.0	7.0	11.0	15.0
9 years......	3.0	3.5	4.0	5.5	8.0	14.0	18.0
10 years......	3.0	3.5	4.0	6.0	9.0	16.0	20.0
11 years......	3.5	4.0	5.0	6.0	10.0	16.0	19.0
NEGRO							
Boys							
6 years......	3.0	3.0	3.0	4.0	4.0	5.5	6.0
7 years......	2.5	3.0	3.0	4.0	4.0	5.0	6.0
8 years......	3.0	3.0	3.0	4.0	4.5	5.0	6.0
9 years......	3.0	3.0	3.0	4.0	4.0	6.0	7.0
10 years......	3.0	3.0	3.0	4.0	5.0	7.0	10.0
11 years......	3.0	3.0	4.0	4.0	5.0	6.5	13.0
Girls							
6 years......	3.0	3.0	3.0	4.0	5.0	6.0	7.0
7 years......	3.0	3.0	3.0	4.0	5.0	7.0	8.0
8 years......	2.5	3.0	3.0	4.0	6.0	9.0	11.0
9 years......	3.0	3.0	4.0	5.0	6.5	8.0	12.0
10 years......	3.0	3.0	4.0	5.0	7.0	10.0	16.0
11 years......	3.0	4.0	4.0	5.0	8.0	13.0	16.0

(data from Johnston et al., 1973)

TABLE   332

MIDAXILLARY SKINFOLD THICKNESS (mm) IN U.S. CHILDREN

Race, sex	Percentiles						
and age	5	10	25	50	75	90	95
WHITE							
Boys							
12 years......	3.0	3.2	3.8	4.8	6.8	14.2	19.6
13 years......	3.2	3.3	4.1	5.0	7.1	13.8	19.5
14 years......	3.2	3.5	4.2	5.2	7.1	14.3	19.8
15 years......	3.4	3.8	4.4	5.5	7.6	13.4	19.7
16 years......	3.5	4.0	4.5	5.6	7.5	13.2	18.3
17 years......	3.6	4.1	4.6	5.8	8.3	15.1	22.4
Girls							
12 years......	3.6	4.2	5.1	6.6	10.6	16.8	23.5
13 years......	4.0	4.5	5.7	7.5	11.5	18.2	23.3
14 years......	4.4	5.1	6.4	8.4	12.8	20.1	23.2
15 years......	4.6	5.2	6.5	8.7	13.4	22.5	27.4
16 years......	4.8	5.6	6.9	9.5	13.7	21.6	28.2
17 years......	4.6	5.4	6.7	8.8	13.8	21.1	26.2
NEGRO							
Boys							
12 years......	3.1	3.2	3.5	4.2	5.1	7.4	13.2
13 years......	3.0	3.2	3.7	4.7	5.8	12.2	15.2
14 years......	2.8	3.2	4.0	4.6	5.5	8.2	10.4
15 years......	3.2	3.5	4.2	4.8	5.7	7.2	10.5
16 years......	3.2	3.5	4.3	5.4	6.6	7.9	9.6
17 years......	3.5	4.0	4.3	4.8	6.7	8.9	18.6
Girls							
12 years......	3.7	4.3	5.4	7.1	9.8	17.3	22.5
13 years......	3.5	4.2	5.4	6.7	9.8	18.3	23.8
14 years......	4.1	4.5	6.2	8.3	12.0	16.7	21.1
15 years......	4.4	4.8	6.1	7.6	11.4	19.2	23.4
16 years......	4.5	5.0	6.1	8.2	11.7	21.4	23.5
17 years......	5.0	5.3	6.2	9.1	13.7	18.0	21.3

(data from Johnston et al., 1974)

TABLE 333

LATERAL THORACIC SKINFOLD THICKNESS (mm) FOR
CHILDREN IN THE PRESCHOOL NUTRITION SURVEY

Age Intervals (months)	Percentiles				
	10	25	50	75	100
BOYS					
12-17	3.6	4.2	4.6	5.2	5.8
18-23	3.2	3.6	4.3	5.0	5.6
24-29	3.2	3.6	4.2	5.0	5.9
30-35	3.4	3.6	4.0	4.6	5.2
36-41	3.4	3.8	4.2	4.8	5.4
42-47	3.2	3.6	4.1	5.0	5.7
48-53	3.2	3.6	4.0	4.8	5.5
54-59	3.2	3.6	4.0	4.8	5.6
60-65	3.4	3.6	4.2	4.7	5.2
66-71	3.2	3.4	4.2	4.8	5.2
GIRLS					
12-17	3.5	4.0	4.6	5.2	6.4
18-23	3.2	3.8	4.4	5.2	6.0
24-29	3.6	4.0	4.4	5.0	5.7
30-35	3.4	3.8	4.3	5.2	6.3
36-41	3.4	4.0	4.6	5.4	5.8
42-47	3.6	4.0	4.6	5.2	5.9
48-53	3.4	3.6	4.2	5.2	6.0
54-59	3.4	3.8	4.2	5.2	6.1
60-65	3.4	4.0	4.6	5.4	6.4
66-71	3.4	4.0	4.8	5.4	6.2

(data from Owen, Kram, Garry, et al., 1974, Pediatrics 53:
597-646. Copyright American Academy of Pediatrics 1974)

TABLE   334

MIDAXILLARY SKINFOLD THICKNESS (mm) IN DENVER CHILDREN

Age (years)	Boys			Girls		
	Median	Mean	S.D.	Median	Mean	S.D.
4.00	3.03	3.23	0.68	3.50	3.64	0.84
4.50	3.12	3.35	0.90	3.92	3.77	0.63
5.00	3.19	3.45	1.01	3.52	3.76	0.99
5.50	3.03	3.21	0.71	3.52	3.69	0.80
6.00	3.52	3.36	0.72	3.51	3.80	1.15
6.50	3.11	3.28	0.80	3.56	3.74	0.94
7.00	3.21	3.40	0.91	3.24	4.14	1.71
7.50	3.65	3.59	0.77	3.50	4.13	1.49
8.00	3.48	3.60	0.86	3.72	4.79	2.67
8.50	3.64	3.68	0.96	4.19	4.85	2.62
9.00	3.80	3.99	1.29	4.43	5.15	2.49
9.50	4.02	4.96	2.47	4.84	6.11	3.24
10.00	4.04	4.82	2.36	5.19	5.85	2.86
10.50	4.08	5.37	3.89	5.04	6.14	3.63
11.00	4.90	5.83	3.43	5.12	6.70	3.73
11.50	4.06	6.33	4.58	5.07	6.32	3.48
12.00	4.44	6.34	5.20	5.08	6.70	3.63
12.50	4.90	6.94	5.53	5.97	6.95	4.08
13.00	4.85	7.12	5.85	5.53	6.61	3.37
13.50	4.65	7.08	6.71	6.50	7.48	3.52
14.00	4.60	6.58	5.95	6.38	7.12	2.98
14.50	4.33	6.63	5.25	5.99	7.66	3.67
15.00	4.76	6.56	5.46	6.43	7.90	3.82
15.50	5.04	7.35	5.67	7.50	8.25	3.57
16.00	5.19	7.30	5.63	6.31	8.80	5.43
16.50	5.45	8.27	7.30	--	--	--
17.00	4.92	6.39	4.42	6.97	8.02	3.75
17.50	4.75	7.10	5.47	--	--	--
18.00	5.07	6.53	3.26	7.53	8.42	4.14

(data from McCammon, unpublished)

TABLE 335

BREADTH OF CHEST FAT AT LEVEL OF NIPPLE (mm)*
MEASURED ON RADIOGRAPHS OF OHIO CHILDREN

Age (Years)	Boys			Girls		
	Mean	S.D.	Median	Mean	S.D.	Median
6.5	3.0	0.8	2.4	3.1	1.0	2.5
7.5	3.1	1.1	2.4	3.4	1.1	2.7
8.5	3.4	1.6	2.6	3.7	1.4	3.5
9.5	3.7	1.6	2.9	4.0	1.7	2.9
10.5	4.1	2.5	3.1	4.5	2.3	3.2
11.5	4.5	2.5	3.5	5.0	2.9	4.9
12.5	4.9	2.8	3.9	5.2	2.7	4.3
13.5	4.9	2.9	3.5	5.5	2.4	4.0
14.5	4.6	2.5	3.3	6.0	3.2	4.7
15.5	4.7	2.4	3.7	7.1	3.7	5.5
16.5	4.9	2.2	3.8	6.7	3.6	4.5
17.5	5.8	2.7	5.0	6.5	2.4	5.5

* lateral edge of chest in posteroanterior view)

(data from Reynolds, 1951)

TABLE 336

MIDAXILLARY SKINFOLD THICKNESS
(mm) FOR PHILADELPHIA CHILDREN

Age (years)	WHITE BOYS		BLACK BOYS	
	Mean	S.D.	Mean	S.D.
6	4.89	3.03	3.88	2.89
7	5.29	3.70	4.20	2.26
8	5.70	3.49	4.28	2.58
9	6.73	5.52	5.22	3.75
10	8.73	7.11	5.93	5.91
11	7.95	5.23	6.41	6.82
12	7.97	5.53	5.13	3.38
13	--	--	5.65	3.97

Age (years)	WHITE GIRLS		BLACK GIRLS	
	Mean	S.D.	Mean	S.D.
6	5.05	1.95	4.61	2.90
7	5.90	3.28	5.35	3.52
8	6.99	4.65	5.79	3.77
9	8.47	5.63	6.31	4.69
10	9.55	5.18	7.34	5.06
11	8.60	4.53	7.27	4.98
12	--	--	7.39	4.64
13	--	--	7.50	4.92

Age is for completed years.

(data from Malina, unpublished a)

TABLE 337

MEAN LATERAL THORACIC SKINFOLDS (mm) IN
BLACK AND WHITE CHILDREN IN
THE PRESCHOOL NUTRITION SURVEY

Age Interval (year)		Boys	Girls
1.00 -	1.49		
	Black	4.72	5.01
	White	4.61	4.80
1.50 -	2.49		
	Black	4.37	4.43
	White	4.37	4.63
2.50 -	3.49		
	Black	4.07	4.47
	White	4.30	4.59
3.50 -	4.49		
	Black	3.95	4.14
	White	4.38	4.66
4.50 -	5.49		
	Black	3.83	4.46
	White	4.24	4.84
5.50 -	6.49		
	Black	4.01	4.86
	White	4.25	5.01

(data from Owen, Kram, Garry, et al.,
1974, Pediatrics 53:597-646. Copyright
American Academy of Pediatrics 1974)

TABLE   338

MEAN LATERAL THORACIC SKINFOLD THICKNESS (mm) IN BLACK AND WHITE
CHILDREN OF SIMILAR SOCIOECONOMIC STATUS (Warner Ranks I and II)
IN THE PRESCHOOL NUTRITION SURVEY

Age Intervals (year)	Race	Boys	Girls
1.50 - 2.49	Black	4.37	4.43
	White	4.46	4.71
2.50 - 3.49	Black	4.07	4.47
	White	4.36	4.71
3.50 - 4.49	Black	3.95	4.14
	White	4.32	4.48
4.50 - 5.49	Black	3.83	4.46
	White	4.28	4.86
5.50 - 6.49	Black	4.01	4.86
	White	4.46	4.78

(data from Owen, Kram, Garry, et al., 1974, Pediatrics 53:597-646.
Copyright American Academy of Pediatrics 1974)

TABLE   339

FAT THICKNESS (mm) OVER THE TENTH RIB IN MICHIGAN INFANTS

Gestational Age (weeks)	-2 S.D.	Mean	+2 S.D.
<30	0	0	0
30 - 31+	0	1.9	2.9
32 - 33+	2.3	3.2	4.1
34 - 35+	2.3	3.7	5.1
36 - 37+	2.7	4.1	5.5
38 - 39+	2.6	5.1	7.7
40 - 41+	3.1	5.8	8.6

(data from Kuhns, Berger, Roloff, et al., 1974, RADIOLOGY
111;665-671)

TABLE 340

FAT THICKNESS AT THE TENTH RIB (mm)
MEASURED ON RADIOGRAPHS OF OHIO CHILDREN

Age (Years)	Boys			Girls		
	Median	Mean	S.D.	Median	Mean	S.D.
0.08	2.71	2.92	0.95	3.20	3.16	0.77
0.25	4.66	4.73	1.43	4.80	4.93	1.79
0.50	4.79	5.04	1.41	5.21	5.61	1.90
0.75	4.72	4.76	1.18	5.11	5.10	1.47
1.00	4.71	4.90	1.32	5.30	5.41	1.63
1.50	4.09	4.42	1.29	4.78	4.85	1.41
2.00	4.04	4.17	1.18	4.31	4.44	1.22
2.50	3.86	4.26	1.12	4.28	4.58	1.48
3.00	3.65	3.93	1.18	4.47	4.42	1.01
3.50	3.70	4.00	0.93	4.11	4.28	1.06
4.00	3.38	3.52	0.77	4.22	4.20	1.02
4.50	3.69	3.83	1.28	4.02	4.30	1.28
5.00	3.13	3.27	0.83	3.50	3.53	0.80
5.50	3.26	3.57	1.03	3.51	3.97	1.54
6.00	2.91	3.11	0.77	3.31	3.61	1.42
6.50	3.09	3.39	1.24	3.50	3.90	1.67
7.00	3.00	3.92	3.33	--	--	--
7.50	3.07	3.37	1.31	3.56	3.89	1.65
8.50	3.18	3.77	2.18	3.94	4.63	2.41
9.50	3.61	4.46	2.93	4.11	4.96	2.72
10.00	3.83	4.07	1.37	--	--	--
10.50	4.27	5.32	3.61	4.96	6.40	4.28
11.50	4.58	5.80	3.73	5.04	6.27	3.58
12.50	4.39	5.82	3.70	5.48	7.18	4.88

TABLE 340 (continued)

FAT THICKNESS AT THE TENTH RIB (mm)
MEASURED ON RADIOGRAPHS OF OHIO CHILDREN

Age (Years)	Boys			Girls		
	Median	Mean	S.D.	Median	Mean	S.D.
13.50	4.30	5.95	3.93	6.10	7.59	5.03
14.50	4.57	5.73	3.52	6.87	8.32	4.44
15.50	4.60	5.61	2.88	7.28	8.42	4.38
16.50	4.95	5.86	2.81	7.70	8.77	3.65
17.50	5.10	5.64	2.01	6.89	7.65	2.84

(data from Roche, unpublished)

TABLE 341

RADIOGRAPHIC FAT THICKNESS (mm) OVER THE ELEVENTH RIB IN NEWLY-BORN WHITE INFANTS
IN OHIO

Sex	Percentiles				
	5	15	50	85	95
Boys	1.8	2.1	2.9	3.9	4.6
Girls	2.0	2.3	3.0	3.9	4.2

(data from "Fat, body size and growth in the new born," Human Biology 30:265-280, 1958, by S. M. Garn, by permission of the Wayne State University Press. Copyright 1958, Wayne State University Press, Detroit, Michigan 48202)

TABLE 342

SUPRAILIAC SKINFOLD THICKNESS (mm) IN U.S. CHILDREN

Race, sex and age	Percentiles						
	5	10	25	50	75	90	95
WHITE							
Boys							
12 years......	3.8	4.5	5.9	9.1	16.5	27.1	31.8
13 years......	4.2	4.8	6.1	9.1	16.2	27.8	34.4
14 years......	4.3	5.0	6.5	9.4	15.6	27.5	33.6
15 years......	5.1	5.6	7.2	10.1	15.7	27.6	34.2
16 years......	5.1	5.7	7.2	10.3	16.8	25.9	32.8
17 years......	5.2	5.8	7.6	11.1	18.3	30.2	38.3
Girls							
12 years......	4.9	5.6	7.7	12.3	18.7	25.5	30.8
13 years......	5.5	6.7	9.1	13.4	19.3	26.1	32.0
14 years......	6.3	7.4	10.2	14.2	20.5	26.8	31.3
15 years......	6.6	7.6	10.5	14.9	22.1	29.7	34.5
16 years......	7.0	8.2	10.8	15.1	20.2	27.2	31.9
17 years......	6.2	7.4	10.2	14.1	20.9	28.6	33.0
NEGRO							
Boys							
12 years......	3.5	4.1	4.9	6.2	8.8	18.2	23.5
13 years......	3.3	3.8	5.0	6.6	11.8	22.4	33.3
14 years......	3.6	4.1	5.2	6.8	9.5	15.6	26.1
15 years......	4.1	4.3	5.2	6.5	10.0	13.4	23.4
16 years......	4.1	4.5	5.5	7.9	10.8	15.4	19.6
17 years......	4.3	4.8	5.5	7.1	10.5	19.8	26.6
Girls							
12 years......	4.0	5.2	7.5	11.3	18.7	25.8	32.3
13 years......	5.0	5.6	6.8	10.9	17.0	27.7	36.6
14 years......	4.7	6.3	8.6	13.7	18.7	25.4	30.2
15 years......	5.5	6.4	8.8	13.2	18.2	23.8	29.5
16 years......	6.0	6.9	9.3	12.6	21.3	30.8	35.5
17 years......	5.5	7.2	9.5	12.8	17.7	24.7	30.4

(data from Johnston et al., 1974)

TABLE   343

SUPRAILIAC SKINFOLD THICKNESS (mm) IN DENVER CHILDREN

Age (years)	Boys			Girls		
	Median	Mean	S.D.	Median	Mean	S.D.
4.00	4.00	4.00	1.57	3.85	4.80	2.58
4.50	3.51	4.15	1.82	4.99	4.98	1.61
5.00	3.99	4.13	1.39	4.03	5.06	2.43
5.50	3.50	4.02	1.83	3.97	4.38	1.54
6.00	4.01	4.42	1.51	4.01	4.45	1.69
6.50	4.02	3.99	1.38	3.99	4.41	1.58
7.00	3.94	4.16	1.38	4.03	4.77	2.25
7.50	3.93	4.22	1.88	4.35	5.54	2.78
8.00	4.10	4.12	1.68	4.47	5.83	3.82
8.50	4.06	4.20	1.33	4.98	6.51	4.58
9.00	4.73	5.14	2.65	5.93	6.90	5.00
9.50	5.18	6.11	3.84	5.36	7.36	5.41
10.00	4.24	6.04	4.17	5.00	7.27	5.35
10.50	6.07	6.70	4.75	6.69	7.94	5.28
11.00	5.64	7.44	5.26	6.92	8.83	5.52
11.50	6.00	8.60	6.62	6.10	7.72	4.77
12.00	6.04	9.77	7.58	6.91	8.34	4.73
12.50	7.04	9.13	6.12	7.91	8.69	3.97
13.00	7.05	8.95	8.17	7.75	8.61	3.37
13.50	7.32	9.85	8.07	9.17	10.12	5.15
14.00	6.93	8.85	5.59	7.60	9.03	4.27
14.50	6.27	8.40	6.55	7.66	8.73	4.04
15.00	6.95	8.87	7.02	7.98	9.93	4.79
15.50	7.65	9.69	6.33	11.46	10.92	4.79
16.00	6.94	9.82	6.95	9.85	10.70	4.60
16.50	7.33	10.72	8.55	--	--	--
17.00	6.45	8.37	5.71	9.00	9.47	3.79
17.50	6.19	9.72	7.58	--	--	--
18.00	7.59	8.67	5.17	9.71	9.96	3.70

(data from McCammon, unpublished)

TABLE   344

SUPRAILIAC SKINFOLD THICKNESS (mm) FOR BOYS IN TECUMSEH (MI)

| Per-centile | \multicolumn{14}{c}{Age (years)} |
	5	6	7	8	9	10	11	12	13	14	15	16	17	18
95	12	18	20	28	26	36	43	47	41	43	43	52	54	45
90	9	12	11	21	17	26	33	40	31	33	34	43	43	36
75	6	7	7	9	10	15	19	23	18	20	21	25	25	23
50	5	5	5	6	6	8	10	11	11	11	13	15	15	16
25	4	4	4	5	5	6	6	7	7	8	10	10	11	11
10	3	3	3	4	4	4	3	4	5	7	7	8	8	8
5	3	3	3	3	3	3	2	4	4	6	6	7	6	7
Mean	5.5	6.9	7.3	9.2	8.9	12.2	13.2	16.8	15.4	16.0	17.4	20.1	20.6	18.6
S.D.	3.0	5.6	7.0	8.7	8.3	9.8	11.1	13.4	12.1	12.4	11.3	13.9	14.8	11.5

(data from Montoye, 1978, 1970. Copyright 1978, 1970 by Allyn & Bacon, Inc., Boston. Originally published by Phi Epsilon Kappa Fraternity, Indianapolis)

TABLE   345

SUPRAILIAC SKINFOLD THICKNESS (mm) FOR GIRLS IN TECUMSEH (MI)

| Per-centile | \multicolumn{14}{c}{Age (years)} |
	5	6	7	8	9	10	11	12	13	14	15	16	17	18
95	15	15	21	26	34	34	36	40	35	40	46	38	39	46
90	12	12	18	21	20	27	29	32	27	32	38	32	32	35
75	8	9	12	13	14	18	19	21	20	22	27	22	23	25
50	6	7	7	9	9	12	12	14	13	15	18	16	16	17
25	5	5	5	5	6	7	8	9	10	11	12	11	11	12
10	4	4	4	4	5	5	6	6	8	8	9	8	8	9
5	3	4	3	4	4	4	4	5	6	7	8	7	7	8
Mean	7.5	7.6	9.2	10.9	11.6	14.4	15.2	16.8	16.1	18.2	21.0	17.7	18.4	21.0
S.D.	4.6	4.0	6.4	7.6	8.8	9.6	9.6	11.0	9.4	10.9	12.6	10.0	10.4	13.5

(data from Montoye, 1978, 1970. Copyright 1978, 1970 by Allyn & Bacon, Inc., Boston. Originally published by Phi Epsilon Kappa Fraternity, Indianapolis)

TABLE   346

PARAUMBILICAL SKINFOLD THICKNESS (mm)
IN DENVER CHILDREN

Age (Years)	Boys			Girls		
	Median	Mean	S.D.	Median	Mean	S.D.
4.00	5.00	5.17	1.67	5.00	5.76	2.84
4.50	4.99	5.19	1.49	5.00	5.52	2.38
5.00	4.83	5.02	1.70	5.02	5.95	3.75
5.50	4.32	4.37	1.12	5.02	5.32	1.56
6.00	4.88	5.05	1.93	5.01	5.79	2.68
6.50	4.62	4.52	1.11	5.49	5.48	1.98
7.00	4.50	4.96	1.94	4.81	5.65	2.65
7.50	4.28	4.74	1.67	5.72	7.01	3.74
8.00	4.37	5.00	2.07	5.68	7.25	5.18
8.50	4.74	5.31	2.53	6.02	8.22	5.76
9.00	5.25	6.76	3.58	6.53	7.86	5.31
9.50	6.25	7.87	4.51	7.84	9.63	5.82
10.00	6.35	7.94	4.09	7.71	9.73	6.15
10.50	7.47	8.39	5.00	7.94	10.80	7.85
11.00	8.36	9.87	6.16	9.92	12.96	8.83
11.50	7.54	10.75	8.50	8.96	11.50	7.24
12.00	7.85	11.57	8.74	11.33	12.81	6.45
12.50	8.21	11.25	8.27	11.98	13.11	6.51
13.00	9.61	12.25	9.24	10.71	12.46	6.24
13.50	8.79	11.06	8.56	11.31	12.72	5.94
14.00	8.42	10.24	7.19	13.00	14.49	6.04
14.50	6.63	10.07	7.89	11.19	12.48	5.02
15.00	7.64	10.03	7.66	12.82	14.79	6.15
15.50	7.42	10.86	7.78	13.82	15.28	5.81

TABLE 346 (continued)

PARAUMBILICAL SKINFOLD THICKNESS (mm)
IN DENVER CHILDREN

Age (Years)	Boys			Girls		
	Median	Mean	S.D.	Median	Mean	S.D.
16.00	9.95	12.17	7.55	14.34	15.98	6.34
16.50	9.24	12.74	9.64	--	--	--
17.00	8.06	10.23	6.95	14.19	14.30	4.67
17.50	9.01	11.21	8.30	--	--	--
18.00	10.02	10.50	5.57	15.40	15.10	5.35

(data from McCammon, unpublished)

TABLE 347

PARAUMBILICAL SKINFOLD THICKNESS (mm) AT
BIRTH IN RELATION TO BIRTH WEIGHT
FOR MONTREAL INFANTS

Weight group (gm)	Double skinfold thickness (mm)	
	Mean	S.D.
501-1,000	2.4	0.44
1,001-1,500	2.8	0.45
1,501-2,000	3.6	0.71
2,001-2,500	3.8	0.63
2,501-3,000	4.4	0.96
3,001-3,500	4.9	0.48
3,501-4,000	5.7	0.81
4,001-4,500	5.8	1.06

(data from Usher and McLean, 1969, The Journal
of Pediatrics 74:901-910)

TABLE   348

SKINFOLD THICKNESSES OF THE ABDOMEN (mm)
(midway between the umbilicus and iliac crest)
IN NEW JERSEY CHILDREN

Age	Boys			Girls		
	Mean	Median	S.D.	Mean	Median	S.D.
Newborn	3.2	3.1	0.6	3.1	3.1	0.4
3 months	5.6	5.5	1.3	6.4	6.4	2.5
6 months	5.3	4.9	1.5	6.1	5.9	1.8
9 months	5.1	4.7	1.4	5.5	5.1	1.4
12 months	4.8	4.7	1.4	5.4	5.2	1.3
18 months	4.8	4.4	1.4	4.8	4.8	0.9
2 years	4.7	4.7	1.2	5.0	4.9	1.4
2½ years	4.2	4.2	0.9	5.2	5.2	1.6
3 years	4.3	4.1	1.2	5.4	5.0	1.7
3½ years	4.4	4.1	1.4	5.1	4.6	1.8
4 years	4.0	3.9	1.0	4.8	4.3	1.8
4½ years	4.0	3.8	1.2	5.5	4.1	4.3
5 years	4.6	3.9	1.4	5.3	4.0	3.0
5½ years	4.5	4.1	1.9	5.1	4.3	2.3
6 years	4.6	3.8	2.5	5.6	4.5	2.7
6½ years	4.9	3.7	3.9	6.0	5.1	2.4
7 years	4.2	3.8	1.8	6.1	5.3	2.8
7½ years	3.9	3.5	1.5	7.4	6.1	4.0
8 years	4.5	4.0	1.8	7.5	6.0	4.1
8½ years	4.6	4.0	2.1	8.7	7.5	4.5
9 years	5.2	3.9	3.4	9.3	8.3	5.2
9½ years	5.6	4.4	3.2	9.7	8.0	5.7
10 years	5.4	4.6	2.9	9.3	8.5	4.8
10½ years	6.8	4.8	5.3	8.0	8.0	6.2
11 years	6.2	4.7	3.7	10.1	8.0	5.8
11½ years	7.8	6.0	5.0	9.9	8.2	4.7
12 years	7.8	6.5	4.5	11.4	10.0	4.9
12½ years	8.7	5.8	5.7	10.5	9.8	3.8
13 years	8.8	5.4	5.8	12.8	12.5	5.7
13½ years	6.9	5.2	4.0	13.4	13.7	4.4
14 years	9.8	6.0	7.2	--	--	--
14½ years	8.0	6.0	--	--	--	--
15 years	9.9	6.0	--	--	--	--

(data from "Typical and atypical changes in the soft tissue distribution during
childhood," Human Biology 29:62-82, 1957, by W. Kornfeld, by permission of the Wayne
State University Press. Copyright 1957, Wayne State University Press, Detroit,
Michigan 48202)

TABLE 349

PARAUMBILICAL SKINFOLD THICKNESS (mm) FOR BOYS OF TECUMSEH (MI)

Per-centile	5	6	7	8	9	Age (years) 10	11	12	13	14	15	16	17	18
95	10	19	19	26	23	35	40	47	44	43	47	53	54	44
90	9	12	11	17	15	26	30	38	34	35	36	42	45	37
75	7	8	7	9	10	15	16	22	20	19	21	26	27	24
50	5	6	6	6	6	8	9	11	11	11	14	15	15	16
25	4	5	4	5	4	5	6	7	7	8	10	11	11	12
10	3	4	4	4	4	4	4	5	5	6	7	9	9	9
5	3	3	3	3	4	3	4	4	5	6	6	8	8	8
Mean	6.0	7.3	7.3	8.6	8.6	12.3	13.8	17.0	16.0	16.0	18.1	20.2	21.0	19.5
S.D.	3.1	5.7	6.6	7.8	8.2	9.9	11.7	13.7	12.4	11.6	12.5	13.9	14.2	11.2

(data from Montoye, 1978, 1970. Copyright 1978, 1970 by Allyn & Bacon, Inc., Boston. Originally published by Phi Epsilon Kappa Fraternity, Indianapolis)

TABLE 350

PARAUMBILICAL SKINFOLD THICKNESS (mm) FOR GIRLS OF TECUMSEH (MI)

Per-centile	5	6	7	8	9	Age (years) 10	11	12	13	14	15	16	17	18
95	18	15	22	27	36	36	46	46	41	45	55	46	43	55
90	11	12	17	23	26	29	38	37	34	38	46	39	37	46
75	9	10	11	14	14	20	23	25	25	28	31	29	28	31
50	7	7	7	9	9	12	14	17	19	19	21	22	21	22
25	5	6	5	6	6	8	9	11	13	13	15	16	15	15
10	4	5	4	4	5	6	6	8	9	9	11	12	11	11
5	4	4	4	4	4	5	5	6	8	7	9	10	9	10
Mean	8.0	8.0	9.2	11.2	12.1	15.6	17.7	19.7	19.8	21.8	25.0	23.8	22.8	25.2
S.D.	4.7	3.8	6.3	7.9	9.2	10.7	11.8	12.3	10.8	12.2	13.7	11.9	11.5	14.6

(data from Montoye, 1978, 1970. Copyright 1978, 1970 by Allyn & Bacon, Inc., Boston. Originally published by Phi Epsilon Kappa Fraternity, Indianapolis)

TABLE 351

ABDOMINAL SKINFOLD THICKNESS (log units) IN
JAPANESE-AMERICAN CHILDREN IN LOS ANGELES

Age	Boys		Girls	
(years)	Mean	S.D.	Mean	S. D.
6	140.7	53.6	163.9	19.2
7	169.9	33.5	173.7	25.9
8	158.7	45.4	175.4	36.6
9	150.8	47.5	184.3	37.9
10	178.0	37.5	194.9	31.4
11	178.9	39.7	189.1	33.0
12	189.1	37.1	196.1	24.6
13	182.0	34.7	199.5	29.7
14	178.1	35.7	205.8	21.4
15	188.5	39.2	206.6	28.8
16	186.3	28.2	--	--
18	--	--	222.2	20.2

(data from Kondo and Eto, 1975, pp. 13-45 in Comparative
Studies on Human Adaptability of Japanese, Caucasians and
Japanese Americans; S. Horvath et al. (eds), University
of Tokyo Press)

TABLE 352

PARAUMBILICAL SKINFOLD THICKNESS (mm) AT BIRTH IN RELATION TO GESTATIONAL AGE
FOR MONTREAL INFANTS

Gestational age	Mean	S.D.
24-26	2.34	0.43
27-28	2.72	0.38
29-30	2.67	0.36
31-32	3.35	0.85
33	3.75	0.80
34	3.89	0.68
35	4.51	1.20
36	3.86	0.72
37	4.58	0.87
38	4.85	0.81
39	5.19	0.98
40	5.41	1.12
41	5.24	1.14
42	4.70	0.94
43	4.84	0.85
44	4.73	1.20

(data from Usher and McLean, 1969, The Journal of Pediatrics
74:901-910)

TABLE   353

BREADTH OF FAT AT WAIST (mm) IN POSTEROANTERIOR RADIOGRAPH
IN OHIO CHILDREN

Age	Boys			Girls		
(Years)	Mean	S.D.	Median	Mean	S.D.	Median
6.5	4.8	1.7	3.8	5.7	1.8	5.2
7.5	5.0	1.6	4.2	5.9	2.2	5.6
8.5	5.6	2.4	4.5	6.5	2.9	5.5
9.5	6.6	3.4	4.9	7.2	3.4	5.8
10.5	7.4	4.2	6.1	8.4	4.4	6.8
11.5	8.0	4.1	7.3	9.1	5.0	7.0
12.5	9.0	4.5	7.7	9.6	4.6	8.5
13.5	9.6	5.5	6.9	10.6	4.5	9.2
14.5	9.2	4.8	7.5	11.3	4.4	9.7
15.5	9.7	4.8	7.8	12.8	5.4	12.2
16.5	10.0	4.5	9.2	12.0	5.5	11.0
17.5	10.9	5.1	8.5	12.0	3.4	11.0

(data from Reynolds, 1951)

TABLE   354

HIP RADIOGRAPHIC FAT THICKNESS (mm)

Age (Years)	Boys			Girls		
	Median	Mean	S.D.	Median	Mean	S.D.
0.13	9.26	9.69	3.10	10.41	10.55	2.84
0.38	16.39	16.61	3.81	17.37	18.05	3.54
0.50	18.30	18.42	4.37	19.16	19.97	4.43
1.00	18.33	18.81	4.87	18.53	19.61	4.17
1.50	17.63	17.37	4.33	18.91	19.24	4.32
2.00	15.77	16.21	4.34	17.34	18.13	3.97
2.50	15.21	15.47	3.67	17.63	17.62	3.68
3.00	14.46	14.65	3.55	17.11	17.22	3.64
3.50	13.88	14.47	3.52	17.42	17.13	4.07
4.00	13.73	13.92	3.80	17.93	17.50	3.90
4.50	13.38	13.60	3.37	17.27	17.40	4.27
5.00	13.03	13.21	3.83	17.12	17.41	4.51
5.50	12.87	12.91	3.95	16.93	17.23	5.08
6.00	12.86	13.43	4.37	17.35	17.70	5.53
6.50	12.76	13.44	5.29	16.94	17.46	5.87
7.00	13.31	13.79	5.40	17.33	17.65	6.43
7.50	13.21	13.47	5.36	17.66	18.52	7.35
8.00	12.85	13.81	6.25	18.86	19.83	6.91
8.50	13.15	14.04	6.44	19.28	20.49	7.42
9.00	14.73	15.46	7.02	19.79	21.29	7.90
9.50	14.96	16.06	7.19	20.62	21.74	7.77
10.00	15.26	16.55	7.54	21.56	21.20	7.50
10.50	15.06	16.08	7.76	20.82	21.90	7.56

TABLE 354 (continued)

HIP RADIOGRAPHIC FAT THICKNESS (mm)

Age (Years)	Boys			Girls		
	Median	Mean	S.D.	Median	Mean	S.D.
11.00	16.09	17.16	7.86	22.15	22.37	6.97
11.50	15.96	16.70	7.75	21.71	22.22	6.92
12.00	16.33	17.19	8.73	22.28	23.40	7.68
12.50	15.36	17.38	8.90	22.76	23.55	7.69
13.00	15.35	17.22	8.72	22.20	24.12	7.39
13.50	13.41	15.84	8.72	26.02	25.38	7.88
14.00	13.08	14.81	9.18	25.72	25.81	8.11
14.50	11.25	14.15	9.49	25.93	25.21	7.45
15.00	10.08	13.49	9.66	26.82	27.94	7.84
15.50	10.41	12.06	7.39	--	--	--
16.00	11.35	13.63	9.86	31.19	30.25	9.89
16.50	11.33	12.86	6.38	--	--	--
17.00	9.78	11.53	8.69	28.06	31.02	10.85
17.50	--	--	--	--	--	--
18.00	9.37	10.44	7.34	--	--	--

(data from McCammon, unpublished)

TABLE 355

BREADTH OF TROCHANTERIC FAT (mm) IN POSTEROANTERIOR RADIOGRAPH
OF OHIO CHILDREN

Age	Boys			Girls		
(Years)	Mean	S.D.	Median	Mean	S.D.	Median
6.5	9.9	2.9	9.2	12.0	3.6	11.2
7.5	10.1	4.4	8.5	12.5	4.5	11.5
8.5	11.0	5.0	9.5	13.6	5.2	12.0
9.5	11.9	5.6	9.5	14.2	5.8	12.2
10.5	12.1	6.8	9.8	14.8	6.8	13.2
11.5	13.6	6.7	11.5	15.7	7.6	13.8
12.5	14.3	7.4	11.0	16.7	7.9	15.0
13.5	14.0	7.6	11.0	17.8	8.2	16.0
14.5	12.5	5.8	10.5	19.5	9.2	17.7
15.5	11.4	5.3	9.0	22.8	11.9	19.0
16.5	11.5	5.9	9.0	22.2	10.8	18.0
17.5	12.9	7.6	10.5	22.2	9.1	18.5

(data from Reynolds, 1951)

TABLE 356

SUPRAILIAC AND LATERAL CHEST (at level of nipple) SKINFOLD
THICKNESS (mm) OF FEMALE INFANTS
OF WESTERN MASSACHUSETTS

Age			Percentiles				
(months)	Mean	S.D.	10	25	50	75	90
Suprailiac							
1	5.5	2.6	0.0	4.0	6.0	7.0	8.7
2	7.8	2.9	4.0	6.2	8.0	10.0	11.1
3	8.8	3.2	5.0	7.0	8.5	10.9	12.2
4	8.8	3.2	5.0	6.4	8.5	10.5	12.9
5	8.9	3.2	5.0	7.0	8.5	10.2	13.0
6	8.7	3.1	5.0	6.0	8.5	10.5	12.0
Chest							
1	3.0	1.9	0.0	2.0	3.0	4.0	5.0
2	4.5	1.6	3.0	3.5	4.5	5.1	6.6
3	4.8	1.9	3.0	4.0	4.5	5.2	6.2
4	4.6	1.6	3.0	3.5	4.2	5.0	6.5
5	4.5	1.6	3.0	3.5	4.0	5.0	6.0
6	4.4	1.5	3.0	3.2	4.2	5.0	6.0

(data from Ferris, Beal, Laus, and Hosmer, 1979, Pediatrics 64:397-401, Copyright
American Academy of Pediatrics 1979)

TABLE 357

TRUNK SKINFOLDS (mm) FOR BLACK BOYS IN TEXAS

Age	Chest				Umbilical			
	Lower Income		Middle Income		Lower Income		Middle Income	
(years)	Mean	S.D.	Mean	S.D.	Mean	S.D.	Mean	S.D.
10.1 - 11	4.9	1.81	4.2	0.75	9.4	6.56	6.5	1.97
11.1 - 11	4.2	0.60	5.0	1.51	8.6	3.66	11.8	6.30
12.1 - 13	6.9	7.29	6.7	5.60	11.5	9.43	12.3	8.59
13.1 - 14	5.6	1.69	7.5	6.20	8.9	3.61	12.5	8.62
14.1 - 15	5.9	2.13	5.6	1.48	9.6	5.03	9.4	3.70
15.1 - 16	4.8	0.89	5.8	2.04	7.2	1.37	10.2	5.42
16.1 - 17	5.3	1.61	7.4	3.03	9.1	3.79	12.3	5.82
17.1 - 18	7.7	6.15	5.2	1.10	12.3	11.9	9.0	2.97

Age	Suprailiac				Subscapular			
	Lower Income		Middle Income		Lower Income		Middle Income	
(years)	Mean	S.D.	Mean	S.D.	Mean	S.D.	Mean	S.D.
10.1 - 11	8.4	6.00	5.6	1.71	6.4	1.90	5.3	0.78
11.1 - 12	6.8	2.26	10.6	5.76	6.0	1.94	6.7	2.43
12.1 - 13	12.4	13.66	13.2	14.72	9.3	8.95	8.9	5.74
13.1 - 14	9.2	4.46	13.4	10.38	7.0	1.60	8.6	4.28
14.1 - 15	9.2	4.69	9.7	4.23	7.8	1.60	8.0	2.92
15.1 - 16	6.5	1.70	9.0	4.81	6.8	1.38	8.8	2.13
16.1 - 17	8.9	5.46	10.1	4.64	8.1	2.28	10.5	3.87
17.1 - 18	12.3	11.88	7.5	3.12	10.8	6.03	9.0	2.24

(data from Schutte, 1979)

TABLE   358

SKINFOLD THICKNESSES (mm) IN CANADIAN
YOUTHS AGED 18 AND 19 YEARS

Percentiles	Subscapular	Triceps	Abdominal	Suprailiac
		BOYS		
95	17	11	21	19
75	10	8	12	9
50	8	6	9	7
25	7	5	7	5
5	6	4	5	4
		GIRLS		
95	17	20	24	15
75	12	16	16	11
50	10	13	12	8
25	9	11	9	6
5	7	8	6	5

Age is that at last completed year. The abdominal measurements were
made to the left of the umbilicus (vertical fold).

(data from The Canadian Association for Health, Physical Education
and Recreation, n.d.)

TABLE 359

SUM OF FOUR SKINFOLDS*(mm) FOR BOYS IN TECUMSEH (MI)

Per-centile	Age (years)													
	5	6	7	8	9	10	11	12	13	14	15	16	17	18
95	40	59	61	85	77	114	119	144	136	128	160	160	174	138
90	36	42	43	63	52	90	91	124	102	98	112	132	142	113
75	30	32	32	36	38	54	57	74	64	64	70	84	91	71
50	24	25	26	28	27	35	37	40	42	42	50	52	62	55
25	20	21	22	22	22	28	28	30	31	33	38	41	51	46
10	18	19	19	19	18	22	22	23	25	26	28	35	39	34
5	17	16	16	17	17	18	19	20	23	23	25	31	34	28
Mean	26.1	29.6	30.6	35.1	34.4	45.5	48.7	58.2	54.5	54.7	60.2	67.2	68.0	63.9
S.D.	8.9	16.6	20.8	24.2	24.8	29.1	33.2	39.5	36.2	34.3	34.1	31.9	42.0	32.0

*Triceps, subscapular, suprailiac and paraumbilical.

(data from Montoye, 1978, 1970. Copyright 1978, 1970 by Allyn & Bacon, Inc., Boston.
Originally published by Phi Epsilon Kappa Fraternity, Indianapolis)

TABLE   360

SUM OF FOUR SKINFOLDS*(mm) FOR GIRLS IN TECUMSEH (MI)

Per-centile	Age (years)													
	5	6	7	8	9	10	11	12	13	14	15	16	17	18
95	54	57	69	92	116	110	131	141	130	146	170	144	140	175
90	45	47	60	75	75	90	106	114	103	119	143	118	118	136
75	37	37	42	52	50	69	70	76	77	82	104	90	87	99
50	31	31	33	37	37	45	48	52	57	61	73	69	66	72
25	26	27	26	26	29	34	36	39	43	48	55	53	50	53
10	21	23	21	23	25	26	28	34	33	37	42	46	41	43
5	19	20	19	21	23	22	25	28	28	32	34	41	37	40
Mean	33.7	33.7	37.5	43.1	45.2	55.2	57.9	64.7	65.9	77.0	87.7	77.2	74.6	86.1
S.D.	14.2	12.3	19.0	23.6	26.5	31.7	31.3	35.5	33.0	35.4	43.0	34.0	33.4	46.8

*Triceps, subscapular, suprailiac and paraumbilical.

(data from Montoye, 1978, 1970. Copyright 1978, 1970 by Allyn & Bacon, Inc., Boston.
Originally published by Phi Epsilon Kappa Fraternity, Indianapolis)

TABLE 361

SKINFOLD THICKNESSES OF THE CHEEKS (mm) IN NEW JERSEY CHILDREN

Age	Boys			Girls		
	Mean	Median	S.D.	Mean	Median	S.D.
Newborn	7.7	7.5	1.9	8.6	8.3	1.9
3 months	9.4	9.7	2.0	9.5	9.5	1.8
6 months	10.1	10.1	1.9	10.1	10.0	1.8
9 months	10.4	10.1	1.8	10.2	10.2	1.3
12 months	9.9	9.9	2.1	9.8	10.1	1.9
18 months	10.0	9.9	1.6	10.8	11.0	1.6
2 years	9.5	9.5	1.6	9.2	9.3	1.6
2½ years	9.5	9.4	1.7	8.4	8.5	1.6
3 years	9.1	9.0	1.9	9.0	9.0	1.5
3½ years	8.7	8.5	1.4	8.7	8.9	1.7
4 years	8.3	8.2	1.6	8.2	8.0	1.6
4½ years	8.0	7.8	1.7	8.2	8.3	2.1
5 years	7.6	7.8	1.6	8.0	8.0	2.0
5½ years	7.7	7.5	1.8	7.5	7.2	1.6
6 years	7.6	7.4	2.0	7.4	7.0	1.9
6½ years	7.7	7.5	1.7	7.2	7.0	1.6
7 years	7.4	7.3	1.3	7.4	7.2	2.0
7½ years	7.4	7.5	1.6	7.3	7.0	2.1
8 years	7.4	7.5	1.5	7.2	6.7	1.8
8½ years	7.6	7.5	1.4	7.2	7.3	1.6
9 years	7.2	7.1	1.4	7.5	7.6	2.1
9½ years	6.8	6.7	1.4	7.4	7.3	2.1
10 years	6.9	6.7	1.7	6.7	6.7	1.9
10½ years	6.5	6.5	1.1	6.5	6.8	1.9
11 years	6.8	6.8	1.7	6.9	6.8	1.3
11½ years	7.1	6.7	1.6	6.9	6.5	2.1
12 years	6.7	6.6	1.7	6.3	6.2	1.1
12½ years	7.4	7.2	1.5	6.9	7.1	1.7
13 years	7.0	6.3	1.8	6.7	6.2	1.5
13½ years	6.4	6.2	1.6	6.5	6.5	1.4
14 years	6.5	6.7	1.4	--	--	--
15 years	6.0	6.0	--	--	--	--

(data from "Typical and atypical changes in the soft tissue distribution during childhood," Human Biology 29:62-82, 1957, by W. Kornfeld, by permission of the Wayne State University Press. Copyright 1957, Wayne State University Press, Detroit, Michigan 48202)

TABLE 362

TRICEPS SKINFOLD THICKNESS (mm) IN U.S. CHILDREN

Race, sex and age	Percentiles						
	5	10	25	50	75	90	95
WHITE							
Boys							
6 years.......	5.0	6.0	6.5	8.0	9.5	12.0	13.0
7 years.......	5.0	6.0	7.0	8.0	10.0	12.0	14.5
8 years.......	5.0	6.0	7.0	8.0	11.0	14.0	17.0
9 years.......	5.0	6.0	7.0	9.0	12.0	17.0	21.0
10 years.......	5.5	6.0	7.5	9.5	13.0	16.0	20.0
11 years.......	5.5	6.0	8.0	10.0	14.0	19.0	22.0
Girls							
6 years.......	6.0	6.5	8.0	10.0	11.0	14.0	16.0
7 years.......	6.5	7.0	8.0	10.0	12.5	16.0	18.0
8 years.......	6.0	7.0	9.0	11.0	14.0	18.0	20.0
9 years.......	7.0	8.0	9.0	11.5	15.0	20.0	22.5
10 years.......	6.0	7.0	9.0	12.0	16.0	20.0	23.0
11 years.......	7.0	7.5	9.0	12.0	16.0	20.1	22.0
NEGRO							
Boys							
6 years.......	4.0	5.0	5.5	7.0	8.0	10.0	11.0
7 years.......	4.0	4.0	5.0	6.0	7.0	9.0	10.0
8 years.......	4.0	4.0	5.0	6.5	8.0	12.0	13.0
9 years.......	4.0	4.0	5.0	6.5	8.0	11.0	14.0
10 years.......	4.0	4.0	5.5	7.0	9.0	11.0	13.0
11 years.......	4.0	4.0	6.0	7.0	9.0	12.0	18.0
Girls							
6 years.......	5.0	5.0	6.0	7.0	9.0	11.0	14.0
7 years.......	5.0	5.0	6.0	7.5	9.0	12.0	16.0
8 years.......	5.0	5.0	6.5	8.0	11.0	14.0	20.0
9 years.......	5.0	6.0	7.0	9.0	12.5	15.5	19.0
10 years.......	5.0	6.0	7.0	9.0	12.0	20.0	20.2
11 years.......	4.0	6.0	7.0	10.0	12.0	20.0	25.0

(data from Johnston et al., 1973)

TABLE   363

TRICEPS SKINFOLD THICKNESS (mm) IN U.S. CHILDREN

	Percentiles						
Race, sex and age	5	10	25	50	75	90	95
WHITE							
Boys							
12 years......	5.2	5.7	7.2	9.7	13.6	19.8	23.2
13 years......	4.8	5.4	7.1	9.4	13.4	19.7	22.6
14 years......	4.3	5.0	6.3	8.2	12.5	17.4	21.2
15 years......	4.3	4.8	6.0	7.8	11.2	16.4	21.3
16 years......	4.2	5.0	5.9	7.6	11.6	16.5	20.5
17 years......	4.1	4.5	5.6	7.7	11.6	15.8	20.7
Girls							
12 years......	6.1	7.1	9.2	12.0	16.0	22.1	25.1
13 years......	6.6	7.6	9.6	12.7	17.2	22.7	25.4
14 years......	7.3	8.5	11.0	14.2	18.7	23.5	26.8
15 years......	7.5	8.8	12.0	15.1	20.0	25.4	29.9
16 years......	8.1	9.7	12.3	16.0	21.2	25.5	29.1
17 years......	8.5	10.1	12.4	16.3	20.8	25.3	29.5
NEGRO							
Boys							
12 years......	3.8	4.7	5.6	7.4	10.3	15.3	23.2
13 years......	3.6	4.2	5.2	7.2	10.3	15.6	25.2
14 years......	3.6	4.1	4.8	6.4	8.4	14.2	19.2
15 years......	3.9	4.2	5.1	6.4	7.7	10.6	14.7
16 years......	4.0	4.3	4.9	6.7	8.9	11.7	12.8
17 years......	4.2	4.6	5.3	6.1	8.5	14.0	15.8
Girls							
12 years......	6.0	6.5	7.7	10.6	16.2	22.5	25.6
13 years......	6.0	6.3	7.7	9.8	13.6	23.9	27.2
14 years......	5.5	6.7	10.0	12.5	17.3	22.1	24.6
15 years......	7.2	8.2	10.2	12.8	17.9	22.7	25.8
16 years......	7.0	7.6	9.6	13.1	17.8	26.4	31.5
17 years......	7.1	7.6	10.7	13.4	18.2	23.3	25.8

(data from Johnston et al., 1974)

TABLE   364

TRICEPS SKINFOLD THICKNESS (mm) IN U.S. CHILDREN

ALL BOYS

Age (years)	Mean Age	Mean	Median	S.E. of Mean
1	1.55	10.7	10.7	0.43
2	2.45	10.1	10.2	0.32
3	3.46	9.9	10.4	0.33
4	4.49	9.5	9.8	0.33
5	5.53	9.3	9.2	0.36
6	6.46	8.8	8.6	0.49
7	7.47	8.7	8.3	0.34
8	8.46	8.9	8.6	0.45
9	9.47	11.0	10.2	0.55
10	10.52	10.4	10.0	0.49
11	11.48	10.8	9.6	0.61
12	12.56	11.2	10.5	0.51
13	13.48	11.3	10.2	0.96
14	14.46	10.2	8.9	0.92
15	15.50	10.9	9.3	0.89
16	16.52	9.8	8.4	0.83
17	17.51	9.0	7.6	0.75

WHITE BOYS

1	--	11.0	11.1	0.44
2	--	9.7	9.9	0.39
3	--	10.1	10.5	0.40
4	--	9.7	10.1	0.36
5	--	9.4	9.4	0.39
6	--	9.2	9.1	0.57

TABLE 364 (continued)

TRICEPS SKINFOLD THICKNESS (mm) IN U.S. CHILDREN

WHITE BOYS

Age (years)	Mean Age	Mean	Median	S.E. of Mean
7	--	9.0	8.6	0.37
8	--	9.1	8.8	0.52
9	--	11.6	10.6	0.75
10	--	10.5	10.3	0.56
11	--	11.5	10.2	0.66
12	--	11.6	10.8	0.60
13	--	11.9	10.4	1.33
14	--	11.0	9.6	1.06
15	--	10.9	9.2	0.87
16	--	10.2	8.8	0.98
17	--	8.9	7.7	0.84

NEGRO BOYS

Age (years)	Mean Age	Mean	Median	S.E. of Mean
1	--	9.5	9.0	0.57
2	--	11.8	10.5	1.57
3	--	9.1	10.0	0.65
4	--	8.4	8.1	0.57
5	--	8.7	7.7	0.87
6	--	7.1	7.5	0.64
7	--	6.2	6.3	0.38
8	--	7.3	6.6	0.76
9	--	8.5	8.0	0.83
10	--	8.7	8.3	0.74
11	--	7.4	6.6	0.60
12	--	8.7	7.6	0.72
13	--	8.4	7.6	1.53

TABLE 364 (continued)

TRICEPS SKINFOLD THICKNESS (mm) IN U.S. CHILDREN

NEGRO BOYS

Age (years)	Mean Age	Mean	Median	S.E. of Mean
14 15		8.7	8.2	1.56
16	--	7.3	6.8	0.66
17	--	9.6	6.4	2.29

ALL GIRLS

1	1.49	10.3	10.6	0.42
2	2.45	10.4	10.5	0.28
3	3.50	11.3	11.8	0.36
4	4.54	10.5	10.7	0.29
5	5.55	10.7	10.3	0.29
6	6.47	10.2	10.7	0.42
7	7.51	10.7	10.9	0.59
8	8.47	12.8	11.6	0.70
9	9.52	13.5	13.4	0.61
10	10.47	13.4	12.6	0.52
11	11.55	13.9	13.1	0.83
12	12.49	14.8	15.1	0.56
13	13.52	16.6	15.7	0.87
14	14.50	17.6	17.1	0.83
15	15.48	17.6	17.2	1.05
16	16.55	17.8	17.1	1.25
17	17.46	19.7	18.9	1.42

TABLE   364 (continued)

TRICEPS SKINFOLD THICKNESS (mm) IN U.S. CHILDREN

WHITE GIRLS

Age (years)	Mean Age	Mean	Median	S.E. of Mean
1	--	10.2	10.6	0.42
2	--	10.6	10.7	0.31
3	--	11.5	12.0	0.37
4	--	10.8	10.9	0.30
5	--	10.7	10.3	0.31
6	--	10.6	11.3	0.51
7	--	10.8	11.3	0.71
8	--	12.7	11.5	0.76
9	--	14.1	14.4	0.70
10	--	13.6	13.1	0.54
11	--	13.8	13.1	0.84
12	--	14.9	15.1	0.61
13	--	16.7	15.7	0.97
14	--	17.9	17.3	0.99
15	--	18.8	18.0	1.16
16	--	18.0	17.1	1.37
17	--	20.2	19.0	1.76

NEGRO GIRLS

Age (years)	Mean Age	Mean	Median	S.E. of Mean
1	--	10.7	10.7	0.73
2	--	9.5	9.9	0.41
3	--	10.1	10.4	0.65
4	--	8.9	8.5	0.60
5	--	10.2	10.0	0.96
6	--	8.9	8.6	0.44
7	--	10.4	9.7	0.90

TABLE 364 (continued)

TRICEPS SKINFOLD THICKNESS (mm) IN U.S. CHILDREN

NEGRO GIRLS

Age (years)	Mean Age	Mean	Median	S.E. of Mean
8	--	13.5	13.1	2.24
9	--	8.7	7.6	0.81
10	--	11.8	10.8	1.13
11	--	14.2	13.1	2.07
12	--	14.5	14.8	1.72
13	--	16.0	15.7	1.04
14	--	16.7	15.8	1.44
15	--	12.8	12.3	1.18
16	--	15.9	13.9	2.26
17	--	16.0	14.6	2.03

(data from Abraham et al., 1975)

TABLE 365

TRICEPS SKINFOLD THICKNESS (mm) IN BALTIMORE (MD) CHILDREN

Race	Sex	Age (years)	Mean	S.D.
Black	Boys	7-9	7.58	4.31
		10-12	10.04	5.83
	Girls	7-9	10.69	5.24
		10-12	12.63	5.24
White	Boys	7-9	9.54	4.07
		10-12	12.60	7.47
	Girls	7-9	12.42	5.46
		10-12	15.5	8.15

(data from Stine, Hepner and Greenstreet, 1975, American Journal of Diseases of Children 129:905-911. Copyright 1975, American Medical Association)

TABLE 366

TRICEPS SKINFOLD THICKNESS (mm)
IN CINCINNATI CHILDREN

Age (years)	White					Non-white				
	Mean	Median	S.D.	Range P_3 -	P_{97}	Mean	Median	S.D.	Range P_3 -	P_{97}
					Boys					
6	7.8	7.8	2.2	4.8	11.2	7.9	6.4	3.4	5.8	9.0
7	7.4	7.2	2.0	4.2	12.4	6.6	6.2	1.6	5.0	8.4
8	8.5	8.8	3.0	4.6	14.0	6.8	6.3	2.7	4.7	11.6
9	9.6	8.2	4.9	5.2	20.8	7.1	6.0	3.7	4.8	17.2
10	9.6	8.6	4.9	5.6	19.2	7.4	7.6	2.8	5.1	15.4
11	10.5	9.6	5.5	5.9	22.0	8.1	7.6	5.5	4.0	23.2
12	9.6	8.6	4.9	4.8	20.6	8.3	7.2	4.7	4.9	19.5
13	10.0	9.2	5.5	5.0	23.0	7.9	6.6	5.1	4.7	19.4
14	9.6	8.4	5.4	4.2	23.4	7.8	7.2	4.3	4.8	18.5
15	8.3	7.6	4.0	4.6	16.5	7.4	6.7	3.1	4.9	15.2
16	9.1	8.7	3.9	4.4	16.2	7.1	6.4	2.4	5.2	12.4
17	9.8	9.2	5.2	4.4	22.0	7.4	6.8	3.2	4.2	14.2
					Girls					
6	8.9	8.5	3.5	6.2	15.4	8.5	8.2	2.9	6.2	11.2
7	9.1	9.1	3.7	5.0	14.4	8.5	8.0	3.1	6.0	15.4
8	9.9	10.0	4.4	5.4	18.2	9.6	9.4	5.8	5.3	22.4
9	12.0	12.4	4.8	6.0	22.2	9.3	8.6	4.5	5.4	17.6
10	11.8	11.4	5.2	5.8	21.8	8.9	8.4	4.9	4.6	18.8
11	10.8	9.8	6.2	4.4	21.0	9.3	8.4	3.6	5.0	14.2
12	11.8	11.4	5.2	5.3	21.5	13.4	11.4	8.5	7.0	32.8
13	12.6	12.4	5.7	7.0	22.2	13.5	14.0	6.0	6.4	21.0
14	13.5	14.2	5.5	6.4	22.8	13.2	12.4	5.5	6.0	20.1
15	15.6	15.2	7.4	7.6	33.0	14.8	13.4	7.1	9.0	23.5
16	14.6	14.1	6.4	7.8	25.4	12.5	11.2	7.5	6.2	33.0
17	14.6	13.0	5.2	9.4	21.5	13.3	11.4	7.7	7.4	27.3

(data from "An evaluation of triceps skinfold measures from urban school children," Human Biology 40:363-374, 1968, by Rauh and Schumsky, by permission of the Wayne State University Press. Copyright 1968, Wayne State University Press, Detroit, Michigan 48202)

TABLE 367

TRICEPS SKINFOLD THICKNESS (mm)
IN DENVER CHILDREN

Age (Years)	Boys			Girls		
	Median	Mean	S.D.	Median	Mean	S.D.
4.00	9.20	8.88	2.45	9.49	9.71	2.25
4.50	8.01	8.47	2.23	9.55	10.11	2.24
5.00	8.02	8.60	2.45	10.02	10.01	2.23
5.50	7.99	8.03	2.04	10.10	10.29	2.20
6.00	7.94	8.48	2.34	10.03	10.02	2.66
6.50	7.93	8.31	2.34	9.50	9.83	2.22
7.00	7.97	8.34	2.14	9.44	10.55	3.18
7.50	7.13	7.97	2.56	10.83	10.76	3.43
8.00	8.26	8.41	3.11	10.71	11.01	3.46
8.50	7.94	8.19	2.88	11.71	11.44	4.47
9.00	9.26	9.51	3.68	10.75	11.07	4.36
9.50	9.32	10.14	4.08	10.95	11.50	3.41
10.00	8.50	10.37	5.09	10.63	11.42	4.01
10.50	9.40	10.22	4.40	12.67	12.14	4.44
11.00	10.02	11.25	5.43	11.89	12.73	4.66
11.50	9.58	11.43	6.08	10.66	11.21	3.92
12.00	9.63	10.48	5.30	11.45	12.42	5.28
12.50	8.56	10.65	5.42	11.89	12.44	5.14
13.00	7.81	10.14	5.49	10.98	11.85	4.87
13.50	8.06	8.84	4.67	11.85	11.78	4.45
14.00	7.05	8.93	4.58	11.76	11.99	4.27
14.50	5.42	7.86	5.44	10.01	11.39	4.05
15.00	5.88	7.17	3.92	12.79	13.45	5.16

TABLE 367 (continued)

TRICEPS SKINFOLD THICKNESS (mm) IN DENVER CHILDREN

Age (Years)	Boys			Girls		
	Median	Mean	S.D.	Median	Mean	S.D.
15.50	6.26	7.92	4.33	13.73	12.56	3.56
16.00	6.90	8.14	4.84	13.87	14.07	5.48
16.50	6.64	8.52	6.09	--	--	--
17.00	6.17	7.04	3.28	14.47	13.47	4.84
17.50	6.43	7.13	2.83	--	--	--
18.00	6.55	7.08	2.84	14.95	14.68	5.12

(data from McCammon, unpublished)

TABLE 368

TRICEPS SKINFOLD THICKNESS (mm) IN OHIO CHILDREN

Age (Years)	Boys			Girls		
	Median	Mean	S.D.	Median	Mean	S.D.
3.00	10.50	10.21	1.21	10.82	11.87	2.23
3.50	9.47	9.90	2.01	11.90	11.99	2.14
4.00	10.01	10.43	1.94	11.80	11.81	2.53
4.50	10.15	10.46	2.65	11.18	11.41	2.52
5.00	9.14	9.70	2.71	11.11	11.28	2.29
5.50	9.22	9.48	1.84	10.12	10.72	2.73
6.00	8.68	8.84	2.01	10.75	11.16	2.66
6.50	9.24	9.10	1.82	10.17	11.14	3.73
7.00	9.11	8.96	2.13	9.52	10.64	2.90

TABLE 368 (continued)

TRICEPS SKINFOLD THICKNESS (mm)
IN OHIO CHILDREN

Age (Years)	Boys			Girls		
	Median	Mean	S.D.	Median	Mean	S.D.
7.50	8.71	9.54	2.89	9.93	10.75	2.76
8.00	8.68	9.65	3.23	10.25	11.26	3.38
8.50	8.52	9.39	2.74	10.49	11.87	3.97
9.00	9.10	9.88	3.13	11.01	12.42	4.33
9.50	9.10	10.21	3.16	10.96	12.23	4.44
10.00	8.68	9.33	2.74	11.21	12.96	4.74
10.50	9.37	10.54	4.11	12.83	13.21	5.03
11.00	9.52	11.21	4.77	12.93	12.93	4.11
11.50	10.05	11.60	5.58	12.40	12.96	4.63
12.00	10.18	11.47	4.87	12.80	14.13	5.47
12.50	9.63	11.10	4.41	10.94	12.90	5.29
13.00	9.63	10.81	4.48	11.84	13.03	4.87
13.50	9.16	11.48	5.58	12.74	13.28	5.44
14.00	8.35	9.65	4.58	13.01	13.17	3.96
14.50	8.07	10.02	5.50	12.21	13.61	4.38
15.00	8.19	9.41	4.93	13.57	14.32	4.40
15.50	8.07	9.59	4.92	13.40	14.02	4.40
16.00	7.74	9.36	4.50	13.47	15.35	6.22
16.50	6.57	8.72	5.08	12.55	13.79	5.45
17.00	7.02	8.28	3.66	13.74	14.37	4.60
17.50	6.10	6.99	2.71	12.60	13.79	5.61
18.00	6.80	7.62	2.94	14.69	14.91	4.64

(data from Roche, unpublished)

TABLE 369

TRICEPS SKINFOLD THICKNESS (mm) FOR BOYS IN TECUMSEH (MI)

Per-centiles	5	6	7	8	9	10	11	12	13	14	15	16	17	18
						Age (years)								
95	14	18	16	21	21	26	28	29	26	24	27	28	30	26
90	13	14	14	16	15	22	22	25	22	20	22	25	24	23
75	11	11	12	13	12	16	16	18	17	15	16	18	17	16
50	9	9	10	10	10	12	12	12	12	12	12	13	12	12
25	7	8	8	8	8	9	9	10	9	9	10	10	9	9
10	6	6	7	6	7	8	7	8	8	7	8	7	7	7
5	6	5	6	6	6	6	6	7	7	6	6	6	6	6
Mean	9.5	9.9	10.3	10.9	10.6	13.4	13.4	14.7	13.6	13.3	13.9	14.8	14.0	13.7
S.D.	2.5	3.9	4.3	4.4	4.7	5.7	5.9	7.2	6.1	6.7	6.2	6.8	7.5	6.4

(data from Montoye, 1978, 1970. Copyright 1978, 1970 by Allyn & Bacon, Inc., Boston. Originally published by Phi Epsilon Kappa Fraternity, Indianapolis)

TABLE 370

TRICEPS SKINFOLD THICKNESS (mm) FOR GIRLS IN TECUMSEH (MI)

Per-centile	5	6	7	8	9	10	11	12	13	14	15	16	17	18
						Age (years)								
95	18	19	19	25	23	27	27	31	31	34	40	34	33	36
90	16	16	17	21	19	23	24	26	27	29	35	30	28	31
75	13	13	14	16	16	18	19	20	21	22	27	25	23	26
50	11	11	12	13	13	14	14	15	16	17	21	20	18	22
25	10	10	10	10	11	11	11	12	13	13	16	16	15	17
10	8	8	8	8	8	8	9	10	11	11	13	14	13	14
5	7	7	7	7	8	7	7	8	9	10	11	12	11	12
Mean	12.0	12.0	12.4	13.5	13.6	15.3	15.5	16.8	17.8	18.6	22.2	21.3	19.4	22.6
S.D.	3.6	3.5	4.2	5.2	4.8	6.6	6.1	6.8	7.0	7.3	9.1	7.2	6.6	8.8

(data from Montoye, 1978, 1970. Copyright 1978, 1970 by Allyn & Bacon, Inc., Boston. Originally published by Phi Epsilon Kappa Fraternity, Indianapolis)

TABLE 371

TRICEPS SKINFOLD THICKNESS (mm) BY ETHNIC GROUPS

Age Mid-Point (years)	Boys Ethnic Group		Girls Ethnic Group	
	Black	White	Black	White
	Mean	Mean	Mean	Mean
1	10.1	9.9	10.5	9.8
2	10.2	10.0	10.4	10.0
3	10.2	9.8	10.0	9.7
4	9.3	9.3	9.4	10.0
5	8.5	9.1	9.0	10.4
6	8.4	8.6	9.2	10.0
7	7.9	8.7	9.8	10.5
8	8.2	9.2	9.9	11.0
9	8.8	10.2	10.6	12.5
10	8.8	11.0	11.8	13.8
11	10.3	11.8	13.2	13.7
12	10.2	12.4	13.1	14.2
13	10.5	11.7	14.1	15.3
14	10.1	11.3	15.9	16.1
15	9.5	11.9	15.8	17.4
16	9.4	12.0	16.5	16.8
17	8.9	9.8	17.0	17.7

(data from Ten-State Nutrition Survey, 1968-1970)

TABLE 372

TRICEPS SKINFOLD THICKNESS (mm) FOR PHILADELPHIA CHILDREN

Age	WHITE BOYS		BLACK BOYS	
(years)	Mean	S.D.	Mean	S.D.
6	10.95	3.83	7.52	4.65
7	11.40	5.18	7.78	3.83
8	12.06	4.96	8.06	3.71
9	12.78	5.58	9.47	4.47
10	14.68	7.40	9.97	6.75
11	13.71	6.39	10.63	8.16
12	13.81	7.64	9.11	5.60
13	--	--	10.13	6.39

Age	WHITE GIRLS		BLACK GIRLS	
(years)	Mean	S.D.	Mean	S.D.
6	12.29	3.07	9.66	4.41
7	13.49	4.43	10.76	5.02
8	14.15	5.43	11.67	5.55
9	15.66	5.51	11.95	6.89
10	17.10	5.50	12.57	6.61
11	16.45	5.25	12.39	6.45
12	17.87	4.57	12.21	7.10
13	--	--	11.50	5.25

Age is for completed years.

(data from Malina, unpublished a)

TABLE   373

TRICEPS SKINFOLD THICKNESS (cm)
FOR NAVAJO CHILDREN IN ARIZONA

Age (years)	Mean	Range
BOYS		
1 - 4	0.92	0.6 - 1.6
5 - 9	0.92	0.5 - 1.7
10 - 14	1.12	0.4 - 2.7
15 - 19	1.45	0.7 - 2.4
GIRLS		
1 - 4	0.94	0.7 - 1.3
5 - 9	1.11	0.6 - 2.2
10 - 14	1.29	0.6 - 2.2

(data from Nutrition Survey of Lower Greasewood Chapter
Navajo Tribe, 1968-1969)

TABLE   374

TRICEPS SKINFOLD THICKNESS (log units) IN
JAPANESE-AMERICAN CHILDREN IN LOS ANGELES

Age (years)	Boys		Girls	
	Mean	S.D.	Mean	S.D.
6	188.0	11.3	194.5	8.2
7	194.5	13.5	192.9	11.0
8	195.1	18.0	198.6	17.4
9	191.1	16.6	199.0	18.0
10	199.9	16.9	201.8	17.1
11	195.2	21.3	199.1	20.6
12	197.8	24.2	201.0	17.8
13	186.7	24.0	205.6	20.3
14	181.3	24.4	210.6	17.6
15	186.1	20.9	210.7	18.3
16	189.1	15.3	--	--
18	--	--	222.8	12.6

(data from Kondo and Eto, 1975, pp. 13-45 in S. Horvath
et al. (eds), Comparative Studies on Human Adaptability
of Japanese, Caucasians and Japanese Americans. University of Tokyo Press)

TABLE 375

TRICEPS SKINFOLD THICKNESS (mm) IN CANADIAN CHILDREN

| Age Group (years) | Sex | Percentiles | | | | | | |
		10	25	40	50	60	75	90
2	M. + F.	--	7.8	8.5	9.4	9.9	10.7	--
3	M. + F.	--	7.6	8.4	9.1	9.8	11.0	--
4	M. + F.	--	7.2	8.0	8.5	9.2	10.2	--
5	M. + F.	--	6.3	7.2	7.7	8.2	9.6	--
6	M. + F.	--	6.1	7.0	7.6	8.1	9.4	--
7	M.	4.2	5.1	5.9	6.4	7.1	8.1	9.8
	F.	5.5	6.6	7.4	8.0	8.7	9.7	12.1
8	M.	4.2	5.1	5.9	6.4	7.0	8.1	9.6
	F.	5.5	6.6	7.8	8.5	9.2	10.5	12.9
9	M.	4.3	5.1	5.9	6.4	7.0	8.1	10.4
	F.	5.6	6.9	7.9	8.5	9.3	11.0	14.6
10	M.	4.4	5.2	6.1	6.7	7.3	8.4	11.3
	F.	5.5	6.8	7.8	8.6	9.6	11.0	14.6
11	M.	4.2	5.4	6.3	7.0	7.8	9.2	12.3
	F.	5.4	7.0	8.3	9.1	10.0	12.0	15.4
12	M.	4.2	5.2	6.2	6.9	7.8	9.7	12.5
	F.	5.7	7.1	8.2	8.9	10.1	12.4	18.1
13	M.	4.1	5.3	6.1	6.7	7.8	9.5	13.9
	F.	5.9	7.5	8.9	10.0	11.5	13.8	17.3
14	M.	3.7	4.6	5.3	6.2	7.1	8.4	13.0
	F.	5.8	7.8	9.4	10.7	12.0	13.4	17.4
15	M.	3.4	4.4	5.0	5.4	6.3	7.8	10.5
	F.	6.2	7.9	9.4	10.2	11.7	14.3	17.5
16-17	M.	3.2	4.0	4.7	5.2	5.8	7.8	10.9
	F.	7.0	9.5	11.9	13.1	14.3	16.4	20.2
18-19	M.	3.1	4.0	4.7	5.3	5.8	6.9	9.5
	F.	6.8	9.6	12.0	13.1	15.2	17.3	20.1

(data from Pett and Ogilvie, 1957, Canadian Bulletin on Nutrition 5:1-81. Reproduced by permission of the Minister of Supply and Services Canada)

TABLE   376

TRICEPS SKINFOLD THICKNESS (mm) IN RELATION TO INCOME LEVEL
IN U.S. CHILDREN

Age (years)	Mean Age	Income Below Poverty Level			Income Above Poverty Level		
		Mean	Median	S.E. of Mean	Mean	Median	S.E. of Mean
				BOYS			
1	1.55	9.8	8.5	1.36	10.9	11.0	0.46
2	2.45	10.6	9.0	1.25	9.9	10.3	0.42
3	3.46	9.2	10.2	0.54	10.2	10.5	0.38
4	4.49	9.6	10.2	0.29	9.5	9.6	0.43
5	5.53	9.1	9.8	1.09	9.4	9.2	0.38
6	6.46	7.8	7.8	0.55	9.4	9.2	0.66
7	7.47	7.8	7.4	0.61	8.8	8.4	0.44
8	8.46	8.1	8.2	0.70	9.1	8.7	0.61
9	9.46	9.9	8.6	1.28	11.9	10.9	0.75
10	10.52	9.4	8.6	0.85	10.7	10.3	0.51
11	11.48	8.8	9.2	0.41	11.5	10.7	0.87
12	12.56	9.4	7.6	1.61	11.6	10.8	0.49
13	13.48	9.8	8.8	1.19	11.5	10.3	1.42
14	14.46	9.3	8.2	1.98	10.5	9.3	1.02
15	15.50	10.5	9.2	1.72	10.8	9.1	0.96
16	16.52	6.9	6.8	0.81	10.4	9.0	1.05
17	17.51	11.0	11.2	2.29	8.9	7.6	0.61
				GIRLS			
1	1.49	10.2	10.6	0.67	10.3	10.6	0.46
2	2.45	9.9	10.0	0.62	10.6	10.7	0.27
3	3.50	10.0	10.4	0.85	11.8	12.0	0.44
4	4.54	10.0	10.7	0.58	10.6	10.7	0.33
5	5.55	10.4	10.0	0.93	10.7	10.3	0.32
6	6.47	9.9	10.2	0.98	10.3	10.7	0.39
7	7.51	9.6	9.5	0.82	11.1	11.2	1.02
8	8.47	10.3	8.7	1.13	13.5	13.2	0.81
9	9.52	11.2	10.4	1.35	13.9	14.4	0.72
10	10.47	10.9	10.8	1.28	13.9	13.1	0.60
11	11.55	14.2	10.8	2.56	13.8	13.4	0.84
12	12.49	18.4	20.2	1.19	13.8	14.2	0.56
13	13.52	15.2	14.4	1.06	16.8	16.1	1.03

TABLE   376   (continued)

TRICEPS SKINFOLD THICKNESS (mm) IN RELATION TO INCOME LEVEL
IN U.S. CHILDREN

		Income Below Poverty Level			Income Above Poverty Level		
Age (years)	Mean Age	Mean	Median	S.E. of Mean	Mean	Median	S.E. of Mean
			GIRLS				
14	14.50	17.9	18.1	2.73	17.6	16.7	0.99
15	15.48	15.9	15.3	0.95	16.8	17.1	1.44
16	16.55	14.9	12.0	2.14	18.9	17.7	1.57
17	17.46	20.4	17.7	3.12	19.6	18.8	1.47

(data from Abraham et al., 1975)

TABLE   377

ADJUSTED MEAN MONTHLY TRICEPS SKINFOLD MEASUREMENTS (mm)
OF FOUR FEEDING GROUPS CLASSIFIED AT AGE 2 MONTHS
IN WESTERN MASSACHUSETTS

				Groups				
Age	Breast		Breast and Food Supplements		Formula		Formula and solids	
(months)	Mean	S.D.	Mean	S.D.	Mean	S.D.	Mean	S.D.
1	5.8	2.1	6.1	1.8	6.0	1.4	5.6	3.3
2	7.3	2.0	8.2	3.2	7.7	2.6	7.5	2.3
3	7.8	2.6	8.1	2.9	7.8	1.7	9.3	2.7
4	7.8	2.4	8.1	3.2	7.6	2.2	9.1	2.9
5	8.1	2.4	8.3	2.9	8.6	3.8	8.8	3.0
6	8.7	2.7	8.5	2.6	8.4	3.3	8.2	1.9

(data from Ferris, Beal, Laus, and Hosmer, 1979, Pediatrics 64:397-401, Copyright
American Academy of Pediatrics 1979)

TABLE 378

UPPER LIMB SKINFOLD THICKNESSES (mm) FOR BLACK BOYS IN TEXAS

| Age (years) | Triceps | | | | Biceps | | | | Forearm | | | |
| | Lower Income | | Middle Income | | Lower Income | | Middle Income | | Lower Income | | Middle Income | |
	Mean	S.D.	Mean	S.D.	Mean	S.D.	Mean	S.D.	Mean	S.D.	Mean	S.D.
10.1-11	9.3	4.22	6.7	2.28	5.4	3.05	3.8	0.96	5.4	2.84	4.5	1.03
11.1-12	9.2	2.49	10.7	3.27	4.5	1.15	5.7	1.75	5.4	0.98	7.1	3.82
12.1-13	11.3	6.58	9.5	3.79	6.8	5.90	5.7	3.46	6.8	3.33	6.0	1.79
13.1-14	7.8	2.24	10.0	4.30	5.0	1.38	5.8	2.50	5.0	1.38	5.8	2.50
14.1-15	8.5	3.06	8.6	2.41	4.9	2.00	4.5	0.99	5.8	2.42	5.3	1.23
15.1-16	6.9	2.20	9.0	3.50	4.0	1.03	5.2	3.44	4.6	1.38	5.2	1.81
16.1-17	6.9	2.25	10.4	4.63	4.7	1.15	4.5	1.57	4.9	1.16	5.2	1.40
17.1-18	9.4	5.49	7.1	2.46	5.1	2.94	4.0	0.80	5.7	3.42	4.5	1.03

(data from Schutte, 1979)

TABLE 379

FREQUENCY DISTRIBUTION OF SKINFOLD THICKNESSES IN ENGLISH NEONATES

Measurements	Triceps		Subscapular	
	Boys	Girls	Boys	Girls
2.0 - 2.9	1	4	5	2
3.0 - 3.9	40	33	37	22
4.0 - 4.9	81	55	65	41
5.0 - 5.9	50	36	50	47
6.0 - 6.9	13	15	28	21
7.0 - 7.9	2	1	2	11
Total	187	144	187	144
Mean	4.67	4.70	4.87	5.22
S.D.	0.87	0.99	1.09	1.20

SKINFOLD THICKNESSES (mm) AND BODY SIZE (cm) OF ENGLISH
NEONATES IN RELATION TO BIRTH WEIGHT

	Under 2.5 kg		2.5 - 2.99 kg		3.0 - 3.49 kg		3.5 and over kg	
	Boys	Girls	Boys	Girls	Boys	Girls	Boys	Girls
Triceps Skinfold								
Mean	3.59	3.60	4.21	4.22	4.54	4.81	5.14	5.47
Subscapular Skinfold								
Mean	3.24	3.73	4.29	4.64	4.73	5.42	5.48	5.99
Recumbent Length								
Mean	44.1	44.81	48.00	48.12	49.91	49.66	51.90	51.25
S.D.	2.18	3.15	1.63	1.39	1.63	1.22	1.46	1.68
Head Circumference								
Mean	30.97	31.10	33.47	32.58	34.47	33.91	35.21	34.47
S.D.	2.04	2.41	0.78	1.70	0.88	0.90	0.90	1.14

(data from "The relation of skinfold thickness in the neonate to sex, length of gestation, size at birth and maternal skinfold," Human Biology 37:29-37, 1965, by B. Gampel, by permission of the Wayne State University Press. Copyright 1965, Wayne State University Press, Detroit, Michigan 48202)

TABLE 380

SKINFOLD THICKNESSES (mm), BIRTHWEIGHT (kg), AND BODY
SIZE (cm) IN ENGLISH FULL TERM NEONATES

	Boys	Girls
Triceps Skinfold		
Mean	4.68	4.90
Subscapular Skinfold		
Mean	4.96	5.35
Birthweight		
Mean	3.38	3.28
S.D.	0.37	0.34
Recumbent Length		
Mean	50.27	49.83
S.D.	1.87	1.53
Head Circumference		
Mean	34.61	33.70
S.D.	1.07	0.92

(data from "The relation of skinfold thickness in
the neonate to sex, length of gestation, size at birth
and maternal skinfold," Human Biology 37:29-37, 1965,
by B. Gampel, by permission of the Wayne State University
Press. Copyright 1965, Wayne State University Press,
Detroit, Michigan 48202)

TABLE 381

TRICEPS AND SUBSCAPULAR SKINFOLD THICKNESS (mm) OF FEMALE INFANTS
IN WESTERN MASSACHUSETTS

Age			Percentiles				
(months)	Mean	S.D.	10	25	50	75	90
Triceps							
1	5.9	2.2	4.0	5.0	6.0	7.0	8.0
2	7.4	2.3	5.0	6.0	7.0	8.5	11.0
3	8.1	2.5	5.0	6.0	8.0	10.0	11.0
4	8.2	2.6	5.2	6.2	8.0	10.0	11.3
5	8.3	2.8	5.0	6.0	8.0	10.0	12.0
6	8.5	2.6	5.5	6.2	8.2	10.0	11.5
Subscapular							
1	5.8	2.4	0.1	5.0	6.1	7.3	8.0
2	7.3	1.8	5.0	6.2	7.5	8.0	9.5
3	7.9	1.9	6.0	6.0	7.5	9.0	10.5
4	7.6	1.7	6.0	6.2	7.0	8.7	10.5
5	7.3	1.8	5.5	6.0	7.0	8.0	10.0
6	7.2	1.8	5.2	6.0	7.0	8.0	10.0

(data from Ferris, Beal, Laus, and Hosmer, 1979, Pediatrics 64:397-401, Copyright
American Academy of Pediatrics 1979)

TABLE   382

SKINFOLD THICKNESSES AND UPPER ARM CIRCUMFERENCE
IN TEXAS INFANTS

Age (months or years)		White			Black			Mexican-American		
		Triceps	Sub-Scapular	Circ.	Triceps	Sub-Scapular	Circ.	Triceps	Sub-Scapular	Circ.
0-2 mos.	Mean	7.79	.72	10.76	6.09	6.56	11.00	7.66	7.51	10.58
	S.D.	1.45	1.43	1.13	1.48	1.53	1.06	0.91	1.00	0.82
2-4 mos.	Mean	8.72	8.43	11.35	8.45	8.33	11.97	8.14	7.96	11.23
	S.D.	0.72	0.61	1.43	0.92	0.73	1.25	0.40	0.46	1.02
4-6 mos.	Mean	8.65	8.29	12.23	8.28	8.21	13.12	8.89	8.46	12.35
	S.D.	0.53	0.41	1.24	0.92	0.79	1.66	0.73	0.92	1.26
6-8 mos	Mean	8.20	8.12	13.47	9.09	8.62	12.35	8.29	8.33	12.46
	S.D.	0.50	0.58	1.51	0.60	0.81	1.69	0.50	0.60	1.36
8-10 mos.	Mean	9.46	9.45	12.98	9.26	8.97	13.94	9.58	9.69	12.99
	S.D.	0.60	0.60	1.46	0.59	0.50	1.13	0.48	0.55	1.89
10-12 mos.	Mean	9.08	9.13	14.42	9.98	9.59	13.33	9.50	9.19	14.75
	S.D.	0.64	0.49	1.55	0.64	0.50	1.61	0.64	0.63	1.10
1-2 yrs.	Mean	9.47	9.19	14.17	8.97	8.36	14.52	9.44	9.32	15.70
	S.D.	0.67	0.57	1.38	1.40	1.39	1.39	1.31	1.14	0.77
2-3 yrs.	Mean	9.75	9.76	14.88	8.78	9.02	14.83	10.17	9.55	15.31
	S.D.	0.70	0.69	1.15	0.92	0.67	1.21	0.66	0.42	1.06

Circ. = upper arm circumference

(data from Fry et al., 1975)

TABLE   383

MEAN SKINFOLDS (mm) FOR BOSTON BOYS AND GIRLS

SAMPLE	MEAN AGE (years)	ARM	FOREARM	WAIST	BACK	CALF
BOYS	6.0	10.5	6.8	4.5	5.2	12.7
	8.0	11.2	6.4	5.6	6.2	14.5
	9.8	11.5	6.9	4.9	6.3	14.4
	11.9	10.8	6.0	6.9	6.8	13.3
	13.9	11.4	6.8	7.3	6.7	13.4
GIRLS	6.0	13.9	8.4	8.4	7.6	14.0
	8.1	13.0	8.4	8.3	7.2	13.4
	10.0	13.4	8.6	9.7	8.0	14.2
	12.2	15.0	8.4	11.8	9.8	15.6
	13.8	15.7	9.1	12.9	11.8	17.2

(data from Baker et al., 1958)

TABLE   384

SKINFOLD THICKNESSES (mm) FOR MEXICAN-
AMERICAN BOYS IN AUSTIN (TX)

Skinfold	Age (years)	Mean	S.D.	Median
Triceps	9	8.4	2.9	7.8
	10	8.1	2.4	7.4
	11	8.7	2.2	7.8
	12	8.4	2.2	8.2
	13	9.4	4.5	7.5
	14	7.7	3.4	6.6
Subscapular	9	6.3	1.6	5.6
	10	6.4	2.0	5.6
	11	7.9	4.0	7.0
	12	7.6	2.2	7.4
	13	8.1	2.7	7.6
	14	7.7	2.9	7.2
Midaxillary	9	5.3	1.7	4.6
	10	5.4	1.6	5.0
	11	7.0	4.5	5.4
	12	6.4	1.9	5.8
	13	7.4	3.7	6.2
	14	6.6	4.1	5.4
Suprailiac	9	6.7	2.7	6.3
	10	6.9	3.1	5.9
	11	9.4	7.7	5.9
	12	8.0	2.8	7.4
	13	10.3	6.0	7.6
	14	7.2	4.1	6.0
Medial Calf	9	8.2	3.1	7.8
	10	7.6	2.6	6.8
	11	9.3	4.7	7.8
	12	8.1	2.7	8.4
	13	9.0	4.3	8.8
	14	7.0	4.6	5.3
Biceps	9	5.8	2.0	5.9
	10	5.1	1.5	4.5
	11	5.9	2.9	4.6
	12	5.1	1.2	4.8
	13	5.6	3.0	4.8
	14	4.5	1.6	4.2

TABLE 384   (continued)

SKINFOLD THICKNESSES (mm) FOR MEXICAN-
AMERICAN BOYS IN AUSTIN (TX)

Skinfold	Age (years)	Mean	S.D.	Median
Thigh	9	10.3	2.9	10.6
	10	11.4	3.7	10.2
	11	11.9	4.6	9.9
	12	10.4	2.9	10.6
	13	10.7	3.8	10.2
	14	9.4	4.8	8.4
Juxtanipple	9	8.4	3.5	7.8
	10	9.1	3.3	8.6
	11	10.0	4.6	7.6
	12	9.7	3.3	8.8
	13	10.8	4.3	10.8
	14	8.2	3.8	7.2
Abdominal	9	8.0	4.1	7.5
	10	8.6	5.4	6.6
	11	11.4	7.4	8.3
	12	10.3	4.4	9.0
	13	13.4	9.0	10.5
	14	9.7	5.6	7.7

(data from Zavaleta, 1976)

TABLE 385

SKINFOLD THICKNESSES (mm) IN WEST GERMAN CHILDREN

Age	BOYS Percentiles					GIRLS Percentiles				
(years)	10	25	50	75	90	10	25	50	75	90

Triceps

Age										
1-2	7,0	8,0	9,8	11,0	12,6	7,2	8,8	10,2	12,4	14,2
3	7,9	9,2	11,0	12,5	13,4	8,2	10,2	11,6	12,9	14,6
4	7,8	9,0	10,6	11,6	13,4	9,1	10,4	11,7	13,1	14,8
5	7,5	8,9	10,4	12,0	13,4	9,6	10,5	12,0	13,6	15,0
6	6,3	7,2	8,7	10,2	11,9	7,5	8,5	10,0	12,0	14,6
7	6,0	6,8	8,4	10,0	11,8	7,1	8,4	10,0	12,0	14,5
8	6,1	7,2	8,8	11,0	13,4	7,1	8,4	10,5	13,0	15,8
9	6,4	7,4	8,9	12,0	15,1	7,5	9,0	11,6	14,1	17,3
10	6,0	7,1	8,3	12,2	16,8	7,3	9,4	10,9	14,8	17,0
11	6,4	7,2	9,1	12,4	17,0	7,2	7,8	10,3	12,5	17,5
12	6,6	7,9	9,8	13,0	18,9	7,6	9,7	11,6	13,8	16,6
13	6,9	8,0	9,3	11,4	15,6	8,0	9,1	11,0	14,6	18,3
14	6,2	7,5	8,7	11,9	20,0	7,3	9,4	14,4	17,0	20,4

Subscapular

Age										
1-2	4,2	4,8	5,4	6,5	8,0	3,9	4,9	5,4	6,4	7,4
3	4,2	4,8	5,5	6,3	7,1	4,3	5,0	5,7	7,0	7,6
4	4,1	4,4	4,9	5,6	6,8	4,4	4,8	5,4	6,3	7,3
5	4,0	4,4	4,9	5,6	6,4	4,2	4,9	5,5	6,5	7,4
6	3,8	4,3	4,9	5,6	6,6	4,2	4,7	5,4	6,4	8,9
7	3,9	4,3	5,0	5,8	6,8	4,2	4,7	5,5	6,9	8,2
8	3,9	4,4	5,2	6,2	8,4	4,2	4,9	6,0	7,8	11,0
9	4,0	4,6	5,6	7,0	10,0	4,5	5,1	6,3	8,6	13,2
10	3,9	4,6	5,4	6,2	11,4	5,0	5,4	6,8	9,3	12,2
11	4,2	4,7	5,8	7,3	10,4	5,3	6,2	6,9	8,7	14,0
12	4,3	5,4	6,4	8,3	12,4	5,4	6,2	7,5	10,8	22,5
13	4,6	5,6	6,2	7,4	12,0	5,3	6,7	7,7	11,2	14,8
14	4,9	5,6	6,4	8,8	20,6	5,6	7,2	9,3	12,8	17,8

Suprailiac

Age										
1-2	3,4	4,2	5,5	7,6	10,0	3,6	4,8	7,3	9,2	10,0
3	3,8	5,1	6,4	7,4	10,1	4,9	6,0	7,3	10,3	12,0
4	3,8	4,6	5,6	6,8	8,3	4,7	5,6	7,8	9,3	12,0
5	4,1	4,6	5,5	7,0	9,3	4,6	5,9	7,4	9,3	12,8
6	3,4	4,0	4,6	5,7	7,2	3,8	4,7	5,6	7,5	12,0
7	3,4	4,0	4,8	6,1	8,4	3,8	4,6	6,1	8,3	12,3
8	3,6	4,1	4,9	7,2	10,5	4,2	5,0	6,8	9,6	14,4
9	3,6	4,5	5,8	8,8	14,6	4,2	5,5	7,3	11,9	19,4
10	3,7	4,6	6,1	8,8	17,8	4,6	5,2	8,2	13,6	17,2
11	4,3	5,0	6,2	11,8	16,7	5,9	6,0	9,3	12,4	18,9
12	4,4	5,6	8,0	13,8	22,6	5,9	7,0	10,3	15,2	22,5
13	4,7	6,4	7,9	11,4	20,0	6,1	7,9	10,2	15,9	22,3
14	6,2	7,0	8,0	11,2	27,0	6,2	8,2	10,4	16,2	18,6

(data from Maaser et al., 1972)

TABLE 386

ULTRASOUND MEASUREMENTS OF SUBCUTANEOUS
FAT THICKNESSES IN GERMAN CHILDREN

Age	Triceps		Subscapular		Suprailiac	
(years)	Mean	S.D.	Mean	S.D.	Mean	S.D.

Girls

Age	Mean	S.D.	Mean	S.D.	Mean	S.D.
3	6.4	1.2	4.1	1.1	5.7	1.7
4	6.3	2.0	4.2	1.2	6.0	2.3
5	6.5	1.7	4.2	1.1	6.2	1.9
6	5.7	1.4	3.6	0.9	4.6	1.7
7	5.2	1.5	3.6	1.2	4.6	2.3
8	5.5	1.6	3.6	1.2	4.8	2.3
9	6.1	1.6	4.0	1.6	5.3	2.6
10	6.9	2.5	4.7	2.2	6.8	4.0
11	6.6	2.1	5.0	2.0	6.8	3.6
12	7.0	2.3	5.3	2.3	7.7	4.3
13	8.1	2.5	5.6	1.9	7.9	3.9
14	8.1	2.6	5.3	1.8	7.2	3.5

Boys

Age	Mean	S.D.	Mean	S.D.	Mean	S.D.
3	6.1	1.4	3.8	0.8	5.2	1.8
4	5.9	1.3	3.9	0.8	4.8	1.2
5	5.9	1.3	3.9	1.2	5.1	1.8
6	4.8	1.3	3.2	1.0	3.8	1.5
7	4.9	1.7	3.2	1.3	4.0	2.4
8	5.1	1.6	3.2	1.2	3.9	1.8
9	5.2	1.6	3.2	1.0	3.9	1.8
10	6.2	2.0	4.5	2.7	6.2	4.7
11	6.1	1.9	4.3	1.9	5.8	2.9
12	7.0	2.4	5.2	3.2	6.2	3.6
13	5.9	1.8	4.6	1.7	5.7	3.1
14	6.2	2.1	4.7	2.3	5.8	3.4

(data from Maaser, 1972)

TABLE 387

BICEPS SKINFOLD THICKNESS (mm)
IN DENVER CHILDREN

Age (Years)	Males			Females		
	Median	Mean	S.D.	Median	Mean	S.D.
4.00	4.02	4.06	1.25	3.00	3.49	1.09
4.50	3.88	3.86	0.95	4.12	4.34	0.91
5.00	3.69	3.62	1.11	4.00	4.11	1.08
5.50	3.51	3.54	0.82	4.01	4.21	1.14
6.00	3.99	3.85	1.06	3.76	4.03	1.36
6.50	3.23	3.40	1.11	4.19	4.23	1.18
7.00	3.20	3.32	0.94	4.03	4.41	1.31
7.50	3.02	3.40	1.17	4.40	4.54	1.72
8.00	3.94	3.62	1.25	4.32	5.01	1.77
8.50	3.07	3.57	1.30	4.90	4.95	1.79
9.00	3.54	3.98	1.47	4.28	4.88	2.02
9.50	3.35	4.19	1.74	5.03	5.41	2.19
10.00	3.94	4.50	1.83	5.01	5.15	2.12
10.50	4.19	4.79	2.24	5.71	5.58	2.07
11.00	4.01	4.99	2.36	5.71	6.04	2.48
11.50	3.81	4.96	2.78	5.11	5.49	2.36
13.00	4.01	5.05	2.84	5.10	6.08	2.55
12.50	4.29	5.31	3.15	5.42	5.88	2.44
13.00	3.95	4.91	3.09	5.70	5.61	2.00
13.50	3.80	4.67	3.40	5.13	5.99	2.72
14.00	3.42	4.29	2.51	5.11	5.76	2.15
14.50	3.08	4.32	2.84	5.91	5.47	1.85
15.00	3.00	3.74	1.90	5.46	5.96	2.69

TABLE   387   (continued)

BICEPS SKINFOLD THICKNESS (mm)
IN DENVER CHILDREN

Age (Years)	Males			Females		
	Median	Mean	S.D.	Median	Mean	S.D.
15.50	3.02	4.16	2.55	5.88	6.24	2.61
16.00	3.01	4.56	3.17	5.74	6.49	3.62
16.50	3.48	4.92	4.75	--	--	--
17.00	3.05	3.72	2.36	6.20	6.02	2.66
17.50	3.01	3.93	2.30	--	--	--
18.00	2.99	3.29	1.20	5.90	6.01	2.42

(data from McCammon, unpublished)

TABLE 388

ARM FAT AREA X 100 (mm^2)
IN DENVER CHILDREN

Age (Years)	Male			Female		
	Median	Mean	S.D.	Median	Mean	S.D.
4.00	901.52	924.45	263.83	948.24	992.28	236.36
4.50	860.37	897.14	241.38	994.50	1050.12	240.27
5.00	895.51	930.51	280.83	1047.81	1063.38	250.65
5.50	877.47	866.34	221.28	1052.15	1109.49	273.81
6.00	896.15	939.11	268.74	1078.61	1091.95	326.62
6.50	881.48	923.51	252.57	1075.78	1088.92	263.25
7.00	900.38	955.14	269.07	1102.41	1198.97	382.32
7.50	835.81	926.16	305.52	1181.46	1235.58	411.85
8.00	943.05	978.10	367.45	1266.89	1306.90	463.88

TABLE 388 (continued)

ARM FAT AREA X 100 (mm^2)
IN DENVER CHILDREN

Age (Years)	Male			Female		
	Median	Mean	S.D.	Median	Mean	S.D.
8.50	904.08	987.27	359.03	1376.56	1377.69	620.98
9.00	1097.40	1185.93	492.92	1257.28	1345.84	581.53
9.50	1245.29	1307.42	572.25	1329.29	1437.25	496.08
10.00	1105.56	1370.47	732.43	1361.15	1444.34	572.39
10.50	1244.01	1370.09	620.40	1485.67	1548.33	627.85
11.00	1336.64	1537.59	753.89	1471.70	1663.27	676.04
11.50	1399.16	1562.65	811.79	1470.76	1506.52	578.76
12.00	1370.51	1480.24	740.57	1511.09	1695.72	754.05
12.50	1377.22	1546.08	814.17	1641.39	1750.32	775.52
13.00	1290.85	1533.40	876.77	1542.71	1682.17	725.13
13.50	1301.36	1365.31	744.83	1576.54	1698.31	691.17
14.00	1133.76	1390.55	736.53	1711.57	1759.34	688.98
14.50	853.46	1244.20	880.69	1538.84	1673.82	582.74
15.00	1006.47	1194.02	702.22	1824.47	1987.47	763.81
15.50	1074.26	1340.59	780.02	2070.18	1921.79	528.26
16.00	1153.08	1441.22	928.47	2081.59	2105.24	838.46
16.50	977.23	1470.81	1114.60	--	--	--
17.00	1091.65	1232.49	588.51	2145.39	2043.84	765.49
17.50	1201.69	1281.79	535.28	--	--	--
18.00	1170.31	1265.70	526.11	2313.59	2253.24	826.25

(data from McCammon, unpublished)

TABLE   389

SELECTED BODY MEASUREMENTS IN WHITE MINNESOTA
INFANTS AT A MEAN AGE OF 32 DAYS

	Male Infants		Female Infants	
	Mean	S.D.	Mean	S.D.
Skinfold (mm)				
Upper arm (ventral)	3.9	0.80	3.7	0.89
Upper arm (dorsal)	5.8	1.32	5.7	1.25
Forearm	5.8	1.52	5.6	1.10
Subscapular	5.2	1.32	5.4	1.37
Thigh	8.0	1.95	8.6	2.23
Calf	7.6	1.65	7.4	1.74
Skeletal Diameters (cm)				
Bicristal	9.2	0.48	8.9	0.48
Humerus	3.0	0.28	2.8	0.30
Radius-ulna	2.6	0.16	2.4	0.19
Femur	4.0	0.30	3.8	0.25
Malleolus	3.0	0.20	2.8	0.27
Skeletal diameters corrected for fat thickness				
Upper arm	3.0	0.25	2.9	0.23
Forearm	3.0	0.27	2.8	0.18
Thigh	4.4	0.43	4.3	0.40
Calf	3.1	0.37	3.0	0.22

(data from Novak, Hamamoto, Orvis, et al., 1970, American Journal of
Diseases of Children 119:419-423. Copyright 1970, American Medical
Association)

TABLE 390

LOG SKINFOLD THICKNESSES IN RELATION TO BIRTHWEIGHT IN ENGLISH INFANTS

Birth-Weight (kg)	1 Month				3 months			
	Biceps	Triceps	Subscap- ular	Supra- iliac	Biceps	Triceps	Subscap- ular	Supra- iliac
BOYS								
<3.2	148.4 (19.1)	169.3 (13.0)	166.3 (12.3)	153.6 (15.9)	168.3 (17.0)	191.8 (13.4)	179.9 (16.4)	178.3 (16.6)
<3.2	148.6 (16.6)	173.3 (13.8)	170.6 (12.4)	150.8 (17.2)	166.1 (15.2)	191.5 (11.4)	182.1 (12.3)	173.2 (14.9)
>3.7	152.8 (14.4)	177.0 (11.6)	175.5 (15.4)	156.2 (15.6)	166.7 (13.8)	193.3 (11.7)	181.8 (17.8)	175.1 (22.8)
ALL BOYS	149.9 (16.6)	173.2 (13.3)	170.8 (13.5)	153.5 (16.5)	167.0 (15.3)	192.2 (12.0)	181.2 (15.0)	175.5 (17.6)
GIRLS								
<3.0	146.2 (13.3)	170.9 (11.8)	166.0 (13.9)	150.3 (15.3)	162.4 (19.3)	185.3 (11.6)	176.8 (12.9)	169.4 (12.9)
<3.0	147.3 (15.7)	173.8 (11.0)	172.5 (11.2)	152.8 (15.2)	169.5 (14.0)	190.0 (10.1)	184.9 (12.0)	179.3 (17.8)
>3.5	149.8 (12.7)	179.0 (9.2)	176.6 (11.6)	158.6 (14.8)	165.7 (17.4)	192.8 (11.2)	185.5 (11.6)	181.7 (13.3)
ALL GIRLS	147.8 (14.1)	174.5 (11.1)	171.7 (12.2)	153.8 (15.3)	165.9 (16.6)	189.4 (11.1)	182.4 (12.6)	176.8 (15.9)

	6 Months				9 Months			
BOYS								
<3.2	173.9 (18.6)	197.6 (8.5)	180.6 (13.8)	180.4 (18.1)	179.7 (12.6)	200.7 (12.4)	182.1 (13.3)	179.9 (19.2)
<3.2	175.2 (14.3)	196.8 (16.0)	184.0 (14.5)	182.6 (18.1	178.3 (14.3)	200.6 (12.0)	182.0 (16.5)	177.6 (21.5)
>3.7	177.1 (12.4)	197.9 (13.0)	183.5 (14.1)	180.3 (23.0)	176.0 (18.1)	200.4 (11.7)	183.9 (14.0)	180.5 (20.3)

TABLE 390   (continued)

LOG SKINFOLD THICKNESSES IN RELATION TO BIRTHWEIGHT IN ENGLISH INFANTS

Birth-Weight (kg)	6 Months				9 Months			
	Biceps	Triceps	Subscap-ular	Supra-iliac	Biceps	Triceps	Subscap-ular	Supra-iliac
ALL BOYS	175.4 (15.2)	197.4 (13.3)	182.7 (14.2)	181.1 (19.3)	178.0 (14.8)	200.6 (11.9)	182.7 (14.9)	179.3 (20.4)
GIRLS								
< 3.0	174.6 (15.0)	191.0 (13.5)	181.3 (14.2)	178.9 (13.4)	173.1 (14.5)	194.4 (13.2)	181.3 (15.4)	181.8 (19.4)
< 3.0	176.9 (12.1)	195.4 (9.4)	186.1 (13.4)	181.5 (16.9)	177.2 (19.3)	197.6 (9.9)	185.2 (13.7)	184.0 (18.4)
> 3.5	174.0 (13.6)	196.7 (13.0)	185.5 (11.8)	187.5 (11.1)	179.2 (15.2)	196.9 (16.4)	183.0 (11.2)	179.9 (9.4)
ALL GIRLS	175.2 (13.3)	194.5 (11.9)	184.3 (13.2)	182.6 (14.7)	176.5 (16.8)	196.2 (13.0)	183.2 (13.5)	181.9 (17.4)

	12 Months			
BOYS				
< 3.2	177.9 (16.0)	196.0 (16.6)	181.5 (13.3)	177.7 (20.4)
< 3.2	176.9 (14.7)	197.5 (15.1)	181.2 (15.4)	180.3 (18.5)
> 3.7	175.8 (14.1)	200.1 (13.5)	185.0 (15.1)	181.8 (18.3)
ALL BOYS	176.8 (14.8)	197.7 (14.9)	182.6 (14.7)	179.9 (18.9)
GIRLS				
< 3.0	175.7 (15.9)	198.0 (11.5)	179.8 (11.2)	181.6 (18.2)
< 3.0	175.7 (15.7)	198.8 (12.7)	181.7 (11.7)	178.3 (18.1)
> 3.5	172.2 (18.3)	198.7 (16.3)	180.9 (15.7)	183.4 (18.1)
ALL GIRLS	174.5 (16.5)	198.5 (13.4)	180.8 (12.8)	181.1 (18.1)

The figures in parentheses are S.D.

(data from Hutchinson-Smith, 1973)

TABLE 391

DELTOID RADIOGRAPHIC FAT THICKNESS (mm)
IN DENVER CHILDREN

Age (Years)	Male			Females		
	Median	Mean	S.D.	Median	Mean	S.D.
0.13	8.68	9.08	1.92	9.57	9.50	1.60
0.38	11.58	11.69	1.78	12.52	12.26	1.56
0.50	12.46	12.45	1.91	12.75	12.76	1.86
1.00	11.93	12.16	2.29	12.91	12.75	2.26
1.50	11.75	11.88	2.04	12.51	12.62	2.18
2.00	11.50	11.28	1.88	12.67	12.44	2.22
2.50	11.30	11.22	1.78	12.32	12.20	2.14
3.00	10.70	10.94	1.78	11.94	12.00	2.08
3.50	10.55	10.61	1.77	11.63	11.77	2.10
4.00	10.19	10.31	2.10	11.87	11.92	2.32
4.50	9.95	9.99	1.77	11.46	11.71	2.34
5.00	10.06	9.86	2.07	10.91	11.64	2.47
5.50	9.48	9.60	1.98	11.35	11.47	2.80
6.00	9.29	9.62	2.30	10.75	11.57	3.33
6.50	9.21	9.50	2.58	10.82	11.39	3.26
7.00	9.06	9.49	2.64	10.70	11.63	3.30
7.50	8.51	9.27	2.57	11.23	12.15	3.87
8.00	9.27	9.79	3.27	11.45	12.32	4.01
8.50	9.34	9.67	3.42	11.78	12.66	4.20
9.00	8.93	10.17	3.70	12.70	13.24	4.49
9.50	10.17	10.54	3.41	12.91	13.49	4.31
10.00	10.05	10.74	3.81	12.96	13.41	4.32
10.50	10.40	10.82	4.08	13.15	13.51	4.41

TABLE 391 (continued)

DELTOID RADIOGRAPHIC FAT THICKNESS (mm)
IN DENVER CHILDREN

Age (Years)	Males			Females		
	Median	Mean	S.D.	Median	Mean	S.D.
11.00	10.43	11.28	4.06	13.31	13.88	4.45
11.50	10.61	11.14	4.12	14.12	14.24	4.41
12.00	9.81	11.73	4.68	13.92	14.86	4.65
12.50	10.01	11.87	4.97	13.82	14.65	4.26
13.00	11.08	12.10	5.42	14.67	14.88	4.18
13.50	8.77	10.74	5.21	13.41	14.77	4.40
14.00	7.89	9.84	4.99	14.66	15.39	4.64
14.50	7.24	9.55	5.52	13.18	14.32	3.98
15.00	8.04	9.40	5.71	15.96	16.02	4.69
15.50	6.61	8.59	5.42	--	--	--
16.00	6.73	9.53	6.76	17.40	17.06	4.79
16.50	7.43	8.42	4.32	--	--	--
17.00	6.05	7.53	4.60	19.67	19.66	5.19
17.50	--	--	--	--	--	--
18.00	5.81	6.86	4.11	--	--	--

(data from McCammon, unpublished)

TABLE   392

BREADTH OF FAT (mm) AT LEVEL OF DELTOID*
IN OHIO CHILDREN

Age	Boys			Girls		
(Years)	Mean	S.D.	Median	Mean	S.D.	Median
6.5	5.2	1.4	4.6	6.3	1.8	5.5
7.5	5.1	1.8	4.0	6.4	2.3	5.7
8.5	5.5	2.2	4.2	6.7	2.6	5.7
9.5	5.6	2.5	4.2	7.0	3.0	5.7
10.5	5.9	2.7	5.0	7.4	3.6	5.9
11.5	6.4	2.8	5.2	7.2	3.8	5.7
12.5	6.3	3.4	5.3	7.3	3.3	6.8
13.5	6.0	3.3	4.4	7.6	3.2	7.1
14.5	5.4	2.4	4.0	8.0	3.5	7.2
15.5	5.1	2.0	3.9	9.3	4.7	8.0
16.5	5.4	1.9	4.5	8.8	4.1	7.3
17.5	5.8	2.6	4.4	8.4	3.2	6.7

(data from Reynolds, 1951)

*measured in anteroposterior view at greatest lateral bulge of the muscle

TABLE 393

FAT THICKNESSES MEASURED ON RADIOGRAPHS (mm) IN COLORADO CHILDREN

	Level of Deltoid Insertion						Level of Max. Hip Bulge					
	Boys			Girls			Boys			Girls		
Age	Percentiles			Percentiles			Percentiles			Percentiles		
(yr.-mo.)	10	50	90	10	50	90	10	50	90	10	50	90
0 - 2	7.50	9.21	11.98	8.67	10.02	11.72	7.45	10.56	15.64	8.81	11.62	15.28
0 - 4	9.18	11.26	13.58	9.82	12.03	13.74	11.77	15.80	21.37	13.50	16.34	21.88
0 - 6	10.25	12.52	14.95	10.31	12.96	15.22	13.18	18.44	25.21	14.34	19.87	26.71
1 - 0	9.68	11.83	14.69	9.75	12.60	15.49	13.19	18.40	25.89	14.46	18.95	25.08
1 - 6	9.53	11.65	14.49	9.62	12.45	15.50	11.82	17.41	21.83	13.02	18.76	25.51
2 - 0	8.62	11.38	13.45	9.23	12.34	15.21	11.01	15.84	21.34	13.13	17.30	24.30
2 - 6	8.76	11.42	13.57	9.36	12.30	14.60	11.21	15.26	20.64	11.88	17.48	22.72
3 - 0	8.77	10.65	13.14	9.14	11.98	14.94	9.95	14.47	19.87	12.21	17.25	21.80
3 - 6	8.27	10.42	12.87	9.36	11.76	14.08	10.40	14.07	19.80	11.11	16.98	22.21
4 - 0	7.57	10.25	13.00	8.82	11.59	15.28	9.71	13.72	19.20	11.42	17.43	22.28
4 - 6	7.82	9.96	12.09	8.78	11.19	15.26	9.33	13.33	18.15	11.34	17.16	22.58
5 - 0	6.94	9.98	12.28	8.85	10.81	15.53	8.70	13.45	17.19	11.20	16.63	23.20
5 - 6	7.30	9.58	12.00	7.96	11.01	16.02	7.85	12.87	17.71	10.48	16.71	23.96
6 - 0	6.75	9.34	12.53	7.91	10.65	16.84	8.07	12.99	17.84	10.28	17.41	24.20
6 - 6	6.49	9.21	12.14	7.29	10.78	15.18	8.52	12.57	18.52	10.11	16.66	24.97
7 - 0	6.50	9.16	12.81	8.06	10.72	16.20	7.83	13.19	18.74	9.85	16.94	24.08
7 - 6	6.45	8.74	12.45	7.73	11.32	17.86	7.45	13.28	17.84	9.72	17.03	27.53
8 - 0	6.28	9.55	14.54	7.76	11.07	18.38	7.00	13.42	20.47	11.14	18.86	29.82
8 - 6	5.94	8.87	13.88	8.26	11.88	18.46	7.14	13.16	20.64	11.36	18.88	31.27
9 - 0	6.05	9.24	15.58	8.15	12.20	19.93	8.06	14.37	25.54	11.36	19.62	32.56
9 - 6	6.32	9.78	15.71	8.69	12.84	19.62	7.78	14.93	24.00	12.78	20.22	32.58
10 - 0	6.62	10.12	16.32	7.99	12.84	20.38	9.14	15.79	25.26	12.22	20.96	32.15
10 - 6	6.66	9.97	16.72	8.57	12.86	19.42	8.40	15.11	28.73	12.57	21.79	32.28
11 - 0	6.65	10.54	16.77	9.01	12.93	19.97	7.13	16.37	26.38	13.02	21.93	30.62
11 - 6	6.34	10.05	16.33	9.05	14.36	19.95	7.93	16.28	26.24	13.37	21.49	30.81
12 - 0	6.69	10.82	18.42	9.26	13.77	20.83	7.92	15.35	29.24	13.74	22.12	34.40
12 - 6	7.22	10.13	19.29	9.47	14.12	21.93	8.24	14.95	30.56	13.73	22.24	33.44
13 - 0	6.90	11.43	20.30	8.76	14.30	21.26	7.29	15.67	29.40	15.14	22.17	33.39
13 - 6	6.38	8.91	17.07	9.14	13.26	21.69	6.86	13.06	25.53	15.38	24.59	35.72
14 - 0	5.29	8.90	16.72	9.78	14.44	21.70	6.88	12.27	24.99	13.63	24.91	34.61
14 - 6	4.58	7.33	20.81	9.57	13.25	20.62	5.82	11.24	28.16	15.25	24.99	33.61
15 - 0	5.72	7.38	18.96	10.31	15.06	22.63	5.64	10.54	29.56	17.64	26.74	36.67
15 - 6	4.24	6.95	16.59	--	--	--	5.42	10.60	18.25	--	--	--
16 - 0	4.00	6.39	20.14	10.99	16.14	23.92	5.46	11.22	33.39	17.43	31.36	40.51
16 - 6	4.82	7.54	14.67	--	--	--	7.77	11.18	19.21	--	--	--
17 - 0	4.18	6.30	13.22	13.58	20.70	27.82	5.44	9.85	15.14	21.35	29.05	52.15
18 - 0	3.89	6.10	9.56	--	--	--	4.24	8.41	14.46	--	--	--

(data from McCammon, Human Growth and Development, 1970. Courtesy of Charles C Thomas, Publisher, Springfield, Illinois)

TABLE 394

FOREARM SKINFOLD THICKNESS (mm)
IN DENVER CHILDREN

Age (Years)	Males			Females		
	Median	Mean	S.D.	Median	Mean	S.D.
4.00	5.94	6.24	1.49	6.97	7.28	2.15
4.50	5.91	6.08	1.61	6.98	7.09	1.74
5.00	5.95	5.92	1.77	7.00	6.84	1.21
5.50	5.29	5.38	1.26	6.54	6.72	1.76
6.00	5.33	5.93	1.79	6.53	6.63	1.50
6.50	5.01	5.32	1.57	6.19	6.32	1.70
7.00	5.23	5.44	1.45	6.02	6.45	1.93
7.50	5.32	5.57	1.63	6.02	6.62	2.23
8.00	5.25	5.74	2.18	7.06	7.08	2.22
8.50	5.34	5.60	1.84	7.13	7.21	2.68
9.00	6.05	6.36	2.16	6.60	7.22	2.73
9.50	5.68	6.33	2.45	7.13	7.46	2.26
10.00	5.94	6.79	2.65	6.77	7.17	2.43
10.50	5.54	6.50	2.23	7.01	7.53	2.89
11.00	7.02	7.16	2.98	7.16	8.02	3.07
11.50	5.58	6.59	3.24	6.59	7.44	3.45
12.00	5.87	6.82	3.54	7.53	7.79	2.86
12.50	6.10	7.07	3.05	6.76	7.30	2.94
13.00	5.52	6.69	3.53	7.15	7.18	2.36
13.50	5.14	6.17	3.15	6.99	7.51	2.50
14.00	5.14	6.02	2.93	7.38	7.67	2.47
14.50	4.49	5.62	2.93	7.04	6.94	2.24
15.00	4.91	5.24	1.94	8.00	7.96	2.73
15.50	4.98	5.99	2.94	6.77	7.53	2.21

TABLE 394   (continued)

FOREARM SKINFOLD THICKNESS (mm)
IN DENVER CHILDREN

Age	Males			Females		
(Years)	Median	Mean	S.D.	Median	Mean	S.D.
16.00	4.98	6.40	3.40	7.06	7.96	2.93
16.50	5.47	6.46	3.98	--	--	--
17.00	4.30	5.31	2.38	6.99	7.45	2.81
17.50	5.09	5.50	2.20	--	--	--
18.00	4.73	4.97	1.70	7.01	7.20	2.64

(data from McCammon, unpublished)

TABLE   395

FOREARM RADIOGRAPHIC FAT THICKNESS (mm)
IN DENVER CHILDREN

Age	Males			Females		
(Years)	Median	Mean	S.D.	Median	Mean	S.D.
0.13	11.66	11.45	3.08	11.84	11.99	2.33
0.38	15.72	15.81	2.81	16.72	16.90	2.65
0.50	17.17	17.31	3.05	18.25	17.96	2.83
1.00	16.70	17.22	3.07	17.61	17.93	2.74
1.50	16.48	17.08	3.03	17.45	17.32	2.79
2.00	15.25	15.66	2.79	16.71	16.73	2.79
2.50	15.25	15.35	2.64	15.46	15.82	2.45
3.00	14.57	14.65	2.65	15.04	15.47	2.59
3.50	13.73	14.08	2.40	14.41	14.79	2.33
4.00	13.17	13.32	2.43	14.64	14.57	2.20
4.50	12.41	12.90	2.47	14.09	13.95	2.14
5.00	11.67	12.41	2.42	13.16	13.46	2.23
5.50	11.58	11.72	2.27	13.25	13.18	2.28
6.00	11.38	11.68	2.40	12.90	12.86	2.45

TABLE 395 (continued)

FOREARM RADIOGRAPHIC FAT THICKNESS (mm)
IN DENVER CHILDREN

Age (Years)	Males			Females		
	Median	Mean	S.D.	Median	Mean	S.D.
6.50	10.57	11.22	2.68	12.38	12.28	2.32
7.00	10.75	11.21	3.19	12.02	12.30	2.45
7.50	10.08	10.52	2.40	12.06	12.39	2.71
8.00	9.96	10.68	2.89	12.26	12.39	2.70
8.50	9.80	10.60	2.70	12.33	12.56	2.93
9.00	10.15	10.85	2.91	12.50	12.81	2.93
9.50	10.35	10.94	2.82	12.33	12.90	2.94
10.00	10.77	11.22	3.10	12.24	12.63	2.76
10.50	10.37	10.83	3.18	12.78	12.60	2.94
11.00	10.82	11.08	2.96	12.28	12.71	3.02
11.50	10.60	10.92	2.98	12.55	12.63	2.68
12.00	10.40	11.07	3.15	13.15	12.95	3.07
12.50	10.24	10.78	3.26	12.64	12.62	3.04
13.00	9.74	10.69	3.38	12.32	12.45	2.64
13.50	9.20	9.65	3.53	13.22	12.52	2.87
14.00	8.14	8.80	2.90	13.65	13.20	2.77
14.50	7.74	8.42	3.21	12.92	13.25	3.06
15.00	7.67	8.19	2.84	13.33	13.49	2.84
15.50	6.72	7.64	2.61	--	--	--
16.00	7.38	8.34	3.08	13.62	14.08	3.10
16.50	6.84	7.66	2.79	--	--	--
17.00	5.87	6.89	2.63	14.90	14.69	3.00
17.50	--	--	--	--	--	--
18.00	5.56	6.22	2.38	--	--	--

(data from McCammon, unpublished)

TABLE 396

BREADTH OF FAT IN FOREARM (POSTEROANTERIOR) AT LEVEL OF PROXIMAL END OF THE RADIUS (mm)
IN OHIO CHILDREN

Age	Boys			Girls		
(Years)	Mean	S.D.	Median	Mean	S.D.	Median
6.5	5.2	1.2	4.6	5.9	1.2	5.3
7.5	5.2	1.3	4.5	5.9	1.5	5.3
8.5	5.3	1.4	4.4	6.2	1.7	5.8
9.5	5.2	1.4	4.1	6.1	1.8	5.5
10.5	5.3	1.6	4.4	6.2	2.0	5.8
11.5	5.8	1.9	4.6	6.2	2.2	6.0
12.5	5.8	2.1	5.0	6.3	2.0	5.6
13.5	5.5	2.2	4.4	6.4	1.9	5.7
14.5	4.8	1.8	3.8	6.6	1.9	6.0
15.5	4.6	1.6	3.9	7.1	2.2	6.4
16.5	4.4	1.3	3.7	6.9	2.3	6.2
17.5	4.6	1.4	4.2	6.7	2.1	6.0

(data from Reynolds, 1951)

TABLE 397

THIGH RADIOGRAPHIC FAT THICKNESS (mm)
IN COLORADO CHILDREN

Age (years)	Males			Females		
	Median	Mean	S.D.	Median	Mean	S.D.
0.13	5.99	6.83	2.27	7.64	7.87	3.07
0.38	12.92	12.66	3.26	13.53	13.86	3.11
0.50	13.93	13.98	3.67	14.69	15.27	3.69
1.00	13.57	14.08	4.69	14.65	14.89	3.65
1.50	12.09	12.74	3.73	12.51	13.95	3.59
2.00	11.11	11.91	3.73	11.96	12.85	3.55
2.50	10.34	11.03	3.14	11.58	11.92	2.90
3.00	9.67	10.11	2.92	10.76	11.23	2.78
3.50	8.86	9.32	2.88	10.06	10.82	2.87
4.00	8.60	8.99	2.69	10.61	10.70	2.60
4.50	8.00	8.43	2.51	10.07	10.51	2.88
5.00	7.64	8.07	2.46	9.66	10.26	2.75
5.50	7.46	7.96	2.51	9.80	10.28	2.95
6.00	7.40	7.89	2.67	9.43	10.06	3.10
6.50	7.08	7.95	3.55	9.53	9.79	3.16
7.00	7.35	8.07	3.55	9.76	10.24	3.70
7.50	7.38	7.77	3.17	8.92	10.24	4.07
8.00	7.38	8.10	3.86	9.98	11.02	4.17
8.50	7.16	8.04	3.88	10.42	11.14	4.19
9.00	7.81	8.41	3.73	11.28	11.50	4.64
9.50	7.89	8.42	3.89	10.66	11.49	4.64
10.00	8.31	8.94	4.38	10.10	11.08	4.15
10.50	7.57	8.43	3.92	10.81	11.46	4.39

TABLE 397 (continued)

THIGH RADIOGRAPHIC FAT THICKNESS (mm)
IN COLORADO CHILDREN

Age (years)	Males			Females		
	Median	Mean	S.D.	Median	Mean	S.D.
11.00	8.53	9.12	4.08	11.67	11.79	3.92
11.50	8.66	9.00	3.89	11.10	11.61	3.84
12.00	8.16	9.27	4.19	11.80	12.42	4.32
12.50	7.99	9.46	4.49	11.12	12.11	4.18
13.00	8.77	9.08	4.03	10.50	12.49	4.43
13.50	7.70	8.41	4.16	12.28	13.52	4.83
14.00	7.07	7.90	3.73	12.40	13.29	4.65
14.50	6.33	7.84	4.46	12.64	13.44	4.48
15.00	6.07	7.76	4.14	13.54	14.61	4.92
15.50	7.07	7.22	3.46	--	--	--
16.00	5.81	7.35	4.43	14.96	15.38	4.89
16.50	6.43	6.87	2.51	--	--	--
17.00	5.40	6.33	3.76	14.56	16.34	6.06
18.00	4.80	5.50	2.98	21.17	19.32	5.74

(data from McCammon, unpublished)

TABLE 398

MEDIAL THIGH SKINFOLD THICKNESS (mm)
IN COLORADO CHILDREN

Age (years)	Males			Females		
	Median	Mean	S.D.	Median	Mean	S.D.
4.00	10.00	10.41	2.46	10.91	10.72	2.48
4.50	10.85	10.39	2.60	12.49	12.06	2.72
5.00	10.91	10.64	2.63	12.56	12.14	2.70
5.50	10.17	10.23	2.44	12.94	12.56	3.58
6.00	10.94	10.91	4.09	12.40	12.37	4.02
6.50	9.90	10.19	2.85	12.05	12.15	3.10
7.00	10.02	10.61	3.17	12.03	13.35	4.88
7.50	11.21	10.87	3.38	13.61	13.76	4.22
8.00	10.95	11.07	3.83	13.82	14.51	5.02
8.50	11.71	12.23	4.23	13.98	14.64	6.14
9.00	12.97	13.34	6.20	13.97	14.19	5.29
9.50	12.97	13.83	6.57	16.13	16.23	6.29
10.00	15.22	15.91	9.04	16.51	16.98	5.33
10.50	15.69	15.59	6.33	17.52	18.06	6.79
11.00	15.99	15.79	6.18	I9.03	20.83	7.78
11.50	14.48	16.43	7.30	17.37	18.99	6.13
12.00	15.07	16.46	6.62	18.89	19.88	6.44
12.50	17.02	17.11	7.79	19.47	19.31	6.26
13.00	17.00	17.41	8.61	19.01	18.74	5.63
13.50	13.67	15.93	8.66	19.80	20.06	6.08
14.00	11.86	15.95	9.22	18.95	20.29	5.79
14.50	13.83	16.79	9.89	18.30	19.21	5.50
15.00	12.31	14.44	7.61	22.15	21.62	5.51

TABLE   398   (continued)

MEDIAL THIGH SKINFOLD THICKNESS (mm)
IN COLORADO CHILDREN

Age (years)	Males			Females		
	Median	Mean	S.D.	Median	Mean	S.D.
15.50	12.78	16.34	9.82	21.70	21.12	5.49
16.00	15.00	16.85	8.60	20.72	21.52	6.89
16.50	13.63	17.40	12.09	--	--	--
17.00	11.95	14.54	7.95	22.30	23.94	8.15
17.50	12.55	14.74	7.90	--	--	--
18.00	11.98	14.07	8.21	23.92	24.83	9.00

(data from McCammon, unpublished)

TABLE 399

BREADTH OF SKIN AND SUBCUTANEOUS TISSUE (cm)
IN THE CALF IN BOSTON CHILDREN

Age (years)	Percentiles					Mean	S.D.
	10	25	50	75	90		
BOYS							
6	0.8	0.9	1.0	1.1	1.3	1.03	0.22
7	0.7	0.9	1.0	1.2	1.4	1.04	0.30
8	0.8	0.9	1.0	1.2	1.5	1.08	0.30
9	0.8	0.8	1.0	1.2	1.4	1.09	0.36
10	0.8	0.9	1.0	1.2	1.4	1.04	0.24
11	0.8	0.8	1.1	1.2	1.5	1.09	0.27
12	0.8	0.9	1.1	1.5	1.7	1.20	0.35
13	0.7	0.9	1.2	1.6	1.9	1.32	0.59
14	0.7	1.0	1.2	1.6	1.9	1.37	0.65
15	0.7	0.9	1.0	1.5	1.8	1.22	0.50
16	0.6	0.8	1.0	1.2	1.5	1.05	0.44
GIRLS							
6	1.0	1.1	1.3	1.5	1.6	1.29	0.26
7	1.0	1.2	1.3	1.5	1.6	1.33	0.28
8	1.0	1.2	1.4	1.5	1.8	1.37	0.31
9	1.0	1.2	1.4	1.6	1.9	1.43	0.35
10	1.1	1.2	1.5	1.7	2.2	1.55	0.41
11	1.1	1.3	1.5	1.7	2.0	1.50	0.33
12	1.1	1.3	1.6	1.8	2.2	1.63	0.39
13	1.3	1.5	1.8	1.8	2.4	1.74	0.42
14	1.4	1.6	1.8	2.0	2.6	1.89	0.47
15	1.5	1.7	1.9	2.1	2.4	1.94	0.45
16	1.7	1.7	1.9	2.1	2.3	1.94	0.24

(data from Lombard, 1950)

TABLE   400

WIDTH OF SKIN + SUBCUTANEOUS TISSUE
IN THE CALF (cm) IN BOSTON (MA) CHILDREN

Age	Boys			Girls		
			Percentiles			
(years)	10	50	90	10	50	90
6	0.78	1.06	1.34	1.02	1.35	1.71
7	0.76	1.04	1.39	1.00	1.32	1.69
8	0.82	1.06	1.45	1.12	1.35	1.79
9	0.81	1.06	1.45	1.12	1.33	1.93
10	0.82	1.11	1.46	1.12	1.38	1.86

(data from Stuart and Dwinell, 1942)

TABLE   401

SKIN + SUBCUTANEOUS AREA IN RADIOGRAPHS OF THE CALF
IN BOSTON (MA) CHILDREN
(cm^2; area of shadows of soft tissues overlying muscles)

Age	Boys			Girls		
			Percentiles			
(years)	10	50	90	10	50	90
6	16.8	23.2	30.1	21.2	30.0	39.2
7	17.6	24.3	33.7	22.8	33.4	41.6
8	19.4	27.5	38.2	26.1	34.0	47.3
9	19.8	27.7	44.9	28.9	38.2	54.6
10	21.3	30.5	45.0	31.4	38.7	63.0

(data from Stuart and Dwinell, 1942)

TABLE   402

WIDTH OF SKIN AND SUBCUTANEOUS TISSUE OF CALF (mm) IN
BOSTON BOYS

Age (years)	Percentiles					Mean	S.D.
	10	25	50	75	90		
A. Three-Foot Tube Distance							
1/4	12	13	15	18	20	15	3.3
1/2	14	17	20	22	24	19	3.4
3/4	16	18	21	24	26	21	4.1
1	15	17	19	22	24	19	3.4
1 1/2	14	15	17	19	21	17	2.7
2	13	14	16	18	19	16	2.8
2 1/2	12	13	14	16	18	14	2.3
3	10	12	14	16	18	14	2.7
3 1/2	10	12	13	15	17	13	2.5
4	10	11	13	14	16	13	2.8
4 1/2	10	11	12	14	16	12	2.3
5	9	10	12	13	16	12	2.6
5 1/2	9	10	11	13	15	12	2.6
6	8	10	11	13	16	11	3.0
B. Six-Foot Tube Distance							
6	8	10	11	12	14	10	2.0
7	8	9	11	12	14	10	3.0
8	8	9	11	12	14	11	3.0
9	8	8	10	12	14	11	3.6
10	8	8	10	12	14	10	2.4
11	8	8	12	13	15	11	3.1
12	8	9	11	16	19	12	3.8
13	8	9	12	16	22	14	7.0

(data from Stuart and Sobel, 1946, The Journal of Pediatrics 28:637-649)

TABLE   403

CALF RADIOGRAPHIC FAT THICKNESS (mm)
IN COLORADO CHILDREN

Age (years)	Males			Females		
	Median	Mean	S.D.	Median	Mean	S.D.
0.13	9.59	10.96	3.09	11.63	11.51	2.38
0.38	17.57	17.77	3.61	18.51	18.60	2.87
0.50	19.99	19.74	3.85	20.31	20.11	3.44
1.00	19.11	19.64	4.34	19.21	19.47	3.49
1.50	16.82	16.99	3.46	17.23	17.38	3.03
2.00	14.97	15.52	3.16	16.50	16.30	3.14
2.50	14.10	14.68	2.79	15.13	15.23	2.74
3.00	13.17	13.58	2.89	13.96	14.31	2.57
3.50	12.76	12.85	2.66	13.68	13.85	2.57
4.00	11.75	12.29	2.74	13.50	13.63	2.54
4.50	11.73	11.95	2.47	12.49	13.05	2.46
5.00	11.55	11.64	2.48	12.59	12.76	2.41
5.50	11.14	11.26	2.61	12.26	12.73	2.56
6.00	11.06	11.43	2.61	12.21	12.60	2.65
6.50	10.91	11.28	2.79	12.10	12.60	2.83
7.00	10.85	11.42	2.78	12.20	12.62	2.84
7.50	10.90	11.16	2.52	12.20	12.75	2.91
8.00	11.13	11.47	3.00	12.32	13.11	3.26
8.50	11.17	11.44	3.03	12.92	13.39	3.13
9.00	11.96	11.88	3.11	13.86	14.12	3.46
9.50	11.46	12.11	3.36	13.10	14.16	3.48
10.00	12.58	12.42	3.29	13.51	14.14	3.37
10.50	11.67	12.24	3.16	13.84	14.62	3.53

TABLE 403 (continued)

CALF RADIOGRAPHIC FAT THICKNESS (mm)
IN COLORADO CHILDREN

Age (years)	Males			Females		
	Median	Mean	S.D.	Median	Mean	S.D.
11.00	12.26	12.80	3.45	13.94	14.68	3.51
11.50	12.76	12.60	3.20	14.50	14.90	3.13
12.00	13.16	12.95	3.73	15.38	15.60	3.46
12.50	13.28	13.08	3.96	15.12	15.65	3.27
13.00	13.17	13.19	4.38	15.29	15.87	3.30
13.50	12.02	12.47	3.83	15.37	16.15	4.02
14.00	11.33	11.41	3.26	15.94	16.58	3.45
14.50	10.57	11.40	3.90	14.82	16.27	3.53
15.00	9.90	11.00	3.79	15.99	16.83	2.92
15.50	10.12	10.31	3.02	--	--	--
16.00	10.99	10.96	3.70	17.06	17.82	4.05
16.50	11.08	10.75	3.27	--	--	--
17.00	9.82	9.81	3.12	14.84	16.55	3.88
18.00	8.59	9.05	3.51	--	--	--

(data from McCammon, unpublished)

TABLE   404

BREADTH OF FAT IN CALF (mm)
IN OHIO CHILDREN

Age	Boys			Girls		
(years)	Mean	S.D.	Median	Mean	S.D.	Median
6.5	5.7	1.2	5.2	6.1	1.2	5.5
7.5	5.7	1.5	5.0	6.2	1.4	5.4
8.5	6.0	1.6	5.2	6.4	1.5	5.4
9.5	6.2	1.6	5.3	6.5	1.5	5.8
10.5	6.3	1.7	5.7	6.7	1.5	6.0
11.5	6.7	1.6	6.0	6.9	1.5	6.4
12.5	6.8	1.8	6.0	7.1	1.7	6.7
13.5	6.8	1.7	6.3	7.3	1.7	6.9
14.5	6.4	1.6	5.7	7.5	2.0	7.1
15.5	5.9	1.6	5.3	8.1	2.0	7.7
16.5	5.5	1.4	4.9	7.7	1.6	7.1
17.5	5.8	1.4	4.6	7.9	1.8	7.3

(data from Reynolds, 1951)

TABLE 405

LEVELS AND INCREMENTS IN RADIOGRAPHIC CALF FAT THICKNESS (mm) AT MAXIMUM BREADTH
(combined medial and lateral thicknesses)
IN OHIO CHILDREN

Fat Thickness Percentiles

Age (months)	Sex	5	15	50	85	95
1	Boys	6.2	7.8	9.5	11.3	13.3
1	Girls	6.3	7.7	9.6	11.8	13.7
3	Boys	9.2	13.2	15.7	17.7	21.0
3	Girls	11.4	13.4	15.4	18.5	19.4
6	Boys	15.3	17.2	20.0	23.3	26.6
6	Girls	14.8	17.2	19.8	23.2	26.0
9	Boys	15.8	18.2	21.0	24.0	27.9
9	Girls	15.6	17.7	20.7	24.5	27.3
12	Boys	14.6	16.9	19.3	22.1	25.6
12	Girls	13.9	16.5	20.5	23.1	27.8

Fat Increment Precentiles

Age Interval (months)	Sex	5	15	50	85	95
1-3	Boys	1.9	3.7	5.7	7.4	9.3
1-3	Girls	2.8	4.2	5.7	7.4	8.8
3-6	Boys	0.5	2.4	4.5	7.2	9.2
3-6	Girls	0.9	2.8	4.7	6.7	10.6
6-9	Boys	-2.0	-0.6	0.8	2.3	4.1
6-9	Girls	-3.3	-0.5	0.7	2.5	4.2
9-12	Boys	-4.1	-2.8	-1.3	0.3	2.5
9-12	Girls	-3.8	-2.3	-0.9	0.8	3.5

(data from "Fat thickness and growth progress during infancy," Human Biology 28:232-250,
1956, by Garn, Greaney, and Young, by permission of the Wayne State University Press.
Copyright 1956, Wayne State University Press, Detroit, Michigan 48202)

TABLE   406

MEDIAL CALF SKINFOLD THICKNESS (mm) IN U.S. CHILDREN

Race, sex,	Percentiles						
and age	5	10	25	50	75	90	95
WHITE							
Boys							
12 years........	5.2	5.8	7.7	11.1	16.2	21.9	26.8
13 years........	5.5	6.3	8.2	11.3	16.1	22.8	27.5
14 years........	5.1	5.6	7.1	10.2	14.9	20.7	25.4
15 years........	4.7	5.5	7.2	10.0	13.5	20.3	25.1
16 years........	4.2	5.0	6.5	8.7	13.3	19.6	22.6
17 years........	4.2	4.7	6.3	8.8	12.5	19.1	23.0
Girls							
12 years........	7.5	8.6	11.3	14.4	19.6	24.7	30.4
13 years........	7.7	9.5	12.3	15.9	21.7	26.5	31.4
14 years........	9.2	10.5	13.1	17.3	23.1	28.4	30.9
15 years........	9.2	11.0	14.5	18.4	23.4	30.2	35.6
16 years........	10.3	11.6	15.0	19.2	23.8	28.8	31.6
17 years........	10.2	11.7	15.5	20.0	24.7	31.2	34.8
NEGRO							
Boys							
12 years........	3.6	4.2	5.2	7.6	10.6	16.1	19.8
13 years........	3.6	4.2	5.2	7.1	10.2	17.4	24.5
14 years........	4.0	4.3	5.2	7.1	9.6	12.4	21.2
15 years........	4.2	4.5	5.3	6.3	7.8	10.7	14.8
16 years........	3.5	4.1	5.1	6.5	9.2	11.7	13.2
17 years........	3.3	3.6	4.9	6.2	8.4	16.2	18.8
Girls							
12 years........	6.5	7.5	9.7	11.4	15.8	20.8	25.5
13 years........	6.3	7.2	9.3	11.9	17.0	24.6	30.6
14 years........	7.1	7.8	10.4	14.5	20.2	25.3	27.1
15 years........	7.4	8.5	10.6	13.8	17.1	24.2	27.8
16 years........	7.6	9.1	11.0	14.3	19.7	25.9	38.5
17 years........	7.9	8.5	11.0	13.4	17.9	21.2	22.7

(data from Johnston et al., 1974)

TABLE 407

MEDIAL CALF SKINFOLD THICKNESS (mm)
IN COLORADO CHILDREN

Age (years)	Males			Females		
	Median	Mean	S.D.	Median	Mean	S.D.
4.00	8.01	8.45	2.31	8.83	8.67	2.44
4.50	7.92	7.62	1.85	8.10	8.72	2.33
5.00	7.95	7.53	2.25	8.47	8.82	2.46
5.50	7.13	6.86	1.88	8.20	8.96	2.97
6.00	7.22	7.29	2.01	7.75	8.66	2.84
6.50	7.00	7.27	2.75	8.72	8.80	2.54
7.00	7.47	7.65	2.24	8.90	9.44	3.07
7.50	7.65	7.57	2.22	8.71	9.34	3.21
8.00	7.50	7.62	2.29	9.90	9.96	3.15
8.50	7.92	7.79	2.50	9.57	9.98	3.51
9.00	8.73	8.79	3.38	10.31	10.29	3.97
9.50	8.83	9.35	3.93	11.29	11.54	4.91
10.00	11.13	10.67	4.82	11.00	11.76	4.33
10.50	9.38	9.62	3.57	12.19	12.47	5.12
11.00	10.53	10.76	4.83	12.07	12.80	4.84
11.50	10.04	11.25	5.54	11.71	12.58	4.56
12.00	9.89	11.14	5.40	12.27	13.19	5.46
12.50	10.05	11.16	5.56	12.28	12.94	4.95
13.00	9.86	11.56	6.95	13.89	12.64	4.04
13.50	9.79	10.62	6.19	13.68	13.66	4.93
14.00	7.95	10.03	4.94	13.53	14.16	4.25
14.50	7.85	10.33	5.98	12.95	13.99	4.03
15.00	7.84	9.43	4.85	15.54	15.17	4.39

TABLE 407 (continued)

MEDIAL CALF SKINFOLD THICKNESS (mm)
IN COLORADO CHILDREN

Age (years)	Males			Females		
	Median	Mean	S.D.	Median	Mean	S.D.
15.50	8.02	10.06	5.79	14.31	15.87	5.07
16.00	9.55	10.39	5.18	13.83	16.19	7.10
16.50	9.89	10.74	5.51	--	--	--
17.00	7.74	8.14	3.58	13.48	15.01	5.34
17.50	8.82	9.11	3.69	--	--	--
18.00	7.99	8.37	4.10	15.55	16.27	6.93

(data from McCammon, unpublished)

TABLE 408

FAT THICKNESSES MEASURED ON RADIOGRAPHS IN COLORADO BOYS (mm)

Age Yr. Mo.	Max. Forearm Percentiles			Mid-thigh (Lateral half) Percentiles			Max. Calf Percentiles		
	10	50	90	10	50	90	10	50	90
0 - 2	8.91	11.58	16.24	5.32	7.56	10.97	8.82	11.00	16.38
0 - 4	11.64	15.05	19.40	7.92	12.21	16.99	12.55	16.61	21.99
0 - 6	14.12	17.54	22.13	9.46	14.04	19.17	15.00	20.16	24.97
1 - 0	13.28	16.83	21.65	9.11	13.27	20.97	15.46	18.62	24.71
1 - 6	12.98	16.35	21.73	8.58	12.10	17.06	12.78	16.41	22.11
2 - 0	12.25	15.24	19.82	7.39	11.34	17.14	11.69	14.99	19.82
2 - 6	12.03	15.13	19.44	7.16	10.42	15.89	11.50	14.26	18.97
3 - 0	11.18	14.31	18.53	6.54	9.85	14.32	10.20	13.09	17.00
3 - 6	11.46	13.62	17.41	6.13	9.03	13.63	9.30	12.56	16.19
4 - 0	10.62	13.17	16.65	5.87	8.59	13.01	9.11	11.95	15.90
4 - 6	9.40	12.25	16.30	5.26	7.99	11.44	8.62	11.76	15.20
5 - 0	9.31	11.86	15.61	5.14	7.83	11.08	8.38	11.66	14.45
5 - 6	9.13	11.50	14.89	4.84	7.52	11.22	8.09	11.19	14.90
6 - 0	8.87	11.42	15.37	5.11	7.56	12.08	8.24	10.98	14.64
6 - 6	8.14	10.91	13.71	4.95	7.26	11.37	8.17	10.82	14.71
7 - 0	8.41	10.80	14.16	4.64	7.57	11.46	8.49	11.04	14.93
7 - 6	7.89	10.21	13.93	4.86	7.27	11.60	8.01	10.99	14.99
8 - 0	7.68	10.22	15.63	4.69	7.30	11.62	8.01	11.20	15.83
8 - 6	7.80	9.84	15.97	4.78	7.18	11.96	7.57	10.95	14.80
9 - 0	7.34	10.12	14.90	4.71	7.78	13.24	8.03	11.58	16.02
9 - 6	7.82	10.64	14.66	4.59	7.59	13.27	7.90	11.73	15.52
10 - 0	7.54	10.90	15.47	4.47	8.37	14.08	8.77	12.17	16.95
10 - 6	7.82	10.25	15.75	3.87	8.51	13.59	8.41	12.38	16.22
11 - 0	7.47	11.14	15.54	4.57	8.41	14.92	8.91	12.21	16.93
11 - 6	7.36	10.68	15.05	4.12	8.55	14.55	8.71	12.50	15.21
12 - 0	7.54	10.48	15.12	4.18	8.16	15.06	8.18	12.84	17.47
12 - 6	7.29	10.38	14.99	4.51	7.87	16.00	8.75	12.88	17.87
13 - 0	7.25	9.95	14.58	4.72	8.75	15.02	7.89	12.97	17.91
13 - 6	5.90	9.22	13.60	4.27	7.39	15.61	7.96	12.03	17.54
14 - 0	5.79	8.00	12.94	4.19	6.92	14.00	7.84	11.12	16.32
14 - 6	5.40	7.78	13.47	3.66	6.31	15.93	7.07	10.39	17.35
15 - 0	5.04	7.22	13.05	3.94	6.60	14.12	6.45	9.31	16.61
15 - 6	5.53	6.82	11.86	4.42	6.33	10.49	6.99	10.12	14.76
16 - 0	5.38	7.45	12.75	3.95	6.09	13.20	6.40	10.55	15.70
16 - 6	4.88	7.02	11.04	4.38	6.52	10.56	6.18	10.74	14.30
17 - 0	4.91	6.27	9.83	3.52	5.45	9.95	7.20	10.15	13.19
18 - 0	4.44	5.71	8.15	3.00	4.68	10.42	5.73	8.32	14.52

(data from McCammon, Human Growth and Development, 1970. Courtesy of Charles C Thomas, Publisher, Springfield, Illinois)

TABLE 409

FAT THICKNESSES MEASURED ON RADIOGRAPHS IN COLORADO GIRLS (mm)

Age (yr.-mo.)	Max. Forearm Percentiles			Mid-thigh (Lateral half) Percentiles			Max. Calf Percentiles		
	10	50	90	10	50	90	10	50	90
0 - 2	10.72	12.57	15.72	6.05	8.58	12.90	10.07	12.86	16.09
0 - 4	12.31	16.26	20.29	9.05	12.50	17.60	14.38	17.56	21.52
0 - 6	14.01	18.23	21.57	10.78	14.72	20.90	14.37	20.32	25.25
1 - 0	14.24	17.32	21.80	10.46	14.12	19.47	15.26	19.16	23.84
1 - 6	13.72	17.27	20.68	9.66	13.06	19.13	13.65	17.09	20.89
2 - 0	12.92	16.73	20.55	8.55	12.01	18.45	12.27	16.10	20.02
2 - 6	12.72	15.31	19.08	8.06	11.50	15.92	11.99	14.88	18.28
3 - 0	12.38	15.00	19.30	7.81	10.83	14.84	11.42	13.96	17.68
3 - 6	11.91	14.74	17.89	7.11	10.17	15.16	10.77	13.67	16.90
4 - 0	11.46	14.60	17.16	7.34	10.39	14.41	10.05	13.16	16.55
4 - 6	11.00	13.99	16.79	6.74	9.89	14.54	10.22	12.64	16.35
5 - 0	10.66	13.12	16.17	6.85	9.45	14.29	9.88	12.28	15.97
5 - 6	10.06	12.91	16.04	6.47	9.42	13.98	9.47	12.26	17.02
6 - 0	9.70	12.70	15.95	6.14	9.44	14.00	9.66	12.14	16.10
6 - 6	9.22	12.36	15.21	5.84	9.06	13.70	9.29	12.09	16.42
7 - 0	9.34	12.06	14.87	5.90	9.67	15.18	9.38	12.01	17.62
7 - 6	8.78	12.06	15.88	5.83	8.82	16.04	9.22	12.18	17.37
8 - 0	8.96	12.09	15.72	6.08	10.22	17.12	9.64	12.28	18.52
8 - 6	8.74	12.21	16.34	6.52	10.70	17.08	9.71	13.02	18.06
9 - 0	9.59	12.49	16.26	6.21	10.75	17.86	9.99	13.80	19.47
9 - 6	9.32	12.22	16.84	6.70	10.77	17.99	9.93	13.26	19.64
10 - 0	9.43	12.12	16.67	6.56	10.18	16.20	10.29	13.15	18.96
10 - 6	8.80	12.45	15.76	6.65	10.76	17.53	10.18	13.86	19.48
11 - 0	8.99	12.30	17.42	7.40	11.24	17.03	10.64	13.59	19.93
11 - 6	8.82	12.54	16.20	6.86	11.09	16.71	11.33	14.36	19.39
12 - 0	8.59	12.89	17.70	7.64	11.49	17.89	11.27	15.03	20.94
12 - 6	8.74	12.58	17.28	7.53	11.29	18.86	11.66	15.28	21.47
13 - 0	9.66	11.90	15.58	7.50	11.11	18.90	12.03	15.02	20.84
13 - 6	8.57	12.50	16.64	7.77	11.98	19.94	11.50	15.02	22.26
14 - 0	9.67	13.20	17.30	7.61	12.30	21.14	12.52	15.44	22.24
14 - 6	9.29	12.48	17.54	8.19	12.38	18.77	12.06	14.94	22.40
15 - 0	8.77	13.42	16.91	8.36	13.57	22.55	13.03	16.13	21.09
16 - 0	10.62	13.54	19.05	9.04	15.60	21.49	13.18	16.85	24.94
17 - 0	10.84	15.07	18.83	11.15	14.65	25.15	13.78	15.07	22.81

(data from McCammon, Human Growth and Development, 1970. Courtesy of Charles C Thomas, Publisher, Springfield, Illinois)

TABLE 410

MEAN INCREMENTS (mm) IN FAT THICKNESSES
IN OHIO CHILDREN

BOYS   Age Intervals (years)

Site	6.5- 7.5	7.5- 8.5	8.5- 9.5	9.5- 10.5	10.5- 11.5	11.5- 12.5	12.5- 13.5	13.5- 14.5	14.5- 15.5	15.5- 16.5	16.5- 17.5
Calf	0.0	0.3	0.2	0.1	0.4	0.1	0.0	-0.4	-0.5	-0.4	0.3
Trochanter	0.2	0.9	0.9	0.2	1.5	0.7	-0.3	-1.5	-1.1	0.1	1.4
Waist	0.2	0.6	1.0	0.8	0.6	1.0	0.6	-0.4	0.5	0.3	0.9
Chest	0.1	0.3	0.3	0.4	0.4	0.4	0.0	-0.3	0.1	0.2	0.9
Forearm	0.0	0.1	-0.1	0.1	0.5	0.0	-0.3	-0.7	-0.2	-0.2	0.2
Deltoid	-0.1	0.4	0.1	0.3	0.5	-0.1	-0.3	-0.6	-0.3	0.3	0.4

GIRLS   Age Intervals (years)

Site	6.5- 7.5	7.5- 8.5	8.5- 9.5	9.5- 10.5	10.5- 11.5	11.5- 12.5	12.5- 13.5	13.5- 14.5	14.5- 15.5	15.5- 16.5	16.5- 17.5
Calf	0.1	0.2	0.1	0.2	0.2	0.2	0.2	0.2	0.6	-0.4	0.2
Trochanter	0.5	1.1	0.6	0.6	0.9	1.0	1.1	1.7	3.3	-0.6	0.0
Waist	0.2	0.6	0.7	1.2	0.7	0.5	1.0	0.7	1.5	-0.8	0.0
Chest	0.3	0.3	0.3	0.5	0.5	0.2	0.3	0.5	1.1	-0.4	-0.2
Forearm	0.0	0.3	-0.1	0.1	0.0	0.1	0.1	0.2	0.5	-0.2	-0.2
Deltoid	0.1	0.3	0.3	0.4	-0.2	0.1	0.3	0.4	1.3	-0.5	-0.4

(data from Reynolds, 1951)

TABLE 411

LOWER LIMB SKINFOLD THICKNESSES (mm) FOR BLACK BOYS IN TEXAS

| Age | Thigh | | | | Calf | | | |
| | Lower Income | | Middle Income | | Lower Income | | Middle Income | |
(years)	Mean	S.D.	Mean	S.D.	Mean	S.D.	Mean	S.D.
10.1 - 11	10.8	4.05	8.5	2.28	7.6	4.94	5.2	1.84
11.1 - 12	11.9	3.23	13.7	3.00	7.6	2.61	9.1	4.42
12.1 - 13	14.3	7.28	12.0	5.33	9.5	6.80	7.7	3.61
13.1 - 14	9.1	2.36	12.0	4.27	6.9	2.52	9.1	4.16
14.1 - 15	10.9	2.67	11.0	3.90	7.4	2.88	7.8	2.44
15.1 - 16	8.8	2.61	11.2	4.28	6.5	2.96	7.4	4.43
16.1 - 17	9.1	3.60	10.8	4.47	7.0	2.87	7.5	3.09
17.1 - 18	9.3	3.89	8.9	2.40	6.8	3.70	6.5	2.43

(data from Schutte, 1979)

TABLE 412

THE RATIO TRANSVERSE DIAMETER OF
HEART/INTERNAL CHEST DIAMETER
IN COLORADO CHILDREN

| Age | Boys | | Girls | |
(yr.-mo.)	Mean	S.D.	Mean	S.D.
Supine				
0 - 1	.499	.034	.484	.032
0 - 2	.484	.034	.475	.028
0 - 3	.491	.035	.480	.037
0 - 6	.469	.041	.463	.041
0 - 9	.454	.034	.455	.037
1 - 0	.449	.039	.445	.039
1 - 6	.439	.031	.439	.038
2 - 0	.444	.033	.442	.040
2 - 6	.439	.032	.444	.033
3 - 0	.444	.031	.441	.032
3 - 6	.446	.028	.446	.029
4 - 0	.446	.026	.442	.021
Erect				
1 - 0	.488	.032	.467	.032
1 - 6	.461	.036	.476	.045
2 - 0	.466	.036	.460	.043
2 - 6	.470	.040	.457	.035
3 - 0	.460	.026	.452	.032
3 - 6	.453	.035	.447	.026
4 - 0	.442	.027	.447	.026
4 - 6	.437	.027	.441	.026
5 - 0	.435	.028	.437	.028
5 - 6	.428	.032	.434	.030
6 - 0	.430	.028	.430	.029
6 - 6	.427	.030	.431	.034
7 - 0	.428	.028	.428	.026
7 - 6	.424	.028	.429	.031
8 - 0	.420	.029	.431	.030
8 - 6	.420	.033	.427	.028
9 - 0	.424	.032	.428	.029
9 - 6	.414	.031	.422	.028
10 - 0	.412	.027	.424	.032
10 - 6	.409	.027	.418	.030
11 - 0	.415	.031	.419	.031
11 - 6	.413	.030	.419	.029
12 - 0	.410	.030	.413	.031
12 - 6	.409	.028	.409	.031

TABLE 412 (continued)

THE RATIO TRANSVERSE DIAMETER OF
HEART/INTERNAL CHEST DIAMETER
IN COLORADO CHILDREN

Age (yr.-mo.)	Boys Mean	Boys S.D.	Girls Mean	Girls S.D.
13 - 0	.406	.030	.413	.029
13 - 6	.406	.028	.408	.032
14 - 0	.402	.028	.404	.032
14 - 6	.403	.035	.401	.030
15 - 0	.399	.030	.404	.028
16 - 0	.394	.031	.402	.030
17 - 0	.397	.030	.401	.029
18 - 0	.399	.030	.405	.033

(data from McCammon, Human Growth and Development, 1970.
Courtesy of Charles C Thomas, Publisher, Springfield,
Illinois)

TABLE 413

REDUCTION IN MEAN TRANSVERSE DIAMETER (TD, mm) OF
THE HEART (mm) FROM BIRTH TO 5 DAYS
FOR INFANTS IN NEW YORK CITY

Group	TD at Birth	TD at 5 days	Mean Difference	S.D. of Difference
Control, clamped	49.4	47.5	1.86	1.75
Control, stripped	51.8	49.5	2.24	1.39
Low Apgar score, clamped	52.3	49.8	3.50	1.83
Low Apgar score, stripped	--	--	--	--
Intermediate, clamped	54.0	50.3	3.71	2.37
Intermediate, stripped	58.3	51.9	6.43	2.69

(data from Burnard and James, 1961, Pediatrics 27:727-739.
Copyright American Academy of Pediatrics 1961)

TABLE   414

ANGLES OF CLEARANCE OF THE LEFT VENTRICLE (degrees)
IN IOWA CHILDREN

Sex	First Angle of Clearance		Second Angle of Clearance	
	Mean	S.D.	Mean	S.D.
Boys	51.4	6.0	63.1	7.0
Girls	52.2	5.3	63.2	8.4

(data from Jackson, Einstein, Blau, et al., 1944, American Journal
of Diseases of Children 68:157-162. Copyright 1944, American
Medical Association)

TABLE   415

PENIS, TESTICULAR, AND BODY MEASUREMENTS
IN TERM NEWBORN ISRAELI INFANTS

Measurement	Mean ± S.D.
Penis length (cm)	3.5 ± 0.4
Penis width (mid shaft)	1.1 ± 0.1
Penis circumference (cm)	3.3 ± 0.3
Left testicular volume (ml)	1.6 ± 0.4
Right testicular volume (ml)	1.6 ± 0.4
Body length (cm)	50 ± 2
Body weight (gr)	3,350 ± 370
Head circumference (cm)	34.5 ± 1.70
Gestational age (days)	279 ± 7

(data from Flatau, Josefsberg, Reisner, et al., 1975, The
Journal of Pediatrics 87:663-664)

TABLE 416

LENGTH AND CIRCUMFERENCE OF THE PENIS AND VOLUME OF THE TESTIS
FOR BOYS IN NEW YORK CITY

Age	Length		Circumference			Volume of Testis	
	Mean	S.D.	Relaxed Mean	Relaxed S.D.	Erect	Mean	S.D.
	cm.	cm.	cm.	cm.	cm.	cc.	cc.
0 - 5 mo.	3.84	.563	3.64	.424	4.93	.56	.192
6 - 11 mo.	4.11	.633	3.74	.286	5.06	.74	.181
1 year	4.63	.644	3.91	.331	5.28	.80	.223
2 years	5.04	.596	4.04	.360	5.45	.87	.228
3 years	5.41	.880	4.08	.416	5.51	.81	.173
4 years	5.63	.704	4.13	.277	5.57	.83	.215
5 years	6.08	.701	4.23	.344	5.70	.80	.154
6 years	5.98	.753	4.23	.453	5.70	.78	.195
7 years	6.16	.895	4.49	.458	6.05	.84	.234
8 years	6.41	.827	4.29	.431	5.78	.87	.205
9 years	6.33	.907	4.47	.437	6.02	1.01	.312
10 years	6.26	.785	4.56	.383	6.14	1.11	.640
11 years	6.63	1.338	4.93	.979	6.62	2.31	2.051
12 years	7.50	2.199	5.41	1.540	7.26	4.06	3.798
13 years	9.05	2.257	6.09	1.858	8.15	5.72	4.101
14 years	10.02	2.534	6.76	1.718	9.03	8.11	4.488
15 years	11.98	2.080	7.64	1.483	10.19	11.85	4.729
16 years	12.76	1.654	8.03	1.159	10.70	13.37	3.353
17 years	13.41	1.639	8.47	1.433	11.28	15.02	3.698
18 years	13.35	2.017	8.42	1.398	11.22	15.72	4.997

(data from Schonfeld and Beebe, 1942, The Journal of Urology 48:759-799. Copyright 1942,
The Williams & Wilkins Co., Baltimore, MD)

TABLE 417

PERCENTAGE OF OHIO BOYS IN EACH CATEGORY OF PENIS SIZE, BY AGE LEVEL

Size	Age (years)								
	10	10.5	11	11.5	12	12.5	13	13.5	14
1.....	28	26	24	25	16	12	8	3	3
2.....	53	54	54	48	56	36	10	13	8
3.....	14	15	16	17	16	40	36	8	5
4.....	5	5	6	8	7	5	26	36	11
5.....	0	0	0	2	0	2	15	20	35
6.....	0	0	0	0	5	2	3	15	22
7.....	0	0	0	0	0	2	3	5	16

Size	Age (years)								
	14.5	15	15.5	16	16.5	17	17.5	18	19-21
1.....	0	0	0	0	0	0	0	0	0
2.....	3	0	0	0	0	0	0	0	0
3.....	3	3	0	0	0	0	0	0	0
4.....	18	16	20	16	15	14	18	27	20
5.....	27	23	20	20	23	36	35	40	60
6.....	33	42	40	48	42	41	41	27	20
7.....	15	16	20	16	19	9	6	7	0

(data from Reynolds and Wines, 1951, American Journal of Diseases of Children 82:529-547. Copyright 1951, American Medical Association)

TABLE 418

KIDNEY SIZE
IN CHILDREN OF BALTIMORE (MD)

Age (years)	Length (cm)	Vertebral Span	Width (cm)	Cortical Thickness			Axial Midline Distance		Axial Angle (Degrees)
				Lateral (cm)	Inferior (cm)	Superior (cm)	Superior (cm)	Inferior (cm)	
Newborn	4.8	5.0	2.9	1.3	1.5	1.3	1.7	3.1	15
1-4 mo.	5.3	4.5	3.2	1.6	1.5	1.7	2.0	3.4	15
4-8 mo.	5.9	4.8	3.3	1.1	1.5	1.6	2.2	3.2	11
8-12 mo.	5.8	4.1	3.6	1.3	1.7	1.4	2.1	3.4	12
1-2	6.9	4.2	4.0	1.4	1.9	2.0	2.3	3.2	7
2-3	8.0	4.4	4.3	2.0	2.2	2.3	2.4	4.0	12
3-4	8.2	4.4	4.5	1.9	1.9	2.4	2.3	4.0	12
4-5	8.2	4.0	4.5	2.1	2.1	2.5	2.5	4.0	10
5-6	8.2	3.9	4.6	2.0	2.1	2.4	2.5	4.0	11
6-7	9.2	4.0	4.7	2.1	2.5	2.7	2.7	4.4	10
7-8	9.1	3.7	4.9	2.3	2.2	2.5	2.9	4.6	10
8-9	9.2	3.5	4.9	2.3	2.3	2.5	2.8	4.6	12
9-10	9.6	3.6	5.0	2.2	2.3	2.5	2.7	5.2	15
10-11	9.5	3.7	5.0	2.1	2.3	2.5	3.2	5.3	13
11-12	10.8	4.0	5.2	2.0	2.9	3.4	3.1	5.8	15
12-13	10.8	3.8	5.4	2.5	2.8	2.7	2.7	6.3	17
13-14	11.5	3.6	5.5	2.6	2.9	2.9	3.4	6.5	16
14-15	11.7	3.4	5.6	2.3	2.8	3.1	3.6	6.7	14

(data from Gatewood, Glasser, and Vanhoutte, 1965, American Journal of Diseases of Children 110:162-165. Copyright 1965, American Medical Association)

TABLE 419

THE NORMAL RANGE (±2 S.D.) OF KIDNEY LENGTH IN MM GIVEN FOR EACH MM OF L1 to L3 LENGTH.
OUTSIDE THE FRAME THE LENGTH OF THE LUMBAR SEGMENT IS GIVEN IN TENS OF MM ALONG THE y-AXIS
AND IN UNITS ALONG THE x-AXIS.   EXAMPLE: LENGTH OF THE LUMBAR SEGMENT 62 mm.   NORMAL
RANGE OF RENAL LENGTH 72 TO 101 mm.   THE NORMAL RANGE OF THE RATIO
RIGHT KIDNEY LENGTH/LEFT KIDNEY LENGTH COVERS THE INTERVAL 1.12 to 0.84
IN SWEDISH CHILDREN

	0	1	2	3	4	5	6	7	8	9
20	--	--	--	--	--	36- 65	37- 66	38- 67	39- 68	40- 69
30	41- 70	42- 71	43- 72	44- 73	45- 74	46- 75	47- 76	48- 77	49- 78	50- 79
40	51- 79	52- 80	53- 81	54- 82	55- 83	56- 84	57- 85	58- 86	59- 87	60- 88
50	61- 89	62- 90	63- 91	64- 92	65- 93	66- 94	67- 95	68- 96	69- 97	70- 98
60	70- 99	71-100	72-101	73-102	74-103	75-104	76-105	77-106	78-107	79-108
70	80-109	81-110	82-111	83-112	84-113	85-114	86-115	87-116	88-117	89-118
80	90-118	91-119	92-120	93-121	94-122	95-123	96-124	97-125	98-126	99-127
90	100-128	101-129	102-130	103-131	104-132	105-133	106-134	107-135	108-136	109-137
100	109-138	110-139	111-140	112-141	113-142	114-143	115-144	116-145	117-146	118-147
110	119-148	120-149	121-150	122-151	123-152	124-153	--	--	--	--

(data from Eklöf and Ringertz, 1976)

TABLE   420

RENAL INDEX (length/width) IN WHITE AND NEGRO MISSOURI
CHILDREN

Category	Left Renal Index		Right Renal Index		Renal Difference (L-R) Index	
	Mean	S.D.	Mean	S.D.	Mean	S.D.
White male children	52.46	7.95	50.61	7.97	1.85	5.29
White female children	52.99	8.64	50.83	7.73	2.16	6.48
Negro male children	54.40	8.92	51.08	9.08	3.32	5.39
Negro female children	54.75	7.04	51.71	8.34	3.04	5.82
Children 2 years old or less	61.04	11.10	59.24	10.28	1.80	8.79
Children of 3-5 years	55.08	6.54	52.54	6.26	2.54	5.99
Children of 6-10 years	51.96	7.23	49.52	6.96	2.45	5.64
Children of 11-15 years	48.46	7.08	46.16	7.69	2.29	5.35
Males of 11-15 years	49.56	7.75	46.26	8.41	3.30	6.52
Females of 11-15 years	48.19	6.51	45.86	7.50	2.33	5.20

(data from Friedenberg, Walz, McAlister, et al., 1965, RADIOLOGY 84:1022-1030)

TABLE 421

DISTRIBUTION (percentage within 0.5 cm intervals)
OF DIFFERENCES IN LENGTH OF THE PAIR OF KIDNEYS
(left kidney minus right) IN MISSOURI CHILDREN

Interval in cm	Percentage
-3.0 to -2.5	0.0
-2.5 to -2.0	0.0
-2.0 to -1.5	0.3
-1.5 to -1.0	1.1
-1.0 to -0.5	8.4
-0.5 to 0.0	30.0
0.0 to 0.5	35.8
0.5 to 1.0	20.3
1.0 to 1.5	3.2
1.5 to 2.0	0.3
2.0 to 2.5	0.0
2.5 to 3.0	0.0

(data from Friedenberg, Walz, McAlister, et al.,
1965, RADIOLOGY 84:1022-1030)

TABLE 422

VOLUME (cc) OF A SWALLOW OF WATER
IN CHILDREN AGED 1.25 TO 3.5 YEARS

per subject		per Kg	
Mean	S.D.	Mean	S.D.
4.6	± 1.9	0.33	± 0.10

(data from Jones and Work, 1961, American
Journal of Diseases of Children 102:427.
Copyright 1961, American Medical Associa-
tion)

TABLE 423

RATIO OF TONGUE AREA TO INTERMAXILLARY
AREA EXPRESSED AS A PERCENTAGE
IN ENGLISH CHILDREN

Age (years)	BOYS		GIRLS	
	Mean	S.D.	Mean	S.D.
4	--	--	71.4	9.37
5	69.3	5.18	70.9	7.81
6	71.3	6.36	71.1	6.84
7	71.5	5.85	70.9	6.43
8	73.3	6.28	71.8	5.10
9	73.8	4.84	73.1	6.61
10	74.9	6.08	76.0	7.47
11	75.4	6.01	77.5	7.54
12	77.0	5.11	79.2	8.54
13	78.4	5.31	78.7	6.53
14	79.8	6.59	79.1	6.52
15	80.8	6.55	79.1	5.48
16	82.4	6.81	79.6	6.02
17	80.6	6.99	79.8	6.21
18	79.4	7.14	82.6	5.72

(data from Cohen and Vig, 1976)

TABLE 424

TONGUE SHADOW AREA (cm^2)
IN ENGLISH CHILDREN

Age (years)	Mean	S.D.
Boys		
5	16.5	2.95
6	17.4	1.14
7	18.3	1.94
8	19.4	1.98
9	20.4	2.74
10	21.4	2.23
11	22.4	2.67
12	23.8	1.92
13	25.2	2.67
14	26.4	4.97
15	27.8	3.75
16	29.2	3.46
17	30.0	4.19
18	30.2	3.88
Girls		
4	15.3	2.02
5	15.9	1.77
6	16.8	1.69
7	17.5	1.50
8	18.3	1.20
9	19.3	1.61
10	20.8	2.01
11	22.2	2.01
12	23.3	2.99
13	24.1	1.20
14	24.7	2.59
15	25.1	2.18
16	25.5	2.28
17	25.7	2.88
18	25.8	2.40

(data from Cohen and Vig, 1976)

TABLE   425

DIMENSIONS OF THE NASOPHARYNX
FOR CHILDREN IN DENVER (CO)

Age (± 6 months)	Np Area (mm)				Np height (mm)			
	Boys		Girls		Boys		Girls	
	Mean	S.D.	Mean	S.D.	Mean	S.D.	Mean	S.D.
0y 9m	246	64	184	90	15.6	1.7	15.1	1.8
1y 9m	257	78	260	71	16.3	1.5	17.6	2.8
2y 9m	303	61	288	82	17.7	1.3	18.5	2.3
3y 9m	301	90	343	36	18.5	2.4	20.2	1.4
4y 9m	334	72	360	42	19.9	1.9	21.0	1.4
5y 9m	332	86	363	70	19.7	2.5	21.6	1.9
6y 9m	355	84	409	55	20.4	2.1	21.8	1.2
7y 9m	364	69	453	68	21.1	1.9	23.9	2.2
8y 9m	395	79	479	55	22.3	2.3	24.4	2.4
9y 9m	435	88	484	50	23.6	2.3	24.9	2.2
10y 9m	451	84	490	56	24.0	2.7	25.3	2.0
11y 9m	490	93	532	30	25.3	2.8	26.2	2.0
12y 9m	507	101	574	31	25.6	2.8	27.3	2.0
13y 9m	522	141	584	49	26.3	3.1	27.3	2.2
14y 9m	558	128	560	70	27.7	2.2	27.2	2.6
15y 9m	567	125	576	77	28.1	2.6	27.4	2.7
16y 9m	544	86	585	28	27.8	2.4	28.0	2.6
17y 9m	638	138	563	87	29.0	2.9	26.8	3.1

Age (± 6 months)	Np depth (mm)				Angle in degrees			
	Boys		Girls		Boys		Girls	
	Mean	S.D.	Mean	S.D.	Mean	S.D.	Mean	S.D.
0y 9m	30.6	2.6	30.3	2.9	26.2	3.7	31.3	3.5
1y 9m	28.2	4.9	32.0	2.9	27.8	3.7	30.5	2.5
2y 9m	29.3	1.6	30.7	2.4	26.7	2.5	30.7	3.4
3y 9m	27.8	2.0	30.3	3.0	29.2	3.4	30.5	1.8
4y 9m	27.7	1.5	27.8	1.0	29.8	3.2	30.1	1.3
5y 9m	27.4	2.1	28.5	2.3	29.3	3.2	32.0	1.9
6y 9m	27.9	1.9	29.5	2.2	29.1	2.7	28.2	4.5
7y 9m	27.7	2.3	30.1	1.9	30.1	3.6	30.1	3.1
8y 9m	27.3	2.5	30.8	1.2	30.3	3.6	29.6	1.8
9y 9m	27.8	3.4	31.2	2.6	30.1	2.8	30.8	3.6
10y 9m	27.8	8.3	31.4	2.0	29.5	4.0	31.6	3.4
11y 9m	28.7	3.6	31.8	2.0	29.8	4.2	30.3	5.0
12y 9m	28.8	3.5	32.1	2.2	29.2	3.9	30.3	4.6
13y 9m	29.0	4.5	31.8	2.4	30.5	3.1	29.2	3.0
14y 9m	29.3	5.3	31.0	2.0	31.0	2.9	30.2	5.3

TABLE 425   (continued)

DIMENSIONS OF THE NASOPHARYNX
FOR CHILDREN IN DENVER (CO)

Age	Np depth (mm)				Angle in degrees			
(± 6 months)	Boys		Girls		Boys		Girls	
	Mean	S.D.	Mean	S.D.	Mean	S.D.	Mean	S.D.
15y 9m	28.9	4.5	31.4	1.7	30.1	2.5	29.8	6.0
16y 9m	28.1	3.0	32.0	0.9	31.2	1.9	31.0	7.8
17y 9m	31.0	4.6	31.2	2.2	29.0	3.6	29.3	2.1

Note: The sample sizes are small. The angle is between the lower margin of the
sphenoid and the palatal plane. np = nasopharynx.

(data from Handelman and Osborne, 1976)

TABLE 426

ANNUAL INCREMENTS OF DIMENSIONS
OF THE NASOPHARYNX
FOR CHILDREN IN DENVER (CO)

Measurement	Sex	Growth Period	Increment/ year
Area	Boys	0y 9m to 17y 9m	22.5 mm^2/y
	Girls	0y 9m to 12y 9m	32.5 mm^2/y
Height	Boys	0y 9m to 17y 9m	0.8 mm/y
	Girls	0y 9m to 12y 9m	1.0 mm/y
Depth	Boys	1y 9m to 17y 9m	0.2 mm/y
	Girls	1y 9m to 12y 9m	0.0 mm/y

Note: The sample sizes are small.

(data from Handelman and Osborn, 1976)

TABLE   427

SIZE OF THE PHARYNGEAL CAVITY (mm) AT THREE AGES
IN MICHIGAN CHILDREN

Linear Dimensions	Sex	7.5 years		10.5 years		14.0 years	
		Mean	S.D.	Mean	S.D.	Mean	S.D.
Length of pharyngeal cavity	Male	73.02	4.37	82.83	4.82	97.36	5.34
	Female	72.44	4.17	81.75	4.21	93.84	5.09
Nasal opening	Male	16.97	2.51	19.39	2.37	21.60	2.31
	Female	15.40	2.42	17.94	1.89	20.16	1.78
Oral opening	Male	51.02	3.42	57.37	3.25	63.72	3.62
	Female	51.28	2.91	57.51	3.34	64.77	3.76
Laryngeal inlet	Male	19.52	1.96	22.52	2.99	27.94	2.93
	Female	19.33	2.10	22.40	2.24	26.79	2.17

(data from Castelli et al., 1973)

TABLE   428

SIZE OF NASOPHARYNX IN CHILDREN
AGED 13 YEARS WITH NORMAL OCCLUSION

Measurement	Mean	S.D.
Nasopharyngeal area (in^2)	.31	.09
Nasopharyngeal depth (mm)	15.8	2.8
Anterior osseous facial convexity (degrees)	3.0	1.9
Midface depth (mm)	78.7	4.1

(data from Mergen and Jacobs, 1970)

TABLE 429

ESOPHAGEAL LENGTH (cm)

Age (Years)	Jackson and Jackson (1950) incisors to hiatus (cm)	Holinger (1975) teeth to cardia (cm)	Strobel et al. (1979) incisors to LES (cm)
Birth	17	18	--
1	18	21	23
3	21	23	26
5	--	25	29
6	24	--	31
10	26	27	38
14	33	--	44
15	--	33	45

LES = lower esophageal sphincter

(data from Jackson and Jackson (1950) and Holinger (1975), cited by Strobel
 et al., 1979, The Journal of Pediatrics 94:81-84)

TABLE 430

SMALL INTESTINE LENGTH (cm) IN RELATION TO BODY LENGTH (cm)
IN COLORADO CHILDREN

Body Length	Intestine Length Mean	S.D.	Average Intestine Length / Average Body Length
32.5-40	190	±63	
41-45	217	±82	
45.5-50	266	±56	5:1
50.5-55	284	±77	
55.5-60	338	±113	
61-70	383	±101	6:1
72-80	383	±61	
81-90	431	±109	5:1
91-98	448	±100	
100-110	428	±101	
112-130	465	±123	
130-140	514	±86	4:1
141-160	554	±100	

(data from Reiquam, Allen, and Akers, 1965, American Journal
of Diseases of Children 109:447-451. Copyright 1965,
American Medical Association)

TABLE   431

MEASUREMENTS OF PREPUBERTAL AND PUBERTAL THYROID CARTILAGES.
DISTANCES MEASURED IN MILLIMETERS, WEIGHT IN GRAMS AND ANGLES IN DEGREES

Group	F'F	AD	A'D'	AB	A'B'	CD	C'D'
BOYS							
Prepubertal							
Mean	10.77	31.18	30.05	11.65	10.96	6.18	5.97
S.D.	1.51	4.45	4.28	2.36	2.79	0.56	0.62
% of adult	59	70+		89	92	85	88
Pubertal							
Mean	14.72	39.05	41.23	13.61	14.39	8.20	8.34
S.D.	1.55	0.30	6.06	3.42	1.93	1.38	0.84
% of adult	80	92+		104	102	113	123
GIRLS							
Prepubertal							
Mean	9.61	28.64	25.92	10.29	8.18	5.84	5.51
S.D.	0.74	2.76	2.82	1.80	1.69	1.00	1.58
% of adult	73	78+		86	62	89	82
Pubertal							
Mean	10.81	35.58	35.64	14.14	15.09	6.81	6.47
S.D.	0.73	1.97	2.25	1.10	1.04	1.28	1.26
% of adult	83	100+		118	115	104	96

Group	A'A	B'B	C'C	D'D	E'E	∡1	Wt
BOYS							
Prepubertal							
Mean	33.76	30.10	21.08	21.21	21.17	88.00	2.26
S.D.	5.36	5.35	4.02	2.33	1.52	12.13	0.58
% of adult	87	76	66	65	58	126	26
Pubertal							
Mean	34.03	37.94	28.73	28.60	30.93	84.20	5.25
S.D.	a	4.06	2.54	3.42	3.06	7.25	0.92
% of adult	a	95	90	88	85	123	61
GIRLS							
Prepubertal							
Mean	33.05	29.46	21.28	20.72	21.00	87.80	2.00
S.D.	2.09	2.14	1.80	1.01	0.80	4.37	0.12
% of adult	97	80	85	80	82	108	48
Pubertal							
Mean	34.50	33.34	24.27	23.59	23.91	92.50	2.81
S.D.	2.62	2.62	1.51	0.61	2.68	6.61	0.28
% of adult	103	101	97	91	94	114	67

a = could not be calculated because only one specimen had both superior horns intact;
F'F = anterior thyroid height; A'D' = left and right posterior thyroid height;
respectively; AB, A'B' = left and right superior thyroid cornua, respectively; CD, C'D' =
left and right inferior thyroid cornua, respectively; A'A = superior interhorn distance;
B'B = superior interroot distance; C'C = inferior interroot distance; D'D = inferior
interhorn distance; E'E = anterioposterior length; 1 = angle of thyroid laminae;
Wt = weight. (data from Kahane, 1978)

TABLE   432

MEASUREMENTS OF PREPUBERTAL AND PUBERTAL CRICOID CARTILAGES.
DISTANCES MEASURED IN MILLIMETERS AND WEIGHT IN GRAMS.

Group	IJ	I'J'	L'L	H'H	J'J	KI	K'I'	Wt
Prepubertal				BOYS				
Mean	3.98	15.21	8.72	14.18	15.08	17.05	17.54	1.39
S.D.	0.83	1.74	1.59	1.56	2.81	2.98	3.40	0.60
% of adult	63	64	54	61	62	57+		26
Pubertal								
Mean	5.30	21.67	13.84	20.28	20.86	25.24	25.62	3.51
S.D.	0.55	3.68	1.72	2.63	2.63	3.55	2.90	1.03
% of adult	84	91	85	85	85	84+		66
Prepubertal				GIRLS				
Mean	3.97	14.78	8.62	13.56	14.54	16.88	17.37	1.25
S.D.	0.23	0.39	0.94	0.86	0.58	1.11	1.11	0.10
% of adult	71	77	71	74	76	72+		43
Pubertal								
Mean	4.30	18.63	10.81	17.23	17.23	20.39	20.75	2.14
S.D.	0.50	1.03	0.96	0.68	0.68	1.05	1.26	0.23
% of adult	77	97	89	90	90	86+		74

+ represents the mean of right and left measurements; IJ, anterior height of cricoid
cartilage; I'J', posterior height of cricoid cartilage; L'L, distance between the
cricoarytenoid facets; H'H, greatest width of cricoid lumen; J'J, midsagittal length
of cricoid lumen; KI, K'I', superiolateral length of the cricoid on left and right,
respectively; wt, weight.

(data from Kahane, 1978)

TABLE   433

MEASUREMENTS OF PREPUBERTAL AND PUBERTAL ARYTENOID CARTILAGES.
DISTÅNCES MEASURED IN MILLIMETERS AND WEIGHT IN GRAMS.

Group	QR	Q'R'	RS	R'S'	QS	Q'S'	Wt$_l$	Wt$_r$
				BOYS				
Prepubertal								
Mean	9.63	10.25	9.63	9.88	11.03	11.08	0.10	0.10
S.D.	0.55	1.30	0.79	1.33	1.00	1.07	0.04	0.04
% of adult	59+		66+		63+		26+	
Pubertal								
Mean	14.36	14.35	12.60	13.16	15.92	15.94	0.26	0.25
S.D.	2.13	2.05	1.11	1.19	1.12	1.77	0.07	0.07
% of adult	81+		87+		90+		64+	
				GIRLS				
Prepubertal								
Mean	9.91	10.28	8.56	8.72	10.64	10.69	0.08	0.09
S.D.	0.88	1.17	0.52	0.49	0.36	0.31	0.02	0.02
% of adult	77+		77+		75+		47+	
Pubertal								
Mean	12.04	11.93	10.08	10.14	12.62	12.74	0.13	0.12
S.D.	1.60	1.77	0.73	0.67	0.83	0.89	0.02	0.01
% of adult	90+		90+		88+		68+	

+ represents mean of left and right measurements; QR, Q'R', anterior ridge
height on left and right sides, respectively; RS,R'S', width on left and right sides,
respectively; QS, Q'S', posterior ridge height on left and right sides, respectively;
wt$_l$, weight of left cartilage; wt$_r$, weight of right cartilage.

(data from Kahane, 1978)

TABLE 434

MEASUREMENTS FOR PREPUBERTAL AND PUBERTAL TOTAL VOCAL CORD
LENGTH AND ANTERIOR CRICOTHYROID DISTANCE. DISTANCES MEASURED IN
MILLIMETERS. (YI', total vocal fold length; F'I, anterior
cricothyroid distance)

Group	YI'	F'I
BOYS		
Prepubertal		
Mean	17.35	4.75
S.D.	4.35	1.03
% of adult	60	49
Pubertal		
Mean	28.21	7.26
S.D.	3.64	1.45
% of adult	98	76
GIRLS		
Prepubertal		
Mean	17.31	5.00
S.D.	3.68	0.39
% of adult	81	64
Pubertal		
Mean	23.15	6.23
S.D.	2.75	0.95
% of adult	108	79

(data from Kahane, 1978)

TABLE 435

LEVELS OF THE LARYNX IN RELATION TO CERVICAL VERTEBRAE
IN AUSTRALIAN CHILDREN

Age (years)	Superior margin of body of hyoid		Tip of Epiglottis		Inferior margin of arytenoid	
	Boys	Girls	Boys	Girls	Boys	Girls
2.25	inf. C3	--	disc C2-3	--	inf. C4	--
2.5	mid. C3	mid. C3	sup. C2	inf. C2	disc C4-5	inf. C4
2.75	--	mid. C3	disc C3-4	inf. C2	sup. C6	inf. C4
3.0	disc C3-4	mid. C3	inf. C2	inf. C2	sup. C5	mid. C4
3.25	inf. C3	inf. C3	inf. C2	disc C2-3	sup. C5	inf. C4
3.5	inf. C3	inf. C3	sup. C3	disc C2-3	disc C4-5	disc C4-5
3.75	disc C3-4	disc.C4	sup. C3	sup. C3	mid. C5	inf. C4
4.0	inf. C3	disc C3 4	disc C2-3	disc C2-3	mid. C5	inf. C4
4.5	mid. C3	disc C3-4	disc C2-3	disc C2-3	disc C4-5	inf. C4
5.0	inf. C3	disc C3-4	disc C2-3	disc C2-3	sup. C5	sup. C5
5.5	inf. C3	disc C3-4	disc C2-3	inf. C2	sup. C5	inf. C4
6.0	disc C3-4	inf. C3	disc C2-3	disc C2-3	--	--
7.0	mid. C3	inf. C3	disc C2-3	inf. C2	mid. C5	--
8.0	sup. C4	inf. C3	disc C2-3	sup. C3	mid. C5	sup. C5
9.0	disc C3-4	inf. C3	mid. C3	sup. C3	mid. C5	disc C4-5
10.0	disc C3-4	disc C3-4	mid. C3	mid. C3	mid. C5	sup. C5
11.0	disc C3-4	inf. C3	mid. C3	sup. C3	inf. C5	sup. C5

Sup. = superior third; mid. = middle third; and inf. = inferior third of body. C2,3, etc. refer to cervical vertebral bodies.

(data from Roche and Barkla, 1965)

TABLE 436

BRONCHIAL ANGLES (in degrees) IN MASSACHUSETTS CHILDREN

Group	RBA[1]	LBA[2]	Difference in Bronchial Angles
Birth to 1 Year			
Mean	27.71	29.00	1.29
S.D.	9.09	7.50	6.24
1 to 5 Years			
Mean			
Total	27.42	31.80	2.33
Without Ao	29.35	30.85	1.00
With Ao	27.00	38.00	11.00
S.D.			
Total	9.46	10.44	8.05
Without Ao	9.95	10.88	7.78
With Ao	7.07	4.24	2.83
5 to 10 Years			
Mean			
Total	28.13	30.13	2
Without Ao	29.71	29.57	-0.14
With Ao	17	34	17
S.D.			
Total	6.42	3.94	6.89
Without Ao	4.96	3.91	3.53
10 to 15 Years			
Mean			
Total	26.33	33.56	7.22
Without Ao	32.33	34.00	1.67
With Ao	14.33	32.67	18.33
S.D.			
Total	10.32	5.68	9.56
Without Ao	5.85	6.42	4.93
With Ao	4.04	4.93	6.51
15 to 18 Years			
Mean			
Total	20.27	38.91	18.64
Without Ao	28.00	31.50	3.50
With Ao	18.56	40.56	22.00
S.D.			
Total	6.40	6.88	9.18
Without Ao	9.90	7.78	2.12
With Ao	4.56	5.90	5.89

[1] = Right bronchial angle in degrees; [2] = Left bronchial angle in degrees; [3] Negative sign denotes difference when right bronchial angle is greater than left bronchial angle.

(data from Cleveland, 1979, RADIOLOGY 133:89-93)

TABLE 437

THE ANTEROPOSTERIOR DIAMETER (mm) OF THE TRACHEAL LUMEN
IN NEWLY BORN INFANTS IN MICHIGAN AND WISCONSIN

	Level of Measurement		
Diameter	5th Cervical Vertebra	1st Thoracic Vertebra	3rd Thoracic Vertebra
	Number of cases	Number of cases	Number of cases
7	3	0	0
6	21	5	0
5	178	67	33
4	120	198	145
3	27	74	166
2	1	4	5
1	0	2	1
Mean	4.5	3.9	3.5

(data from Donaldson and Tomsett, "Tracheal diameter in the
normal newborn infant," American Journal of Roentgenology 67:
785-787. Copyright 1952, American Roentgen Ray Society.
Reproduced by permission.)

TABLE 438

THE HILAR HEIGHT RATIO IN A BOSTON (MA) SAMPLE
(AGE RANGE 15-30 YEARS)

	Right		Left	
Sex	Mean	S.D.	Mean	S.D.
Males	1.19	0.132	0.842	0.073
Females	1.32	0.160	0.824	0.085

Ratio = apex to hilum
        dome to hilum

(data from Homer, 1978, Radiology 129:11-16)

TABLE 439

PLACENTA: MACRO-AND MICROMORPHOMETRY (%)

	Boston		Guatemala	
Tissue Component	Mean	±1 S.E.M.	Mean	±1 S.E.M.
Macroscopic components				
Parenchyma	72.06	1.39	67.90	1.28
Nonparenchyma	27.61	1.37	29.66	1.10
Pathology				
Old infarct	0.09	0.06	1.60	0.52
New infarct	0.21	0.11	0.53	0.22
Subchorionic fibrin	0.01	0.006	0.26	0.08
Conversion Factor, Fresh-to- Processed Volumes	0.787	0.007	0.829	0.006
Microscopic components of parenchyma				
Intervillous space	32.32	0.89	30.04	0.79
Fibrin	3.58	0.28	8.64	0.61
Stem villi	13.39	0.96	15.65	0.85
Peripheral villi	49.95	0.87	45.60	1.05
Miscellaneous, nonvillous	0.91	0.36	0.24	0.10
Peripheral villi				
Capillaries	8.43	1.0	8.18	0.57
Connective tissue	24.60	0.76	21.94	0.87
Fibroblast component	7.61	0.36	7.80	0.16
Cytotrophoblast	0.19	0.014	0.16	0.03
Syncytiotrophoblast	16.73	0.59	15.30	0.42
Stem villi				
Stem vessels	1.74	0.31	1.84	0.24
Capillaries	0.45	0.10	0.58	0.07
Connective tissue	10.25	0.75	12.17	0.77
Fibroblast component	1.65	0.12	1.88	0.12
Trophoblast	0.83	0.12	1.07	0.10

(data from Laga, Driscoll, and Munro, 1972a, Pediatrics 50:24-32. Copyright American Academy of Pediatrics, 1972)

TABLE 440

MASS COMPONENTS (gm/placenta) FOR PLACENTAS OBTAINED IN
BOSTON AND GUATEMALA

Mass Component	Boston		Guatemala	
	Mean	S.E.M.	Mean	S.E.M.
Fresh placental weight	571.4	21.20	485.40	18.34
Trimmed placental weight	469.0	18.48	404.2	15.72
Nonparenchyma and pathology	130.2	6.82	129.8	5.07
Effective placental mass	338.8	16.34	274.6	13.86
Peripheral villous mass	167.6	8.80	124.9	7.64
Peripheral trophoblastic mass	57.4	2.97	42.3	2.71
Peripheral fibroblast mass	25.8	1.85	21.2	1.59
Peripheral connective tissue mass	84.0	5.45	59.0	4.02
Stem villous mass	46.1	3.37	42.9	3.42
Stem villous trophoblastic mass	2.7	0.37	3.0	0.33
Stem villous fibroblast mass	5.7	0.53	5.2	0.37
Stem villous connective tissue mass	34.8	3.19	27.5	2.46

(data from Laga, Driscoll, and Munro, 1972a, Pediatrics 50:24-32. Copyright American
Academy of Pediatrics, 1972)

TABLE 441

BIOCHEMICAL CHARACTERISTICS OF PLACENTA

Parameter	Boston		Guatemala	
	Mean	+1 S.E.M.	Mean	+1 S.E.M.
DNA (mg/placenta)	1353	148	1057	96
RNA (mg/placenta)	1637	127	1411	168
RNA/DNA ratio (mg/mg)	1.42	0.15	1.51	0.18
Protein (gm/placenta)	58.3	4.1	49.7	3.3
Protein/DNA ratio (gm/gm)	57.8	9.9	52.4	4.1
Heat-Stable alkaline phosphatase (moles/min/placenta)	5.37	0.35	4.77	0.41
Polysome/Monosome ratio	2.96	0.32	1.4	0.13
Membrane-Bound ribosomes/ total ribosomes (%)	21.1	1.3	20.7	0.8
Cell-Free ^{14}C-Leucine incorporation (cpm/mg rRNA)	2454	254	2277	159
Cell-Free ^{14}C-Leucine incorporation (10^3Xcpm/placenta)	4061	511	3198	402

(data from Laga, Driscoll, and Munro, 1972b, Pediatrics 50:33-39. Copyright American Academy of Pediatrics, 1972)

TABLE   442

HUMAN PLACENTA – VILLOUS SURFACE ANALYSIS (M^2)
FOR PLACENTAS OBTAINED IN BOSTON AND GUATEMALA

Characteristic	Boston Mean	Boston ±1 S.E.M.	Guatemala Mean	Guatemala ±1 S.E.M.
Peripheral villous surface	16.09	0.51	12.62	0.64
Peripheral villous capillary surface	12.04	1.29	8.06	0.68
Stem villous surface	0.650	0.074	0.798	0.069
Stem villous capillary surface	0.363	0.053	0.522	0.068

(data from Laga, Driscoll, and Munro, 1972a, Pediatrics 50:24-32. Copyright American
Academy of Pediatrics, 1972)

TABLE   443

TRANSILLUMINATION DISTANCE (cm) AS MEASURED FROM CENTER OF RIGHT BEAM
IN TEXAS INFANTS

Age	Frontal Mean	Frontal S.D.	Parietal Mean	Parietal S.D.	Occipital Mean	Occipital S.D.
Newborn	5.4	0.4	4.8	0.4	4.5	0.5
Two months	5.6	0.6	5.1	0.5	5.4	0.7
Four months	5.7	0.7	5.0	0.3	5.1	0.5
Six months	5.5	0.6	5.0	0.5	5.1	0.5
Nine months	5.7	0.7	5.0	0.7	5.0	0.5
Twelve months	5.7	0.3	5.1	0.4	5.2	0.5
Eighteen months	5.4	0.4	4.9	0.5	4.7	0.5

(data from Cheldelin, Davis, and Grant, 1975, The Journal of Pediatrics 87:
937-938)

TABLE 444

SPINAL CORD/SAS RATIO
IN ENGLISH CHILDREN

Level	Ratio of sagittal diameters		Ratio of transverse diameters	
	Mean	S.D.	Mean	S.D.
C-1	0.56	0.06	--	--
C-2	0.57	0.05	--	--
C-3	0.61	0.05	--	--
C-4	0.61	0.05	--	--
C-5	0.59	0.05	--	--
C-6	0.58	0.06	--	--
C-7	0.55	0.06	--	--
Th-1	0.51	0.06	--	--
Th-2	0.52	0.07	--	--
Th-3	0.53	0.07	--	--
Th-4	0.53	0.08	--	--
Th-5	0.53	0.09	--	--
Th-6	0.54	0.08	0.67	0.06
Th-7	0.57	0.08	0.65	0.06
Th-8	0.56	0.08	0.65	0.07
Th-9	0.55	0.07	0.65	0.07
Th-10	0.56	0.07	0.64	0.06
Th-11	0.56	0.06	0.64	0.07
Th-12	0.56	0.06	0.59	0.08

C = cervical; Th = thoracic; SAS ratio = spinal cord width/
subarachnoid space width.

(data from Boltshauser and Hoare, 1976)

TABLE 445

BODY AND LIVER MEASUREMENTS IN IOWA CHILDREN AGED 5 TO 12 YEARS*

	BOYS		GIRLS	
Age (years)	9.2		8.9	
	+1.7		+1.9	
Investigator	A	B	A	B
Body measurements: Stature (cm)	107.7	108.3	109.5	110.4
	+21.5	+21.3	+18.6	+18.9
Weight (kg)	52.3	51.8	53.3	53.0
	+4.1	+4.3	+4.2	+4.3
Liver measurement at right mid-clavicular line: Span (height, cm)	8.9 +1.2	8.5 +1.3	9.2 +1.0	9.0 +1.1
Edge below costal margin (cm)	4.3	4.0	4.4	3.8
	+1.8	+1.5	+1.2	+1.1

*Numbers shown are mean + S.D. values.

(data from Younoszai and Mueller, 1975, Clinical Paediatrics 14:378-380. Copyright Lippincott/Harper & Row)

TABLE   446

WEIGHTS OF ORGANS OF MALE NEWBORN INDIANA INFANTS

Age Group		Body Wt. gm.	Body Length cm.	Brain gm.	Thymus gm.	Heart gm.
Stillborn	Mean	3,277	49.4	409	12	23
	S.D.	653	4.5	60	5	6
Liveborn to 1 week	Mean	3,171	49.3	406	8.6	23
	S.D.	488	2.8	55	5.5	5

Age Group		Lungs Combined gm.	Liver gm.	Spleen gm.	Pancreas gm.	Adrenal Glands Combined gm.	Kidneys Combined gm.
Stillborn	Mean	59	130	12	5.2	8.5	30
	S.D.	21	39	4	2.7	3.5	9
Liveborn to 1 week	Mean	66	129	10	5.3	7.0	29
	S.D.	25	33	3	3.2	2.8	8

(data from Schulz, Giordano, and Schulz, 1962, Archives of Pathology 74:244-250.
Copyright 1962, American Medical Association)

TABLE 447

WEIGHTS OF ORGANS OF FEMALE NEWBORN INDIANA INFANTS

Age Group		Body Wt. gm.	Body Length cm.	Brain gm.	Thymus gm.	Heart gm.
Stillborn	Mean	3,172	49.6	394	10	21
	S.D.	617	3.3	81	5	5
Liveborn to 1 wk.	Mean	3,085	49.7	401	7.2	21
	S.D.	491	3.7	41	3.6	5

Age Group		Lungs Combined gm.	Liver gm.	Spleen gm.	Pancreas gm.	Adrenal Glands Combined gm.	Kidney Combined gm.
Stillborn	Mean	52	129	10	5.0	8.0	28
	S.D.	13	44	4	3.1	3.2	9
Liveborn to 1 wk.	Mean	65	131	11	5.6	6.9	29
	S.D.	17	43	4	2.8	3.1	7

(data from Schulz, Giordano, and Schulz, 1962, Archives of Pathology 74:244-250.
Copyright 1962, American Medical Association)

TABLE 448

WEIGHTS OF ORGANS OF MALE INDIANA INFANTS

Age (months)		Body Length cm.	Brain gm.	Thymus gm.	Heart gm.	Lungs Combined gm.
1	Mean	51.4	460	7.8	23	64
	S.D.	3.2	47	5.3	7	21
2	Mean	54.0	506	9.4	27	74
	S.D.	2.9	67	4.4	7	26
3	Mean	57.7	567	10	30	89
	S.D.	2.9	81	5	7	23
4	Mean	60.4	620	10	31	96
	S.D.	4.1	71	6	7	27
5	Mean	62.0	746	12	35	93
	S.D.	3.1	91	7	5	18
6	Mean	64.2	762	10	40	115
	S.D.	3.9	73	6	8	31
7	Mean	66.7	767	12	43	118
	S.D.	5.0	32	9	8	33
8	Mean	68.2	774	10	44	104
	S.D.	3.4	95	6	8	32
9	Mean	69.4	820	10	45	109
	S.D.	4.2	49	4	7	33
10	Mean	69.7	850	9	46	110
	S.D.	3.9	96	5	6	34
11	Mean	70.5	875	19	48	130
	S.D.	4.3	89	4	7	31
12	Mean	73.8	954	12	50	116
	S.D.	4.1	35	5	6	23

TABLE 448 (continued)

WEIGHTS OF ORGANS OF MALE INDIANA INFANTS

Age (months)		Liver gm.	Spleen gm.	Pancreas gm.	Adrenal Glands Combined gm.	Kidneys Combined gm.
1	Mean	140	12	6.2	5.1	34
	S.D.	40	4	3.6	1.7	9
2	Mean	160	15	7.2	5.0	39
	S.D.	46	5	4.4	1.6	9
3	Mean	179	16	7.7	5.0	45
	S.D.	41	5	3.1	1.3	10
4	Mean	195	17	11	4.9	47
	S.D.	41	5	5	2.0	12
5	Mean	228	18	11	5.3	54
	S.D.	47	7	4	1.9	11
6	Mean	259	20	11	5.2	62
	S.D.	58	7	5	2.0	14
7	Mean	276	23	12	5.5	69
	S.D.	54	10	6	2.1	14
8	Mean	285	20	13	5.4	66
	S.D.	57	7	7	2.3	14
9	Mean	288	22	16	5.4	67
	S.D.	47	5	7	2.0	16
10	Mean	300	24	14	5.7	72
	S.D.	69	11	6	2.1	17
11	Mean	305	28	16	6.1	76
	S.D.	81	10	3	1.8	19
12	Mean	325	28	14	6.3	76
	S.D.	39	7	6	2.2	13

(data from Schulz, Giordano, and Schulz, 1962, Archives of Pathology 74:244-250.
Copyright 1962, American Medical Association)

TABLE   449

WEIGHTS OF ORGANS OF FEMALE INDIANA INFANTS

Age (months)		Body Length cm	Brain gm	Thymus gm	Heart gm	Lungs Combined gm
1	Mean	51.9	433	6.6	21	64
	S.D.	4.5	59	4.9	5	27
2	Mean	54.0	490	5.8	26	74
	S.D.	3.7	51	4.7	6	23
3	Mean	57.0	525	9.7	28	81
	S.D.	3.7	89	6.9	4	14
4	Mean	59.0	595	9.0	30	91
	S.D.	3.7	80	7.3	6	24
5	Mean	62.2	725	13	36	102
	S.D.	3.3	62	5	5	22
6	Mean	63.0	730	10	37	111
	S.D.	3.0	85	6	7	30
7	Mean	65.4	750	10	40	111
	S.D.	4.2	92	8	9	38
8	Mean	66.5	770	8	41	109
	S.D.	4.5	96	5	7	35
9	Mean	68.3	810	9	41	105
	S.D.	4.7	82	5	5	28
10	Mean	67.5	830	12	43	105
	S.D.	4.2	117	7	7	21
11	Mean	70.5	875	15	44	125
	S.D.	3.1	64	8	8	31
12	Mean	71.5	886	11	49	115
	S.D.	4.7	64	8	6	34

TABLE 449   (continued)

WEIGHTS OF ORGANS OF FEMALE INDIANA INFANTS

Age (months)		Liver gm	Spleen gm	Pancreas gm	Adrenal Glands Combined gm	Kidneys Combined gm
1	Mean	139	11	5.0	4.8	31
	S.D.	31	4	1.8	1.9	8
2	Mean	159	14	7.1	4.7	36
	S.D.	31	5	2.9	1.4	10
3	Mean	183	15	8.5	4.8	42
	S.D.	39	5	3.2	1.4	12
4	Mean	204	17	9.0	4.6	50
	S.D.	49	5	3.0	2.1	11
5	Mean	227	19	11	4.8	52
	S.D.	38	5	3	2.2	13
6	Mean	242	18	11	4.6	58
	S.D.	58	8	4	1.5	20
7	Mean	272	22	10	5.5	65
	S.D.	51	8	3	2.2	14
8	Mean	276	20	11	5.3	60
	S.D.	54	9	5	2.3	13
9	Mean	288	18	14	5.4	62
	S.D.	67	6	5	1.5	10
10	Mean	284	25	13	5.7	66
	S.D.	48	11	6	1.7	10
11	Mean	292	23	14	6.2	68
	S.D.	36	9	7	2.0	14
12	Mean	315	27	15	6.0	72
	S.D.	38	9	8	1.4	19

(data from Schulz, Giordano, and Schulz, 1962, Archives of Pathology 74:244-250.
Copyright 1962, American Medical Association)

TABLE 450

RECUMBENT LENGTHS (cm) OF NEWBORN INFANTS AND THEIR ORGAN
WEIGHTS (gm) IN RELATION TO BODY WEIGHT (gm; means ± S.D.)

Body Weight	Recumbent Length	Heart	Lungs combined	Spleen	Liver	Adrenal Glands combined	Kidneys combined	Thymus	Brain	Gestational Age Wks.	Days
500	29.4	5.0	12	1.3	26	2.6	5.4	2.2	70	23.	5
	±2.5	±1.6	±5	±0.8	±10	±1.7	±2.1	±0.8	±18	±2.	3
750	32.9	6.3	19	2.0	39	3.2	7.8	2.8	107	26.	0
	±3.0	±1.8	±6	±1.2	±12	±1.5	±2.6	±1.3	±27	±2.	6
1000	35.6	7.7	24	2.6	47	3.5	10.4	3.7	113	27.	5
	±3.1	±2.0	±8	±1.5	±12	±1.6	±3.4	±2.0	±34	±3.	1
1250	38.1	9.6	30	3.1	56	4.0	12.9	4.9	174	29.	0
	±3.0	±3.3	±9	±1.8	±21	±1.7	±3.9	±2.1	±38	±3.	0
1500	41.0	11.5	34	4.3	65	4.5	14.9	6.1	219	31.	3
	±2.7	±3.3	±11	±2.0	±18	±1.8	±1.2	±2.7	±52	±2.	3
1750	42.6	12.8	40	5.0	74	5.2	17.4	6.8	247	32.	4
	±3.1	±3.2	±13	±2.5	±20	±2.0	±4.7	±3.0	±51	±2.	6
2000	44.9	14.9	44	6.0	82	5.3	18.8	7.9	281	34.	6
	±2.8	±4.2	±13	±2.7	±23	±2.0	±5.0	±3.4	±56	±3.	2
2250	46.3	16.0	48	7.0	88	6.0	20.2	8.2	308	36.	1
	±2.9	±4.3	±15	±3.3	±24	±2.3	±1.9	±3.4	±19	±3.	0
2500	47.3	17.7	48	8.5	105	7.1	22.6	8.3	339	38.	0
	±2.3	±4.2	±14	±3.5	±21	±2.8	±5.5	±4.4	±50	±3.	2
2750	18.7	19.1	51	9.1	117	7.5	24.0	9.6	362	39.	2
	±2.9	±3.8	±15	±3.6	±26	±2.7	±5.4	±3.8	±18	±2.	2
3000	50.0	20.7	53	10.1	127	8.3	24.7	10.2	380	40.	0
	±2.9	±5.3	±13	±3.3	±30	±2.9	±5.3	±4.3	±55	±2.	1
3250	50.7	21.5	59	11.0	145	9.2	27.3	11.6	395	40.	4
	±2.6	±4.3	±18	±4.0	±33	±3.4	±6.6	±4.4	±53	±1.	6
3500	51.8	22.8	63	11.3	153	9.8	28.0	12.8	411	40.	4
	±3.0	5.9	±17	±3.6	±33	±3.5	±6.5	±5.1	±55	±1.	5
3750	52.1	23.8	65	12.5	159	10.2	29.5	13.0	413	40.	6
	±2.3	±5.1	±15	±4.1	±40	±3.3	±6.8	±4.8	±55	±2.	3
4000	52.1	25.8	67	14.1	180	10.8	30.2	11.4	420	41.	4
	±2.7	±5.3	±20	±4.0	±39	±3.1	±6.2	±3.2	±62	±4.	3
4250	53.2	26.5	68	13.0	197	12.0	30.7	11.7	415	41.	2
	±2.5	±5.3	±16	±2.5	±12	±3.7	±5.8	±3.7	±38	±2.	1

(data from Gruenwald and Minh, 1960, American Journal of Clinical Pathology 34:247-253.
By permission of Lippincott/Harper & Row)

TABLE 451

STATURE (cm), WEIGHT (kg) AND CARDIAC
DIMENSIONS (cm) IN GERMAN CHILDREN

Age (years)	Stature Mean	S.D.	Weight Mean	S.D.	Heart Length Mean	S.D.	Heart Width Mean	S.D.	Heart Depth Mean	S.D.	HV (ml) Mean	S.D.	HV kg^{-1} Mean	S.D.
8	132.4	6.4	28.4	4.3	11.2	0.7	9.1	0.6	7.8	0.3	281.8	35.2	10.0	0.5
9	138.2	7.1	30.8	5.8	11.8	0.7	9.3	0.6	8.1	0.4	312.3	45.3	10.3	1.3
10	144.8	7.0	32.3	4.1	11.8	0.7	9.6	0.6	8.2	0.7	328.2	50.6	10.1	0.9
11	149.6	5.5	35.6	5.3	12.2	0.6	9.8	0.5	8.6	0.9	362.1	41.9	10.3	1.5
12	154.6	6.4	38.6	6.1	12.4	0.7	10.0	0.6	9.0	0.7	394.9	49.7	10.3	1.0
13	158.8	10.1	44.8	12.1	13.2	1.0	10.4	0.8	9.1	0.8	443.5	85.8	10.1	1.3
14	163.6	8.8	49.0	8.5	13.7	0.9	10.9	0.9	9.5	0.8	503.0	85.7	10.3	0.9
15	170.5	8.3	56.1	10.5	14.1	1.1	11.5	1.0	9.6	0.9	551.3	117.2	9.8	1.3
16	172.9	6.9	63.0	9.7	14.8	0.7	11.9	0.7	9.7	0.8	603.0	82.6	9.6	1.0
17	177.5	5.3	66.7	7.9	14.8	1.0	12.2	0.8	10.2	0.7	646.1	104.5	9.7	1.0
18	177.5	7.6	66.8	7.9	15.3	0.7	12.3	0.7	10.1	1.2	671.2	97.1	10.1	1.2

HV = heart volume (cc)

(data from Bouchard et al., 1977)

TABLE   452

OCULAR REFRACTION, THE OPTICAL COMPONENTS, STATURE AND WEIGHT IN RELATION
TO AGE: MEAN VALUES AND STANDARD DEVIATIONS FOR ENGLISH CHILDREN

Mean Age (years and months)		Stature (cm)		Weight (kg)		Ocular refraction in vertical meridian (D)		Corneal power in vertical meridian (D)	
y.	m.	Mean	S.D.	Mean	S.D.	Mean	S.D.	Mean	S.D.
					BOYS				
3	6	98	3.8	16.4	1.6	+2.2	0.7	42.6	1.5
4	6	104	3.4	18.4	1.7	+2.2	0.8	42.8	1.5
5	6	110	5.0	19.8	2.2	+2.1	1.02	43.3	1.8
6	5	116	5.1	22.1	2.3	+1.5	0.83	43.3	1.5
7	5	124	5.6	25.4	3.5	+1.6	1.2	43.1	1.4
8	6	128	5.6	27.4	3.4	+1.5	1.1	42.5	1.4
9	6	133	5.7	30.7	5.4	+1.2	1.2	42.8	1.4
10	5	139	6.2	33.9	5.1	+1.3	1.07	42.7	1.5
11	8	142	6.9	35.5	5.7	+1.3	1.04	42.8	1.6
12	5	147	7.2	39.2	6.3	+1.0	1.1	42.6	1.4
13	5	153	7.9	44.0	8.7	+1.1	0.91	42.6	1.4
14	6	160	10.0	50.6	8.8	+1.0	1.0	42.9	1.5
					GIRLS				
3	6	97	3.6	16.2	1.6	+2.7	0.8	43.8	1.5
4	6	103	3.6	17.8	1.8	+2.2	0.9	43.5	1.2
5	6	110	4.2	20.1	2.2	+2.1	1.1	43.7	1.2
6	6	116	5.7	22.2	3.5	+1.7	1.0	43.7	1.6
7	6	121	5.7	24.6	3.5	+1.8	1.2	43.8	1.3
8	5	127	5.5	26.9	3.7	+1.5	0.92	43.8	1.4
9	6	132	6.7	30.2	5.4	+1.7	1.5	43.6	1.4
10	5	139	6.8	33.8	5.1	+1.1	1.1	43.6	1.5

TABLE 452 (continued)

OCULAR REFRACTION, THE OPTICAL COMPONENTS, STATURE AND WEIGHT IN RELATION
TO AGE: MEAN VALUES AND STANDARD DEVIATIONS FOR ENGLISH CHILDREN

Mean Age (years and months)		Stature (cm)		Weight (kg)		Ocular refraction in vertical meridian (D)		Corneal power in vertical meridian (D)	
y.	m.	Mean	S.D.	Mean	S.D.	Mean	S.D.	Mean	S.D.
GIRLS									
11	6	144	8.1	38.1	7.8	+1.2	1.2	43.6	1.4
12	6	150	6.8	41.9	7.4	+1.1	1.1	43.5	1.5
13	6	155	6.2	46.9	8.4	+0.9	0.62	43.8	1.3
14	6	158	6.1	51.1	9.0	+0.7	1.5	43.5	1.3
15	4	159	5.8	52.0	8.3	+0.8	0.59	43.8	1.3

Mean Age (years and months)		Depth of anterior chamber (mm)		Equivalent power of lens (D)		Total power of eye (D)		Axial length (mm)	
y.	m.	Mean	S.D.	Mean	S.D.	Mean	S.D.	Mean	S.D.
BOYS									
3	6	3.4	0.20	21.8	1.2	60.4	1.8	23.2	0.19
4	6	3.5	0.21	21.5	1.1	60.3	1.7	23.2	0.63
5	6	3.4	0.21	21.5	1.5	60.7	2.3	23.1	0.69
6	5	3.4	0.22	21.4	1.5	60.6	2.2	23.4	0.83
7	5	3.5	0.23	21.1	2.0	60.1	2.1	23.6	0.92
8	6	3.5	0.20	20.6	1.4	59.1	1.9	23.9	0.73
9	6	3.5	0.27	20.6	1.5	59.4	1.9	23.8	0.72
10	5	3.6	0.21	20.5	1.4	59.3	2.0	23.9	0.92
11	8	3.5	0.20	20.3	1.6	59.1	2.3	24.0	0.92
12	5	3.6	0.21	20.1	1.5	58.8	2.0	24.2	0.81
13	5	3.5	0.20	20.1	1.2	59.0	1.5	24.1	0.65
14	6	3.5	0.19	19.9	1.2	59.1	1.9	24.0	0.72

TABLE   452 (continued)

OCULAR REFRACTION, THE OPTICAL COMPONENTS, STATURE AND WEIGHT IN RELATION
TO AGE: MEAN VALUES AND STANDARD DEVIATIONS FOR ENGLISH CHILDREN

GIRLS

Mean Age (years and months)		Depth of anterior chamber (mm)		Equivalent power of lens (D)		Total power of eye (D)		Axial length (mm)	
y.	m.	Mean	S.D.	Mean	S.D.	Mean	S.D.	Mean	S.D.
3	6	3.3	0.19	22.4	1.5	62.0	2.0	22.5	0.74
4	6	3.3	0.21	22.1	1.3	61.6	1.8	22.7	0.62
5	6	3.3	0.22	22.3	1.7	61.7	2.0	22.8	0.61
6	6	3.4	0.16	21.7	1.5	61.3	2.0	23.0	0.70
7	6	3.3	0.22	21.3	1.4	61.0	1.6	23.1	0.69
8	5	3.4	0.22	21.2	1.4	60.9	1.9	23.2	0.68
9	6	3.4	0.19	21.3	1.5	60.8	1.8	23.2	0.81
10	5	3.5	0.26	20.4	1.3	60.1	1.9	23.6	0.75
11	6	3.5	0.20	20.5	1.7	60.1	2.1	23.6	0.86
12	6	3.4	0.21	20.6	1.4	60.1	2.0	23.6	0.84
13	6	3.5	0.21	20.7	1.5	60.4	1.8	23.6	0.61
14	6	3.5	0.24	20.5	1.4	60.1	1.8	23.7	0.76
15	4	3.5	0.16	20.7	1.4	60.6	1.7	23.5	0.63

The values for 43 boys, 44 girls, whose ocular refraction was more than
2 standard deviations from the mean for their age groups, are not included.
D = diopters.

(data from Sorsby, Benjamin and Sheridan, 1961, MRC Special Report Series No.
301, British Crown copyright. Reproduced with the permission of the Controller
of Her Britannic Majesty's Stationery Office)

TABLE 453

OCULAR REFRACTION IN ENGLISH CHILDREN

Age* (years)	BOYS Mean ocular refraction (D)	S.E.	GIRLS Mean ocular refraction (D)	S.E.
3	+2.33	0.24	+2.96	0.19
4	+2.24	0.12	+2.33	0.15
5	+2.21	0.16	+2.20	0.17
6	+1.71	0.16	+1.83	0.19
7	+1.92	0.20	+1.98	0.20
8	+1.76	0.19	+1.63	0.15
9	+1.52	0.24	+2.03	0.24
10	+1.43	0.22	+1.33	0.20
11	+1.63	0.23	+1.50	0.20
12	+1.19	0.23	+1.04	0.18
13	+1.38	0.20	+0.96	0.13
14	+0.93	0.38	+0.62	0.26
15	--	--	+0.64	0.18

*Age is at last birthday.   D = diopters.

(data from Sorsby, Benjamin and Sheridan, 1961, MRC Special Report Series No. 301, British Crown copyright. Reproduced with the permission of the Controller of Her Britannic Majesty's Stationery Office)

TABLE 454

MEAN QUOTIENT OF AXIAL LENGTH (mm) OVER POWER
OF LENS (D) FOR ALL REFRACTIONS IN ENGLISH CHILDREN
COMPARED WITH MEAN QUOTIENTS FOR THE ADULT EMMETROPIC EYE

Age (years)	Boys	Girls
3	1.06	1.00
4	1.08	1.03
5	1.07	1.02
6	1.09	1.06
7	1.12	1.08
8	1.16	1.09
9	1.16	1.09
10	1.17	1.16
11	1.18	1.15
12	1.20	1.15
13	1.20	1.14
14	1.21	1.16
15	--	1.14

D = diopters

(data from Sorsby, Benjamin and Sheridan, 1961, MRC Special
Report Series No. 301, British Crown copyright. Reproduced
with the permission of the Controller of Her Britannic
Majesty's Stationery Office)

TABLE 455

DISTRIBUTION OF DEGREES OF AXIAL ELONGATION IN
ENGLISH CHILDREN RE-EXAMINED AFTER 2-6 YEARS

No. of children
Annual axial elongation (mm)

Age (years)	0-0 M.	0-0 F.	0-1 M.	0-1 F.	0-2 M.	0-2 F.	0-3 M.	0-3 F.	0-4 M.	0-4 F.	0-5 M.	0-5 F.	0-6 M.	0-6 F.	0-7 M.	0-7 F.	Total M.	Total F.
3	--	--	3	1	6	3	2	2	3	1	--	--	--	--	--	--	14	7
4	1	--	--	1	2	5	3	3	3	--	3	--	--	--	--	--	12	9
5	2	--	7	4	2	3	10	4	2	5	1	2	--	1	1	--	25	19
6	--	1	1	5	6	5	9	6	1	4	2	3	--	1	--	--	19	25
7	1	2	4	2	5	13	8	5	5	1	1	--	--	1	--	--	24	24
8	3	3	7	5	6	9	8	2	--	1	--	--	--	--	--	--	24	20
9	2	1	3	7	3	3	--	4	1	--	--	--	--	--	--	--	9	15
10	--	1	5	3	4	4	4	2	2	4	1	--	--	--	--	1	16	15
11	--	1	7	4	7	12	1	2	--	--	1	--	--	--	--	--	16	19
12	3	4	3	9	1	2	2	1	--	1	--	2	--	--	--	--	9	19
13	--	4	5	2	3	4	1	1	--	--	--	--	--	--	--	--	9	11
14	3	2	2	5	--	2	--	--	--	--	--	--	--	--	--	--	5	9
15	--	--	1	4	--	3	--	--	--	1	--	--	--	--	--	--	1	8
16	--	--	--	1	--	2	--	--	--	--	--	--	--	--	--	--	--	3
Totals	15	19	48	53	45	70	48	32	17	18	9	7	--	3	1	1	183	203

(data from Sorsby, Benjamin and Sheridan, 1961, MRC Special Report Series No. 301, British Crown copyright. Reproduced with the permission of the Controller of Her Britannic Majesty's Stationery Office)

TABLE 456

ANNUAL RATES OF AXIAL ELONGATION IN RELATION TO THE RATE OF CHANGE IN
OCULAR REFRACTION AND IN THE POWERS OF THE CORNEA AND OF THE LENS IN ENGLISH CHILDREN

No. of children
Annual axial elongation (mm)

Annual change	0-0 M.	0-0 F.	0-1 M.	0-1 F.	0-2 M.	0-2 F.	0-3 M.	0-3 F.	0-4 M.	0-4 F.	0-5 M.	0-5 F.	0-6 M.	0-6 F.	0-7 M.	0-7 F.
	15	19	48	53	45	70	48	32	16	18	10	7	0	3	1	1
Ocular refraction (D)																
0.0 - 0.2	13	16	42	51	32	47	28	13	5	5	1	1	–	–	–	–
0.3 - 0.5	2	3	6	2	12	21	19	10	5	9	6	3	–	1	–	–
0.6 - 0.8	–	–	–	–	1	2	1	8	6	4	2	3	–	2	–	1
0.9 and over	–	–	–	–	–	–	–	1	–	–	1	–	–	–	1	–
Corneal power (D)																
0.0 - 0.2	14	18	45	50	41	59	27	26	9	14	4	1	–	1	–	–
0.3 - 0.5	–	1	3	3	4	11	21	6	6	4	6	5	–	2	1	1
0.6 - 0.8	1	–	–	–	–	–	–	–	1	–	–	1	–	–	–	–
0.9 and over	–	–	–	–	–	–	–	–	–	–	–	–	–	–	–	–
Lens power (D)																
0.0 - 0.2	14	18	41	47	23	55	21	20	8	4	1	3	–	1	1	–
0.3 - 0.5	1	1	6	6	22	13	26	9	7	10	8	4	–	2	–	1
0.6 - 0.8	–	–	1	–	–	1	–	2	–	3	1	–	–	–	–	–
0.9 and over	–	–	–	–	–	1	1	1	1	1	–	–	–	–	–	–

D = diopters

(data from Sorsby, Benjamin and Sheridan, 1961, MRC Special Report Series No. 301, British Crown copyright. Reproduced with the permission of the Controller of Her Britannic Majesty's Stationery Office)

TABLE   457

COLLATED DATA FOR STANDARD BMR VALUES

Age (years)	BOYS Albritton (1954) Mean	BOYS Albritton (1954) 95% range	kcal/m²/h Fleisch (1951) Mean	GIRLS Albritton (1954) Mean	GIRLS Albritton (1954) 95% range	Fleisch (1951) Mean
1	--	--	53.0	--	--	53.0
2	--	--	52.4	--	--	52.4
3	60.1	51.8 - 68.3	51.3	54.5	47.0 - 62.0	51.2
4	57.9	49.9 - 65.9	50.3	53.9	46.5 - 61.3	49.8
5	56.3	48.5 - 64.1	49.3	53.0	45.7 - 60.3	48.4
6	54.0	46.5 - 61.5	48.3	51.2	44.1 - 58.3	47.0
7	52.3	45.1 - 59.5	47.3	49.7	42.8 - 56.6	45.4
8	50.8	43.8 - 57.8	46.3	48.0	41.4 - 54.6	43.8
9	49.5	42.7 - 56.3	45.2	46.2	39.8 - 52.6	42.8
10	47.7	41.1 - 54.3	44.0	44.9	38.7 - 51.1	42.5
11	46.5	40.1 - 52.9	43.0	44.1	38.0 - 50.2	42.0
12	45.3	39.0 - 51.6	42.5	42.0	36.2 - 47.8	41.3
13	44.5	38.4 - 50.6	42.3	40.5	34.9 - 46.1	40.3
14	43.8	37.8 - 49.8	42.1	39.2	33.8 - 44.6	39.2
15	43.7	37.7 - 49.7	41.8	38.3	33.0 - 43.6	37.9
16	42.9	37.0 - 48.8	41.4	37.7	32.5 - 42.9	36.9
17	41.9	36.1 - 47.7	40.8	36.2	31.2 - 41.2	36.3
18	40.5	34.9 - 46.1	40.0	35.7	30.8 - 40.6	35.9

TABLE 457 (continued)

COLLATED DATA FOR STANDARD BMR VALUES

Age (years)	BOYS Albritton (1954) Mean	BOYS Albritton (1954) 95% range	kJ/m²/h* Fleisch (1951) Mean	GIRLS Albritton (1954) Mean	GIRLS Albritton (1954) 95% range	Fleisch (1951) Mean
1	--	--	222	--	--	222
2	--	--	219	--	--	219
3	252	217 – 286	215	228	197 – 260	214
4	242	209 – 276	211	226	195 – 257	208
5	236	203 – 268	206	222	191 – 252	203
6	226	195 – 257	202	214	185 – 244	197
7	219	189 – 249	198	208	179 – 237	190
8	213	183 – 242	194	201	173 – 229	183
9	207	179 – 236	189	193	167 – 220	179
10	200	172 – 227	184	188	162 – 214	178
11	195	168 – 221	180	185	159 – 210	176
12	190	163 – 216	178	175	152 – 200	173
13	186	161 – 212	177	170	146 – 193	169
14	183	158 – 208	176	164	141 – 187	164
15	182	158 – 208	175	160	138 – 182	159
16	180	155 – 204	173	158	136 – 180	154
17	175	151 – 200	171	152	131 – 172	152
18	170	146 – 193	167	149	129 – 170	150

*The values in kJ have been calculated by using the relationship 1 $kcal_{15}$ = 4.185 kJ.

(data from Diem and Lentner, 1972. Reproduced from DOCUMENTA GEIGY Scientific Tables, 7th edition, with kind permission of CIBA-GEIGY Ltd., Basel, Switzerland)

TABLE 458

METABOLISM TESTS ON COLORADO BOYS, FED AND ASLEEP,
BETWEEN 1 AND 37 MONTHS OF AGE, IN
RELATION TO AGE, SURFACE AREA, WEIGHT AND STATURE

| Age | cal/hr | | cal/hr/m^2 | | cal/hr/kg | | cal/hr/cm | |
(months)	Mean	S.D.	Mean	S.D.	Mean	S.D.	Mean	S.D.
1 - 3	10.8	1.5	42.2	3.7	2.32	0.22	0.192	0.022
3 - 5	14.6	2.3	45.9	3.4	2.27	0.16	0.234	0.026
5 - 7	17.9	1.5	49.9	4.1	2.33	0.26	0.269	0.020
7 - 9	19.7	1.9	49.6	3.6	2.21	0.18	0.281	0.025
9 - 11	21.4	2.6	51.3	5.2	2.27	0.26	0.294	0.032
11 - 13	22.8	2.7	51.8	4.4	2.27	0.22	0.302	0.031
13 - 15	23.9	1.9	52.0	3.8	2.24	0.22	0.308	0.022
15 - 19	25.2	1.8	51.9	3.6	2.23	0.24	0.313	0.020
19 - 25	27.1	2.2	52.2	3.3	2.24	0.18	0.318	0.023
25 - 31	28.8	2.0	51.7	3.3	2.20	0.17	0.321	0.021
31 - 37	30.6	2.4	51.9	4.4	2.22	0.23	0.326	0.027

(data from Lee and Iliff, 1956, Pediatrics 18:739-749. Copyright American Academy of
Pediatrics 1956)

TABLE 459

METABOLISM TESTS ON COLORADO GIRLS, FED AND ASLEEP,
BETWEEN 1 AND 37 MONTHS OF AGE, IN
RELATION TO AGE, SURFACE AREA, WEIGHT AND STATURE

Age	cal/hr		cal/hr/m^2		cal/hr/kg		cal/hr/cm	
(months)	Mean	S.D.	Mean	S.D.	Mean	S.D.	Mean	S.D.
1 - 3	10.0	1.8	39.9	4.4	2.22	0.20	0.180	0.027
3 - 5	13.6	2.0	44.9	4.0	2.30	0.22	0.221	0.025
5 - 7	16.3	1.4	48.8	2.9	2.37	0.19	0.253	0.018
7 - 9	18.1	2.1	48.8	4.2	2.29	0.23	0.264	0.026
9 - 11	19.5	1.9	49.5	4.1	2.28	0.21	0.274	0.025
11 - 13	20.9	2.2	50.4	5.3	2.29	0.31	0.284	0.028
13 - 15	21.6	1.9	50.3	4.1	2.28	0.28	0.286	0.021
15 - 17	23.6	1.9	51.9	3.9	2.30	0.21	0.303	0.023
17 - 21	22.4	2.3	48.0	2.2	2.15	0.13	0.279	0.021
21 - 25	26.6	2.5	52.1	4.5	2.28	0.24	0.313	0.025
25 - 31	26.3	2.5	49.5	3.8	2.18	0.19	0.298	0.025
31 - 37	28.6	2.8	49.5	4.0	2.16	0.19	0.306	0.028

(data from Lee and Iliff, 1956, Pediatrics 18:739-749. Copyright American Academy of
Pediatrics 1956)

TABLE   460

BASAL METABOLIC RATES IN BOYS AGED 13 TO 16 YEARS
IN DENVER (CO)

	Age (years)	Mean	S.D.
Calories per hour	13-14	57.6	6.71
	14-15	61.6	6.65
	15-16	64.0	5.54
Calories per hour per square meter	13-14	41.5	2.46
	14-15	41.0	2.44
	15-16	39.9	3.16
Calories per hour per kilogram	13-14	1.33	0.133
	14-15	1.28	0.116
	15-16	1.22	0.111
Calories per hour per centimeter	13-14	0.367	0.0306
	14-15	0.375	0.0302
	15-16	0.376	0.0314

(data from Lewis, Duvall and Iliff, 1943, American Journal of Diseases of
Children 65:845-856. Copyright 1943, American Medical Association)

TABLE   461

BASAL METABOLIC RATES IN GIRLS AGED 13 TO 16 YEARS
IN DENVER (CO)

	Age (years)	Mean	S.D.
Calories per hour	13-14	56.5	5.04
	14-15	56.8	3.98
	15-16	56.2	4.07
Calories per hour per square meter	13-14	37.8	2.48
	14-15	36.0	2.23
	15-16	34.7	1.66
Calories per hour per kilogram	13-14	1.16	0.148
	14-15	1.06	0.136
	15-16	1.02	0.103
Calories per hour per centimeter	13-14	0.350	0.0262
	14-15	0.346	0.0224
	15-16	0.336	0.0218

(data from Lewis, Duvall and Iliff, 1943, American Journal of Diseases of
Children 65:845-856. Copyright 1943, American Medical Association)

TABLE 462

VALUES OF BASAL METABOLISM IN CALIFORNIA ADOLESCENTS
(calories per square meter per hour)

Age	Boys		Girls	
(years)	Mean	S.D.	Mean	S.D.
11.5	43.63	3.46	41.65	3.78
12.0	45.03	4.18	40.99	2.95
12.5	44.37	3.86	40.43	2.54
13.0	44.13	3.09	39.90	2.92
13.5	43.22	3.76	38.82	3.21
14.0	43.46	3.71	37.96	3.03
14.5	42.91	3.34	36.45	3.03
15.0	42.82	3.32	35.68	2.61
15.5	41.41	4.32	34.35	3.19
16.0	41.13	3.92	34.23	3.08
16.5	40.96	4.16	34.62	2.43
17.0	40.96	4.58	33.44	2.79
17.5	40.63	3.54	33.39	1.78

(data from Shock, 1942, American Journal of Diseases of Children 64:19-32. Copyright 1942, American Medical Association)

TABLE   463

BASAL OXYGEN CONSUMPTION FOR CALIFORNIAN ADOLESCENTS
(cubic centimeters of oxygen consumed per square meter per minute)

Age	Boys		Girls	
(years)	Mean	S.D.	Mean	S.D.
11.5	149.85	12.15	143.83	14.70
12.0	154.47	14.76	140.75	10.58
12.5	150.83	10.92	138.64	8.48
13.0	151.48	10.40	137.01	10.11
13.5	148.30	13.28	132.64	10.13
14.0	149.70	13.27	130.05	10.31
14.5	148.07	11.30	124.69	10.83
15.0	148.43	11.65	123.12	9.97
15.5	143.32	16.00	118.35	11.49
16.0	140.92	12.75	117.70	10.64
16.5	141.00	14.34	118.75	8.93
17.0	140.86	15.54	114.83	9.45
17.5	140.25	12.42	114.50	6.19

(data from Shock, N. W., 1942, American Journal of Diseases of Children
64:19-32. Copyright 1942, American Medical Association)

TABLE   464

COEFFICIENT OF VARIATION OF BASAL METABOLISM IN CALIFORNIA ADOLESCENTS

Age (years)	Total cc. oxygen per min.	cc. oxygen per kg. per min.	Calories per kg. per hour	cc. oxygen per sq. m. per min.	Calories per sq. m. per hour
			BOYS		
11.5	9.1	10.9	11.0	8.1	7.9
12.0	11.7	11.0	10.9	9.6	9.3
12.5	12.0	9.8	9.8	7.3	8.7
13.0	11.8	9.0	9.0	6.9	7.0
13.5	13.8	11.1	11.1	9.0	8.7
14.0	13.1	10.1	10.3	8.9	8.5
14.5	12.3	10.0	9.6	7.7	7.8
15.0	11.1	9.7	9.6	7.8	7.8
15.5	14.2	12.3	10.8	11.2	10.4
16.0	11.2	11.6	9.1	9.0	9.5
16.5	11.6	9.0	8.6	10.2	11.7
17.0	12.1	11.6	10.2	11.0	11.2
17.5	13.3	7.1	6.2	8.9	8.7
			GIRLS		
11.5	17.3	9.4	9.0	10.2	9.1
12.0	13.7	12.1	11.7	7.5	7.2
12.5	12.0	11.4	11.4	6.1	6.3
13.0	11.4	13.0	12.8	7.4	7.3
13.5	11.6	11.5	11.6	7.6	8.3
14.0	11.6	11.4	11.6	7.9	8.0
14.5	11.1	12.0	12.0	8.2	8.3
15.0	12.9	10.3	10.2	8.1	7.3
15.5	11.4	13.1	11.0	9.7	9.3
16.0	10.7	11.8	9.6	9.0	9.0
16.5	11.7	9.7	9.9	7.5	7.0
17.0	10.7	7.8	8.6	8.2	8.3
17.5	11.5	9.6	10.4	5.4	5.3

(data from Shock, 1942, American Journal of Diseases of Children 64:19-32. Copyright 1942, American Medical Association)

TABLE 465

BASAL METABOLIC RATES (calories per hour per m2 of body surface)
FOR DENVER (CO) CHILDREN 2 TO 15 YEARS OF AGE, INCLUSIVE

Age (years)	Calories per hour per square meter		Age (years)	Calories per hour per square meter	
	Boys	Girls		Boys	Girls
2.00	56.9	52.9	9.00	45.0	42.7
2.25	56.2	52.4	9.25	44.6	42.4
2.50	55.6	52.0	9.50	44.3	42.1
2.75	55.0	51.6	9.75	44.0	41.7
3.00	54.5	51.3	10.00	43.6	41.4
3.25	53.9	50.9	10.25	43.2	41.1
3.50	53.4	50.6	10.50	42.8	40.8
3.75	53.0	50.2	10.75	42.5	40.6
4.00	52.6	49.9	11.00	42.2	40.4
4.25	52.2	49.5	11.25	41.9	40.3
4.50	51.8	49.1	11.50	41.7	40.2
4.75	51.4	48.7	11.75	41.6	39.9
5.00	51.0	48.4	12.00	41.5	39.7
5.25	50.6	48.0	12.25	41.5	39.4
5.50	50.3	47.6	12.50	41.4	39.1
5.75	50.0	47.2	12.75	41.4	38.7
6.00	49.6	46.9	13.00	41.4	38.4
6.25	49.2	46.5	13.25	41.3	38.0
6.50	48.9	46.2	13.50	41.2	37.6
6.75	48.6	45.8	13.75	41.2	37.2
7.00	48.2	45.5	14.00	41.1	36.8
7.25	47.8	45.1	14.25	41.0	36.4
7.50	47.4	44.7	14.50	40.9	36.0
7.75	47.0	44.3	14.75	40.7	35.6
8.00	46.6	44.0	15.00	40.5	35.2
8.25	46.2	43.6	15.25	40.2	34.8
8.50	45.8	43.3	15.50	39.8	34.5
8.75	45.4	43.0	15.75	39.5	34.1

(data from Lewis, Duval, and Iliff, 1943, The Journal of Pediatrics 23:1-18)

TABLE 466

BASAL CALORIES PER HOUR REFERRED
TO SURFACE AREA (mm^2)
IN DENVER (CO) CHILDREN

Surface Area	Calories per hour Boys	Calories per hour Girls	Surface Area	Calories per hour Boys	Calories per hour Girls
0.500	--	26.2	1.060	47.1	45.2
0.520	--	27.0	1.080	47.6	45.8
0.540	30.8	28.1	1.100	48.1	46.4
0.560	31.6	29.5	1.120	48.5	47.0
0.580	32.4	30.4	1.140	49.0	47.6
0.600	33.1	31.3	1.160	49.5	48.2
0.620	33.8	32.2	1.180	50.0	48.8
0.640	34.5	33.0	1.200	50.4	49.4
0.660	35.2	33.8	1.220	51.0	49.9
0.680	35.9	34.5	1.240	51.7	50.5
0.700	36.6	35.2	1.260	52.5	51.1
0.720	37.3	35.8	1.280	53.3	51.6
0.740	38.0	36.4	1.300	54.1	52.2
0.760	38.7	37.0	1.320	54.8	52.7
0.780	39.4	37.6	1.340	55.5	53.3
0.800	40.1	38.2	1.360	56.2	53.9
0.820	40.7	38.8	1.380	56.9	54.5
0.840	41.3	39.4	1.400	57.6	55.1
0.860	42.0	39.9	1.420	58.3	55.6
0.880	42.6	40.4	1.440	59.0	56.2
0.900	43.2	40.9	1.460	59.7	56.8
0.920	43.8	41.4	1.480	60.4	57.4
0.940	44.4	41.9	1.500	61.1	57.9
0.960	45.0	42.4	1.520	61.8	58.5
0.980	45.5	42.9	1.540	62.5	59.1
1.000	46.0	43.4	1.560	--	59.7
1.020	46.3	44.0	1.580	--	60.3
1.040	46.7	44.6	--	--	--

(data from Lewis, Duval, and Iliff, 1943, The Journal of Pediatrics 23:1-18)

TABLE 467

BASAL CALORIES PER HOUR REFERRED TO WEIGHT (kg)
IN DENVER (CO) CHILDREN

Weight	Calories per hour		Weight (kg)	Calories per hour	
	Boys	Girls		Boys	Girls
10.0	--	24.3	30.0	47.6	45.8
11.0	--	26.2	31.0	48.3	46.7
12.0	30.5	28.0	32.0	49.0	47.3
13.0	31.9	29.8	33.0	49.6	48.0
14.0	33.2	31.2	34.0	50.3	48.5
15.0	34.5	32.7	35.0	51.0	49.0
16.0	35.7	34.0	36.0	51.7	49.5
17.0	37.0	35.2	37.0	52.4	50.0
18.0	38.2	36.1	38.0	53.1	50.5
19.0	39.2	37.2	39.0	53.9	51.0
20.0	40.2	38.0	40.0	54.6	51.7
21.0	41.2	39.0	41.0	55.4	52.5
22.0	42.1	39.8	42.0	56.2	53.5
23.0	43.0	40.5	43.0	57.0	54.6
24.0	43.8	41.2	44.0	57.8	56.0
25.0	44.5	42.0	45.0	58.6	--
26.0	45.2	42.7	46.0	59.4	--
27.0	45.8	43.5	47.0	60.3	--
28.0	46.4	44.4	48.0	61.2	--
29.0	47.0	45.2	49.0	62.0	--
--	--	--	50.0	62.8	--

(data from Lewis, Duval, and Iliff, 1943, The Journal of Pediatrics 23:1-18)

TABLE 468

BASAL CALORIES PER HOUR REFERRED TO STATURE
(cm) FOR BOYS AND GIRLS
IN DENVER (CO)

Stature	Calories per hour		Stature (cm)	Calories per hour	
	Boys	Girls		Boys	Girls
84	--	26.8	130	45.3	43.0
86	--	27.5	132	46.0	43.8
88	30.7	28.2	134	46.6	44.6
90	31.4	28.9	136	47.2	45.4
92	32.1	29.6	138	47.9	46.2
94	32.8	30.3	140	48.6	47.0
96	33.5	31.0	142	49.3	47.8
98	34.2	31.7	144	50.0	48.7
100	34.9	32.4	146	50.7	49.6
102	35.6	33.1	148	51.6	50.5
104	36.3	33.8	150	52.5	51.4
106	37.0	34.5	152	53.5	52.3
108	37.7	35.2	154	54.9	53.2
110	38.4	35.9	156	56.3	54.1
112	39.1	36.6	158	57.7	55.0
114	39.8	37.3	160	59.1	55.9
116	40.5	38.0	162	60.5	56.8
118	41.2	38.7	164	61.9	57.7
120	41.8	39.4	166	63.2	58.6
122	42.5	40.1	168	64.3	--
124	43.2	40.8	170	65.3	--
126	43.9	41.5	172	66.0	--
128	44.6	42.2	174	66.6	--

(data from Lewis, Duval, and Iliff, 1943, The Journal of Pediatrics 23:1-18)

MATURATION

TABLE 469

PERCENTAGE DISTRIBUTION OF BEDWETTERS (adjusted for birth weight)
ACCORDING TO FREQUENCY OF WET BEDS
FOR INFANTS IN BALTIMORE, MD

Frequency of Wet Beds	Age (years)			
	3-5	6-7	8-10	12
Less than once per week, but as often as once per month	37	27	29	33
At least once per week, but not every night	43	63	59	52
Every night	21	10	12	15

PERCENTAGE RELAPSES BY DURATION OF
RELAPSE FOR BEDWETTERS AND DAYWETTERS
FOR INFANTS IN BALTIMORE, MD

Relapses		Length of Relapse in Years								
		Less Than 1	1-1.9	2-2.9	3-3.9	4-4.9	5-5.9	6-6.9	7-7.9	8 or More
Bedwetters	Percent	27	16	15	12	10	8	4	5	3
Daywetters	Percent	45	22	13	12	3	3	1	0	0

(data from Oppel, Harper and Rider, 1968, Pediatrics 42:614-626. Copyright 1968,
American Academy of Pediatrics)

TABLE 470

BLADDER CONTROL IN CHILDREN IN ROCHESTER (MN)
AGED 2.5 YEARS; PREVALENCE OF DRYNESS AND WETNESS

Condition	Percent		
	Boys	Girls	Total
Day			
Dry	50.8	67.6	59.9
Wet	49.2	32.4	40.1
Night			
Dry	48.1	63.8	56.6
Wet	51.9	36.2	43.4
Nap			
Dry	78.9	88.5	84.1
Wet	21.1	11.5	15.9

Frequency of Lack of Urinary Control During the Day, Night, and Nap

Periods When Wetting Occurs	Percent	
	Boys	Girls
Dry only	16.2	15.4
Night only	14.9	14.3
Day and night	15.5	10.2
Day, nap and night	16.4	6.5
Night and nap	4.7	4.4
Day and nap	0.0	0.3
Nap only	0.0	0.3
Total wet at some time	67.7	51.4
Total dry all the time	32.3	48.6

(data from Roberts and Schoellkopf, 1951b, American Journal of
Diseases of Children 82:144-152. Copyright 1951, American
Medical Association)

TABLE 471

PERCENT OF NIGHTTIME RELAPSES BY RACE AND
AGE CATEGORIES OF FIRST ATTAINING NIGHTTIME DRYNESS
FOR INFANTS IN BALTIMORE, MD

Age (years) of First Attaining Dryness	Percent	
	Relapsing at Least Once	Never Relapsing
Caucasian		
Less than 3	15	85
3 to 6	19	81
6 or older	22	78
Total	17	83
Negro		
Less than 3	35	65
3 to 6	28	72
6 or older	18	82
Total	31	69

(data from Oppel, Harper and Rider, 1968, Pediatrics
42:614-626. Copyright 1968, American Academy of
Pediatrics)

TABLE   472

MOTHERS' REPORTS OF THUMB-SUCKING,
THE UNIVERSITY OF CALIFORNIA NURSERY SCHOOL
1947 to 1956 and 1960

Age (years)	Boys		Girls	
	Thumb-suckers	%	Thumb-suckers	%
2	27	34.2	39	50.0
3	93	41.0	116	50.4
4	19	28.4	29	46.8

(data from Honzik and McKee, 1962, The Journal of Pediatrics
61:726-732)

TABLE   473

MEAN AGE (months) OF STANDING WITH SUPPORT, AND COMPARISON
BETWEEN GROUPS IN RELATION TO TYPE OF ARTIFICIAL FEEDING
IN PHILADELPHIA CHILDREN

Group	Mean	S.D.
All white	8.9	±1.69
I	8.8	±1.73
II	8.7	±1.66
III	9.3	±1.58
IV	9.1	±1.70
All Negro	9.2	±1.95
I	8.9	±1.52
II	9.5	±2.96
III	9.1	±1.38
IV	9.2	±1.45

  I = irradiated evaporated milk,
 II = non-irradiated evaporated milk plus cod-liver oil,
III = irradiated evaporated milk plus carotene, and
 IV = irradiated evaporated milk plus carotene and yeast.

(data from Rhoads, Rapaport, Kennedy, et al., 1945, The Journal of
Pediatrics 26:415-454)

TABLE 474

MEAN AGE (weeks) OF WALKING IN NEGRO INFANTS IN WASHINGTON (DC) IN RELATION
TO MATERNAL EDUCATION, NUMBER OF SIBLINGS, AND WEIGHT (kg) OF THE INFANT

Factors	Mean	S.D.
Maternal Education		
≤ 9th grade	50.8	5.6
10th – 12th grade	48.0	13.2
≤ 13th grade	49.2	7.0
Number of Siblings		
0	49.6	6.5
1	50.2	7.0
2	50.2	5.8
3 +	51.0	6.9
Weight (Boys)		
< 9.5	47.6	--
9.5 – 11.3	49.8	--
> 11.3	54.1	--
Weight (Girls)		
< 9.5	46.8	--
9.5 – 11.3	50.6	--
> 11.3	50.0	--

(data from Ferguson, Cutter, and Scott, 1956, The Journal of Pediatrics 48:
308-313)

TABLE 475

AGE OF WALKING BY SEX AND SOCIAL CLASS IN SAMPLES OF EUROPEAN CHILDREN
ALL VALUES ARE IN TERMS OF LOGARITHM OF AGE IN MONTHS EXCEPT THE LAST TWO COLUMNS,
WHICH ARE IN MONTHS (S.D.'s x 100)

	Social Class	Boys		Girls		Total Sample			
		Mean	S.D.	Mean	S.D.	Mean	S.D.	Mean (months)	Median (months)
Brussels	1 & 2	1.118	6.99	1.108	6.46				
	3	1.109	6.75	1.089	5.49	1.102	6.34	12.65	12.48
	4 & 5	1.089	5.50	1.095	5.76				
London	1 & 2	1.162	6.16	1.128	12.57				
	3	1.111	6.59	1.106	6.27	1.124	7.40	13.31	13.23
	4 & 5	1.121	6.51	1.131	7.78				
Paris	1 & 2	1.132	4.39	1.146	6.92				
	3	1.144	6.13	1.138	6.92	1.140	7.33	13.81	13.58
	4 & 5	1.138	8.01	1.141	8.63				
Stockholm	1 & 2	1.104	5.01	1.100	5.92				
	3	1.099	6.97	1.095	6.38	1.097	6.45	12.51	12.44
	4 & 5	1.096	7.02	1.093	6.70				
Zurich	1 & 2	1.122	6.65	1.142	5.06				
	3	1.136	5.54	1.132	5.52	1.134	5.65	13.59	13.63
	4 & 5	1.125	4.90	1.150	6.10				

The samples are those included in studies conducted by the International Children's Centre (Paris).

(data from "Differences in age of walking in five European longitudinal samples," Human Biology 38:364-379, 1966, by Hindley, Filliozat, Blackenberg, et al., by permission of the Wayne State University Press. Copyright 1966, Wayne State University Press, Detroit, Michigan 48202)

TABLE   476

TIME OF FIRST VOIDING
BY FULL-TERM INFANTS
IN BALTIMORE (MD)

Time	%	Cumulative %
Delivery room	17.0	17.0
Hours		
12	50.6	67.6
12-24	24.8	92.4
24-48	7.0	99.4
Over 48	* 0.6	100.0

*These patients voided at 50, 50, and 51 hours.

(data from Sherry and Kramer, 1955, The Journal
of Pediatrics 46:158-159)

TABLE 477

STOOL FREQUENCY, WEIGHT (g), AND WATER
CONTENT (%) IN NORMAL ENGLISH
CHILDREN

Variables	1st week	8-28 days	1-12 months	13-24 months
Stool frequency				
Mean interval between stools (hours) $\pm$ 1 S.D.	5.2 $\pm$ 1.9	9.9 $\pm$ 6.5	13.2 $\pm$ 9.2	14.9 $\pm$ 8.3
Range (hours)	0.5 - 22	0.7 - 38.5	0.5 - 53	2 - 120
No. of stools per 24 h $\pm$ 1 S.D.	4 $\pm$ 1.8	2.2 $\pm$ 1.6	1.8 $\pm$ 1.2	1.7 $\pm$ 0.6
Range	1 - 12	0 - 6	0 - 5	0 - 3
Weight of individual stools (g)				
Mean	4.3	11.0	17	35
Range	0.5 - 48	0.3 - 40	2 - 98	4 - 180
Stool water content (%)				
Mean $\pm$ 1 S.D.	72.8 $\pm$ 5.0	73.3 $\pm$ 2.7	75.0 $\pm$ 3.0	73.8 $\pm$ 3.2
Range	65 - 84	72 - 77	72 - 80	66 - 81

(data from Lemoh and Brooke, 1979, Archives of Disease in Childhood 54:719-720)

TABLE 478

TIME OF PASSAGE OF FIRST
STOOL BY FULL-TERM INFANTS

	%	Cumula- tive %
Delivery room	27.2	27.2
Hours		
12	41.8	69.0
12-24	25.0	94.0
24-48	5.8	99.8
Over 48	*0.2	100.0

*Sixty-two hours.

(data from Sherry and Kramer, 1955, The Journal
of Pediatrics 46:158-159)

TABLE 479

FREQUENCY OF BOWEL MOVEMENTS
IN CHILDREN AGED 2.5 YEARS

Frequency	Children (percent)	
1 time daily	61.2	⎫
1 time, occasionally twice	17.6	⎬ 88.8
2 or 3 times daily	10.0	⎭
Variable pattern	7.6	
Every other day	3.5	
Extremely infrequent	0.1	

Regularity and Variability of Bowel Movements

	Percent	
Regularity	Boys	Girls
Regular	53.1	45.1
Irregular	41.5	51.8
Variable	5.4	3.1

(data from Roberts and Schoellkopf, 1951a, American Journal of
Diseases of Children 82:144-152. Copyright 1951, American
Medical Association)

TABLE 480

PERCENTAGE DISTRIBUTIONS OF STAGING SEXUAL
MATURATION IN U.S. WHITE BOYS

Pubic Hair	Age 12 Years Genital Stage					Age 13 Years Genital Stage				
Stage	I	II	III	IV	V	I	II	III	IV	V
V	0.0	0.0	0.7	0.6	0.4	0.0	0.0	0.4	5.8	6.1
IV	0.0	0.2	3.3	5.2	0.0	0.0	0.4	5.4	16.7	0.6
III	0.2	9.4	12.0	0.2	0.0	1.1	9.5	15.6	2.4	0.0
II	11.3	26.9	3.1	0.0	0.0	5.6	18.2	3.0	0.0	0.0
I	20.7	5.4	0.2	0.0	0.2	5.8	3.3	0.2	0.0	0.0

Pubic Hair	Age 14 Years Genital Stage					Age 13 Years Genital Stage				
Stage	I	II	III	IV	V	I	II	III	IV	V
V	0.0	0.0	0.6	11.8	27.2	0.0	0.0	0.2	15.1	49.8
IV	0.0	0.4	7.6	24.2	2.7	0.0	0.2	2.9	20.3	3.4
III	0.2	1.9	9.1	1.0	0.2	0.0	0.4	3.6	1.1	0.0
II	2.5	6.7	1.3	0.0	0.0	0.4	1.9	0.2	0.0	0.0
I	1.7	0.8	0.0	0.0	0.2	0.4	0.0	0.0	0.0	0.0

Pubic Hair	Age 16 Years Genital Stage					Age 17 Years Genital Stage				
Stage	I	II	III	IV	V	I	II	III	IV	V
V	--	--	0.2	8.9	72.3	--	--	--	5.8	88.2
IV	--	--	1.0	10.1	4.4	--	--	--	2.6	2.9
III	--	--	0.8	0.6	0.4	--	--	--	0.2	0.2
II	--	--	0.0	0.2	0.0	--	--	--	0.0	0.0
I	--	--	0.0	0.0	0.4	--	--	--	--	--

(data from Harlan, Grillo, Cornoni-Huntley, et al., 1979, The Journal of Pediatrics 95:
293-297)

TABLE 481

PERCENTAGE OF OHIO BOYS AT EACH STAGE OF GENITALIA MATURATION, BY AGE LEVEL

	Age (years)									
Stage	9	9.5	10	10.5	11	11.5	12	12.5	13	13.5
1.....	100	97	95	83	65	36	23	9	3	0
2.....	0	3	5	17	33	55	63	42	18	13
3.....	0	0	0	0	2	4	9	42	41	23
4.....	0	0	0	0	0	4	5	7	38	64
5.....	0	0	0	0	0	0	0	0	0	0

	Age (years)									
Stage	14	14.5	15	15.5	16	16.5	17	17.5	18	19-21
1.....	0	0	0	0	0	0	0	0	0	0
2.....	8	3	0	0	0	0	0	0	0	0
3.....	5	3	3	0	0	0	0	0	0	0
4.....	87	94	97	97	96	88	77	56	25	0
5.....	0	0	0	3	4	12	23	44	75	100

(data from Reynolds and Wines, 1951, American Journal of Diseases of Children 82: 529-547. Copyright 1951, American Medical Association)

TABLE 482

PERCENTAGE OF OHIO BOYS AT EACH STAGE
OF PUBIC HAIR MATURATION, BY AGE LEVEL

	Age (years)										
Stage	8.5	9	9.5	10	10.5	11	11.5	12	12.5	13	13.5
1.....	100	98	98	98	91	86	69	60	48	28	10
2.....	0	2	2	2	9	14	29	31	43	38	33
3.....	0	0	0	0	0	0	2	7	7	20	20
4.....	0	0	0	0	0	0	0	2	2	12	33
5.....		0	0	0	0	0	0	0	0	2	3

	Age (years)									
Stage	14	14.5	15	15.5	16	16.5	17	17.5	18	19-21
1.....	3	3	3	0	0	0	0	0	0	0
2.....	14	3	0	0	0	0	0	0	0	0
3.....	22	9	3	0	0	0	0	0	0	0
4.....	59	74	76	73	56	46	32	12	0	0
5.....	3	12	18	27	44	54	68	88	100	100

(data from Reynolds and Wines, 1951, American Journal of Diseases of Children 82: 529-547. Copyright 1951, American Medical Association)

TABLE   483

MEDIAN AGE (years) FOR FIRST EJACULATION, FIRST VOICE-CHANGE,
AND FIRST APPEARANCE OF PUBIC HAIR

	Negro	White
Median age of first ejaculation	13.8	13.8
Median age of first appearance of pubic hair	13.3	13.6
Median age for first recognition of voice-change	13.7	13.4

The Negro group is from the north eastern part of the
U.S. and the white group from the mid-west.

(data from "Sexual growth of Negro and white boys," Human
Biology 22:146-149, 1950, by G. V. Ramsey, by permission of
the Wayne State University Press. Copyright 1950, Wayne State
University Press, Detroit, Michigan 48202)

TABLE 484

PREVALENCE OF GYNECOMASTIA AND ASSOCIATED
GENITAL STAGE IN ADOLESCENT U.S. MALES*

| Group | Age (years) | | | | | |
	12	13	14	15	16	17
All males (%)	3.9	8.2	5.0	2.6	2.9	1.6
White males (%)	4.3	8.2	5.1	2.5	2.4	1.2
Genitial stage (Tanner) all males						
Without gynecomastia	2.2	2.5	4.0	4.5	4.8	4.9
With gynecomastia	2.4	3.1	3.6	3.9	4.5	5.0

*Gynecomastia classified as Class I or II with only 6% of the 147 cases being
class II. Class I is benign adolescent gynecomastia presenting as firm
plaque of subareolar tissue ranging from 1 to 10 cm in diameter. Class II
is diffuse gynecomastia with large breast mass and puffy areolae, usually
bilateral.

(data from Harlan, Grillo, Cornoni-Huntley, et al., 1979, The Journal of Pediatrics
95:293-297).

TABLE 485

AREOLAR DIAMETERS (mm) FOR BLACK MALES IN TEXAS

Age (years)	Maximum Diameter				Minimum Diamater			
	Lower Income		Middle Income		Lower Income		Middle Income	
	Mean	S.D.	Mean	S.D.	Mean	S.D.	Mean	S.D.
10.1 - 11	18.3	4.15	18.6	2.43	12.5	2.88	14.3	3.53
11.1 - 12	20.8	3.00	21.3	3.26	15.8	3.65	16.3	4.03
12.1 - 13	21.5	5.54	21.3	3.63	14.1	3.48	16.1	3.58
13.1 - 14	20.3	5.40	20.5	3.31	15.9	3.22	13.9	2.92
14.1 - 15	22.8	4.48	22.5	5.21	15.9	3.45	16.8	4.60
15.1 - 16	24.8	4.45	26.6	3.81	19.1	3.27	19.8	4.92
16.1 - 17	25.8	5.13	29.2	3.89	21.1	5.08	20.7	5.00
17.1 - 18	25.9	4.30	26.6	2.23	17.9	6.24	18.8	3.79

(data from Schutte, 1979)

TABLE 486

DISTRIBUTION OF STAGES OF PUBIC HAIR AT DEVELOPMENTAL POINTS
IN CALIFORNIA GIRLS

Developmental Point	Stages of Pubic Hair				
	1	2	3	4	5
Prepuberal	21	47	8	0	0
Puberal onset	7	58	19	4	0
Puberal end	0	1	3	46	40
Postpuberal	0	0	0	4	87

(data from Faust, 1977)

TABLE 487

AGE (years) AT TRANSITION TO STAGES IN
DEVELOPMENT OF SECONDARY
SEXUAL CHARACTERISTICS
IN CALIFORNIA GIRLS

Stage	Mean	S.D.
P_2	9.03	0.39
A_2	10.24	0.79
B_2	10.58	1.18
P_3	10.97	0.99
B_3	11.63	1.20
A_3	11.87	1.16
P_4	11.90	1.04
A_4	12.63	0.95
P_5	13.08	1.05
B_4	13.18	1.08
A_5	13.50	1.13

P = pubic hair; A = axillary hair; B = breast

(data from Faust, 1977)

TABLE 488

PREVALENCE OF STAGES OF PUBIC HAIR DEVELOPMENT IN OHIO GIRLS

						Age (years)								
Stage	8	8.5	9.0	9.5	10	10.5	11	11.5	12	12.5	13	13.5	14	14.5
1	48	48	47	44	40	28	21	12	5	3	2	1	1	0
2	1	1	2	4	7	16	14	13	12	6	3	3	1	0
3	0	0	0	1	1	3	8	11	8	7	3	1	2	2
4	0	0	0	0	1	2	5	6	8	14	17	12	7	5
5	0	0	0	0	0	0	0	1	0	1	4	6	6	7

(data from Reynolds and Wines, 1948, American Journal of Diseases of Children 75:329-350. Copyright 1948, American Medical Association)

TABLE 489

DISTRIBUTION OF STAGES IN AXILLARY HAIR AT DEVELOPMENTAL POINTS
IN CALIFORNIA GIRLS

	Stage				
Developmental Point	1	2	3	4	5
Prepuberal	70	13	4	0	0
Puberal onset	43	38	9	1	0
Puberal end	0	11	27	33	20
Postpuberal	0	0	3	17	71

(data from Faust, 1977)

TABLE 490

PERCENTAGE OF GIRLS MENSTRUATING IN THE UNITED STATES

					Age (years)				
Race	6-17	6-9	10	11	12	13	14	15	16-17
				Percentage of Girls					
White	42.9	0.2	0.8	11.6	41.7	72.9	91.4	98.2	99.6
Black	45.2	0.2	4.0	21.3	51.2	74.1	93.5	98.7	100.0
				Standard Error					
White	0.75	0.09	0.44	1.73	2.60	2.28	1.99	0.55	0.21
Black	2.15	0.22	2.30	5.89	7.41	6.65	2.00	0.86	--

(data from MacMahon, 1973)

TABLE 491

MEDIAN AGE OF MENARCHE IN THE UNITED STATES IN RELATION TO REGION,
SIZE OF PLACE OF RESIDENCE AND FAMILY INCOME

Characteristic	Median age at menarche		
	All races	White	Black
	12.76	12.80	12.52
Geographic region			
Northeast	12.71	12.79	12.21
Midwest	12.88	12.90	12.66
South	12.68	12.69	12.67
West	12.72	12.75	12.43
Population size of place of residence			
Giant SMSA*	12.57	12.65	12.23
Other large SMSA	12.96	13.01	12.76
Other SMSA	12.70	12.69	---
Other areas	12.80	12.83	---
Annual family income			
Less than $3,000	12.61	12.65	12.49
$3,000-$4,999	12.81	12.91	12.43
$5,000-$6,999	12.81	12.86	12.39
$7,000-$9,999	12.87	12.88	12.72
$10,000-$14,999	12.68	12.69	---
$15,000 or more	12.79	12.81	---

SMSA = standard metropolitan statistical area
(data from MacMahon, 1973)

TABLE   492

REPORTED AGES AT MENARCHE FOR UNITED STATES WOMEN
BY SUM OF SKINFOLDS* AT TIME OF SURVEY

Race and age		Sum of skinfolds (cm) at time of survey				
	Total	Less than 2.5	2.5-3.4	3.5-4.4	4.5-4.9	5.0 or more
			Mean age at menarche			
All races 18-79 years	13.08	13.15	13.05	13.11	13.13	13.03
18-34 years	12.76	13.03	12.70	12.69	12.82	12.40
35-54 years	13.06	13.17	13.14	13.07	12.90	13.00
55-79 years	13.53	13.60	13.78	13.53	13.62	13.37
White 18-79 years	13.09	13.11	13.04	13.10	13.18	13.09
18-34 years	12.72	13.00	12.66	12.64	12.81	12.38
35-54 years	13.06	13.15	13.13	13.03	12.90	13.07
55-79 years	13.55	13.49	13.76	13.58	13.71	13.40
Black 18-79 years	12.94	13.17	13.20	13.06	12.51	12.66
18-34 years	12.85	13.11	13.04	12.43	12.75	12.48
35-54 years	12.92	13.21	13.25	13.47	12.53	12.55
55-79 years	13.10	13.34	13.79	13.08	12.11	13.05
			Standard error			
All races 18-79 years	0.02	0.06	0.04	0.06	0.08	0.06
18-34 years	0.03	0.08	0.06	0.10	0.14	0.12
35-54 years	0.04	0.09	0.07	0.11	0.15	0.06
55-79 years	0.06	0.21	0.15	0.18	0.21	0.11
White 18-79 years	0.02	0.05	0.04	0.07	0.09	0.06
18-34 years	0.03	0.08	0.06	0.12	0.14	0.14
35-54 years	0.05	0.12	0.07	0.12	0.16	0.06
55-79 years	0.06	0.18	0.15	0.21	0.22	0.12
Black 18-79 years	0.08	0.20	0.22	0.18	0.32	0.15
18-34 years	0.17	0.23	0.44	0.18	0.34	0.44
35-54 years	0.12	0.49	0.30	0.27	0.46	0.20
55-79 years	0.17	0.40	0.58	0.41	0.79	0.34

*Sites not described

(data from MacMahon, 1973)

TABLE 493

AGE AT MENARCHE (years) OF FRENCH CANADIAN GIRLS
OF DIFFERENT SOCIOECONOMIC STRATA IN MONTREAL

	MEAN	S.D.
Total samples (all socioeconomic classes plus group of undertermined status)		
	13.08	1.10
Socioeconomic classes combined (I + II + III)		
	13.13	1.10
Socieconomic class:		
Upper (I)	13.43	1.10
Middle (II)	13.03	1.10
Lower (III)	13.12	1.08
Undetermined	12.92	1.11
Category of family income:		
1.	13.48	1.04
2.	13.19	1.12
3.	13.37	1.09
4.	13.02	1.11
5.	12.95	1.10
6.	13.04	1.09
7.	13.45	1.11
8.	13.20	1.05
Category of father's occupation:		
1.	13.59	1.05
2.	13.06	1.12
3.	13.42	1.09
4.	13.01	1.09
5.	12.97	1.12
Category of father's education:		
1.	13.28	1.11
2.	12.97	1.11
3.	13.20	1.09
4.	13.10	1.10
5.	12.90	1.09
Category of mother's education:		
2.	13.21	1.09
3.	13.13	1.09
4.	13.13	1.11
5.	12.69	1.12
Girls having working mothers		
Yes	13.12	1.10
No	13.13	1.10

TABLE 493 (continued)

AGE AT MENARCHE (years) OF FRENCH CANADIAN GIRLS
OF DIFFERENT SOCIO-ECONOMIC STRATA IN MONTREAL

	Mean	S.D.
Body frame (laterality):		
According to the index		
biacromial width (cm) X 100/stature		
large ($\geqslant P_{75}$)	13.28	1.09
medium (P_{25}-P_{75})	13.08	1.10
small ($< P_{25}$)	12.92	1.11
According to the index		
bicristal width (cm) X 100/stature		
large ($\geqslant P_{75}$)	13.45	1.09
medium (P_{25}-P_{75})	13.13	1.09
small ($< P_{25}$)	12.72	1.12
Family size (total offspring):		
1 child	12.68	1.09
2-3 children	12.94	1.10
4-5 children	13.20	1.09
6 and more children	13.45	1.12

(data from Jeniček and Demirjian, 1974, Annals of Human Biology $\underline{1}$:
339-346)

TABLE   494

DISTRIBUTION OF STAGES IN BREAST DEVELOPMENT
AT DEVELOPMENTAL POINTS IN CALIFORNIA GIRLS

Developmental	Stage			
Point	1	2	3	4
Prepuberal	79	8	2	0
Puberal onset	60	28	3	0
Puberal end	0	1	59	31
Postpuberal	0	2	14	78

(data from Faust, 1977)

TABLE   495

PREVALENCE OF STAGES OF BREAST DEVELOPMENT IN OHIO GIRLS

Stage	Age (years)										
	8	8.5	9	9.5	10	10.5	11	11.5	12	12.5	13
1	49	47	45	40	33	25	15	10	7	2	0
2	0	2	2	6	12	14	14	10	5	4	3
3	0	0	2	2	2	6	12	11	8	6	8
4	0	0	0	1	2	4	6	9	12	16	15
5	0	0	0	0	0	0	0	0	0	2	3

(data from Reynolds and Wines, 1948, American Journal of Diseases of Children 75:329-350.
Copyright 1948, American Medical Association)

TABLE   496

AREOLAR DIAMETER (mm) AT DIFFERENT STAGES OF DEVELOPMENT IN BOSTON CHILDREN

Group	Age Range (years)	Areolar Stage		Mean	S.D.
Boys	9-10	0	(immature)	13.1	3.3
Girls	9-10	0	(immature)	11.9	3.0
Girls	9-13	I	(bud)	23.6	3.9
Girls	9-14	II	(mound)	26.8	3.2
Girls	10-16	III	(mature)	33.1	9.9
Girls	17-19	III	(mature)	35.6	8.3
Boys	17-20	A	(adult)	21.5	5.5

(data from Garn, 1952)

TABLE 497

AREOLAR DIAMETERS (mm) AND AREAS (mm^2)
IN AUSTRALIAN GIRLS

Age (years)	Horizontal Diameter Percentiles			Vertical Diameter Percentiles			Areolar Area
	10	50	90	10	50	90	
Boys							
9.0	11.10	14.00	18.80	10.00	11.33	19.40	124.74
9.5	10.53	14.75	18.77	9.37	12.40	18.40	143.81
10.0	11.47	15.25	20.30	9.60	13.83	16.96	165.72
10.5	10.54	16.00	19.07	9.90	13.33	18.30	167.70
11.0	12.30	16.50	20.45	10.55	14.25	17.88	184.61
11.5	11.77	16.90	20.90	9.27	14.30	18.47	189.88
12.0	12.17	17.46	20.75	10.37	14.37	18.87	197.00
12.5	13.07	17.92	22.35	10.07	15.28	19.74	215.14
13.0	13.60	18.75	23.20	11.40	16.00	20.73	235.59
13.5	14.20	19.17	21.97	11.73	17.00	19.97	255.92
14.0	13.90	18.37	23.66	12.85	16.70	20.92	240.91
14.5	16.30	20.87	24.35	14.53	18.90	22.35	309.77
15.0	14.00	20.00	25.60	13.00	17.00	23.60	267.14
Girls							
9.0	11.70	14.75	21.65	11.17	13.83	18.15	160.27
9.5	10.53	15.20	21.70	8.87	12.69	19.20	151.45
10.0	12.27	16.20	22.20	9.80	13.71	20.80	174.38
10.5	11.96	17.31	23.55	10.90	15.25	21.60	207.16
11.0	13.28	18.65	27.30	11.23	17.55	27.15	256.89
11.5	15.07	21.70	31.90	13.37	20.62	31.35	351.57
12.0	15.45	24.06	34.10	13.45	22.12	36.55	418.16
12.5	18.90	27.50	38.10	15.90	26.50	42.20	572.59
13.0	21.15	31.62	44.57	22.30	30.50	48.70	757.75
13.5	22.55	32.50	44.90	23.10	34.07	49.90	868.37
14.0	22.00	37.67	49.00	22.00	37.33	52.00	1104.88
14.5	20.40	34.25	56.80	20.80	37.00	59.60	995.41
15.0	25.30	36.75	54.70	25.60	36.67	58.50	1058.27

(data from "Areolar size during pubescence," Human Biology 43:210-223, 1971, by Roche, French and Davila, by permission of the Wayne State University Press. Copyright 1971, Wayne State University Press, Detroit, Michigan 48202)

TABLE 498

PERCENTILES FOR 6-MONTHLY INCREMENTS IN AREOLAR DIAMETERS (mm)
IN RELATION TO AGE AT PEAK HEIGHT VELOCITY
IN AUSTRALIAN GIRLS

	Horizontal Diameter Percentiles			Vertical Diameter Percentiles		
Intervals	10	50	90	10	50	90
Boys						
-1.75 to -1.25 years	-1.40	1.14	3.70	-1.63	0.08	2.30
-1.25 to -0.75 years	-3.25	0.29	3.34	-2.72	-0.05	1.97
-0.75 to -0.25 years	-2.90	0.30	3.60	-1.98	-0.07	1.85
-0.25 to +0.25 years	-1.91	1.00	3.95	-1.29	0.19	2.48
+0.25 to +0.75 years	-2.12	0.65	3.40	-1.18	0.08	1.92
+0.75 to +1.25 years	-3.43	0.00	3.10	-2.97	-0.21	1.70
+1.25 to +1.75 years	-2.20	-0.58	2.20	-1.50	-0.58	1.60
Girls						
-1.75 to -1.25 years	-2.85	1.25	5.40	-1.73	0.13	4.17
-1.25 to -0.75 years	-1.46	2.23	4.95	-0.48	1.41	4.90
-0.75 to -0.25 years	-1.97	2.22	6.23	-0.80	1.75	5.97
-0.25 to +0.25 years	-0.97	2.46	7.82	-0.30	2.00	5.98
+0.25 to +0.75 years	-1.55	3.00	7.55	-0.19	2.95	9.57
+0.75 to +1.25 years	-3.35	2.94	8.40	-4.20	1.62	8.85
+1.25 to +1.75 years	-3.60	3.25	9.87	-2.23	1.75	10.05

-indicates an age before peak height velocity; +indicates an age after
peak height velocity.

(data from "Areolar size during pubescence," Human Biology 43:210-223,
1971, by Roche, French and Davila, by permission of the Wayne State University
Press. Copyright 1971, Wayne State University Press, Detroit, Michigan
48202)

TABLE 499

PERCENTILES FOR 6-MONTH INCREMENTS OF AREOLAR DIAMETERS
IN RELATION TO MENARCHE (mm) IN AUSTRALIAN GIRLS

	Horizontal Diameter Percentiles			Vertical Diameter Percentiles		
Intervals	10	50	90	10	50	90
-4.25 to -3.75 years	-1.80	-0.50	2.34	-1.40	0.50	1.50
-3.75 to -3.25 years	-2.15	1.25	5.70	-1.30	1.25	4.30
-3.25 to -2.75 years	-2.45	0.79	4.42	-2.26	0.75	4.90
-2.75 to -2.25 years	-2.37	1.00	5.03	-1.43	1.21	4.40
-2.25 to -1.75 years	-0.33	2.32	6.75	-0.20	2.14	5.37
-1.75 to -1.25 years	-1.86	2.12	7.26	-0.40	3.62	8.13
-1.25 to -0.75 years	-2.45	3.00	7.47	-0.49	4.08	10.13
-0.75 to -0.25 years	0.65	4.12	10.55	0.80	4.31	10.90
-0.25 to +0.25 years	-2.20	2.87	9.85	-1.40	3.67	12.90
+0.25 to +0.75 years	-6.60	0.87	7.60	-7.05	1.00	9.63
+0.75 to +1.25 years	-3.20	3.67	10.30	-2.30	4.67	10.95
+1.25 to +1.75 years	-8.00	1.00	10.50	-4.00	1.75	9.50
+1.75 to +2.25 years	-5.70	0.67	10.10	-7.10	0.00	13.60

-indicates before menarche; +indicates after menarche

(data from "Areolar size during pubescence," Human Biology 43:210-223,
1971, by Roche, French and Davila, by permission of the Wayne State University
Press. Copyright 1971, Wayne State University Press, Detroit, Michigan
48202)

TABLE 500

PERCENTILES FOR AREOLAR DIAMETERS (mm) IN RELATION TO PUBIC HAIR STAGES
IN AUSTRALIAN GIRLS

	Pubic Hair Stage	Horizontal Diameter Percentiles			Vertical Diameter Percentiles		
		10	50	90	10	50	90
Boys	1	11.53	16.62	19.95	9.77	14.40	18.30
	2	13.13	18.17	22.53	10.60	15.62	19.89
	3	14.00	19.25	24.50	11.00	16.75	22.25
	4	--	17.67	--	--	16.25	--
	5	--	22.50	--	--	19.00	--
Girls	1	13.40	19.00	27.00	11.75	17.80	27.00
	2	14.18	21.50	34.10	12.18	19.79	31.40
	3	17.87	27.17	35.70	16.60	27.19	38.85
	4	20.00	30.50	41.00	17.00	29.80	49.00
	5	20.00	34.00	56.40	21.20	35.00	61.20

(data from "Areolar size during pubescence," Human Biology 43:210-223,
1971, by Roche, French and Davila, by permission of the Wayne State University
Press. Copyright 1971, Wayne State University Press, Detroit, Michigan
48202)

ENDOCRINES

TABLE 501

SERUM LH CONCENTRATIONS (mIU/ml)
FROM 8 THROUGH 15 YEARS OF AGE
IN CLEVELAND (OH) CHILDREN

Age	Girls		Boys	
(years)	Mean	S.E.	Mean	S.E.
8	1.87	0.14	1.1	.29
9	2.28	0.17	0.63	.04
10	2.70	0.40	0.86	.31
11	5.6	1.5	1.2	.11
12	7.3	1.7	2.1	.27
13	8.0	1.7	4.2	.83
14	11.9	3.5	4.1	.30

LH = luteinizing hormone.

(data from Yen, Vicic, and Kearchner, 1969, Journal of Clinical
Endocrinology 29:382-385, by permission of Lippincott/Harper Company,
Philadelphia, PA)

TABLE 502

SERUM LH CONCENTRATIONS (mIU/ml)
FROM 11 THROUGH 13 YEARS OF AGE
IN CLEVELAND (OH) GIRLS

Age	Premenarchial		Post-menarchial	
(years)	Mean	S.E.	Mean	S.E.
11	5.5	1.6	6.2	.96
12	3.8	.65	11.8	3.3
13	9.9	3.6	6.8	1.8

LH = luteinizing hormone.

(data from Yen, Vicic, and Kearchner, 1969, Journal of Clinical
Endocrinology 29:382-385, by permission of Lippincott/Harper Company,
Philadelphia, PA)

TABLE 503

SERUM LUTEINIZING AND FOLLICLE-STIMULATING HORMONES IN BOYS

LH (mI.U. of 2nd IRP-HMG/ml.)

Groups	Ann Arbor	Baltimore	Cleveland	Philadelphia	San Francisco	Winnipeg
Age (years)						
1-2	4.0	--	--	} 2.2	--	5.4
3-4	4.8	--	--		--	5.9
5-6	4.4	} 3.4	--		--	6.8
7-8	4.4		1.1	} < 2.0	--	7.7
9-10	4.8	4.8	0.9		--	8.6
11-12	3.5	6.8	2.1	2.7	--	9.0
13-14	5.8	9.4	4.1	3.7	--	11.3
15-16	8.4	9.0	4.0	4.3	--	14.0
17-18	7.6	14.1	--	12.8	--	18.5
19-20	7.5	--	--	--	--	16.2
Stages						
P-1	--	3.9	--	--	2.5	7.4
P-2	--	6.8	--	--	3.5	10.1
P-3	--	8.5	--	--	3.8	14.0
P-4	--	9.5	--	--	} 3.9	18.5
P-5	--	11.8	--	--		16.2

FSH (mI.U. of 2nd IRP-HMG/ml.)

Groups	Ann Arbor	Baltimore	Cleveland	San Francisco	Winnipeg
Age (years)					
1-2	2.2	--	--	--	3.5
3-4	2.3	--	--	--	3.8
5-6	3.0	} 4.2	--	--	3.7
7-8	2.5		3.2	--	5.0
9-10	2.8	5.4	5.2	--	5.2
11-12	3.0	5.6	8.4	--	6.3
13-14	4.1	8.1	10.2	--	6.8
15-16	8.0	8.7	--	--	7.9
17-18	7.8	9.2	--	--	7.9
19-20	8.0	--	--	--	10.1
Stages					
P-1	--	4.5	--	1.4	4.6
P-2	--	5.9	--	1.7	6.5
P-3	--	8.1	--	3.1	7.9
P-4	--	8.5	--	} 4.5	7.9
P-5	--	7.2	--		10.1

(data compiled from various authors by Root, 1973, The Journal of Pediatrics 83:1-19)

TABLE 504

URINARY LUTEINIZING AND FOLLICLE-STIMULATING HORMONES
IN PREPUBERTAL AND PUBERTAL BOYS AND GIRLS

Groups	I.U. of 2nd IRP-HMG/24 hr.	
	LH	FSH
Age (years)	BOYS	
Newborn-0.5	3.10	--
0.5-6	0.48-1.04	0.72
5-8	1.37-2.30	1.38-2.0
9-10	2.05-2.9	1.83-2.3
11-12	7.8	4.7
13-14	17.4	5.8
15-16	25.4	7.8
17-18	38.3	7.9
Age (years)	GIRLS	
0.25-0.5	1.70	--
0.5-6	1.10-1.37	1.05
7-8	1.09-1.52	2.31
9-10	0.98-1.85	2.00
Pubertal Stages		
1	2.6	2.2
2	8.6	8.4
3	12.7	7.1
4	23.0	7.8
5	31.1	6.9

The pubertal stages are those of Tanner (1962)

(data compiled from various authors by Root, 1973, The
Journal of Pediatrics 83:1-19)

TABLE 505

SERUM LUTEINIZING AND FOLLICLE-STIMULATING HORMONES IN GIRLS

LH (mI.U. of 2nd IRP-HMG/ml.)

Groups	Ann Arbor	Baltimore	Cleveland	Philadelphia	San Francisco
Age (years)					
1-2	6.5	--	--	--	--
3-4	5.4	2.8	--	< 2.0	--
5-6	6.0	2.6	--	--	--
7-8	4.8	2.6	1.9	2.3	--
9-10	4.3	4.0	2.7	3.7	--
11-12	8.6	8.7	7.3	4.4	--
13-14	10.0	9.6	11.9	4.4	--
15-16	8.0	9.6	--	11.3	--
17-18	8.8	15.3	--	--	--
19-20	9.5	--	--	--	--
Stages					
P-1	--	2.9	--	--	1.8
P-2	--	3.9	--	--	2.6
P-3	--	--	--	--	2.5
P-4	--	8.4	--	--	3.9
P-5	--	11.3	--	--	5.5

FSH (mI.U. of 2nd IRP-HMG/ml.)

Groups	Ann Arbor	Baltimore	Cleveland	San Francisco
Age (years)				
1-2	2.8	--	--	--
3-4	2.0	3.7	--	--
5-6	3.6	4.5	--	--
7-8	2.8	4.5	4.4	--
9-10	3.0	5.4	8.1	--
11-12	5.3	7.5	10.7	--
13-14	11.6	8.0	9.3	--
15-16	8.0	8.0	--	--
17-18	11.0	8.6	--	--
19-20	9.4	--	--	--
Stages				
P-1	--	4.2	--	1.3
P-2	--	5.5	--	3.2
P-3	--	--	--	4.1
P-4	--	8.0	--	4.7
P-5	--	8.0	--	6.6

Stages P-1 to P-5 are the pubertal stages of Tanner (1962).

(data compiled from various authors by Root, 1973, The Journal of Pediatrics 83:1-19)

TABLE 506

MEAN CONCENTRATIONS OF LH AND FSH IN SERUM, LH/FSH RATIO,
AND TESTICULAR VOLUME INDEX AT DIFFERENT CHRONOLOGICAL AGES IN
CALIFORNIA BOYS

Variable	Measurement	Chronological age (years)						
		7	8	9	10	11	12	13
LH	Mean conc, ng/ml	3.97	4.27	4.08	4.42	4.44	4.8	4.55
	S.D.	0.76	0.45	0.33	1.0	0.7	0.58	0.90
FSH	Mean conc, ng/ml	1.27	1.24	1.30	1.67	1.77	2.09	3.11
	S.D.	0.20	0.32	0.20	0.51	0.64	0.64	0.68
LH/FSH ratio	Mean	3.15	3.63	3.20	2.76	2.76	2.46	1.68
	S.D.	0.61	0.89	0.56	0.72	0.93	0.68	0.61
Testicular volume index	Mean	2.65	2.70	3.18	2.74	3.4	7.22	8.39
	S.D.	0.58	0.50	0.71	0.57	0.67	2.21	2.67

LH = luteinizing hormone; FSH = follicle stimulating hormone. The testicular volume
index is length x breadth.

(data from Burr et al., 1970)

TABLE 507

SIZE (mm) OF THE PITUITARY FOSSA IN WHITE CHILDREN
OF CLEVELAND (OH)

Age	Boys		Girls	
	Average Length	Average Depth	Average Length	Average Depth
3 months	6.16	4.32	6.30	4.88
6 months	6.97	4.85	7.20	5.22
9 months	7.57	5.16	7.70	5.45
1 year	7.86	5.41	7.99	5.65
18 months	8.01	5.72	8.27	5.83
2 years	8.31	5.93	8.46	6.01

TABLE 507 (continued)

SIZE (mm) OF THE PITUITARY FOSSA IN WHITE CHILDREN
OF CLEVELAND (OH)

Age	Boys		Girls	
	Average Length	Average Depth	Average Length	Average Depth
30 months	8.49	6.13	8.64	6.12
3 years	8.57	6.22	8.74	6.29
42 months	8.75	6.45	8.95	6.38
4 years	8.84	6.58	9.04	6.42
54 months	8.98	6.66	9.17	6.47
5 years	9.03	6.74	9.23	6.55
6 years	9.21	6.88	9.41	6.68
7 years	9.22	6.90	9.44	6.74
8 years	9.26	6.95	9.52	6.80
9 years	9.27	6.97	9.56	6.83
10 years	9.29	6.98	9.65	6.84
11 years	9.41	6.99	9.76	6.87
12 years	9.54	7.00	10.03	6.96
13 years	9.68	7.03	10.06	7.02
14 years	9.90	7.05	10.18	7.26
15 years	10.14	7.10	10.31	7.27
16 years	10.23	7.20	10.37	7.31
17 years	10.49	7.23	10.55	7.36
18 years	10.62	7.32	10.66	7.41

(data from "Growth of the human pituitary fossa," Human Biology 20:1-20, 1948, by C. C. Francis, by permission of the Wayne State University Press. Copyright 1948, Wayne State University Press, Detroit, Michigan 48202)

TABLE   508

CONCENTRATIONS OF LH AND FSH IN SERUM, LH/FSH RATIO
AND TESTICULAR VOLUME INDEX AT DIFFERENT BONE AGES IN CALIFORNIA BOYS

Variable	Measurement	Bone age (years)							
		6	7	8	9	10	11	12	13
LH	Mean conc, ng/ml	3.97	4.27	4.25	3.67	4.64	4.65	4.57	4.59
	S.D.	1.00	0.73	0.42	0.78	0.58	0.53	0.85	0.95
FSH	Mean conc, ng/ml	1.29	1.37	1.35	1.31	1.64	1.5	1.97	3.08
	S.D.	0.27	0.59	0.32	0.36	0.64	0.27	0.28	1.22
LH/FSH ratio	Mean	3.21	3.39	3.30	2.95	2.99	3.19	2.35	1.74
	S.D.	1.02	0.84	0.76	0.93	0.72	0.68	0.28	0.86
Testicular volume index	Mean	2.93	2.71	3.08	2.64	3.15	3.54	6.31	8.23
	S.D.	0.54	0.48	0.59	0.47	1.02	0.94	1.89	2.70

(data from Burr et al., 1970)

TABLE   509

MEAN CONCENTRATIONS OF LH AND FSH IN SERUM, LH/FSH RATIO,
AND TESTICULAR VOLUME INDEX AT DIFFERENT STAGES OF PUBERTY IN
CALIFORNIA BOYS

Variable	Measurement	Stages of Puberty			
		P_1	P_2	P_3	P_{4-5}
LH	Mean conc, ng/ml	4.19	4.42	4.82	5.32
	S.D.	0.78	0.51	0.91	0.67
FSH	Mean conc, ng/ml	1.36	1.93	2.93	2.69
	S.D.	0.37	0.53	1.14	1.13
LH/FSH ratio	Mean	3.26	2.43	1.94	2.29
	S.D.	0.86	0.61	0.91	0.93
Testicular volume index	Mean	2.77	4.80	8.99	13.16
	S.D.	0.62	1.71	2.12	2.09

LH = luteinizing hormone; FSH = follicle stimulating hormone. Testicular volume index
is length x breadth (cm) calculated as the mean of the values for the right and left
side. The pubertal stages are according to the criteria of Tanner (1962).

(data from Burr et al., 1970)

TABLE   510

CONCENTRATION OF PLASMA LH, FSH, AND TESTOSTERONE WITH SEXUAL DEVELOPMENT
IN CALIFORNIA BOYS

Stage of puberty	Testosterone (ng/100 ml)		FSH (ng/ml)		LH (ng/ml)		TVI (cm^2)		Bone Age (years)	
	Mean	S.E.	Mean	S.E.	Mean	S.E.	Mean	S.E.	Mean	S.E.
P1	18	4.4	0.8	0.05	1.1	0.06	1.7	0.1	7.7	0.3
P2	71	19	0.96	0.15	1.5	0.15	4.3	0.6	12	0.2
P3	248	46	1.7	0.30	1.6	0.07	8.2	0.5	13.7	0.2
P4-5	482	27	2.5	0.50	1.7	0.11	10.4	0.7	15.7	0.3

FSH = follicle stimulating hormone; LH = luteinizing hormone; TVI = testicular volume
index (length x breadth in cm calculated as the mean for the two sides). The pubertal
stages are based on the criteria of Tanner (1962). Bone age is by the method of
Greulich and Pyle (1959).

(data from August, Grumbach, and Kaplan, 1972, Journal of Clinical Endocrinology and
Metabolism 34:319-326, by permission of Lippincott/Harper Company, Philadelphia, PA)

TABLE   511

MEAN CONCENTRATIONS OF LH AND FSH IN SERUM AND LH/FSH
RATIO AT DIFFERENT STAGES OF TESTICULAR DEVELOPMENT IN
CALIFORNIA BOYS

Variable	Measurement	Testicular volume index					
		<2	2.1-4	4.1-6	6.1-8	8.1-10	10.1-12
LH	Mean conc, ng/ml	4.41	4.14	4.80	4.72	4.55	5.05
	S.D.	0.63	0.73	0.38	0.44	1.18	0.72
FSH	Mean conc, ng/ml	1.30	1.44	2.20	2.55	2.46	3.25
	S.D.	0.29	0.41	0.86	1.00	1.05	1.23
LH/FSH ratio	Mean	3.56	3.04	2.42	2.07	2.29	1.76
	S.D.	1.01	0.82	0.73	0.61	1.13	0.67

The testicular volume index is length x breadth averaged for the two sides.

(data from Burr et al., 1970)

TABLE 512

SERUM TSH VALUES (μU/ml) FOR NEWLY BORN DANISH
INFANTS IN RELATION TO SEASON

Season	Serum TSH (μU/ml)		Age (hours)		Weight (g)	
	Mean	S.D.	Mean	S.D.	Mean	S.D.
Summer 1971	9.6	8.7	38.3	12.9	3,540	370
Winter 1971-72	10.1	10.8	41.8	12.2	3,293	507
Summer 1972	8.4	13.3	47.8	13.8	3,370	554
Winter 1972-73	7.0	6.4	40.9	12.2	3,409	360

TSH = thyroid stimulating hormone

(data from Rogowski, Siersbaek-Nielsen, and Mølholm-Hansen, 1974, Journal
of Clinical Endocrinology and Metabolism 39:919-922, by permission of
Lippincott/Harper Company, Philadelphia, PA)

TABLE   513

SERUM GONADOTROPIN CONCENTRATIONS IN INFANTS
IN WINNIPEG (MAN)*

	5-7 days	8-59 days	2-3.9 months	4-11.9 months
FSH				
Boys	3.1 (0-21.5)**	8.6 (0-55.0)	7.5 (0-35.0)	<2.5 (0-9.1)
Girls	5.3 (0-70.0)	18.1 (2.6-88.3)	12.7 (0-160.0)	13.6 (0-70.5)
LH				
Boys	4.2 (3.1-5.2)	5.2 (2.2-15.6)	2.5 (1.2-4.0)	1.7 (0-3.3)
Girls	2.3 (1.3-3.5)	3.3 (1.3-8.5)	3.9 (1.0-7.3)	1.8 (0-3.2)

	1-1.9 years	2-4 years
FSH		
Boys	4.0 (0-16.0)	6.2 (0-9.0)
Girls	9.4 (3.3-24.6)	9.2 (0-16.6)
LH		
Boys	1.9 (0-3.3)	1.5 (0-2.1)
Girls	2.0 (1.6-3.5)	1.5 (0-2.1)

*   Values in µg LER-907/100 ml.
**   Median (range in brackets). Undetectable values are shown as 0.
     FSH = follicle stimulating hormone; LH = luteinizing hormone.

(data from Winter, Faiman, Hobson, et al., 1975, Journal of Clinical Endocrinology and
Metabolism 40:545-551, by permission of Lippincott/Harper Company, Philadelphia, PA)

TABLE 514

TESTICULAR VOLUME (cc) FOR SWISS BOYS

	Age (years)									
	12.5	13	13.5	14	14.5	15	15.5	16	16.5	17
Median	5.3	6.1	7.8	10.2	12.0	13.8	13.6	13.9	14.8	15.4
Mean	6.8	8.0	9.2	11.6	13.8	15.6	15.9	16.0	17.3	17.8
S.D.	3.6	4.4	4.6	5.1	5.6	5.1	4.7	4.4	4.3	4.1

(data from Zachmann et al., 1974)

TABLE 515

TESTICULAR VOLUME (cc) IN RELATION TO PUBIC HAIR
STAGES FOR SWISS BOYS

	Pubic hair stages					
	1	2	3	4	5	6
Mean	6.0	6.8	9.3	12.6	16.3	18.9
S.D.	2.6	3.6	3.8	4.2	4.6	4.0

Pubic hair stages are according to Tanner (1962)

(data from Zachmann et al., 1974)

TABLE   516

COLLATED DATA FOR MEAN SERUM CONCENTRATIONS OF
GONADOTROPINS AND SEX STEROIDS IN CHILDREN AND ADOLESCENTSab

Age	FSH (µg LER907/dl)	LH (µg LER907/dl)	Testosterone (ng/dl)
Males			
5-7 days	3.1 (3.1-21.5)	4.2 (3.1-5.2)	20 (14-25)
8-60 days	8.6 (3-55.0)	5.2 (2.2-15.6)	165 (53-360)
2-12 months	5.0 (3-35.0)	2.1 (0.8-4.0)	60 (3-308)
1-4 years	6.0 (3-16.0)	1.5 (0.8-3.3)	8 (3-12)
4-8	9.0 (3-13.5)	1.6 (0.8-2.9)	5 (3-13)
8-10	10.3 (6.0-21.3)	1.9 (1.1-3.0)	9 (3-26)
10-12	12.6 (6.1-22.6)	2.0 (1.0-3.0)	13 (3-48)
12-14	13.6 (8.9-23.5)	2.5 (1.4-4.9)	78 (10-437)
14-16	15.8 (11.5-25.6)	3.1 (2.1-4.9)	340 (150-772)
16-18	15.7 (11.5-27.0)	4.1 (2.4-5.8)	532 (264-900)
Females			
5-7 days	5.3 (3-70.0)	2.3 (1.3-3.5)	14 (11-28)
8-60 days	18.1 (3-88.3)	3.3 (1.3-8.5)	11 (3-24)
2-12 months	13.3 (3-160.0)	2.8 (0.8-7.3)	5 (3-17)
1-4 years	9.3 (3-24.6)	1.7 (0.8-3.5)	8 (3-15)
4-8	6.9 (4.0-17.4)	2.0 (0.8-3.3)	6 (3-14)
8-10	7.0 (4.1-13.1)	2.3 (1.2-5.0)	8 (4-20)
10-12	11.2 (5.6-21.8)	2.5 (1.0-6.3)	18 (5-47)
12-14	13.8 (7.2-22.4)	4.5 (1.1-12.3)	26 (11-60)
14-16	18.9 (6.5051.3)	8.7 (2.4-45.0)	34 (16-72)
16-18	22.2 (8.7-53.0)	9.4 (5.3-45.0)	41 (20-65)

Age	Dihydrotes- tosterone (ng/dl)	Androstene- dione (ng/dl)	Dehydroepian- drosterone (ng/dl)
Males			
5-7 days	--	36 (17-44)	--
8-60 days	--	27 (3-98)	230 (54-685)
2-12 months	--	14 (2-68)	70 (10-330)
1-4 years	2 (0-7)	5 (3-27)	31 (12-79)
4-8	3 (0-7)	12 (7-19)	47 (12-233)
8-10	3 (0-7)	18 (13-31)	85 (21-283)
10-12	4 (3-7)	45 (31-65)	135 (46-505)
12-14	13 (9-20)	62 (45-99)	229 (84-583)
14-16	32 (20-52)	80 (48-140)	333 (175-634)
16-18	50 (20-75)	108 (70-140)	380 (180-600)
Females			
5-7 days	--	26 (16-57)	--
8-60 days	--	25 (4-68)	344 (99-696)
2-12 months	--	13 (2-23)	95 (20-220)
1-4 years	1 (0-3)	8 (2-16)	29 (19-42)
4-8	3 (0-5)	18 (12-27)	54 (12-187)

TABLE 516 (continued)

COLLATED DATA FOR MEAN SERUM CONCENTRATIONS OF
GONADOTROPINS AND SEX STEROIDS IN CHILDREN AND ADOLESCENTS[a][b]

Age	Dihydrotes-tosterone (ng/dl)	Androstene-dione (ng/dl)	Dehydroepian-drosterone (ng/dl)
8-10	6 (4-10)	32 (22-47)	82 (24-289)
10-12	8 (5-12)	65 (42-100)	261 (50-916)
12-14	12 (7-19)	123 (80-190)	473 (93-2000)
14-16	14 (10-30)	133 (77-224)	555 (250-2000)
16-18	15 (10-32)	151 (80-300)	550 (250-2000)

Age	Estrone (ng/dl)	Estradiol (ng/dl)	Progesterone (ng/dl)	17OH-Progesterone (ng/dl)
Males				
5-7 days	1.2 (0-2.0)	1.9 (0-2.3)	--	98 (63-144)
8-60 days	1.3 (0-2.4)	<1 (0-3.1)	--	138 (22-210)
2-12 months	1.0 (0-1.6)	<1 (0-1.9)	--	51 (7-137)
1-4 years	3.0 (1.8-5.3)	<1 (0-2.7)	--	16 (4-76)
4-8	2.9 (1.7-4.8)	<0.5 (0-0.9)	--	30 (12-70)
8-10	3.3 (2.0-5.4)	<0.5 (0-1.1)	--	36 (18-75)
10-12	3.2 (2.1-4.9)	<0.5 (0-1.5)	--	39 (12-75)
12-14	2.7 (1.7-4.4)	1.0 (0-2.5)	39 (20-60)	42 (15-219)
14-16	3.9 (2.1-7.0)	1.5 (0-2.4)	34 (20-50)	80 (12-200)
16-18	4.6 (2.5-7.0)	1.8 (0.9-3.0)	56 (30-75)	97 (42-200)
Females				
5-7 days	1.2 (0-2.0)	1.0 (0-3.2)	--	105 (58-128)
8-60 days	1.2 (0-2.0)	1.5 (0-5.0)	--	86 (19-214)
2-12 months	1.0 (0-1.6)	1.1 (0-7.5)	--	43 (4-147)
1-4 years	3.0 (1.9-4.6)	1.1 (0-2.0)	--	*26 (20-39)
4-8	2.7 (1.7-4.4)	<0.5 (0-1.7)	5 (3-23)	28 (9-63)
8-10	4.6 (3.1-7.0)	<0.5 (0-3.1)	6 (3-32)	35 (15-90)
10-12	4.4 (2.8-6.8)	1.6 (0-6.3)	8 (3-55)	42 (18-111)
12-14	5.9 (3.7-14.0)	4.2 (0.5-12.2)	37 (10-450)	84 (24-234)
14-16	8.9 (2.1-14.0)	8.6 (1.0-25.2)	189 (5-1400)	94 (27-234)
16-18	6.7 (1.4-14.0)	8.4 (1.0-28.4)	161 (23-1500)	108 (34-240)

[a]Data summarized as mean (range in brackets). The data in this table are derived mainly from cross-sectional studies of healthy children carried out in Winnipeg (Winter and Faiman, 1972, 1973; Faiman and Winter, 1974; Winter et al., 1975, 1976; Hughes and Winter, 1976). These have been supplemented by data from the literature (Blizzard et al., 1972; Forest et al., 1974; Jenner et al., 1972; Bidlingmaier et al., 1973; Nankin et al., 1975; Sizonenko and Paunier, 1975; Hopper and Yen, 1975; Lee and Migeon, 1975; Gupta et al., 1975; Angsusingha et al., 1974; Penny et al., 1970; Korth-Schutz et al., 1976; de Peretti and Forest, 1976).
[b]Values of FSH and LH can be converted to IU/liter by multiplying by 0.5 and 4.5, respectively. Steroid levels can be converted to nmol/liter by multiplying by the following factors: testosterone (0.0347); dihydrotestosterone (0.0344); androstenedione (0.0349); dehydroepiandrosterone (0.0347); estrone (0.0370); estradiol (0.0367); progesterone (0.0318); and 17OH-progesterone (0.0303).
FSH = follicle stimulating hormone; LH = luteinizing hormone.

(data from Winter, 1978)

TABLE 517

COLLATED DATA FOR MEAN SERUM CONCENTRATIONS OF
GONADOTROPINS AND SEX STEROIDS BY PUBERTAL STAGE[a][b]

Stage of puberty	FSH µg LER907/dl	FSH IU/liter	LH µg LER907/dl	LH IU/liter
Boys				
G1	9.7 + 0.4	4.5 (2.5-7.0)	1.7 + 0.1	3.9 (2.5-5.8)
G2	12.8 + 0.3	5.9 (3.0-9.0)	2.2 + 0.1	6.8 (4.0-12.0)
G3	13.5 + 0.7	8.1 (2.5-14.0)	2.5 + 0.2	8.5 (6.0-11.0)
G4	17.2 + 1.0	8.5 (3.5-15.0)	3.5 + 0.2	9.5 (4.0-15.5)
G5	17.0 (8.6-25.5)	8.0 (2.9-14.3)	4.2 (2.7-6.5)	11.5 (4.4-19.0)
Girls				
B1/PH1	8.2 + 0.6	4.2 (3.1-5.7)	2.1 + 0.1	2.9 (2.0-7.5)
B2/PH2	9.4 + 0.9	5.5 (4.6-7.1)	2.1 + 0.1	3.9 (2.5-11.5)
B3/PH3	14.2 + 1.1	8.0 (5.0-12.0)	3.7 + 0.3	8.4 (2.5-14.0)
B4/PH4	19.5 + 1.2	8.0 (3.5-13.0)	7.6 + 1.0	11.3 (3.0-29.0)
Adult				
Follicular	21.9 + 2.2	12.3 (10.4-14.2)	6.9 + 0.7	18.9 (15.2-22.6)
Midcycle	Peak 25-50	18.8 (14.3-22.8)	Peak 20-60	80.3 (62.4-98.2)
Luteal	8.2 + 1.0	7.4 (6.6-8.3)	5.3 + 0.7	14.2 (11.9-16.5)

Stage of puberty	Testosterone (ng/dl)	Dihydro-testosterone (ng/dl)	Andro-stenedione (ng/dl)	DHEA (ng/dl)
Boys				
G1	10 + 1	8 (2-20)	20 + 2	63 (10-280)
G2	85 + 5	16 (5-28)	40 + 3	175 (50-560)
G3	121 + 17	22 (6-36)	66 + 5	266 (120-590)
G4	493 + 42	36 (15-51)	111 + 4	339 (180-640)
G5	605 (260-1000)	52 (21-76)	107 (50-200)	370 (180-700)
Girls				
B1/PH1	7 + 1	5 (2-16)	26 + 2	75 (19-300)
B2/PH2	17 + 3	8 (3-24)	60 + 3	293 (45-1600)
B3/PH3	30 + 2	14 (9-25)	70 + 5	465 (125-1700)
B4/PH4	38 + 2	16 (9-32)	130 + 3	471 (153-1620)
Adult				
Follicular	35 (15-65)	15 (10-32)	121 (80-300)	430 (200-1500)
Midcycle	42	e	160	e
Luteal	30	e	130	e

TABLE 517 (continued)

COLLATED DATA FOR MEAN SERUM CONCENTRATIONS OF
GONADOTROPINS AND SEX STEROIDS BY PUBERTAL STAGE[ab]

Stage of puberty	Estrone (ng/dl)	Estradiol (ng/dl)	Progesterone (ng/dl)	17OH-Progesterone (ng/dl)
Boys				
G1	1.1 + 0.3	0.8 + 0.3	30	33 + 3
G2	1.6 + 0.3	1.1 + 0.3	36	37 + 3
G3	2.1 + 0.2	1.6 + 0.5	40	57 + 5
G4	3.3 + 0.6	2.2 + 0.8	40	85 + 8
G5	4.2 (2.5-6.9)	2.5 (1.4-3.8)	35 (16-60)	97 (38-200)
Girls				
B1/PH1	1.4 + 0.3	0.8 + 0.3	10	35 + 3
B2/PH2	2.0 + 0.3	1.3 + 0.5	16	45 + 7
B3/PH3	3.0 + 0.5	2.5 + 0.5	23	62 + 7
B4/PH4	4.9 + 0.6	7.6 + 0.8	161	114 + 13
Adult				
Follicular	2.7 (1.4-3.8)	8.7 + 1.5	40 (29-75)	42 (11-80)
Midcycle	6.2 (3.8-18.0)	peak 15-35	--	112 (70-160)
Luteal	3.9 (1.8-7.6)	6.2 + 1.4	940 (250-2000)	160 (30-285)

[a]Data shown as means + SEM or means (range in brackets). Pubertal staging
of males is based upon changes in testicular and penile size, independent
of pubic hair growth. Pubertal staging of females is based upon
breast development (B1 to B5) for FSH, LH, estrogen, and progestin
results, but upon pubic hair growth (PH1 to PH5) for androgen results.
[b]Factors for conversion of steroid levels to the less commonly used
nmol/liter are shown in Table II. Values of FSH and LH may be
converted from µg LER907/dl to µg LER907/dl to µg pure standard/liter
by multiplying by 0.069 and 0.227, respectively.
[c]No significant change with menstrual cycle.
FSH = follicle stimulating hormone; LH = luteinizing hormone; G = genital
stage; B = breast stage; PH = pubic hair stage.

(data from Winter, 1978; collated from same sources as in Table 516)

TABLE 518

PLASMA ANDROGENS IN BOYS

Groups	ng/100 ml.			
	T	DHT	Δ	DHA

Age (years)

2-3	5-11	--	--	--
4-5	13-20	--	40	28
6-7	14-30	--	48	27
8-9	21-34	--	59-71	120
10-11	41-60	--	32-76	142
12-13	131-249	--	57-90	154
14-15	328-643	--	87-140	537

Stages

P-1	17-25	⎫	--	--
P-2	25-85	⎬ 4	⎫	--
P-3	52-328	10	⎬ 60-62	--
P-4	134-532	25	43	--
P-5	293-564	38	78	--

Abbreviations: T = testosterone; DHT = dihydrotestosterone; Δ = androstenedione; DHA = dehydroepiandrosterone.

Stages P-1 to P-5 are the pubertal stages of Tanner (1962)

(data compiled from various authors by Root, 1973, The Journal of Pediatrics 83:1-19)

TABLE 519

URINARY ANDROGENS IN PREPUBERTAL AND PUBERTAL BOYS

Age (years)	17-KS (mg./24 hr.)	T (μg/24 hr.)	EpiT (μg/24 hr.)	DHA (mg./24 hr.)	A (mg./24 hr.)	E (mg./24 hr.)
3-5	0.6-1.6	1.0- 2.0	1.7- 5.3	0 -0.5	0.13-0.19	0 -0.06
6-9	0.9-1.8	3.5- 5.0	5.0- 6.0	0.02-0.10	0.02-0.26	0.09-0.19
10-11	1.9-2.5	5.8- 9.0	7.3- 9.8	0.02-0.12	0.26-0.50	0.24-0.35
12-13	3.3-4.5	10.7-16.0	13.2-14.6	0.20-0.30	0.76-1.12	0.36-0.54
14-15	5.2	25.8-32.3	15.3-15.6	1.03	1.84	1.80

Abbreviations: 17-KS = 17-ketosteroids; T = testosterone; EpiT = epitestosterone; DHA = dehydroepiandrosterone; A = androsterone; E = etiocholanolone. Composite data from recent literature.

(data compiled from various authors by Root, 1973, The Journal of Pediatrics 83:1-19)

TABLE 520

SERUM HORMONE CONCENTRATIONS ACCORDING TO
AGE (YEARS) IN GIRLS BEFORE MENARCHE

Hormone		10	12	13
LH	Mean	4.6	6.1	7.2
(mlU/ml)	S.E.	0.7	2.1	1.9
FSH	Mean	8.8	12.6	11.1
(mlU/ml)	S.E.	1.6	2.5	1.0
E_2	Mean	--	--	123.3
(pg/ml)	S.E.	--	--	31.9
Prog	Mean	--	263	333
(pg/ml)	S.E.	--	35	77
17-OH-P	Mean	--	--	790
(pg/ml)	S.E.	--	--	127
T	Mean	--	--	395
(pg/ml)	S.E.	--	--	10
A	Mean	--	--	937
(pg/ml)	S.E.	--	--	105

LH = luteinizing hormone; FSH = follicle stimulating hormone;
E_2 = estradiol; Prog = progesterone; 17-OH-P = 17-hydroxyprogesterone,
T = testosterone and A = androstenedione.

(data from Lee, Xenakis, Winer, et al., 1976, Journal of Clinical
Endocrinology and Metabolism 43:775-784. Copyright 1976, Lippin-
cott/Harper Co., Philadelphia, PA)

TABLE   521

ESTROGEN AND ANDROGEN LEVELS IN URINE OF GIRLS

Age (years)	E_1 (µg/24 hr.)	E_2 (µg/24 hr.)	E_3 (µg/24 hr.)			
3-4	3.9	0.9	1.8			
5-6	2.4	0.8	1.1			
7-8	4.8	1.2	2.1			
9-10	4.1	1.4	2.2			
11-12	6.8	2.1	2.4			
13-14	6.3	1.8	2.8			

Age (years)	17-KS (mg./24 hr.)	T (µg/24 hr.)	EpiT (µg/24 hr.)	DHA (mg./24 hr.)	A (mg./24 hr.)	E (mg./24 hr.)
3-5	--	0 -0.8	1.0- 5.4	--	--	--
6-9	0.87	3.6-4.3	6.0- 6.7	0.02-0.05	0.14-0.22	0.09-0.2
10-12	3.34	6.8-7.9	8.1- 8.8	0.07-0.22	0.38-0.67	0.11-0.55
13-15	4.34	8.6-9.8	11.8-12.0	0.43	1.14	0.81

Abbreviations: 17-KS = 17-ketosteroids; T = testosterone; EpiT = epitestosterone; DHA = dehydroepiandrosterone; A = androsterone; E = etiocholanolone. Composite data from recent literature; E_1 = estrone; E_2 = estradiol; E_3 = estriol.

(data compiled from various authors by Root, 1973, The Journal of Pediatrics 83:1-19)

TABLE   522

ESTROGEN AND ANDROGEN LEVELS IN PLASMA OF GIRLS

Groups	Plasma
Stages	E_2 (ng,/100 ml.)
P-1	< 0.7
P-2	1.3
P-3	2.5
P-4	4.4
P-5	5.8

Age (years)	T (ng./100 ml.)	Δ (ng./100 ml.)	DHA (ng./100 ml.)
2-12	7-19	30-65	74-142

Abbreviations: Stages P-1 to P-5 are the pubertal stages of Tanner (1962); E_2 = estradiol; T = testosterone; Δ = androstenedione; DHA = dehydroepiandrosterone. Composite date from recent literature

(data compiled from various authors by Root, 1973, The Journal of Pediatrics 83:1-19)

TABLE 523

PLASMA 17β-ESTRADIOL, FSH, AND LH IN CALIFORNIA GIRLS

Pubertal stage	17β -Estradiol				FSH (LER 869) ng/ml		LH (LER 960) ng/ml	
	Uncorrected E_2 (pg)		Corrected E_2 (pg/ml)					
	Mean	S.E.	Mean	S.E.	Mean	S.E.	Mean	S.E.
P_1	15	0.98	--	--	.73	.09	0.8	.08
P_2	25	2.8	13	2.4	1.78	.51	1.12	.24
P_3	--	--	25	4.5	2.25	.53	1.08	.20
P_4	--	--	44	12	2.62	.41	1.68	.23
P_5	--	--	58	5.7	2.53	.36	2.37	.63

Menstrual cycle: P5 subjects

Phase I (days 1-9)			36	6.2				
Phase II (days 10-20)			59	15				
Phase III (days 20-30+)			66	7.3				

FSH = follicle stimulative hormone; LH = luteinizing hormone; pubertal stages are according to the method of Marshall and Tanner (1969)

(data from Jenner et al., 1972)

TABLE   524

SERUM STEROID CONCENTRATIONS IN BOYS

Steroids		Age (years)				
		9-11	12	13	14	15-18
DHA	Mean	--	3,967	3,958	4,515	--
(pg/ml)	S.E.	--	413	542	618	--
DHAS	Mean	--	725	940	1,005	--
(ng/ml)	S.E.	--	108	108	152	--
P	Mean	--	--	325	338	557
(pg/ml)	S.E.	--	--	40	26	77
17-OH-P	Mean	695	725	734	857	1,208
(pg/ml)	S.E.	136	107	115	127	108
E_1	Mean	--	51.5	51.0	48.8	56.0
(pg/ml)	S.E.	--	5.6	5.6	5.3	6.9
E_2	Mean	21.8	24.6	35.3	32.3	34.3
(pg/ml)	S.E.	5.0	2.8	3.8	7.0	6.0
A	Mean	--	623	976	799	1,082
(pg/ml)	S.E.	--	100	146	103	140

DHA = dehydroepiandrosterone; DHAS = dehydroepiandrosterone sulfate;
P = progresterone; 17-OH-P = 17=hydroxyprogesterone; E_1 = estrone;
E_2 = estradiol; A = androstenedione.

(data from Lee and Migeon, 1975, Journal of Clinical Endocrinology
and Metabolism 41:556-562 with permission of Lippincott/Harper & Row).

TABLE 525

PLASMA FREE STEROID LEVELS (ng/100 ml) IN FRENCH CHILDREN

			Age (years)			
	<6	6-8	8-10	10-12	12-14	>14
Dehydroepiandrosterone (DHA)						
Girls						
Mean*	28.6	109.3	116.4	351.7	370.1	465.9
CL†	19.4-42.1	72.6-164.5	74.4-179.6	233.7-529.4	224.3-610.6	282.4-770.5
Boys						
Mean	43.4	43.8	84.7	264.8	352.9	505.9
CL	26.3-71.7	29.1-66.0	53.3-134.6	183.0-383.3	239.5-520.3	306.7-834.8
Androstenedione (Δ)						
Girls						
Mean	14.8	17.7	32.1	64.5	123.2	131.4
CL	9.6-22.9	11.5-27.4	21.6-47.4	41.7-99.6	79.8-190.4	77.2-223.9
Boys						
Mean	11.6	16.1	20.2	44.7	66.6	82.0
CL	7.1-19.0	10.4-24.8	13.1-31.3	30.7-65.2	44.9-98.7	48.2-139.7
Testosterone (T)						
Girls						
Mean	8.4	9.8	8.7	15.1	33.0	37.3
CL	5.2-13.4	6.1-15.8	5.7-13.3	9.4-24.2	20.6-53.0	20.9-66.5
Boys						
Mean	6.2	6.9	7.2	13.1	50.7	204.2
CL	3.6-10.6	4.3-11.1	4.5-11.5	8.6-20.1	33.1-77.8	119.4-349.1
Dihydrotestosterone (DHT):						
Girls						
Mean	1.4	2.8	6.1	7.7	11.7	7.1
CL	0-2.6	1.5-5.4	3.9-9.6	4.7-12.4	7.2-19.0	4.0-12.6
Boys						
Mean	2.0	3.4	2.6	3.9	13.4	31.8
CL	0-7.2	1.6-7.1	1.6-4.2	2.6-5.9	9.1-19.7	19.6-51.6
Estrone (E_1)						
Girls						
Mean	3.0	2.7	4.6	4.4	8.9	5.9
CL	1.9-4.6	1.7-4.4	3.1-7.0	2.8-6.8	5.7-14.0	3.4-10.2
Boys						
Mean	3.0	2.9	3.3	3.2	2.7	3.9
CL	1.8-5.3	1.7-4.8	2.0-5.4	2.1-4.9	1.7-4.4	2.1-7.1
Estradiol (E_2)						
Girls						
Mean	1.9	1.2	2.1	4.9	7.0	6.4
CL	1.3-3.1	0-2.0	1.4-3.3	3.1-7.8	4.4-11.1	3.6-11.3
Boys						
Mean	3.0	2.3	2.0	2.0	2.3	4.5
CL	1.7-5.3	1.3-3.8	1.2-3.5	1.3-3.1	1.4-3.9	2.4-8.4
Cortisol (F)						
Girls						
Mean	7.7	9.3	10.0	7.8	9.8	12.6
CL	5.3-11.2	6.6-12.9	9.4-13.6	5.5-11.2	6.9-14.0	8.4-19.0
Boys						
Mean	11.3	7.4	6.1	5.7	10.1	7.6
CL	7.5-17.1	5.2-10.5	4.3-8.6	4.2-7.6	7.5-13.6	5.2-11.1

*Mean concentration transformed back from logarithms; †95% confidence levels

(data from Ducharme, Forest, de Peretti, et al., 1976, Journal of Clinical Endocrinology and Metabolism 42:468-476, by permission of Lippincott/Harper Company, Philadelphia, PA)

TABLE 526

STATURE, SERUM HGH, LH, FSH, AND SERUM TESTOSTERONE LEVELS,
ACCORDING TO PUBERTAL STAGES, IN OHIO BOYS

| | Pubertal Stages | | | | | |
| | 3 | | 4 | | 5 | |
	Mean	S.D.	Mean	S.D.	Mean	S.D.
Age (years)	13.1	(1.2)	14.6	(1.2)	15.5	(1.3)
Stature (in)	59.2	(3.1)	62.6	(3.7)	65.5	(3.6)
Serum testosterone (ng/100 ml)	95	(39)	193	(106)	298	(131)
HGH (ng/ml)	3.9	(3.6)	3.6	(2.1)	4.8	(4.2)
LH (m I.U./ml)	7.7	(2.8)	8.7	(1.9)	12.5	(3.7)
FSH (m I.U./ml)	12.8	(2.4)	11.8	(2.8)	12.9	(3.8)

HGH = human growth hormone
LH = luteinizing hormone
FSH = follicle stimulating hormone
The pubertal stages are according to the criteria of Tanner (1962)

(data from Wieland, Chen, Zorn, et al., 1971, Journal of Pediatrics 79:999-1002)

TABLE 527

PLASMA DHEA AND T (ng/100 ml) CONCENTRATIONS AND
TESTICULAR VOLUME INDEX IN NORMAL SWISS BOYS

| | Testicular Volume Index | | | | | |
	2-3	3.1-4	4.1-5	5.1-7	7.1-10	10
DHEA mean	80.5	81.5	171.0	247.1	194.8	286.5
C.L. {	254.8	322.8	475.1	509.0	429.1	659.4
{	21.7	20.6	33.4	119.9	88.5	124.5
T mean	7.6	8.5	19.1	21.1	50.3	294.7
C.L. {	16.6	20.4	27.0	28.8	114.3	796.2
{	3.5	1.5	14.3	15.5	22.2	109.1

C.L. = 95% confidence limits
Testicular volume index = length x breadth
DHEA = dehydroepiandrosterone
T = testosterone

(data from Sizonenko and Paunier, 1975, Journal of Clinical Endocrinology and
Metabolism 41:894-904, by permission of Lippincott/Harper, Philadelphia)

TABLE 528

MEAN PLASMA DHEA AND T (ng/100 ml), FSH (mU, MRC 68-39/ml), AND
LH (mU, MRC 69-104/ml) CONCENTRATIONS AND BONE AGE IN NORMAL SWISS BOYS

		Bone Age (years)									
		5	6	7	8	9	10	11	12	13	14-15
DHEA	mean	31.1	46.6	77.1	71.0	98.6	107.1	163.2	221.2	237.8	333.4
	C.L.	78.8	131.5	233.2	243.2	283.5	247.6	606.0	680.4	583.3	634.2
		12.2	16.5	26.2	20.7	34.8	46.4	52.8	84.4	97.1	176.8
T	mean	6.6	5.6	6.9	7.4	9.6	11.5	14.8	54.8	102.8	340.0
	C.L.	18.0	9.5	11.5	16.8	25.7	43.7	48.2	294.6	937.6	772.2
		2.2	3.3	4.1	3.3	3.6	3.0	4.9	10.2	11.3	150.0
FSH	mean	1.8	1.8	1.9	2.4	2.1	2.1	2.4	2.4	3.7	4.2
	C.L.	3.4	2.9	3.4	4.7	5.7	4.4	7.1	4.4	6.5	6.4
		1.8	1.1	1.0	1.2	0.8	1.0	0.8	1.3	2.0	1.0
LH	mean	3.2	3.4	3.9	4.2	3.2	3.7	6.2	7.3	9.1	9.0
	C.L.	5.2	9.2	8.6	7.8	6.6	9.3	12.9	18.8	13.7	19.3
		1.4	1.1	1.7	2.3	1.5	1.5	8.1	2.9	6.1	4.3

C.L. = 95% confidence limits
DHEA = dehydroepiandrosterone
T = testosterone
FSH = follicle stimulating hormone
LH = luteinizing hormone
MRC = Medical Research Council pituitary standard

(data from Sizonenko and Paunier, 1975, Journal of Clinical Endocrinology and Metabolism
41:894-904, by permission of Lippincott/Harper, Philadelphia)

TABLE 529

PLASMA DHEA AND T (ng/100 ml) CONCENTRATIONS AND PUBIC AND
AXILLARY HAIR STAGES IN NORMAL SWISS BOYS

		Pubic Hair Stages					Axillary Hair Stages			
		1		2	3	4	1		2	3
		TVI<4	TVI>4				TVI<4	TVI>4		
DHEA	mean	77.5	129.3	252.5	262.4	419.3	81.2	147.1	345.8	338.6
	C.L.	286.9	394.8	507.0	651.6	--	323.5	609.8	433.5	641.7
		20.9	42.4	125.8	105.6	--	20.3	35.5	275.9	178.6
T	mean	7.6	17.4	49.2	288.8	478.2	8.3	25.9	164.3	269.5
	C.L.	17.2	48.9	201.6	788.6	--	22.1	87.3	807.5	864.9
		3.4	6.2	12.0	105.7	--	3.1	7.7	33.4	84.0

C.L. = 95% confidence limits
TVI = testicular volume index (length x breadth)
DHEA = dehydroepiandrosterone
T = testosterone

(data from Sizonenko and Paunier, 1975, Journal of Clinical Endocrinology and Metabolism
41:894-904, by permission of Lippincott/Harper, Philadelphia)

TABLE   530

MEAN PLASMA DHEA (ng/100 ml) AND T LEVELS IN NORMAL SWISS BOYS

Chronological Age (years)

		7	8	9	10	11	12	13	14
DHEA	mean	52.7	57.2	79.7	112.0	157.0	188.1	230.2	324.3
	C.L.	132.2	231.4	315.5	312.7	481.6	621.5	545.4	776.4
		20.6	13.8	20.1	40.2	51.3	57.5	97.4	135.6
T	mean	5.4	7.3	6.9	9.9	16.8	41.8	159.0	264.0
	C.L.	12.9	16.7	11.5	24.8	56.7	580.5	2,310.3	1,850.6
		2.3	3.2	4.1	3.9	5.0	4.6	10.8	37.7

CL = 95% confidence limits
DHEA = dehydroepiandrosterone
T = testosterone

(data from Sizonenko and Paunier, 1975, Journal of Clinical Endocrinology and Metabolism 41:894-904, by permission of Lippincott/Harper, Philadelphia)

TABLE   531

PLASMA DHEA AND T (ng/100 ml), FSH (mU, MRC 68-39/ml), AND
LH (mU, MRC 69-404/ml) CONCENTRATIONS AND STAGES OF PUBERTY IN NORMAL SWISS BOYS

Stages of Puberty

		1	2	3	4
DHEA	mean	82.6	174.9	265.7	338.6
	C.L.	286.7	558.3	592.9	641.7
		23.8	49.6	119.1	178.6
T	mean	7.8	23.3	152.8	269.5
	C.L.	15.5	65.9	863.3	814.6
		4.0	8.2	26.3	84.0
FSH	mean	2.4	3.0	3.1	3.2
	C.L.	6.0	5.4	4.5	6.7
		1.0	1.7	2.2	1.5
LH	mean	3.5	5.8	6.7	6.6
	C.L.	9.1	11.7	11.6	9.3
		1.3	2.9	3.8	4.3

C.L. = 95% confidence limits
DHEA = dehydroepiandrosterone
T = testosterone
FSH = follicle stimulating hormone
LH = luteinizing hormone
MRC = Medical Research Council pituitary standard

(data from Sizonenko and Paunier, 1975, Journal of Clinical Endocrinology and Metabolism 41:894-904, by permission of Lippincott/Harper, Philadelphia)

TABLE 532

MEAN PLASMA DHEA AND T LEVELS (ng/100 ml) IN NORMAL SWISS GIRLS

Chronological Age (years)

		6	7	8	9	10	11	12	13
DHEA	mean	44.7	63.7	80.9	83.5	185.5	335.9	439.9	507.1
	C.L. {	168.7	187.3	246.0	289.1	690.6	916.3	2,085.6	2,070.0
		11.8	21.6	26.6	24.1	49.8	123.2	92.8	121.4
T	mean	5.9	6.4	7.7	9.1	12.8	23.5	26.0	25.5
	C.L. {	12.4	14.2	14.3	20.1	31.4	47.3	56.9	60.3
		2.8	2.8	4.2	4.2	5.2	11.6	11.9	10.8

C. L. = 95% confidence limits
DHEA = dehydroepiandrosterone
T = testosterone

(data from Sizonenko and Paunier, Journal of Clinical Endocrinology and Metabolism, 41: 894-904, by permission of Lippincott/Harper, Philadelphia)

TABLE 533

PLASMA DHEA AND T (ng/100 ml), FSH (mU, MRC 68-39/ml), AND
LH (mU, MRC 69-104/ml) CONCENTRATIONS AND STAGES OF PUBERTY IN NORMAL SWISS GIRLS

Stages of Puberty

		1	2	3	4	5
DHEA	mean	75.0	293.1	465.3	450.5	510.7
	C.L. {	301.8	1,904.2	1,729.5	1,321.3	1,619.7
		18.7	45.1	125.2	153.4	161.7
T	mean	7.4	17.3	30.1	35.7	25.7
	C.L. {	18.4	46.8	52.8	65.0	48.3
		3.0	6.4	17.1	19.6	13.6
FSH	mean	2.2	3.2	4.3	4.1	4.2
	C.L. {	5.1	7.0	7.7	11.2	9.3
		0.9	1.4	2.4	1.5	1.9
LH	mean	4.1	5.9	8.7	10.0	8.8
	C.L. {	9.2	16.6	13.6	14.4	17.5
		1.8	2.0	5.6	7.0	4.4

C.L. = 95% confidence limits
DHEA = dehydroepiandrosterone
T = testosterone
FSH = follicle stimulating hormone
LH = luteinizing hormone
MRC = Medical Research Council pituitary standard

(data from Sizonenko and Paunier, 1975, Journal of Clinical Endocrinology and Metabolism 41:894-904, by permission of Lippincott/Harper, Philadelphia)

TABLE 534

PLASMA DHEA AND T (ng/100 ml) CONCENTRATIONS AND PUBIC AND
AXILLARY HAIR STAGES IN NORMAL SWISS GIRLS

		Pubic Hair Stages					Axillary Hair Stages			
		Bl [1]	B2	2	3	4	Bl [1]	B2	2	3
DHEA	mean	70.9	230.4	395.0	531.7	638.3	67.2	246.5	490.3	506.9
	C.L. {	308.2	966.7	1,142	1,484.9	1,970.8	282.3	657.8	1,549	1,577
		16.3	54.9	136.6	190.3	217.3	16.0	92.4	155.2	162.9
T	mean	8.1	12.1	21.4	26.9	37.8	7.4	13.8	20.9	29.9
	C.L. {	25.5	34.4	43.0	40.5	68.1	20.2	41.9	45.0	58.8
		2.6	4.3	10.6	17.9	20.9	2.7	4.5	9.7	15.2

C.L. = 95% confidence limits
DHEA = dehydroepiandrosterone
T = testosterone
Bl - no breast development
B2 = first budding of the breasts

(data from Sizonenko and Paunier, 1975, Journal of Clinical Endocrinology and Metabolism
41:894-904, by permission of Lippincott/Harper, Philadelphia)

TABLE 535

MEAN PLASMA DHEA AND T (ng/100 ml), FSH (mU, MRC 68-39/ml), AND
LH (mU, MRC 69-104/ml) CONCENTRATIONS AND BONE AGE IN NORMAL SWISS GIRLS

		Bone Age (years)									
		5	6	7	8	9	10	11	12	13	14-15
DHEA	mean	30.9	52.3	58.6	85.5	97.9	191.1	325.7	308.1	485.6	755.1
	C.L. {	74.0	196.4	153.3	216.3	583.0	629.9	994.5	844.7	1,827.0	2,310.0
		12.5	13.9	22.3	33.8	16.5	58.0	106.6	112.4	114.4	247.9
T	mean	5.6	8.1	7.1	10.2	10.1	12.0	16.6	23.9	25.0	33.5
	C.L. {	9.1	28.2	16.1	32.5	27.9	32.1	30.1	45.2	43.8	71.5
		3.5	2.3	3.2	3.2	3.6	4.8	10.1	12.6	14.3	15.7
FSH	mean	2.4	2.4	2.0	2.5	2.2	2.6	3.2	3.8	4.3	4.1
	C.L. {	5.1	5.0	4.5	6.5	4.7	5.6	8.8	8.0	9.3	6.4
		1.0	1.1	0.9	1.0	1.0	1.2	1.2	1.8	1.9	2.6
LH	mean	3.7	3.5	4.1	4.9	4.4	4.1	6.6	9.3	9.8	9.6
	C.L. {	6.0	9.1	7.4	8.6	9.7	11.0	14.5	13.2	16.1	15.7
		2.1	1.4	3.0	2.8	2.0	2.0	3.0	6.6	5.9	6.1

C.L. = 95% confidence limits
DHEA = dehydroepiandrosterone
T = testosterone
FSH = follicle stimulating hormone
LH = luteinizing hormone
MRC = Medical Research Council pituitary standard

(data from Sizonenko and Paunier, 1975, Journal of Clinical Endocrinology and Metabolism
41:894-904, by permission of Lippincott/Harper, Philadelphia)

TABLE 536

COLLATED DATA FOR SERUM CONCENTRATIONS OF
THYROID HORMONES IN HEALTHY SUBJECTS[a]

Age	TSH (mIU/liter)	Thyroxine[d] (μg/dl)	Triiodothyronine[d] (ng/dl)
cord	8.5(1.5-31.5)	11.0 (713-15.3)[b]	48 + 3[c]
1-5 days	7.3(<1.5-12.1)	15.5(10.1-20.9)	140 + 16
1-8 wks	<6	12.4(7.1-16.6)	163 + 6
2-12 mos	<6	9.6 (5.5-15.0)	124 + 8
1-6 yrs	<6	9.1(5.6-12.6)	138 + 3
6-10 yrs	<6	8.3(4.9-11.7)	141 + 7
10-16 yrs	<6	7.2(3.8-10.6)	131 + 6
16-20 yrs	<6	7.5(4.1-10.9)	120 + 5

Age	T3 resin uptake (% of standard)	Free T4 (ng/dl)	Free T3 (pg/dl)
cord	0.9 (0.75-1.05)	2.2 + 0.1	130 + 10
1-5 days	1.15 (0.9-1.4)	4.9 + 0.3	410 + 20
1-8 wks	0.95 (0.85-1.1)	2.1 + 0.1	400 + 20
2-12 mos	0.9 (0.8-1.1)		
1-6 yrs	0.95 (0.8-1.1)		
6-10 yrs	0.95 (0.8-1.1)		
10-16 yrs	0.95 (0.8-1.1)		
16-20 yrs	0.95 (0.8-1.1)		

[a]Data summarized from Abuid et al., 1974; Rubinstein et al., 1973;
Avruskin et al., 1973; O'Halloran and Webster, 1972; Hays and Mullard,
1974; Fisher, 1973; Mayberry et al., 1971; Hayek et al., 1973.
[b]Mean (range in brackets).
[c]Mean ± standard error.
[d]Levels of thyroxine and triiodothyronine are given in the units that are
presently in general clinical use; to convert these to SI units, multiply
thyroxine values by 0.0129 (μmol/liter) and triiodothyronine values by
0.015 (nmol/liter). Recent data of Corcoran et al. (1977) suggest that
serum concentration of thyroid hormones in children may be higher than
has been previously reported, with mean serum thyroxine levels of 10.0 ±
2.5 μg/dl (S.D.) and serum triiodothyronine levels of 194 ± 35 ng/dl
below the age of 10 years.
TSH = pituitary thyroid stimulating hormone; T3 = reverse triiodothyronine;
T4 = tetraiodothyronine (thyroxine).

(data from Winter, 1978)

TABLE 537

MEANS OF SERUM THYROXINE IODINE LEVELS (B₄) IN CALIFORNIA CHILDREN

MEANS OF SERUM THYROXINE IODINE LEVELS (B_4)
IN CALIFORNIA CHILDREN

Age (years)	BOYS μg/100 ml		GIRLS μg/100 ml	
	Mean	S.E.	Mean	S.E.
4.1-6	5.734	0.200	--	--
6.1-8	5.798	0.409	5.481	0.167
8.1-10	5.261	0.143	5.497	0.192
10.1-12	4.938	0.149	5.032	0.119
12.1-14	4.804	0.102	4.785	0.131
14.1-16	4.686	0.102	4.960	0.107
16.1-18	4.800	0.110	5.236	0.159
18.1-20	4.929	0.112	5.125	0.178

(data from Oddie and Fisher, 1967, Journal of Clinical Endocrinology 27:89-97, by permission of Lippincott/Harper Company, Philadelphia, PA)

TABLE 538

THYROID FUNCTION IN NEWLY BORN INFANTS IN RELATION TO SEASON

	Serum T₄ μg 100 ml		T₃ test %		FT₄ index units		Age in hours		Weight g	
	Mean	S.D.	Mean	S.D.	Mean	S.D.	Mean	S.D.	Mean	S.D.
Summer 1971	17.2	4.2	52.9	5.8	904	217	37.9	11.2	3,396	484
Winter 1971-72	20.6	2.6	55.7	4.8	1,137	117	41.8	12.1	3,362	455
Summer 1972	18.9	3.0	55.3	4.1	1,003	143	46.2	11.4	3,535	538
Winter 1972-73	21.5	4.1	56.0	6.4	1,209	254	41.6	14.2	3,427	340

T_4 = thyroxine; T_3 = serum triiodothyronine; FT_4 = free T_4 index which is the product of the total T_4 and the T_4 test values.

(data from Rogowski, Siersbaek-Nielsen, and Mølholm-Hansen, 1974, Journal of Clinical Endocrinology and Metabolism 39:919-922, by permission of Lippincott/Harper Company)

TABLE   539

COMPARISON OF MEANS OF CPR AND 17-OHCS VS. WEIGHT, SURFACE AREA AND
CREATININE EXCRETION IN ADOLESCENT BOYS
IN PITTSBURGH (PA)

	24-Hr cortisol production rate, mg per:			24-Hr urinary 17-OHCS, mg per:			
Matura-tion	m^2 body surface area	kg body weight	g creatinine	m^2 body surface area	kg body weight	g creatinine	Creatinine mg/kg body weight
Stage 1	10.2	.324	16.1	1.7	.0553	2.8	20
Stage 3	10.4	.337	15.1	1.6	.0519	2.3	22
Stage 5	11.3	.302	14.3	2.4	0.0629	3.0	21

CPR = cortisol production rate; OCHS = 17-hydroxycorticosteroids.

(data from Kenny, Gancayco, Heald, et al., 1966, Journal of Clinical Endocrinology 26:
1232-1236, by permission of Lippincott/Harper Company, Philadelphia, PA)

TABLE   540

MEAN CORTISOL PRODUCTION RATE AND
URINARY 17-OHCS IN ADOLESCENT BOYS
IN WASHINGTON, D.C.

Maturation	Age (years)	Stature (cm)	Weight (kg)	Surface area (m^2)	CPR mg/24 hr	Urinary 17-OHCS mg/24 hr	Urine creatinine g/day
Stage 1	$12 \frac{10}{12}$	146.6	39.8	1.27	12.9	2.2	0.800
Stage 3	$13 \frac{5}{12}$	156.6	42.4	1.37	14.3	2.2	0.944
Stage 5	$15 \frac{11}{12}$	172.3	66.8	1.78	20.2	4.2	1.411

CPR = cortisol production rate; 17-OCHS = 17.21-dihydroxy-20-ketosteroids;
maturity stages are 1. early sexual maturation; 3. period of rapid growth;
and 5 . end of adolescent growth spurt.

(data from Kenny, Gancayco, Heald, et al., 1966, Journal of Clinical Endocrinology 26:
1232-1236, by permission of Lippincott/Harper Company, Philadelphia, PA)

CARDIORESPIRATORY SYSTEM

TABLE 541

BASAL SYSTOLIC BLOOD PRESSURE (mm Hg)
FOR CHILDREN IN DENVER (CO)

Age (years)	Boys		Girls	
	Mean	S.D.	Mean	S.D.
3.50	91.80	5.97	89.54	4.80
4.00	89.13	4.32	91.56	6.49
4.50	89.38	6.42	--	--
5.00	96.12	9.79	94.60	4.73
5.50	94.93	5.47	93.36	5.38
6.00	93.96	7.32	98.81	7.39
6.50	98.95	6.52	97.45	8.18
7.00	92.71	5.62	95.76	9.15
7.50	--	--	98.83	8.29
8.00	--	--	99.28	7.34
8.50	94.34	4.91	96.91	7.57
9.00	--	--	99.20	9.67
9.50	--	--	98.42	10.34
10.00	98.67	7.40	98.19	11.64
10.50	99.67	8.42	100.26	9.95
11.00	100.80	8.32	100.33	8.82
11.50	100.53	8.31	100.54	6.57
12.00	103.30	7.14	99.99	7.33
12.50	103.25	9.88	101.55	7.39
13.00	103.53	5.95	101.95	6.52
13.50	105.96	9.31	103.25	6.73
14.00	108.60	9.63	105.43	5.60
14.50	106.59	7.98	104.96	6.18
15.00	107.21	6.19	103.52	6.91
15.50	108.11	6.44	108.66	7.19
16.00	109.09	8.08	102.83	8.01
16.50	113.99	4.95	106.49	6.97
17.00	114.14	6.30	106.17	8.38
17.50	110.84	7.98	107.98	6.90
18.00	115.02	10.08	105.34	6.63

(data from Roche, Eichorn, McCammon, et al., 1979)

TABLE   542

SUPINE SYSTOLIC BLOOD PRESSURE (mm Hg) FOR CHILDREN IN DENVER (CO)

Age (years)	Boys Mean	S.D.	Girls Mean	S.D.
2.00	85.96	13.79	85.44	17.94
2.50	88.06	8.64	88.44	10.01
3.00	89.46	8.24	90.05	7.65
3.50	91.24	7.83	92.32	8.67
4.00	94.40	7.45	91.39	5.91
4.50	93.44	6.74	93.11	6.68
5.00	93.67	7.67	93.72	6.22
5.50	94.81	7.09	93.75	7.73
6.00	94.68	6.65	94.96	7.62
6.50	96.61	7.58	95.19	8.69
7.00	95.80	6.40	96.20	7.78
7.50	98.25	7.79	97.19	6.92
8.00	97.63	8.28	96.88	7.61
8.50	97.50	8.95	99.21	8.35
9.00	98.33	8.02	97.73	7.84
9.50	99.33	7.23	99.05	8.70
10.00	99.57	7.64	98.76	7.66
10.50	100.31	8.55	98.89	9.44
11.00	100.46	7.72	100.55	8.41
11.50	101.27	6.93	101.70	9.10
12.00	101.66	9.37	102.61	7.12
12.50	103.45	8.22	102.49	9.02
13.00	104.20	10.96	104.66	7.92
13.50	105.46	7.46	103.85	6.28
14.00	107.85	10.61	105.56	8.29
14.50	108.56	7.43	106.07	8.39
15.00	112.20	8.20	106.57	8.79
15.50	113.69	11.28	102.91	9.06
16.00	115.60	12.46	107.33	8.54
16.50	115.97	10.18	--	--
17.00	116.76	10.77	109.36	7.67
17.50	116.92	7.72	--	--
18.00	117.80	9.25	108.55	7.58

(data from Roche, Eichorn, McCammon, et al., 1979)

TABLE   543

SEATED SYSTOLIC BLOOD PRESSURES (mm Hg; supine to 1.5 years)
FOR CHILDREN IN THE BERKELEY (CA) GROWTH STUDY

Age	Boys		Girls	
(years)	Mean	S.D.	Mean	S.D.
0.08	68.64	5.18	64.68	6.21
0.17	67.89	5.66	68.36	5.53
0.25	69.90	4.76	70.32	3.69
0.33	70.71	3.90	70.26	3.83
0.42	72.56	3.79	71.26	4.76
0.50	75.16	5.05	73.70	4.90
0.58	76.50	4.63	74.90	4.63
0.67	77.59	4.01	75.94	5.56
0.75	78.91	4.83	76.23	5.52
0.83	79.94	4.30	78.44	5.41
0.92	80.51	4.58	80.03	4.78
1.00	79.87	4.30	80.30	5.38
1.08	80.40	4.78	81.68	5.86
1.17	80.96	4.41	81.59	8.41
1.25	81.53	4.44	83.82	9.92
1.50	81.96	5.40	82.02	6.61
1.75	83.90	6.30	81.48	6.86
2.00	85.60	8.63	81.60	7.79
3.00	86.20	6.43	83.72	8.85
4.00	87.83	8.46	87.92	9.26
5.00	91.42	8.59	93.16	11.52
6.00	95.64	6.66	97.23	6.66
7.00	99.32	7.26	99.82	7.90
8.00	91.50	10.81	96.72	9.19
9.00	90.51	8.06	94.96	12.60
10.00	94.16	8.99	94.82	9.45
11.00	101.87	8.49	95.71	9.90
12.00	98.21	8.28	96.70	10.84
13.00	101.80	6.47	100.71	8.99
14.00	103.24	10.44	101.62	9.02
15.00	105.26	8.69	100.81	12.05
16.00	113.54	12.16	106.73	12.17
17.00	113.06	10.22	102.47	11.76
18.00	--	--	97.71	13.85

(data from Roche, Eichorn, McCammon, et al., 1979)

TABLE 544

SEATED SYSTOLIC BLOOD PRESSURE (mm Hg)
FOR CHILDREN IN BOSTON (MA)

Age (years)	BOYS		GIRLS	
	Mean	S.D.	Mean	S.D.
2.50	91.30	7.77	95.02	7.31
3.00	92.70	9.14	94.75	6.98
3.50	94.74	7.72	97.79	7.62
4.00	96.43	9.79	97.47	7.50
4.50	99.60	9.13	98.92	7.43
5.00	97.87	10.17	97.57	8.94
5.50	99.39	6.88	99.55	6.10
6.00	99.49	8.39	98.08	7.61
6.50	101.46	7.80	99.74	7.75
7.00	99.90	8.35	98.60	7.21
7.50	100.77	6.23	100.51	6.70
8.00	99.87	8.68	100.14	7.18
8.50	101.35	6.04	99.93	7.60
9.00	99.21	10.47	99.59	7.60
9.50	102.66	7.92	102.98	6.12
10.00	100.17	8.96	99.30	8.76
11.00	100.82	9.00	101.51	8.67
12.00	102.45	10.87	104.05	9.39
13.00	103.05	10.82	105.86	9.81
14.00	105.95	9.98	105.51	9.06
15.00	109.05	9.66	106.62	7.77
16.00	111.11	9.47	105.00	8.03
17.00	114.20	10.31	104.53	9.03
18.00	114.66	9.22	106.11	9.24

(data from Roche, Eichorn, McCammon, et al., 1979)

TABLE 545

SEATED SYSTOLIC BLOOD PRESSURE (mm Hg)
FOR CHILDREN IN DENVER, CO

Age (years)	Boys Mean	S.D.	Girls Mean	S.D.
2.50	88.39	8.31	85.91	8.81
3.00	92.28	9.09	88.48	8.52
3.50	90.09	9.55	92.22	8.11
4.00	92.95	7.90	92.37	7.97
4.50	93.82	8.77	94.76	8.28
5.00	94.95	7.97	94.75	8.97
5.50	94.75	7.89	97.27	8.17
6.00	96.81	8.40	97.22	8.06
6.50	99.87	7.86	96.74	8.86
7.00	96.42	8.47	96.63	7.62
7.50	100.44	9.62	98.30	7.00
8.00	100.71	8.54	98.72	7.84
8.50	100.44	7.98	101.08	8.56
9.00	100.88	8.74	101.07	7.22
9.50	101.05	7.20	101.62	8.35
10.00	101.07	8.32	100.37	7.17
10.50	102.53	9.02	102.72	8.29
11.00	101.88	8.45	102.62	8.71
11.50	104.04	8.44	102.15	8.93
12.00	105.46	9.65	104.21	7.74
12.50	107.37	9.04	103.75	7.69
13.00	109.41	11.04	106.80	9.08
13.50	110.61	8.95	105.41	8.35
14.00	110.30	9.20	108.98	9.35
14.50	111.61	9.26	108.03	5.80
15.00	114.25	10.80	108.92	9.47
15.50	115.47	9.56	104.25	10.13
16.00	119.44	8.72	109.65	8.22
16.50	118.91	10.25	--	--
17.00	119.90	10.51	107.26	6.84
17.50	121.29	7.69	--	--
18.00	121.94	9.51	111.05	11.66

(data from Roche, Eichorn, McCammon, et al., 1979)

TABLE 546

SEATED SYSTOLIC BLOOD PRESSURE (mm Hg; physical examinations)
IN CHILDREN IN THE FELS STUDY (Yellow Springs, OH)

| Age | Boys | | Girls | |
(year)	Mean	S.D.	Mean	S.D.
2.00	82.90	14.35	87.54	15.46
2.50	88.32	15.31	91.00	15.65
3.00	87.78	16.86	90.50	14.80
3.50	89.75	17.85	92.14	16.72
4.00	89.91	14.43	95.98	12.50
4.50	91.63	15.04	96.05	13.61
5.00	94.55	15.26	94.98	13.70
5.50	94.37	14.65	96.64	13.88
6.00	92.92	15.61	95.03	14.93
6.50	92.52	14.93	96.67	17.62
7.00	89.24	11.15	92.64	11.39
7.50	91.89	9.99	91.61	12.03
8.00	91.97	9.88	91.62	10.22
8.50	91.07	10.75	90.32	9.30
9.00	90.69	7.51	91.67	9.22
9.50	91.50	7.87	92.41	7.77
10.00	92.78	8.00	90.70	7.96
10.50	92.14	7.66	92.94	8.67
11.00	93.25	7.23	93.15	8.77
11.50	93.99	7.68	94.90	9.63
12.00	94.63	8.26	97.46	10.20
12.50	94.58	8.60	97.16	8.60
13.00	97.12	9.32	98.82	9.12
13.50	98.43	9.90	100.40	10.17
14.00	101.63	10.58	100.57	11.02
14.50	102.09	10.31	99.94	8.66
15.00	104.50	9.72	99.99	9.69
15.50	106.01	9.73	99.67	9.73
16.00	106.18	10.82	100.37	9.61
16.50	107.95	9.92	99.58	8.65
17.00	108.04	10.58	101.79	8.44
17.50	111.18	10.41	99.46	8.05
18.00	111.19	9.49	99.85	8.76

(data from Roche, Eichorn, McCammon, et al., 1979)

TABLE 547

SEATED SYSTOLIC BLOOD PRESSURE (mm Hg)
FOR CHILDREN IN BERKELEY (CA) ENROLLED IN THE GUIDANCE STUDY

Age (years)	Boys		Girls	
	Mean	S.D.	Mean	S.D.
1.75	83.93	6.62	83.32	5.55
3.00	90.18	7.99	89.61	8.41
4.00	91.41	7.67	92.80	7.50
5.00	94.88	8.38	94.62	8.97
6.00	98.92	8.19	98.21	8.56
7.00	100.63	7.28	101.61	9.56
8.00	98.21	11.16	97.50	9.62
8.50	95.85	10.56	--	--
9.00	91.03	9.06	91.85	9.29
10.00	95.54	8.62	97.39	9.74
11.00	101.71	9.90	98.78	9.35
12.00	104.16	8.88	99.32	9.74
13.00	108.73	9.50	105.81	9.64
14.00	109.46	10.30	104.76	10.62
15.00	110.08	9.99	102.06	9.23
16.00	114.30	9.90	103.54	10.26
17.00	117.03	10.78	104.43	9.62
18.00	117.00	16.20	102.10	7.52

(data from Roche, Eichorn, McCammon, et al., 1979)

TABLE 548

SEATED SYSTOLIC BLOOD PRESSURE (mm Hg)
FOR CHILDREN IN OAKLAND, CA

Age (years)	BOYS		GIRLS	
	Mean	S.D.	Mean	S.D.
11.00	101.62	7.65	103.66	8.62
11.50	102.34	7.60	101.86	6.44
12.00	103.25	7.19	102.37	9.06
12.50	104.68	7.07	101.17	7.57
13.00	106.27	7.42	101.79	7.81
13.50	107.38	7.90	102.56	5.67
14.00	108.34	7.21	101.94	5.53
14.50	109.29	7.51	101.91	6.34
15.00	110.73	7.92	102.39	6.02
15.50	111.77	7.27	102.65	6.81
16.00	112.52	7.37	104.84	8.11
16.50	113.04	7.24	107.64	9.25
17.00	113.64	7.10	106.90	9.39
17.50	115.25	9.65	105.99	10.64

(data from Roche, Eichorn, McCammon, et al., 1979)

TABLE 549

SYSTOLIC BLOOD PRESSURE (mm Hg) IN LOS ANGELES CHILDREN

Age	Wrist		Ankle	
	Mean	S.D.	Mean	S.D.
1-7 days	41	8	37	7
1-3 months	67	11	61	10.5
4-6 months	73	9.5	68	10
7-9 months	76	9	74	8.5
10-12 months	76	14	79	8.5
1-12 months	72	10.5	68	12

(data from Moss and Adams, Problems of Blood Pressure in Childhood, 1962. Courtesy of Charles C Thomas, Publisher, Springfield, Illinois)

TABLE 550

MEDIAN SYSTOLIC BLOOD PRESSURE (mm Hg) IN THE
UPPER EXTREMITY IN LOS ANGELES CHILDREN

Age (months)	Boys	Girls
1	61.6	61.7
2	69.0	65.0
3	68.8	69.0
4	72.6	74.0
5	73.0	73.0
6	70.2	71.5
7	77.5	73.0
8	79.0	77.0
9	80.0	76.0
10	74.0	79.0
11	73.0	73.0
12	76.0	77.0

(data from Moss and Adams, Problems of Blood Pressure in Childhood, 1962. Courtesy of Charles C Thomas, Publisher, Springfield, Illinois)

TABLE 551

SYSTOLIC BLOOD PRESSURE (mm Hg)
IN ST. LOUIS (MO) CHILDREN

| Age | Mean | | S.D. | | 90% Range | |
(years)	Boys	Girls	Boys	Girls	Boys	Girls
3	99	99	10.1	11.0	83-114	83-120
4	98	98	8.5	9.7	85-114	83-114
5	101	102	9.6	10.1	83-115	84-119
6	105	105	10.5	11.0	91-124	91-125
7	106	107	9.8	10.9	91-122	92-125
8	108	108	10.6	10.3	91-125	94-127
9	111	112	11.3	9.9	94-130	94-129
10	114	114	10.5	11.3	95-133	91-134
11	114	121	11.1	12.2	98-134	103-143
12	116	117	10.3	11.9	100-133	101-136
13	120	121	11.6	12.1	97·144	101-141
14	120	119	10.3	12.0	105-135	93-138
15	125	115	9.8	11.0	112-142	103-140

(data from Londe, 1968, Clinical Pediatrics 7:400-403, by permission of Lippincott/Harper Company, Philadelphia, PA)

TABLE 552

STANDING SYSTOLIC BLOOD PRESSURE (mm Hg)
FOR CHILDREN IN OAKLAND (CA)

Age (years)	BOYS		GIRLS	
	Mean	S. D.	Mean	S.D.
14.00	107.85	6.24	--	--
15.00	112.01	16.27	108.86	9.58
15.50	109.45	7.84	107.23	9.84
16.00	113.68	9.55	108.29	10.67
16.50	109.61	10.18	113.03	10.34
17.00	113.83	13.02	111.12	13.54
17.50	112.93	12.27	106.69	12.35

(data from Roche, Eichorn, McCammon, et al., 1979)

TABLE 553

SYSTOLIC BLOOD PRESSURE (mm Hg) AFTER RECOVERY
FROM EXERCISE FOR CHILDREN IN DENVER (CO)

Age (years)	Boys		Girls	
	Mean	S.D.	Mean	S.D.
7.50	102.70	9.33	--	--
8.00	103.64	8.92	96.64	5.30
8.50	100.93	9.85	100.61	8.76
9.00	101.40	8.90	98.15	9.23
9.50	102.74	9.06	100.70	8.04
10.00	104.43	9.34	100.79	10.90
10.50	100.31	7.86	100.27	8.14
11.00	104.39	8.55	104.43	10.33
11.50	104.22	10.16	101.05	10.58
12.00	107.95	16.33	103.49	10.60
12.50	104.40	8.79	106.74	9.95
13.00	110.26	12.47	110.31	8.55
13.50	109.96	10.07	107.28	9.72
14.00	108.96	10.64	112.88	12.31
14.50	113.19	13.05	108.66	13.06
15.00	113.04	11.82	119.30	18.88
15.50	118.08	13.54	--	--
16.00	115.86	11.05	119.20	12.79
16.50	122.45	14.25	--	--
17.00	120.34	10.90	114.26	7.10
17.50	118.94	14.17	--	--
18.00	121.75	12.41	114.94	7.17

(data from Roche, Eichorn, McCammon, et al., 1979)

TABLE 554

SYSTOLIC BLOOD PRESSURE (mm Hg) AFTER RECOVERY
FROM EXERCISE FOR CHILDREN IN OAKLAND (CA)

| Age | BOYS | | GIRLS | |
(years)	Mean	S.D.	Mean	S.D.
13.50	109.54	7.79	108.35	5.95
14.00	111.11	6.25	107.64	5.17
14.50	113.26	8.22	108.05	7.92
15.00	113.26	7.69	108.74	7.80
15.50	114.42	6.34	105.01	6.97
16.00	115.42	6.98	109.40	11.13
16.50	115.18	7.47	108.12	7.67
17.00	118.26	6.04	109.45	7.81
17.50	120.60	8.18	111.59	7.68

(data from Roche, Eichorn, McCammon, et al., 1979)

TABLE 555

CALF SYSTOLIC BLOOD PRESSURE, BLOOD FLOW, AND PERIPHERAL
RESISTANCE IN 15 TERM INFANTS (MEAN $\pm$ 1 S.D.)
IN TORONTO (ONT)

| Variables | Hours after Birth | | | | | |
	12-24	25-48	49-72	73-96	97-120	121-144
	Full-Term					
Blood Pressure (mm Hg)	62 ± 6	67 ± 6	71 ± 6	76 ± 1	76 ± 5	82 ± 10
Blood Flow (ml/100 ml Tissue/min)	9.2 ±3.6	9.7 ±1.7	8.2 ±1.8	9.4 ±3.2	10.3 ±2.8	9.5 ± 0.9
Resistance (Units)	6.7	6.9	8.7	8.1	7.4	8.6

(data from Kidd, Levison, Gemmel, et al., 1966, American Journal of Diseases of Children
112:402-407. Copyright 1966, American Medical Association)

TABLE 556

SEATED DIASTOLIC BLOOD PRESSURE (mm Hg; fourth phase; supine to 1.5 years)
FOR CHILDREN IN THE BERKELEY (CA) GROWTH STUDY

Age (years)	Boys		Girls	
	Mean	S.D.	Mean	S.D.
0.08	37.47	4.40	34.18	5.65
0.17	38.07	4.03	37.42	4.76
0.25	39.95	3.39	39.63	3.27
0.33	40.02	3.25	39.76	3.14
0.42	42.03	2.62	41.11	2.12
0.50	42.76	4.09	42.19	2.48
0.58	43.71	2.84	42.86	2.41
0.67	44.96	2.19	43.35	2.65
0.75	45.88	2.95	43.74	2.92
0.83	46.91	2.44	45.05	2.08
0.92	47.65	2.48	46.20	2.25
1.00	48.39	2.75	47.20	2.57
1.08	48.26	2.28	48.82	3.52
1.17	48.91	1.99	49.24	3.68
1.25	49.06	2.05	49.69	4.82
1.50	49.33	3.55	49.17	3.86
1.75	48.57	4.83	49.01	4.71
2.00	50.83	6.63	48.77	7.33
3.00	51.69	8.29	53.11	9.97
4.00	54.94	7.96	55.25	8.52
5.00	60.90	8.28	62.22	12.71
6.00	67.98	5.76	68.77	6.27
7.00	70.56	7.52	70.02	8.23
8.00	61.80	9.76	62.35	8.90
9.00	57.37	6.43	59.55	7.88
10.00	61.64	7.87	62.74	6.52
11.00	65.30	5.53	63.12	6.12
12.00	60.75	6.13	65.71	7.51
13.00	61.27	5.55	65.09	6.11
14.00	60.05	8.85	57.72	6.58
15.00	69.42	8.68	59.14	8.36
16.00	69.85	8.05	68.11	11.19
17.00	67.50	7.65	69.92	6.64
18.00	--	--	66.22	12.44

(data from Roche, Eichorn, McCammon, et al., 1979)

TABLE 557

BASAL DIASTOLIC (fourth phase) BLOOD PRESSURE (mm Hg)
FOR CHILDREN IN DENVER (CO)

Age	BOYS		GIRLS	
(years)	Mean	S.D.	Mean	S.D.
3.50	63.25	6.34	60.71	7.59
4.00	61.50	8.34	64.72	9.26
4.50	64.10	6.35	--	--
5.00	65.83	9.82	67.44	5.68
5.50	--	--	64.32	5.87
6.00	66.42	8.36	67.55	5.69
6.50	64.77	7.58	65.48	7.36
7.00	60.97	8.77	67.29	10.01
7.50	--	--	66.35	4.91
8.00	--	--	68.59	4.94
8.50	63.45	6.04	66.05	4.02
9.00	--	--	65.60	7.53
9.50	--	--	65.87	9.32
10.00	67.17	6.68	64.05	8.74
10.50	68.40	10.37	69.45	7.65
11.00	68.40	7.96	68.54	8.04
11.50	67.78	12.59	70.14	6.23
12.00	66.76	8.99	68.07	7.74
12.50	70.48	7.77	69.35	6.43
13.00	70.41	6.86	70.58	6.43
13.50	70.41	8.16	71.20	7.82
14.00	69.73	6.75	72.63	7.52
14.50	70.60	4.43	71.72	8.06
15.00	70.43	6.49	70.87	8.28
15.50	72.13	7.65	74.40	8.84
16.00	72.23	7.64	71.34	10.91
16.50	72.75	8.25	71.04	9.22
17.00	73.28	8.10	71.41	8.67
17.50	73.01	7.05	73.36	7.02
18.00	75.82	10.10	71.98	7.14

(data from Roche, Eichorn, McCammon, et al., 1979)

TABLE 558

BASAL DIASTOLIC PRESSURE (mm Hg; fifth phase)
FOR CHILDREN IN THE DENVER STUDY

| Age | Boys | | Girls | |
(years)	Mean	S.D.	Mean	S.D.
3.50	53.45	11.32	52.59	9.01
4.00	54.13	8.17	57.49	7.14
5.00	55.73	10.89	56.58	8.38
5.50	50.55	13.12	54.60	8.13
6.00	61.98	7.51	57.17	7.46
6.50	52.74	10.78	56.29	7.21
7.00	--	--	58.41	8.94
7.50	--	--	58.36	6.86
8.00	59.60	8.79	57.99	8.21
8.50	52.05	8.41	57.29	7.50
9.00	--	--	59.06	7.64
9.50	--	--	56.56	10.49
10.00	--	--	56.02	9.70
10.50	--	--	60.64	8.57
11.00	--	--	57.71	10.02
11.50	59.65	14.25	62.04	8.19
12.00	59.43	11.53	60.96	8.78
12.50	60.71	7.13	59.38	9.44
13.00	60.52	9.66	62.68	9.98
13.50	64.02	7.51	63.65	8.02
14.00	60.78	7.80	65.06	7.44
14.50	59.32	9.78	64.33	8.92
15.00	60.02	9.89	62.11	10.70
15.50	62.31	9.85	66.78	10.30
16.00	63.48	9.38	65.49	9.90
16.50	63.41	11.32	60.61	8.34
17.00	62.06	11.79	64.11	11.93
17.50	62.81	7.50	67.68	5.65
18.00	64.08	11.00	--	--

(data from Roche, Eichorn, McCammon, et al., 1979)

TABLE 559

SUPINE DIASTOLIC BLOOD PRESSURE (mm Hg; fourth phase)
FOR CHILDREN IN DENVER (CO)

Age (years)	BOYS		GIRLS	
	Mean	S.D.	Mean	S.D.
3.00	--	--	46.43	19.13
3.50	--	--	51.71	11.25
4.00	--	--	47.30	10.58
4.50	--	--	49.87	7.62
5.00	--	--	50.91	6.72
5.50	--	--	54.37	7.54
6.00	--	--	54.99	8.87
6.50	--	--	55.84	7.29
7.00	--	--	54.44	6.59
7.50	--	--	56.18	9.62
8.00	--	--	56.49	10.32
8.50	--	--	59.76	7.83
9.00	--	--	54.85	11.04
9.50	--	--	54.23	11.40
10.00	--	--	55.78	10.02
10.50	55.67	7.58	56.65	8.28
11.00	--	--	58.87	7.96
11.50	--	--	55.80	12.64
12.00	--	--	56.74	9.68
12.50	58.86	7.03	58.05	11.89
13.00	55.80	10.74	57.59	10.26
13.50	--	--	59.26	8.33
14.00	60.64	11.31	60.80	9.34
14.50	56.19	9.99	56.22	7.68
15.00	56.41	12.45	57.30	9.29
15.50	58.34	16.66	--	--
16.00	65.82	11.01	61.05	7.74
17.00	--	--	62.64	9.42
18.00	--	--	65.64	5.78

(data from Roche, Eichorn, McCammon, et al., 1979)

TABLE 560

SEATED DIASTOLIC BLOOD PRESSURE (mm Hg; fifth phase)
FOR CHILDREN IN BOSTON (MA)

Age (years)	BOYS		GIRLS	
	Mean	S.D.	Mean	S.D.
2.50	62.49	2.83	63.82	7.28
3.00	61.59	6.81	62.13	7.13
3.50	62.80	5.08	64.13	6.48
4.00	63.87	7.42	66.16	6.23
4.50	65.43	6.21	66.24	6.56
5.00	66.91	9.83	65.21	6.55
5.50	67.09	7.84	67.21	5.89
6.00	66.30	6.96	65.59	7.02
6.50	67.88	6.84	67.59	5.63
7.00	68.88	7.50	66.38	6.81
7.50	68.04	6.72	69.03	6.16
8.00	67.90	7.81	68.43	7.59
8.50	70.18	6.69	68.43	6.31
9.00	68.13	8.22	67.97	8.41
9.50	71.72	8.95	71.37	5.74
10.00	69.54	7.49	66.47	8.35
11.00	69.79	9.41	66.65	8.51
12.00	68.76	8.10	67.07	8.47
13.00	68.24	9.17	67.38	9.77
14.00	70.34	9.58	64.84	7.05
15.00	73.01	6.74	68.12	7.35
16.00	74.62	7.56	67.74	6.79
17.00	75.61	7.66	66.76	7.29
18.00	74.76	8.34	68.79	7.05

(data from Roche, Eichorn, McCammon, et al., 1979)

TABLE 561

SEATED DIASTOLIC BLOOD PRESSURE (mm Hg; fourth phase)
FOR CHILDREN IN DENVER (CO)

Age (years)	Boys		Girls	
	Mean	S.D.	Mean	S.D.
2.50	53.77	11.80	58.08	13.72
3.00	61.59	11.44	57.59	11.82
3.50	55.97	10.00	58.07	9.18
4.00	60.75	11.88	59.92	7.95
4.50	58.30	10.89	61.00	10.76
5.00	60.45	9.35	64.55	9.24
5.50	59.73	7.74	64.93	8.36
6.00	61.86	7.96	64.37	7.84
6.50	63.35	12.47	65.90	8.70
7.00	64.80	8.22	66.32	7.25
7.50	67.65	8.78	67.75	7.33
8.00	68.74	6.29	67.66	7.26
8.50	68.00	7.77	69.70	6.64
9.00	67.40	7.40	69.19	6.74
9.50	67.29	7.20	68.34	6.92
10.00	69.64	7.83	70.19	8.57
10.50	69.74	9.50	70.09	6.99
11.00	69.66	7.66	69.81	7.28
11.50	71.71	8.38	69.92	6.47
12.00	71.02	7.28	71.68	7.06
12.50	74.18	6.85	70.46	7.89
13.00	71.70	8.97	72.44	6.74
13.50	73.19	9.07	73.21	7.24
14.00	74.67	7.34	75.39	8.91
14.50	72.79	7.25	75.23	6.03
15.00	74.08	6.92	73.02	8.19
15.50	76.13	7.60	73.87	8.23
16.00	76.37	7.69	73.66	5.90
16.50	75.16	9.26	--	--
17.00	77.60	7.59	72.95	7.03
17.50	82.46	26.09	--	--
18.00	79.49	9.33	74.55	10.04

(data from Roche, Eichorn, McCammon, et al., 1979)

TABLE   562

SEATED DIASTOLIC BLOOD PRESSURE (mm Hg; fifth phase)
FOR CHILDREN IN THE FELS STUDY (Yellow Springs, OH)

Age (years)	Boys		Girls	
	Mean	S.D.	Mean	S.D.
2.00	54.10	13.60	55.69	24.10
2.50	56.26	10.63	53.57	13.56
3.00	59.67	16.41	56.21	12.81
3.50	57.53	12.47	61.59	17.96
4.00	58.22	11.98	61.69	16.13
4.50	58.43	13.76	64.68	14.00
5.00	60.56	14.77	62.49	11.85
5.50	59.56	13.94	64.17	12.59
6.00	58.92	14.64	63.10	13.01
6.50	60.60	13.46	64.13	17.53
7.00	58.59	12.42	62.29	13.75
7.50	59.41	10.52	60.18	11.91
8.00	60.87	10.97	59.16	12.44
8.50	58.62	10.41	59.03	9.60
9.00	58.54	8.15	59.62	9.56
9.50	58.54	8.10	57.37	7.95
10.00	57.94	7.10	57.51	7.03
10.50	58.21	6.18	58.55	7.37
11.00	60.19	7.37	58.55	7.72
11.50	59.74	8.27	59.30	6.92
12.00	59.60	8.24	61.27	8.59
12.50	59.65	8.45	61.67	7.30
13.00	60.76	8.13	61.47	8.39
13.50	61.51	7.82	63.33	8.32
14.00	62.13	8.77	64.10	8.57
14.50	63.55	7.50	64.79	6.94
15.00	64.45	8.78	63.81	8.37
15.50	64.93	8.87	63.81	7.07
16.00	66.12	8.79	63.23	7.15
16.50	66.73	9.06	63.99	6.20
17.00	66.45	9.20	64.73	6.45
17.50	67.82	7.41	63.73	6.53
18.00	68.06	7.44	64.70	7.41

(data from Roche, Eichorn, McCammon, et al., 1979)

TABLE 563

SEATED DIASTOLIC BLOOD PRESSURE (mm Hg; fourth phase)
FOR CHILDREN IN BERKELEY (CA) ENROLLED IN THE GUIDANCE STUDY

Age (years)	BOYS		GIRLS	
	Mean	S.D.	Mean	S.D.
1.75	47.82	8.91	46.79	6.12
3.00	50.41	6.11	50.89	6.65
4.00	56.54	8.08	56.71	8.32
5.00	61.54	7.77	62.94	9.87
6.00	68.51	8.17	69.32	8.63
7.00	71.52	6.75	73.14	7.91
8.00	68.46	11.57	68.23	11.21
8.50	64.63	7.95	69.54	5.40
9.00	59.37	8.92	60.02	9.14
10.00	60.90	7.66	62.63	7.75
11.00	66.40	8.08	64.22	7.53
12.00	63.54	8.31	65.35	8.05
13.00	60.95	6.50	67.45	7.61
14.00	62.25	7.26	62.27	7.57
15.00	67.94	8.24	59.78	7.96
16.00	69.80	8.31	64.69	8.24
17.00	71.24	9.11	69.68	6.73
18.00	69.71	6.88	69.60	7.28

(data from Roche, Eichorn, McCammon, et al., 1979)

TABLE   564

SEATED DIASTOLIC BLOOD PRESSURE (mm Hg; fifth phase)
FOR CHILDREN IN OAKLAND, CA

Age (years)	Boys		Girls	
	Mean	S.D.	Mean	S.D.
11.00	67.71	7.86	60.32	9.57
11.50	69.00	8.10	59.28	8.78
12.00	68.04	6.20	58.61	9.06
12.50	67.95	5.93	56.98	7.71
13.00	66.86	5.58	56.82	5.91
13.50	65.71	6.50	55.82	5.38
14.00	65.79	6.30	55.61	4.80
14.50	66.70	6.06	56.80	5.14
15.00	67.76	6.51	57.97	6.48
15.50	68.55	4.99	59.97	6.46
16.00	68.56	6.39	64.38	7.87
16.50	68.68	6.29	68.09	7.94
17.00	68.95	5.69	70.43	7.39
17.50	69.85	6.21	71.23	7.25

(data from Roche, Eichorn, McCammon, et al., 1979)

TABLE 565

STANDING DIASTOLIC BLOOD PRESSURE
(mm Hg; fifth phase) FOR CHILDREN IN OAKLAND (CA)

| Age | Boys | | Girls | |
(years)	Mean	S.D.	Mean	S.D.
14.00	73.79	6.51	--	--
15.00	76.57	8.24	78.40	9.27
15.50	77.16	8.16	76.89	5.21
16.00	81.95	6.57	81.80	9.46
16.50	82.75	8.21	81.84	8.27
17.00	87.43	5.92	83.75	6.07
17.50	83.71	7.45	80.66	7.26

(data from Roche, Eichorn, McCammon, et al., 1979)

TABLE   566

DIASTOLIC (fifth phase) BLOOD PRESSURES (mm Hg) IN ST. LOUIS (MO) CHILDREN

Age (years)	Mean		S.D.		90% Range	
	Boys	Girls	Boys	Girls	Boys	Girls
3	57	58	14.2	10.9	37-80	44-77
4	57	60	9.6	10.2	40-71	41-76
5	60	60	10.2	9.2	43-75	43-75
6	60	64	10.1	9.2	42-79	44-78
7	63	63	9.4	9.5	45-77	46-79
8	61	65	11.5	8.0	43-76	53-79
9	65	67	9.7	9.5	51-79	52-85
10	66	64	8.6	8.9	51-84	51-79
11	65	69	9.3	8.2	51-81	59-90
12	67	65	7.1	8.0	54-81	52-80
13	65	69	8.4	9.8	57-77	53-88
14	68	67	8.1	9.8	53-80	51-86
15	67	67	8.0	8.3	54-82	53-82

(data from Londe, 1968, Clinical Pediatrics 7:400-403, by permission of Lippincott/Harper Company, Philadelphia, PA)

TABLE 567

DIASTOLIC BLOOD PRESSURE (mm Hg; fourth phase)
AT RECOVERY AFTER EXERCISE FOR CHILDREN IN DENVER (CO)

| Age | BOYS | | GIRLS | |
(years)	Mean	S.D.	Mean	S.D.
7.50	60.47	11.01	--	--
8.00	63.14	8.11	59.72	11.25
8.50	61.42	6.35	65.75	9.56
9.00	61.47	8.18	59.62	10.21
9.50	66.81	8.10	64.24	6.85
10.00	65.87	9.00	64.29	9.39
10.50	65.82	9.07	63.05	7.80
11.00	65.61	10.14	61.69	11.47
11.50	66.09	9.38	65.12	10.87
12.00	65.84	8.39	63.66	11.40
12.50	68.09	10.93	64.46	10.76
13.00	70.14	8.54	63.97	14.66
13.50	69.58	10.61	65.95	12.05
14.00	70.59	7.10	68.87	11.26
14.50	71.35	12.57	63.79	11.71
15.00	69.70	9.92	64.12	12.37
15.50	70.39	11.20	--	--
16.00	71.60	10.33	64.00	12.99
16.50	72.57	9.17	--	--
17.00	71.95	9.37	62.56	10.26
17.50	73.15	11.41	--	--
18.00	72.66	9.13	65.66	9.32

(data from Roche, Eichorn, McCammon, et al., 1979)

TABLE   568

DIASTOLIC BLOOD PRESSURE (mm Hg; fifth phase) AT RECOVERY AFTER
EXERCISE FOR CHILDREN IN OAKLAND (CA)

Age	BOYS		GIRLS	
(years)	Mean	S.D.	Mean	S.D.
13.50	77.90	9.73	79.59	8.01
14.00	79.50	9.79	78.88	6.25
14.50	74.99	8.04	72.54	5.94
15.00	70.04	5.13	71.82	6.74
15.50	71.56	5.09	69.53	4.57
16.00	71.34	4.48	71.99	5.20
16.50	71.67	4.85	71.97	4.28
17.00	72.73	4.88	72.83	4.49
17.50	75.28	6.73	72.93	4.95

(data from Roche, Eichorn, McCammon, et al., 1979)

TABLE 569

DIFFERENCE BETWEEN BASAL SYSTOLIC AND DIASTOLIC BLOOD PRESSURES
(mm Hg; fifth phase diastolic) FOR CHILDREN IN DENVER (CO)

Age	Boys		Girls	
(years)	Mean	S.D.	Mean	S.D.
3.50	38.34	10.55	36.38	9.29
4.00	35.00	7.51	34.32	7.87
5.00	40.39	11.13	38.03	8.13
5.50	44.38	14.36	38.82	11.01
6.00	33.37	9.05	42.40	11.09
6.50	45.75	11.71	41.57	8.22
7.00	--	--	37.12	10.00
7.50	--	--	40.96	11.14
8.00	--	--	41.54	11.37
8.50	42.11	10.95	40.07	7.64
9.00	--	--	43.19	9.78
9.50	--	--	42.76	11.29
10.00	--	--	43.05	10.27
10.50	--	--	40.45	10.34
11.00	--	--	43.15	12.11
11.50	41.94	11.10	38.91	10.51
12.00	42.99	11.93	38.84	10.90
12.50	44.70	11.52	43.00	9.27
13.00	42.08	10.01	39.78	10.23
13.50	43.33	6.42	39.48	8.58
14.00	47.80	10.00	40.67	8.48
14.50	47.68	11.04	40.98	9.37
15.00	47.59	10.44	43.08	9.91
15.50	44.71	10.26	41.89	9.28

TABLE 569 (continued)

DIFFERENCE BETWEEN BASAL SYSTOLIC AND DIASTOLIC BLOOD PRESSURES
(mm Hg; fifth phase diastolic) FOR CHILDREN IN DENVER (CO)

Age	Boys		Girls	
(years)	Mean	S.D.	Mean	S.D.
16.00	45.71	11.05	38.91	11.39
16.50	50.59	11.99	44.57	8.69
17.00	49.04	6.86	41.72	9.75
17.50	48.03	8.00	40.87	7.25
18.00	50.50	9.50	--	--

(data from Roche, Eichorn, McCammon, et al., 1979)

TABLE 570

DIFFERENCE BETWEEN BASAL SYSTOLIC AND DIASTOLIC BLOOD PRESSURES
(mm Hg; fourth phase diastolic) FOR CHILDREN IN DENVER, CO

Age	Boys		Girls	
(years)	Mean	S.D.	Mean	S.D.
3.50	28.97	6.70	28.83	8.41
4.00	27.74	6.42	27.25	7.18
4.50	25.28	5.44	32.64	5.53
5.00	29.82	8.53	27.16	6.22
5.50	--	--	29.19	7.24
6.00	27.31	7.20	31.26	7.18
6.50	33.07	6.50	31.97	7.47
7.00	31.74	5.54	28.48	7.41
7.50	--	--	32.48	8.55
8.00	--	--	31.16	7.91

TABLE 570   (continued)

DIFFERENCE BETWEEN BASAL SYSTOLIC AND DIASTOLIC BLOOD PRESSURES
(mm Hg; fourth phase diastolic)FOR CHILDREN IN DENVER, CO

Age	Boys		Girls	
(years)	Mean	S.D.	Mean	S.D.
8.50	31.01	7.25	31.31	6.72
9.00	--	--	33.60	8.20
9.50	--	--	32.55	10.37
10.00	31.49	7.87	34.14	9.84
10.50	31.27	9.05	30.81	8.02
11.00	32.75	5.18	32.36	10.00
11.50	32.69	12.64	30.40	6.58
12.00	36.58	10.82	31.92	8.79
12.50	33.24	8.51	32.37	6.19
13.00	33.12	8.99	31.38	7.59
13.50	35.14	5.83	32.05	8.23
14.00	38.82	7.43	32.80	7.52
14.50	36.36	8.19	33.23	7.82
15.00	37.16	7.47	33.23	6.91
15.50	35.98	7.58	34.26	7.72
16.00	36.86	7.16	32.28	10.48
16.50	41.25	8.89	36.04	8.35
17.00	40.86	8.19	34.76	8.09
17.50	37.82	4.45	34.62	7.19
18.00	39.20	5.38	33.36	6.17

(data from Roche, Eichorn, McCammon, et al., 1979)

TABLE 571

DIFFERENCE BETWEEN SEATED SYSTOLIC AND DIASTOLIC BLOOD PRESSURES
(mm Hg; supine to 1.5 years; fourth phase diastolic)
FOR CHILDREN IN THE BERKELEY (CA) GROWTH STUDY

Age	Boys		Girls	
(years)	Mean	S.D.	Mean	S.D.
0.08	31.17	3.80	30.50	3.44
0.17	29.82	4.52	30.69	3.26
0.25	29.94	4.29	30.67	2.57
0.33	30.69	2.37	30.39	3.42
0.42	30.53	3.29	30.15	4.76
0.50	32.40	3.16	31.51	4.11
0.58	32.79	2.78	32.03	3.16
0.67	32.63	2.77	32.60	3.62
0.75	33.03	2.88	32.49	4.31
0.83	33.36	2.77	33.39	4.57
0.92	32.99	4.02	33.83	3.88
1.00	31.48	3.73	33.10	3.88
1.08	32.14	3.11	32.86	3.53
1.17	31.94	2.85	32.35	5.49
1.25	32.47	3.56	34.13	5.86
1.50	33.08	3.82	32.39	4.29
1.75	35.33	3.71	31.98	3.96
2.00	34.77	4.52	32.83	5.27
3.00	34.51	7.11	30.76	6.22
4.00	32.90	7.00	32.67	5.47
5.00	30.53	6.25	31.33	7.34
6.00	27.66	4.77	28.46	5.35
7.00	28.76	5.19	29.80	5.40
8.00	29.70	5.97	33.80	6.97

TABLE 571 (continued)

DIFFERENCE BETWEEN SEATED SYSTOLIC AND DIASTOLIC BLOOD PRESSURES
(mm Hg; supine to 1.5 years; fourth phase diastolic)
FOR CHILDREN IN THE BERKELEY (CA) GROWTH STUDY

Age	Boys		Girls	
(years)	Mean	S.D.	Mean	S.D.
9.00	33.14	8.31	35.41	9.27
10.00	32.52	7.40	32.08	6.95
11.00	36.57	9.03	32.59	7.93
12.00	37.46	6.87	30.99	8.81
13.00	40.53	6.16	35.62	8.46
14.00	43.19	10.54	43.89	8.16
15.00	35.84	7.64	41.68	10.51
16.00	43.69	11.16	38.63	14.84
17.00	45.56	9.10	32.56	9.17
18.00	--	--	31.49	7.76

(data from Roche, Eichorn, McCammon, et al., 1979)

TABLE 572

DIFFERENCE BETWEEN SUPINE SYSTOLIC AND DIASTOLIC (fifth phase)
BLOOD PRESSURES (mm Hg) FOR CHILDREN IN DENVER (CO)

Age	Boys		Girls	
(years)	Mean	S.D.	Mean	S.D.
3.00	--	--	46.72	17.65
3.50	--	--	44.64	12.32
4.00	--	--	46.39	10.65
4.50	--	--	45.56	12.29
5.00	--	--	42.10	10.10
5.50	--	--	39.51	10.41
6.00	--	--	40.00	11.08
6.50	--	--	39.66	8.84
7.00	--	--	43.04	7.78
7.50	--	--	41.62	10.26
8.00	--	--	41.47	12.00
8.50	--	--	42.42	10.44
9.00	--	--	44.39	12.18
9.50	--	--	46.50	14.67
10.00	--	--	43.28	9.55
10.50	--	--	44.31	7.33
11.00	--	--	45.56	8.25
11.50	--	--	47.82	13.61
12.00	--	--	45.98	8.88
12.50	50.06	8.43	46.17	10.58
13.00	50.65	8.83	47.41	9.53
13.50	--	--	44.51	11.62
14.00	49.24	8.53	44.86	8.09
14.50	52.76	10.68	51.56	10.03

TABLE 572 (continued)

DIFFERENCE BETWEEN SUPINE SYSTOLIC AND DIASTOLIC (fifth phase)
BLOOD PRESSURES (mm Hg) FOR CHILDREN IN DENVER (CO)

Age (years)	Boys		Girls	
	Mean	S.D.	Mean	S.D.
15.00	61.45	14.58	51.11	12.97
15.50	62.71	15.50	--	--
16.00	54.37	11.14	47.02	7.91
17.00	--	--	45.79	10.28
18.00	--	--	42.98	6.85

(data from Roche, Eichorn, McCammon, et al., 1979)

TABLE   573

DIFFERENCE BETWEEN SEATED SYSTOLIC AND DIASTOLIC BLOOD PRESSURES
(mm Hg; fifth phase diastolic) FOR CHILDREN IN BOSTON (MA)

| Age | Boys | | Girls | |
(years)	Mean	S.D.	Mean	S.D.
2.50	28.81	5.92	13.20	5.03
3.00	31.12	3.97	32.62	5.85
3.50	31.94	5.79	33.66	6.21
4.00	32.57	6.37	31.31	6.15
4.50	34.16	7.56	32.70	6.92
5.00	31.11	6.73	32.36	7.76
5.50	32.34	5.52	32.27	5.36
6.00	33.45	7.64	32.75	6.70
6.50	33.58	5.51	32.14	6.99
7.00	31.02	6.30	32.23	7.04
7.50	32.73	6.31	31.39	7.10
8.00	31.98	6.92	31.71	7.48
8.50	31.27	6.74	31.50	6.46
9.00	31.74	7.74	31.61	7.38
9.50	31.15	8.55	31.61	5.65
10.00	31.04	7.45	33.16	7.83
11.00	31.26	10.61	35.19	9.11
12.00	34.00	7.47	36.57	8.07
13.00	34.52	9.44	38.12	10.61
14.00	36.02	9.62	40.42	9.37
15.00	36.45	9.03	38.71	6.90
16.00	36.49	7.83	37.40	7.88
17.00	38.59	9.13	37.87	9.86
18.00	40.15	6.68	37.35	7.48

(data from Roche, Eichorn, McCammon, et al., 1979)

TABLE   574

DIFFERENCE BETWEEN SEATED SYSTOLIC AND DIASTOLIC (fourth phase)
BLOOD PRESSURES (mm Hg) FOR CHILDREN IN DENVER (CO)

| Age | Boys | | Girls | |
(years)	Mean	S.D.	Mean	S.D.
2.50	34.06	9.10	28.36	9.66
3.00	30.48	8.38	30.96	9.39
3.50	33.91	6.77	33.73	7.16
4.00	32.44	9.87	31.75	7.43
4.50	35.23	8.94	33.62	13.23
5.00	34.53	6.20	31.17	7.60
5.50	35.28	8.12	32.33	7.54
6.00	34.55	9.62	33.00	8.16
6.50	36.73	11.81	31.08	9.40
7.00	31.62	6.90	30.46	6.47
7.50	32.79	8.90	30.68	5.81
8.00	31.98	7.19	31.35	6.96
8.50	32.42	7.53	31.49	9.31
9.00	33.49	7.48	32.07	8.38
9.50	33.55	7.24	33.13	7.70
10.00	30.80	7.79	30.48	8.97
10.50	32.80	9.18	32.36	7.87
11.00	32.41	8.23	33.44	6.82
11.50	32.32	7.03	32.25	8.37
12.00	34.70	11.04	32.51	7.45
12.50	33.20	8.18	33.66	7.18
13.00	38.30	11.21	34.34	8.56
13.50	37.45	7.85	32.33	6.81
14.00	35.81	7.84	33.68	7.05

TABLE   574   (continued)

DIFFERENCE BETWEEN SEATED SYSTOLIC AND DIASTOLIC (fourth phase)
BLOOD PRESSURES (mm Hg) FOR CHILDREN IN DENVER (CO)

Age	Boys		Girls	
(years)	Mean	S.D.	Mean	S.D.
14.50	38.82	10.91	32.71	7.22
15.00	40.17	9.27	35.73	8.37
15.50	39.34	8.22	30.91	8.10
16.00	43.07	11.14	36.58	7.73
16.50	43.76	12.30	--	--
17.00	42.16	9.18	34.72	9.60
17.50	38.69	26.12	--	--
18.00	42.37	6.96	36.98	8.90

(data from Roche, Eichorn, McCammon, et al., 1979)

TABLE   575

DIFFERENCE BETWEEN SUPINE SYSTOLIC AND DIASTOLIC (fourth phase)
BLOOD PRESSURES (mm Hg) FOR CHILDREN IN DENVER (CO)

Age	Boys		Girls	
(years)	Mean	S.D.	Mean	S.D.
2.00	36.90	13.19	33.36	10.82
2.50	34.58	10.98	33.03	7.50
3.00	35.36	7.32	31.81	8.52
3.50	36.16	5.46	33.72	8.27
4.00	37.01	7.77	32.91	7.09
4.50	34.24	8.13	34.40	7.87
5.00	36.33	6.84	34.74	8.57
5.50	36.39	7.70	34.25	8.80
6.00	36.70	10.53	34.17	7.63
6.50	37.26	8.11	32.95	8.88

TABLE 575 (continued)

DIFFERENCE BETWEEN SUPINE SYSTOLIC AND DIASTOLIC (fourth phase)
BLOOD PRESSURE (mm Hg) FOR CHILDREN IN DENVER (CO)

Age	Boys		Girls	
(years)	Mean	S.D.	Mean	S.D.
7.00	37.08	7.28	35.70	7.59
7.50	36.68	7.65	34.19	7.53
8.00	36.38	8.42	33.53	7.50
8.50	36.72	8.71	35.57	9.25
9.00	36.42	7.75	33.68	8.08
9.50	36.22	6.94	35.49	10.24
10.00	36.97	8.11	32.62	8.04
10.50	37.63	8.39	33.69	8.20
11.00	37.38	5.91	36.59	8.21
11.50	39.05	6.77	36.06	8.62
12.00	38.70	8.53	36.41	6.28
12.50	37.73	9.89	37.18	8.22
13.00	39.17	11.33	38.10	7.33
13.50	40.47	7.59	38.25	6.99
14.00	41.50	9.79	37.31	7.77
14.50	42.76	8.22	38.30	11.60
15.00	44.46	9.19	39.08	9.23
15.50	45.63	8.12	37.62	9.66
16.00	45.88	10.54	42.28	11.47
16.50	45.88	9.86	40.48	8.25
17.00	47.00	10.05	40.04	13.16
17.50	47.83	8.80	44.06	5.69
18.00	45.63	9.93	39.71	8.72

(data from Roche, Eichorn, McCammon, et al., 1979)

TABLE 576

DIFFERENCE BETWEEN SEATED SYSTOLIC AND DIASTOLIC BLOOD PRESSURES
(mm Hg; fifth phase diastolic) FOR CHILDREN IN THE FELS STUDY
(Yellow Springs, OH)

Age (years)	Boys		Girls	
	Mean	S.D.	Mean	S.D.
2.00	28.80	7.69	32.08	20.00
2.50	32.05	11.97	37.43	15.99
3.00	28.17	7.57	34.32	9.54
3.50	31.90	18.59	30.83	15.72
4.00	31.70	10.34	34.26	13.82
4.50	32.98	8.78	31.54	9.67
5.00	34.08	9.58	32.51	8.18
5.50	34.81	9.89	32.51	9.65
6.00	34.51	10.06	31.79	8.21
6.50	32.12	7.89	32.89	8.92
7.00	30.66	8.13	30.74	10.51
7.50	32.75	7.76	31.31	7.92
8.00	31.13	8.28	32.55	7.95
8.50	32.47	8.01	31.26	7.32
9.00	32.26	7.77	32.02	6.95
9.50	33.01	7.64	35.05	7.52
10.00	34.85	7.88	33.28	7.40
10.50	33.89	8.22	34.38	7.95
11.00	33.11	7.56	34.63	8.61
11.50	34.04	9.00	35.58	7.80
12.00	35.03	8.96	36.18	9.32
12.50	35.09	8.60	35.78	7.60
13.00	36.36	10.58	37.42	9.17
13.50	37.00	9.02	37.07	8.98

TABLE 576   (continued)

DIFFERENCE BETWEEN SEATED SYSTOLIC AND DIASTOLIC BLOOD PRESSURES
(mm Hg; fifth phase diastolic) FOR CHILDREN IN THE FELS STUDY
(Yellow Springs, OH)

Age	Boys		Girls	
(years)	Mean	S.D.	Mean	S.D.
14.00	39.50	12.09	36.52	8.42
14.50	38.55	9.65	35.08	8.04
15.00	40.11	9.67	36.13	8.07
15.50	41.24	10.87	36.92	7.93
16.00	40.01	9.44	37.20	7.74
16.50	41.31	10.65	35.70	7.89
17.00	41.59	10.62	37.22	8.40
17.50	43.50	11.20	36.00	7.81
18.00	43.27	10.42	35.31	8.13

(data from Roche, Eichorn, McCammon, et al., 1979)

TABLE   577

DIFFERENCE BETWEEN SEATED SYSTOLIC AND DIASTOLIC BLOOD PRESSURES
(mm Hg; fifth phase diastolic) FOR CHILDREN IN BERKELEY (CA)
ENROLLED IN THE GUIDANCE STUDY

Age	Boys		Girls	
(years)	Mean	S.D.	Mean	S.D.
1.75	36.11	7.56	36.53	5.69
3.00	39.76	7.20	38.71	6.50
4.00	34.87	6.79	36.09	5.79
5.00	33.34	6.77	31.68	6.02
6.00	30.41	6.23	38.89	5.69
7.00	29.11	6.32	28.47	7.26
8.00	29.75	6.53	29.29	6.70
8.50	29.28	4.04	--	--
9.00	31.66	6.23	31.83	6.78
10.00	34.64	7.56	34.76	7.05
11.00	35.31	6.22	34.56	6.57
12.00	40.61	8.02	33.97	7.22
13.00	47.78	8.09	38.36	9.44
14.00	47.22	11.28	42.49	8.37
15.00	42.13	9.17	42.28	8.19
16.00	44.50	11.42	38.85	8.88
17.00	45.79	11.62	34.74	7.37
18.00	47.29	14.14	32.50	6.97

(data from Roche, Eichorn, McCammon, et al., 1979)

TABLE 578

DIFFERENCE BETWEEN SEATED SYSTOLIC AND DIASTOLIC BLOOD PRESSURES
(mm Hg; fifth phase diastolic) FOR CHILDREN IN OAKLAND (CA)

Age (years)	Boys		Girls	
	Mean	S.D.	Mean	S.D.
11.00	33.91	10.11	43.34	10.51
11.50	33.33	7.49	42.58	7.88
12.00	35.21	6.77	43.77	8.98
12.50	36.73	8.21	44.19	8.19
13.00	39.41	7.59	44.97	7.43
13.50	41.67	8.78	46.67	6.18
14.00	42.54	7.78	46.33	5.89
14.50	42.59	7.80	45.11	5.86
15.00	42.98	7.73	44.42	6.56
15.50	43.22	7.33	42.68	6.63
16.00	42.96	6.58	40.47	8.30
16.50	44.36	7.00	39.66	8.49
17.00	44.69	7.66	36.45	8.03
17.50	45.40	9.52	34.76	8.63

(data from Roche, Eichorn, McCammon, et al., 1979)

TABLE 579

DIFFERENCE BETWEEN STANDING SYSTOLIC AND DIASTOLIC BLOOD PRESSURES
(mm Hg; fifth phase diastolic) FOR CHILDREN IN OAKLAND (CA)

Age (years)	Boys		Girls	
	Mean	S.D.	Mean	S.D.
14.00	34.07	11.29	--	--
15.00	35.44	12.72	30.46	11.91
15.50	32.29	9.57	30.34	9.94
16.00	31.74	8.75	26.49	7.53
16.50	26.86	9.83	31.19	10.22
17.00	26.40	11.18	27.37	14.28
17.50	29.22	13.77	26.03	9.22

(data from Roche, Eichorn, McCammon, et al., 1979)

TABLE 580

BLOOD PRESSURE (mm Hg) IN BALTIMORE CHILDREN

Race	Sex	Age (years)	Systolic Blood Pressure		Diastolic Blood Pressure	
			Mean	S.D.	Mean	S.D.
Black	M	7- 9	94.4	11.7	56.5	10.9
		10-12	97.2	14.7	58.9	12.4
	F	7- 9	92.3	10.9	57.2	11.0
		10-12	99.0	15.9	59.9	12.7
White	M	7- 9	98.9	10.2	61.9	13.7
		10-12	104.1	14.9	62.8	13.9
	F	7- 9	101.0	12.0	65.2	12.9
		10-12	102.1	21.2	64.3	13.4

(data from Stine, Hepner and Greenstreet, 1975, American Journal of Diseases of Children 129:905-911. Copyright 1975, American Medical Association)

TABLE 581

BLOOD PRESSURE (mm Hg) IN MISSOURI CHILDREN 14 TO 18 YEARS OF AGE

Age (year)		BOYS				GIRLS		
		Mean ± S.D.	90th percentile	95th percentile		Mean ± S.D.	90th percentile	95th percentile
14	S	114 ± 11.8	130	133	S	111 ± 10.4	125	128
	D	64 ± 11.0	78	82	D	66 ± 9.7	79	82
15	S	117 ± 12	133	138	S	112 ± 10.7	126	130
	D	67 ± 11	81	86	D	68 ± 9.7	81	84
16	S	120 ± 12.2	136	140	S	111 ± 10.5	125	128
	D	69 ± 10.5	83	86	D	68 ± 9.7	81	84
17	S	121 ± 11.8	137	140	S	110 ± 10.2	124	127
	D	70 ± 9.9	83	86	D	68 ± 9.7	81	84
18	S	121 ± 11.3	136	140	S	109 ± 10.5	123	126
	D	72 ± 9.9	85	88	D	67 ± 9.5	80	83

S = systolic pressure; D = diastolic pressure.

(data from Goldring, Londe, Sivakoff, et al., 1977, The Journal of Pediatrics 91:884-889)

TABLE   582

BLOOD PRESSURE (mm Hg) IN LOS ANGELES (CA) CHILDREN

Age (years)	Systolic		Diastolic 4th		Diastolic 5th	
	Mean	S.D.	Mean	S.D.	Mean	S.D.
BOYS						
3.50 - 4.49	100.2	10.0	57.4	23.0	37.0	22.1
4.50 - 5.49	102.3	6.1	61.3	10.1	41.3	18.5
5.50 - 6.49	106.1	7.1	68.2	9.8	49.0	14.8
6.50 - 7.49	106.2	6.4	70.2	7.7	50.3	15.2
7.50 - 8.49	108.2	10.5	68.0	7.2	47.0	11.8
8.50 - 9.49	108.0	9.3	65.7	8.1	48.1	16.2
9.50 - 10.49	107.8	9.2	69.1	10.2	55.0	16.2
10.50 - 11.49	108.7	13.0	64.7	16.7	48.2	22.9
11.50 - 12.49	111.3	11.5	67.9	15.0	49.0	22.7
12.50 - 13.49	112.0	8.8	67.5	15.9	47.5	23.5
13.50 - 14.49	112.2	10.7	70.4	14.4	51.0	24.0
14.50 - 15.49	109.9	7.7	69.0	9.5	48.2	22.9
15.50 - 16.49	110.8	11.5	75.2	9.5	59.2	21.2
16.50 - 17.49	115.7	9.4	79.2	11.9	62.4	14.3
GIRLS						
3.50 - 4.49	99.3	8.4	66.0	9.8	46.6	14.9
4.50 - 5.49	99.2	5.9	62.2	8.9	39.0	16.4
5.50 - 6.49	103.2	5.6	63.5	8.9	37.2	20.6
7.50 - 8.49	105.1	8.9	68.7	10.6	42.9	28.5
8.50 - 9.49	108.0	9.1	68.0	16.2	53.5	23.5
9.50 - 10.49	108.3	7.0	71.3	8.7	60.5	11.0
10.50 - 11.49	114.0	11.2	67.1	16.4	56.5	22.1
11.50 - 12.49	117.1	10.9	67.0	15.0	55.4	15.9
12.50 - 13.49	112.0	7.2	65.5	9.0	41.4	25.3

(data from Moss and Adams, Problems of Blood Pressure in Childhood, 1962. Courtesy of
Charles C Thomas, Publisher, Springfield, Illinois)

TABLE 583

MERCURY SPHYGMOMANOMETRIC BLOOD PRESSURES (mm Hg) OF LOUISIANA CHILDREN

Age (years)	Mean	S.D.	Percentiles						
			5	10	25	50	75	90	95
SYSTOLIC									
5	96.2	8.2	84	87	90	95	102	105	112
6	95.4	8.6	83	85	89	95	100	108	112
7	94.6	8.4	82	85	88	94	100	105	109
8	96.3	8.7	83	85	91	96	101	107	114
9	97.2	8.2	85	87	92	97	102	107	111
10	98.6	9.4	85	87	93	98	104	111	116
11	101.0	9.2	86	89	94	100	107	114	117
12	103.8	9.3	89	92	98	104	109	116	119
13	105.4	9.8	91	94	99	105	111	116	122
14	107.4	8.5	94	97	101	107	113	119	123
15	108.8	9.9	94	96	102	108	115	121	125
DIASTOLIC (fourth phase)									
5	59.8	6.8	49	50	56	60	64	69	72
6	58.6	7.9	47	48	53	58	64	68	72
7	58.2	7.1	46	50	54	58	63	67	69
8	59.4	6.6	48	51	55	60	64	67	70
9	60.7	6.5	50	52	56	61	65	69	72
10	61.3	7.9	49	51	56	61	66	71	75
11	63.0	7.6	50	54	58	63	68	74	76
12	65.3	7.6	53	55	60	65	70	75	78
13	67.5	7.9	55	57	62	68	73	77	80
14	67.4	7.0	55	58	62	68	72	77	79
15	67.6	7.2	56	58	62	67	72	77	81

(data from Voors, Foster, Frerichs, et al., Studies of blood pressures in children, ages 5-14 years, in a total biracial community. The Bogalusa Heart Study. Circulation 54:319-327, 1976, by permission of the American Heart Association, Inc.)

TABLE   584

BLOOD PRESSURE (mm Hg) IN SCHOOL CHILDREN
OF RICHMOND (VA)

Race				Sex			
Black		White		Boys		Girls	

Systolic Blood Pressure

114.65	113.70	113.96	114.49	115.78	116.03	113.57	112.42
(15.36)	(13.79)	(15.20)	(14.55)	(16.32)	(14.33)	(14.55)	(13.54)

Diastolic Blood Pressure (fourth phase)

72.14	70.42	72.24	71.23	72.32	70.63	72.04	70.56
(11.91)	(10.45)	(10.78)	(9.85)	(12.11)	(11.36)	(11.26)	(9.55)

Age (years)							
≤14		15		16		≥17	

Systolic Blood Pressure

113.27	113.85	114.25	113.47	115.63	114.52	116.42	114.34
(15.34)	(14.70)	(15.25)	(13.75)	(15.59)	(14.01)	(14.24)	(12.13)

Diastolic Blood Pressure (fourth phase)

71.57	70.73	71.65	70.19	73.26	70.86	73.81	71.91
(11.42)	(10.45)	(11.48)	(10.16)	(11.58)	(10.47)	(12.90)	(10.18)

(Standard deviations in parentheses)

(data from McCue et al., 1979)

TABLE 585

BLOOD PRESSURE (mm Hg; Techumseh, MI)

Age (years)	Systolic		Diastolic (muffling)		Diastolic (disappearance)	
	Mean	S.D.	Mean	S.D.	Mean	S.D.
BOYS						
0- 2	107	16	71	12	67	11
3	104	13	69	11	68	11
4	106	11	72	11	67	11
5	109	13	74	10	71	9
6	109	12	74	12	68	10
7	110	12	75	12	69	10
8	111	12	77	10	69	10
9	113	13	77	12	72	11
10	115	13	76	13	69	9
11	116	10	77	11	70	11
12	120	12	79	11	71	10
13-14	124	14	77	10	69	12
15-16	129	16	78	11	70	11
17-19	130	14	82	12	74	12
GIRLS						
0- 2	106	15	71	12	67	10
3	108	14	72	12	69	10
4	106	14	71	11	68	10
5	108	13	75	11	69	9
6	107	13	73	12	67	10
7	112	15	76	10	72	10
8	112	13	74	12	69	11
9	114	12	77	10	70	10
10	116	11	76	9	69	9
11	119	13	77	10	70	10
12	120	12	79	9	72	10
13-14	125	12	79	11	74	11
15-16	124	13	81	11	74	11
17-19	123	13	79	11	73	11

(data from Johnson, Epstein, and Kjelsberg, 1965, printed with permission of Journal of Chronic Diseases 18:147-160, Copyright 1965, Pergamon Press, Ltd.)

TABLE   586

DIFFERENCES BETWEEN SYSTOLIC AND DIASTOLIC BLOOD PRESSURES
(mm Hg; fifth phase diastolic) AT RECOVERY FROM EXERCISE
FOR CHILDREN IN DENVER (CO)

Age	Boys		Girls	
(years)	Mean	S.D.	Mean	S.D.
7.50	42.23	13.85	--	--
8.00	40.50	11.75	36.92	11.50
8.50	39.51	9.15	34.86	10.50
9.00	39.93	12.00	38.79	9.48
9.50	36.03	8.46	36.87	10.20
10.00	38.55	10.04	36.50	9.37
10.50	34.49	10.25	27.23	8.56
11.00	38.78	12.00	42.74	13.39
11.50	38.72	10.17	35.96	11.37
12.00	42.11	16.03	39.83	12.07
12.50	36.31	10.99	42.28	13.29
13.00	40.49	13.69	46.34	15.64
13.50	40.61	12.55	41.33	13.32
14.00	38.37	11.05	44.91	13.53
14.50	41.84	16.89	44.86	13.62
15.00	43.34	10.03	55.18	19.28
15.50	47.70	16.65	--	--
16.00	44.26	11.27	54.84	16.04
16.50	49.88	16.18	--	--
17.00	48.39	12.46	51.71	12.89
17.50	45.79	15.26	--	--
18.00	49.09	16.05	49.28	12.79

(data from Roche, Eichorn, McCammon, et al., 1979)

TABLE 587

DIFFERENCE BETWEEN RECOVERY SYSTOLIC AND DIASTOLIC BLOOD PRESSURES
(mm Hg; fifth phase) FOR CHILDREN IN THE OAKLAND STUDY

Age (years)	Boys		Girls	
	Mean	S.D.	Mean	S.D.
13.50	31.64	8.22	28.76	8.35
14.00	31.61	10.77	28.76	6.69
14.50	38.28	8.21	35.51	10.18
15.00	43.22	6.93	36.92	8.94
15.50	42.86	6.73	35.48	5.59
16.00	44.08	6.17	37.41	9.16
16.50	43.50	6.49	36.15	7.52
17.00	45.53	5.48	36.62	7.81
17.50	45.32	6.24	38.65	5.77

(data from Roche, Eichorn, McCammon, et al., 1979)

TABLE 588

HEART RATE (beats/min) AND ELECTROCARDIOGRAPHIC DATA
FOR NORMAL CHILDREN IN CHICAGO (IL)

	Age (years)					
	5 or less		6 to 10		11 to 15	
Heart Rate						
60 or less	--		2		--	
61 to 80	--		3		9	
81 to 100	3		9		13	
101 to 120	6		4		2	
121 and greater	1		--		--	
Average	112		87		84	
Shift of Electrical Axis	No.	%	No.	%	No.	%
No shift	5	50	7	39	12	50
Right axis shift	4	40	9	50	8	33
Left axis shift	1	10	2	11	4	17
Total	10	--	18	--	24	--
Electrical Position						
Vertical	4	40	11	61	8	33
Semivertical	4	40	4	22	9	38
Intermediate	2	20	3	17	5	21
Semihorizontal	--	--	--	--	1	4
Horizontal	--	--	--	--	1	4
Total	10	--	18	--	24	--
Transition Zone						
V_2 to V_3	4	40	3	17	5	22
V_2 to V_4	--	--	2	11	3	12
V_2 to V_5	--	--	2	11	1	4
V_3 to V_4	4	40	9	50	9	39
V_3 to V_5	--	--	--	--	2	8
V_4 to V_5	1	10	1	5.5	3	13
Others	1	10	1	5.5	1	4
Total	10	--	18	--	21	--

(data from Switzer and Besoain, 1950, American Journal of Diseases of Children 79:449-466.
Copyright 1950, American Medical Association)

TABLE 589

BASAL PULSE RATE (beats/min)
FOR CHILDREN IN DENVER, CO

Age (years)	BOYS		GIRLS	
	Mean	S.D.	Mean	S.D.
0.13	144.73	17.45	136.24	21.68
0.25	126.50	16.83	123.57	27.37
0.38	129.03	14.97	124.42	18.37
0.50	117.17	17.37	115.91	20.56
0.63	105.58	15.84	105.00	14.91
0.75	103.65	17.42	98.32	12.20
0.88	103.29	13.58	106.45	18.27
1.00	102.03	15.87	105.47	19.95
1.50	98.97	16.10	96.44	16.84
2.00	91.58	9.17	91.64	10.13
2.50	91.19	10.50	89.55	11.73
3.00	88.78	8.84	87.95	8.40
3.50	85.58	7.26	85.41	8.47
4.00	84.13	7.62	84.79	8.83
4.50	81.18	7.35	84.14	8.67
5.00	79.47	7.33	82.56	8.61
5.50	77.38	7.30	78.14	7.03
6.00	76.10	6.85	76.27	7.17
6.50	74.83	6.74	76.37	7.73
7.00	74.06	8.05	75.49	7.75
7.50	73.16	7.68	73.93	6.79
8.00	72.52	7.43	72.66	6.81
8.50	70.15	7.51	71.73	6.57
9.00	70.07	6.42	70.14	5.85
9.50	68.68	7.23	68.80	6.43
10.00	67.46	7.49	68.80	6.24
10.50	65.92	6.71	68.48	6.98
11.00	65.48	6.09	67.60	6.60
11.50	65.01	6.24	67.26	5.99
12.00	64.99	6.39	66.75	6.07
12.50	64.54	6.02	66.71	6.36

TABLE   589   (continued)

BASAL PULSE RATE (beats/min)
FOR CHILDREN IN DENVER, CO

Age (years)	BOYS		GIRLS	
	Mean	S.D.	Mean	S.D.
13.00	63.76	6.24	66.06	6.11
13.50	64.28	7.07	65.81	7.45
14.00	62.82	6.40	64.66	6.32
14.50	60.81	6.26	64.12	7.09
15.00	60.03	6.42	64.27	6.21
15.50	59.19	6.58	63.29	5.61
16.00	59.00	6.39	60.93	6.09
16.50	58.92	5.92	61.99	7.06
17.00	57.49	7.28	61.48	6.09
17.50	56.95	6.69	63.06	5.23
18.00	58.60	6.54	60.75	6.33

(data from Roche, Eichorn, McCammon, et al., 1979)

TABLE 590

PULSE RATE (beats/min.) FOR CHILDREN IN DENVER (CO)
(SUPINE AND SEATED*)

Age (years)	BOYS		GIRLS	
	Mean	S.D.	Mean	S.D.
0.08	136.53	15.83	137.67	9.58
0.13	137.67	12.12	131.55	16.51
0.25	131.83	14.63	140.66	15.53
0.38	131.09	14.75	134.60	18.94
0.50	128.66	18.28	139.81	19.54
0.63	--	--	149.11	17.15
0.75	127.33	15.10	123.76	18.04
1.00	131.52	17.02	123.91	12.49
1.50	121.05	16.22	123.92	18.63
2.00	117.48	15.59	119.23	14.39
2.50	114.12	11.70	113.11	17.13
3.00	108.55	12.18	108.35	12.13
3.50	106.35	13.01	107.96	11.69
4.00	103.40	10.76	102.66	10.01
4.50	99.92	10.63	104.95	11.62
5.00	96.50	9.51	103.04	10.86
5.50	96.08	11.29	96.76	11.46
6.00	93.49	10.96	95.99	10.99
6.50	91.45	9.92	93.66	11.48
7.00	88.00	9.89	92.27	11.08
7.50	86.31	10.16	92.73	11.15
8.00	84.92	8.91	90.23	10.51
8.50	83.34	11.67	88.95	11.20
9.00	81.68	8.27	88.17	12.49
9.50	80.98	9.96	86.83	8.98
10.00	80.78	8.43	83.85	9.65
10.50	77.99	8.02	82.73	11.51
11.00	78.06	9.24	84.57	10.38
11.50	76.21	8.60	83.23	9.64
12.00	77.45	8.10	83.57	9.83
12.50	76.93	8.13	84.78	11.13

TABLE 590   (continued)

PULSE RATE (beats/min.) FOR CHILDREN IN DENVER (CO)
(SUPINE AND SEATED*)

| Age | BOYS | | GIRLS | |
(years)	Mean	S.D.	Mean	S.D.
13.00	76.88	9.86	81.02	12.20
13.50	75.57	9.93	79.58	8.25
14.00	74.61	9.08	81.54	11.49
14.50	76.41	9.49	79.66	8.15
15.00	74.18	9.35	81.33	9.84
15.50	71.84	10.01	74.49	10.25
16.00	71.39	11.29	76.67	9.55
16.50	72.28	9.67	--	--
17.00	71.23	10.79	75.51	12.09
17.50	72.58	9.19	--	--
18.00	72.81	9.54	75.17	9.58

*Supine until 3 years of age; seated thereafter.

(data from Roche, Eichorn, McCammon, et al., 1979)

TABLE 591

HEART RATE, RESPIRATORY RATE, AND INDICES OF HEART RATE AND
RESPIRATORY RATE VARIABILITY IN FULL TERM NEWBORN CALIFORNIA INFANTS

Group	Heart rate (beats/min)		Interval Index[1]		Differential Index[2]		Respiratory Rate (breaths/min)		Respiratory variability/ min	
	Mean	S.D.	Mean	S.D.	Mean	S.D.	Mean	S.D.	Mean	S.D.
QS	130	12.0	3.5	1.46	8.2	2.40	36	13.0	48	28
AS	132	11.9	4.8	1.80	8.3	2.42	35	14.7	72	35

Interval index = coefficient of variation of heart rate intervals.
Differential index = S.D. of differences between succession heart rate intervals.
AS = action stage.
QS = quiet stage.

(data from Siassi et al., 1979)

TABLE 592

SEATED PULSE RATE (beats/min; supine to 1.5 years)
FOR CHILDREN IN THE BERKELEY STUDY (CA)

Age (years)	BOYS		GIRLS	
	Mean	S.D.	Mean	S.D.
0.08	145.89	11.89	143.68	13.10
0.17	141.29	9.70	140.00	7.91
0.25	138.73	6.94	139.94	7.93
0.33	134.90	5.30	138.70	8.20
0.42	135.02	6.73	137.39	8.52
0.50	134.20	8.11	136.46	10.49
0.58	133.04	9.37	136.07	6.57
0.67	132.48	7.51	133.44	6.76
0.75	131.52	6.28	134.30	9.00
0.83	131.54	6.04	133.94	6.02
0.92	131.30	6.94	133.97	7.48
1.00	130.92	7.11	131.93	6.84
1.08	131.01	10.54	132.52	5.76
1.17	130.90	10.40	131.82	7.74
1.25	133.92	14.29	131.47	11.81
1.50	129.35	12.98	131.34	16.24
1.75	130.14	14.61	132.44	18.46
2.00	129.84	22.14	131.05	20.96
3.00	109.55	10.55	112.27	15.99
4.00	104.89	10.98	104.28	6.89
5.00	99.63	9.74	104.95	13.06
6.00	97.47	8.31	103.83	8.66
7.00	93.13	10.32	94.98	9.29
14.00	77.37	12.42	78.73	15.32
15.00	78.69	10.63	81.39	13.17
16.00	79.89	12.70	81.77	8.39
17.00	77.04	10.76	78.78	9.74
18.00	--	--	81.59	7.92

(data from Roche, Eichorn, McCammon, et al., 1979)

TABLE   593

SEATED PULSE RATE (beats/min)
FOR CHILDREN IN BERKELY (CA)
ENROLLED IN THE GUIDANCE STUDY

Age (years)	BOYS		GIRLS	
	Mean	S.D.	Mean	S.D.
1.75	117.61	12.04	120.42	12.36
3.00	106.24	10.23	108.16	9.39
4.00	100.22	8.77	103.23	8.95
5.00	98.69	12.55	101.11	11.97
6.00	94.97	8.61	98.06	9.71
7.00	92.34	10.48	94.52	10.88
13.00	--	--	79.98	13.45
14.00	76.96	9.15	78.84	9.76
15.00	75.01	7.78	79.02	8.02
16.00	75.66	7.79	81.09	9.21
17.00	77.12	13.27	82.89	8.45
18.00	74.64	6.92	79.50	10.34

(data from Roche, Eichorn, McCammon, et al., 1979)

TABLE 594

SEATED PULSE RATE (beats/min) FOR CHILDREN IN THE FELS STUDY
(Yellow Springs, OH)

Age (years)	BOYS		GIRLS	
	Mean	S.D.	Mean	S.D.
0.00	133.46	22.61	136.28	24.62
0.08	136.54	17.84	144.26	21.17
0.25	141.71	21.77	145.19	15.21
0.50	135.33	21.35	136.35	16.72
0.75	133.67	20.15	134.05	17.79
1.00	134.85	21.77	139.14	23.23
1.50	129.19	22.15	136.94	20.21
2.00	129.33	21.36	118.34	17.45
2.50	112.82	17.20	123.47	18.39
3.00	114.15	16.70	116.42	17.95
3.50	110.68	15.88	111.46	18.74
4.00	103.59	13.89	106.97	15.33
4.50	109.22	19.87	105.80	15.53
5.00	99.90	13.28	102.81	16.76
5.50	100.31	15.14	102.50	11.12
6.00	93.08	14.40	98.75	15.04
6.50	94.02	11.80	95.91	11.85
7.00	89.67	13.33	95.77	15.44
7.50	88.90	13.09	94.98	12.35
8.00	86.75	11.67	87.12	10.84
8.50	85.81	11.50	89.95	11.60
9.00	84.23	11.78	85.97	13.11
9.50	84.55	11.78	86.64	12.57
10.00	82.97	11.63	83.00	10.98
10.50	81.10	11.11	85.66	13.14
11.00	78.68	11.59	82.21	10.55
11.50	79.96	11.83	83.88	13.19
12.00	79.96	12.03	83.13	11.64
12.50	79.63	10.04	82.34	12.93
13.00	78.08	10.30	82.96	13.20
13.50	77.53	9.96	80.94	11.94

TABLE 594 (continued)

SEATED PULSE RATE (beats/min) FOR CHILDREN IN THE FELS STUDY
(Yellow Springs, OH)

Age (years)	BOYS		GIRLS	
	Mean	S.D.	Mean	S.D.
14.00	77.19	11.25	78.38	11.22
14.50	75.08	10.95	77.94	9.45
15.00	76.62	12.03	78.37	11.12
15.50	72.50	10.51	78.70	12.95
16.00	72.49	11.40	77.93	11.10
16.50	72.95	10.94	80.07	12.59
17.00	72.10	11.40	78.66	11.15
17.50	71.82	9.41	76.93	11.22
18.00	71.32	11.05	76.99	10.88

(data from Roche, Eichorn, McCammon, et al., 1979)

TABLE 595

STANDING PULSE RATE (beats/min) FOR CHILDREN IN OAKLAND (CA)

Age (years)	BOYS		GIRLS	
	Mean	S.D.	Mean	S.D.
14.00	71.19	8.27	--	--
15.00	88.03	13.22	88.85	18.54
15.50	89.35	11.12	89.59	9.95
16.00	85.87	8.54	92.69	8.85
16.50	82.90	7.06	84.97	8.50
17.00	78.44	9.40	81.46	8.91
17.50	74.08	7.44	77.04	7.01

(data from Roche, Eichorn, McCammon, et al., 1979)

TABLE 596

SUPINE PULSE RATE (beats/min) FOR CHILDREN IN DENVER (CO)

Age (years)	BOYS		GIRLS	
	Mean	S.D.	Mean	S.D.
2.00	--	--	111.13	22.03
2.50	--	--	111.00	14.52
3.00	--	--	111.64	13.30
3.50	--	--	108.98	14.72
4.00	--	--	102.64	12.41
4.50	--	--	102.45	11.84
5.00	--	--	102.20	10.88
5.50	98.73	16.99	96.24	10.64
6.00	--	--	93.54	12.19
6.50	87.89	10.37	92.81	10.61
7.00	84.97	13.77	92.08	12.30
7.50	91.55	12.25	89.88	11.29
8.00	88.10	9.90	88.93	10.53
8.50	85.64	13.08	87.73	11.57
9.00	83.90	8.42	86.57	11.85
9.50	81.31	7.89	87.45	9.64
10.00	79.73	6.78	83.36	11.68
10.50	76.47	8.44	83.14	10.08
11.00	80.31	12.32	83.55	9.79
11.50	79.02	9.10	82.64	7.74
12.00	76.57	10.69	81.75	9.82
12.50	76.83	5.49	82.26	11.71
13.00	79.10	9.18	81.96	9.80
13.50	78.57	10.86	78.02	13.31
14.00	76.74	8.71	79.09	7.22
14.50	77.56	9.59	77.36	7.81
15.00	77.14	11.76	79.60	11.46
15.50	72.16	6.72	78.25	11.37
16.00	72.27	5.51	76.54	10.57
16.50	72.96	10.48	--	--
17.00	74.05	6.13	74.61	10.32
18.00	69.86	7.81	75.53	11.61

(data from Roche, Eichorn, McCammon, et al., 1979)

TABLE 597

PULSE RATE (beats/min) AFTER RECOVERY FROM EXERCISE
FOR CHILDREN IN OAKLAND (CA)

Age (years)	BOYS		GIRLS	
	Mean	S.D.	Mean	S.D.
13.50	82.83	10.70	90.34	7.51
14.00	82.71	6.98	89.52	9.12
14.50	79.60	9.74	90.92	10.38
15.00	81.68	10.61	85.13	10.40
15.50	76.37	8.96	83.36	10.87
16.00	74.34	7.94	85.81	9.03
16.50	73.97	8.05	89.61	6.71
17.00	74.90	11.93	86.52	9.81
17.50	76.14	9.11	84.74	9.94

(data from Roche, Eichorn, McCammon, et al., 1979)

TABLE 598

PERCENTAGE OF ELECTROCARDIOGRAMS OF HEALTHY INFANTS AND CHILDREN
IN CALIFORNIA IN WHOM PARTICULAR FINDINGS WERE OBSERVED

Variables	Age (years)							
	½–1½	1½–2½	2½–3½	3½–4½	4½–5½	5½–6½	6½–7½	7½–9
Sinus	98	91	77	61	53	57	50	36
Sinus arry.	2	9	23	39	47	43	50	64
Flat, diphasic or inverted								
T3 only	29	38	35	25	21	18	23	21
P3 only	4	4	8	18	19	17	20	16
T3 and P3	11	10	9	7	9	15	12	3
P2 and P3	--	--	--	2	--	1	1	2
None	56	47	48	48	51	49	45	58
QRS Interval								
.03 sec.	1	--	1	--	--	--	--	--
.04 sec.	14	6	5	8	11	7	8	5
.05 sec.	45	36	22	26	24	22	23	15
.06 sec.	34	48	45	43	51	39	37	41
.07 sec.	5	9	22	20	16	23	25	26
.08 sec.	--	--	4	3	8	7	7	12
.09 sec.	--	--	--	--	--	1	1	--
Average QRS	.053 sec.	.056 sec.	.059 sec.	.058 sec.	.059 sec.	.060 sec.	.060 sec.	.062 sec.
PR Interval								
<.10 sec.	1	3	1	2	2	--	1	--
<.10 sec.	26	17	14	9	7	7	--	4
<.11 sec.	25	26	17	20	8	16	13	11
<.12 sec.	27	27	26	28	29	23	25	26
<.13 sec.	14	14	22	16	25	27	24	30
<.14 sec.	5	10	16	15	14	18	20	17
<.15 sec.	2	3	4	6	11	5	11	7
<.16 sec.	--	--	--	3	1	4	4	4
<.17 sec.	--	--	--	--	1	--	3	--
<.18 sec.	--	--	--	--	1	--	1	--
Average PR interval	.115 sec.	.117 sec.	.122 sec.	.123 sec.	.128 sec.	.126 sec.	.131 sec.	.128 sec.

Arry. = Arrhythmia

(data from Maroney and Rantz, 1950, Pediatrics 5:396-407. Copyright American Academy of Pediatrics, 1950)

TABLE 599

AMPLITUDES OF R AND S WAVES IN PRECORDIAL LEADS

Age	R Wave			S Wave		
	5%	Mean	95%	5%	Mean	95%
Amplitudes in V_4R						
30 hr.	4.0	8.6	14.2	0.2	3.8	13.0
1 mo.	3.3	6.3	8.5	0.8	1.8	4.6
Amplitudes in V_1						
30 hr.	4.3	11.9	21.0	1.1	9.7	19.1
1 mo.	3.3	11.1	18.7	0.0	6.1	15.0
Amplitudes in V_5						
30 hr.	3.1	9.4	16.6	2.4	9.5	18.5
1 mo.	3.8	15.0	24.2	2.8	8.3	16.3
Amplitudes in V_6						
30 hr.	1.5	5.4	11.3	1.0	5.6	13.8
1 mo.	1.0	10.8	16.2	0.0	4.8	9.5

AMPLITUDE OF T WAVES IN PRECORDIAL LEADS V_4 to V_6

Age	V_4			V_5			V_6		
	Mean	95%	S.D.	Mean	95%	S.D.	Mean	95%	S.D.
0-24 hr.	4.3	7.2	0.95	3.3	6.8	1.62	2.4	3.9	0.63
1-7 days	4.4	7.7	1.39	4.9	7.3	1.44	2.9	4.2	0.67
8-30 days	5.3	8.1	1.49	5.3	7.5	1.50	3.5	5.3	1.01

(data from Namin, 1966, and Liebman, 1968; cited by Hastreiter and Abella, 1971, in Journal of Pediatrics 78:146-156)

TABLE 600

DEPTH OF Q WAVE (mm) FOR INFANTS OF CHICAGO (IL)

Lead	Birth - 1 wk.		1 wk. - 3 mo.		3 - 6 mo.		6 mo. - 1 yr.	
	Max.	Min.	Max.	Min.	Max.	Min.	Max.	Min.
1	0	0	1.5	0	0.25	0	0	0
2	2	0	1	0	1	0	1.5	0
3	1.5	0	5	0	4	0	4	0
R	3	0	11	0	13	0	10	0
L	3	0	1	0	3	0	0	0
F	2	0	1.5	0	3	0	2.5	0
V_4R	0	0	0	0	0	0	0	0
V_1	0	0	6	0	0	0	0	0
V_2	0	0	12	0	0	0	0	0
V_4	0	0	2	0	1	0	0	0
V_5	0	0	1.5	0	1	0	0	0
V_6	0	0	0.5	0	1	0	0	0

Lead	1 - 2 yr.		2 - 3 yr.		3 - 4 yr.		4 - 5 yr.	
	Max.	Min.	Max.	Min.	Max.	Min.	Max.	Min.
1	2.5	0	0.5	0	2	0	0.25	0
2	1.5	0	1	0	3	0	0.25	0
3	3	0	1.5	0	1.5	0	1	0
R	12	0	8	0	11	0	11	7
L	5	0	4	0	1.1	0	1	0
F	1.5	0	1.5	0	2.5	0	0.25	0
V_4R	4	0	3	0	0	0	0	0
V_1	1	0	0	0	0	0	0	0
V_2	1.5	0	0	0	0	0	0	0
V_4	2	0	0	0	5	0	0.5	0
V_5	4	0	1	0	3	0	1	0
V_6	3	0	1	0	3	0	1	0.5

(data from Gros, Gordon, and Miller, 1951, Pediatrics 8:349-361. Copyright American Academy of Pediatrics 1951)

TABLE 601

HEIGHT OF R-WAVE (mm) FOR INFANTS OF CHICAGO (IL)

Lead	Birth – 1 wk.			1 wk. – 3 mo.			3 – 6 mo.			6 mo. – 1 yr.		
	Max.	Min.	Av.	Max.	Min.	Av.	Max.	Min.	Av.	Max.	Min.	Av.
1	4	0	1.2	9	4	6.4	11	5	7.8	8	4	6.3
2	10	T	4.5	11	3	7.3	13	5	9.2	12	7	10.0
3	12	T	6.3	7	0	3.0	10	0.25	5.2	9	4	6.7
R	7	T	4.2	3	0	1.8	3	0	1.3	7	0.5	3.0
L	3	0	1.6	8	2	4.5	9	2	4.7	7	2	4.7
F	13	T	5.4	10	2	5.2	9	T	5.9	11	5	8.0
V_4R	14	2.5	5.3	5	0.5	3.0	7	0.5	3.1	8	1	4.0
V_1	11	5	8.0	13	7	10.3	12	5	7.6	10	5	7.0
V_2	15	6	9.5	16	12	13.1	20	11	13.4	15	8	11.7
V_4	12	1.5	7.0	15	6	12.3	19	10	13.5	15	9	12.6
V_5	7	2	3.0	1.5	5	12.5	17	7	12.0	12	8	10.6
V_6	3.5	1	2.0	14	2	9.5	11	4	7.9	7	6	6.3

	1 – 2 yr.			2 – 3 yr.			3 – 4 yr.			4 – 5 yr.		
	Max.	Min.	Av.	Max.	Min.	Av.	Max.	Min.	Av.	Max.	Min.	Av.
1	10	2.5	7.0	7	1	4.0	5	3	4.7	7	5	6.0
2	18	2	11.6	14	4.5	8.9	13	8	9.7	14	7	8.7
3	10	T	8.0	12	1.5	6.6	8	3	6.0	13	0.25	4.3
R	2.5	0	0.7	1.5	T	0.5	2	0	0.7	1.5	0	0.8
L	5	0	4.1	3.5	0.25	1.5	2.5	1	1.8	4	1	2.3
F	14	1.5	9.8	12	3.5	7.5	10	5	7.3	15	8	9.3
V_4R	10	1	3.7	3	0.5	2.4	5	1	2.8	4	2	2.2
V_1	11	3	7.5	10	2.5	5.9	8	T	5.2	9	3	5.7
V_2	20	6	13.2	16	5	11.6	15	9	12.7	10	6	8.0
V_4	17	8	13.5	10	5	8.6	18	10	13.0	20	15	17.0
V_5	17	7	12.5	11	4	7.6	13	8	11.0	19	13	13.5
V_6	15	4	8.4	8	4	7.5	12	8	10.0	15	9	9.3

0 = absent; T = less than 0.5.

(data from Gros, Gordon, and Miller, 1951, Pediatrics 8:349-361. Copyright American Academy of Pediatrics 1951)

TABLE 602

HEIGHT OF P-WAVE (mm) FOR INFANTS OF CHICAGO (IL)

Lead	Birth to 1 wk.			1 wk. to 3 mo.			3 mo. to 1 yr.			1 to 5 yr.		
	Max.	Min.	Av.	Max.	Min.	Av.	Max.	Min.	Av.	Max.	Min.	Av.
1	+1	+0.5	+0.8	+1	+0.5	+0.8	+1	+0.25	+0.6	+1	+Tiny	+0.5
2	+3	+1	+2.0	+1	+0.50	+0.75	+2	+Tiny	+1.0	+1.5	+0.25	+1.0
3	+2	+0.25	+1.0	+Tiny	-1	Ind.	+1	-Tiny	+0.3	+0.75	Ind.	+0.3
R	-1.5	-0.5	-1.0	-1.5	-0.50	-1	-2	-0.50	-0.9	-1	-Tiny	-0.6
L	+0.75	Ind.	+0.3	+1.5	+0.25	+0.4	+0.50	Ind.	+0.2	+0.25	-Tiny	+0.1
F	+2	+0.25	+1.1	+1	+Tiny	+0.5	+1.50	-0.25	+0.5	+2	+0.25	+0.5
V_4R	+1	-0.25	+0.5	-1.50	-0.25	-0.4	+1	-Tiny	+0.1	+1	Ind.	+0.1
V_1	+2	-Tiny	+1.1	+1	Ind.	+0.5	+2	-Tiny	+0.1	+1.25	-Tiny	+0.4
V_2	+1.5	+0.25	+1.1	+1	+Tiny	+0.5	+2	+0.25	+0.5	+1	-0.25	+0.5
V_4	+1.5	+0.50	+1.0	+1	+0.25	+0.6	+1	Ind.	+0.3	+1	+Tiny	+0.3
V_5	+1	+0.50	+0.4	+1	+0.50	+0.7	+1	Ind.	+0.4	+1	+Tiny	+0.2
V_6	+1.5	Ind.	+0.5	+1	+0.25	+0.4	+0.50	Ind.	+0.2	+1	Ind.	+0.2

+ = Upright; - = Inverted; Ind. = Indiscernible.

(data from Gros, Gordon, and Miller, 1951, Pediatrics 8:349-361. Copyright American
Academy of Pediatrics 1951)

TABLE 603

DURATION OF P-WAVE (sec) FOR INFANTS OF CHICAGO (IL)

Lead	Birth - 1 wk.			1 wk. - 3 mo.			3 mo. - 1 yr.			1 - 5 yr.		
	Max.	Min.	Av.	Max.	Min.	Av.	Max.	Min.	Av.	Max.	Min.	Av.
1	0.05	0.04	0.04	0.04	0.04	0.04	0.08	0.04	0.06	0.08	0.04	0.06
2	0.06	0.04	0.04	0.06	0.04	0.04	0.08	0.04	0.06	0.08	0.04	0.06
3	0.05	0.04	0.04	0.04	0.04	0.04	0.04	0.04	0.04	0.08	0.04	0.06
R	0.04	0.04	0.04	0.06	0.04	0.04	0.08	0.04	0.06	0.08	0.04	0.06
L	0.04	0.04	0.04	0.06	0.04	0.04	0.04	0.04	0.04	0.08	0.04	0.06
F	0.07	0.04	0.05	0.04	0.04	0.04	0.08	0.04	0.06	0.08	0.04	0.06
V_4R	0.04	0.04	0.04	0.04	0.04	0.04	0.08	0.04	0.06	0.08	0.04	0.06
V_1	0.06	0.04	0.05	0.04	0.04	0.04	0.06	0.04	0.05	0.08	0.04	0.06
V_2	0.06	0.04	0.04	0.04	0.04	0.04	0.08	0.04	0.06	0.08	0.04	0.06
V_4	0.08	0.04	0.05	0.04	0.04	0.04	0.08	0.04	0.06	0.08	0.04	0.06
V_5	0.08	0.04	0.05	0.04	0.04	0.04	0.04	0.04	0.04	0.08	0.04	0.06
V_6	0.06	0.04	0.04	0.04	0.04	0.04	0.04	0.04	0.04	0.08	0.04	0.06

(data from Gros, Gordon, and Miller, 1951, Pediatrics 8:349-361. Copyright American
Academy of Pediatrics 1951)

TABLE   604

DURATION QRS COMPLEX (sec) FOR INFANTS OF CHICAGO (IL)

Lead	Birth - 1 wk.			1 wk. - 3 mon.			3 mo. - 1 yr.			1 - 5 yr.		
	Max.	Min.	Av.	Max.	Min.	Av.	Max.	Min.	Av.	Max.	Min.	Av.
1	0.04	0.04	0.04	0.04	0.04	0.04	0.04	0.04	0.04	0.04	0.04	0.04
2	0.04	0.04	0.04	0.04	0.04	0.04	0.04	0.04	0.04	0.04	0.04	0.04
3	0.04	0.04	0.04	0.04	0.04	0.04	0.06	0.04	0.04	0.06	0.04	0.05
R	0.04	0.04	0.04	0.04	0.03	0.04	0.04	0.04	0.04	0.04	0.04	0.04
L	0.04	0.04	0.04	0.04	0.03	0.04	0.06	0.04	0.05	0.06	0.04	0.05
F	0.04	0.04	0.04	0.04	0.03	0.04	0.06	0.04	0.05	0.04	0.04	0.04
V_4R	0.04	0.04	0.04	0.04	0.03	0.04	0.04	0.04	0.04	0.07	0.04	0.05
V_1	0.06	0.04	0.05	0.06	0.04	0.04	0.06	0.04	0.05	0.08	0.04	0.06
V_2	0.08	0.04	0.05	0.06	0.04	0.04	0.06	0.04	0.05	0.08	0.04	0.06
V_4	0.06	0.04	0.05	0.04	0.04	0.04	0.06	0.04	0.05	0.06	0.04	0.05
V_5	0.04	0.04	0.04	0.04	0.04	0.04	0.04	0.04	0.04	0.06	0.04	0.05
V_6	0.04	0.04	0.04	0.04	0.04	0.04	0.04	0.04	0.04	0.04	0.04	0.04

(data from Gros, Gordon, and Miller, 1951, Pediatrics 8:349-361. Copyright American
Academy of Pediatrics 1951)

TABLE   605

DURATION OF PR INTERVALS (sec.) FOR INFANTS OF CHICAGO (IL)

Lead	Birth - 1 wk.			1 wk. - 3 mo.			3 mo. - 1 yr.			1 yr. - 5 yr.		
	Max.	Min.	Av.	Max.	Min.	Av.	Max.	Min.	Av.	Max.	Min.	Av.
1	0.16	0.10	0.12	0.12	0.10	0.11	0.14	0.10	0.12	0.14	0.10	0.12
2	0.16	0.10	0.12	0.12	0.10	0.10	0.14	0.10	0.12	0.16	0.10	0.13
3	0.16	0.10	0.12	0.12	0.10	0.10	0.14	0.10	0.12	0.16	0.10	0.13
R	0.16	0.10	0.12	0.12	0.10	0.11	0.14	0.10	0.12	0.14	0.10	0.13
L	0.12	0.10	0.11	0.12	0.10	0.10	0.14	0.10	0.12	0.16	0.10	0.13
F	0.16	0.10	0.12	0.12	0.10	0.11	0.14	0.10	0.12	0.16	0.10	0.13
V_4R	0.16	0.10	0.12	0.12	0.10	0.11	0.14	0.10	0.12	0.16	0.12	0.13
V_1	0.16	0.10	0.12	0.12	0.12	0.12	0.14	0.10	0.12	0.16	0.12	0.13
V_2	0.16	0.10	0.12	0.12	0.12	0.12	0.14	0.10	0.12	0.16	0.12	0.13
V_4	0.16	0.10	0.11	0.12	0.10	0.11	0.14	0.10	0.12	0.16	0.12	0.13
V_5	0.16	0.10	0.11	0.12	0.10	0.11	0.14	0.10	0.12	0.16	0.12	0.13
V_6	0.12	0.10	0.11	0.12	0.10	0.11	0.14	0.10	0.12	0.16	0.12	0.13

(data from Gros, Gordon, and Miller, 1951, Pediatrics 8:349-361. Copyright American
Academy of Pediatrics 1951)

TABLE 606

DEPTH OF S-WAVE (mm) FOR INFANTS OF CHICAGO (IL)

Lead	Birth - 1 wk.			1 wk. - 3 mo.			3 - 6 mo.			6 mo. - 1 yr.		
	Max.	Min.	Av.	Max.	Min.	Av.	Max.	Min.	Av.	Max.	Min.	Av.
1	8	3.5	5.0	4	0	1.6	6	0	3.0	8	1.5	6.5
2	6	0	3.5	5	0	1.4	3	0	0.95	7	0	3.0
3	3	0	1.0	10	0	2.1	6	0	1.0	2	0	3.3
R	2	0	1.0	6	0	1.4	13	0	4.3	9	0	3.0
L	9	3.5	5.1	2	0.25	1.0	7	0	3.9	6	1	4.0
F	4	0	2.0	6	0	1.2	2	0	0.5	3	0	1.3
V_4R	2	0	0.1	1.5	0	0.9	7	0.5	2.8	2	0.5	1.2
V_1	10	3	5.5	6	0	2.1	11	2	6.1	5	2	4.0
V_2	15	7	11.4	10	0	6.5	20	8	11.6	15	4	8.3
V_4	14	11	13.0	7	2	4.3	15	2	8.5	12	4	7.0
V_5	8	3	5.5	8	1	2.6	9	1	4.4	7	2	4.4
V_6	7.5	3	4.3	5	0	1.4	7	0	1.8	4	1	2.7

Lead	1 - 2 yr.			2 - 3 yr.			3 - 4 yr.			4 - 5 yr.		
	Max.	Min.	Av.	Max.	Min.	Av.	Max.	Min.	Av.	Max.	Min.	Av.
1	4	0	1.5	2	0	0.6	1	0	0.6	2.5	1	1.1
2	2	0	0.6	3	0	1.4	1.5	0	0.6	2	0	1.0
3	8.5	0	0.8	2.5	0	0.9	0.5	0	0.3	8	0	0.6
R	12	0	7.7	11	0	6.0	11	6	8.0	8	0	2.6
L	5	0	2.5	4	0	2.8	2	0	1.0	3	0.5	1.8
F	7	0	0.8	2.5	0	1.0	1	0	0.3	1.5	0	0.7
V_4R	7	0	3.6	10	0	4.2	4	1	2.3	4	3.5	3.8
V_1	17	0	6.8	16	2.5	9.9	9	3	5.7	11	5	8.7
V_2	19	4.5	11.6	17	4	14.0	16	10	12.7	16	11	13.0
V_3	7	0	3.4	8	3	5.6	5	0	2.7	6	3	4.0
V_5	4	0	1.6	3	0.5	1.6	1	0	0.7	2	1.5	1.8
V_6	1.5	0	0.8	2	0	0.8	0.5	0	0.2	1.5	0	0.7

0 = absent

(data from Gros, Gordon, and Miller, 1951, Pediatrics 8:349-361. Copyright American Academy of Pediatrics 1951)

TABLE 607

R/S RATIO FOR INFANTS OF CHICAGO (IL)

Lead	Birth - 1 wk.			1 wk. - 3 mo.			3 - 6 mo.			6 mo. - 1 yr.		
	Max.	Min.	Av.	Max.	Min.	Av.	Max.	Min.	Av.	Max.	Min.	Av.
1	0.4	0.1	0.2	6	1.5	4.8	20	1.4	5.7	3.5	1	2.4
2	∞	0.8	1.3	∞	0.6	8.3	∞	3	6.0	∞	1.7	2.6
3	∞	3.3	1.4	∞	0.3	0.5	∞	1.1	3.0	∞	4	4.3
R	∞	0.5	4.5	∞	1.6	1.8	∞	0	0.3	∞	0.1	1.0
L	0.3	0.1	0.1	16	0.5	8.0	28	0.5	6.3	2.0	0.8	1.3
F	∞	1.3	3.0	∞	0.4	7.1	∞	2.5	4.3	∞	2.7	3.8
V_4R	13.0	1.5	8.2	∞	2.7	2.8	2.0	0.16	0.8	4.0	2.0	3.0
V_1	2.3	0.9	1.4	6.5	1.8	3.7	4.0	0.9	1.7	2.5	1.2	1.7
V_2	4.6	0.8	1.7	6.5	1.1	2.4	1.4	1.0	1.2	2.0	1.0	1.7
V_4	1.0	0.1	5.0	6.5	1.3	3.8	1.7	0.8	1.4	2.8	1.3	2.1
V_5	1.0	0.3	6.0	13.5	1.0	8.0	4.3	1.0	2.8	4.0	1.7	2.9
V_6	0.7	0.3	5.0	∞	1.8	7.7	22	1.0	14.0	6.0	1.5	3.3

Lead	1 - 2 yr.			2 - 3 yr.			3 - 4 yr.			4 - 5 yr.		
	Max.	Min.	Av.	Max.	Min.	Av.	Max.	Min.	Av.	Max.	Min.	Av.
1	∞	5	1.2	∞	1.3	1.4	∞	5	5.0	28	3	12.0
2	∞	2.5	7.2	∞	1.7	5.5	∞	8.76	23.3	∞	5	7.0
3	∞	0.7	7.0	∞	2.8	7.3	∞	16	24.0	22	0.5	9.8
R	0.14	0	0.08	∞	0.0	0.2	0.28	0	0.1	∞	0	0.3
L	∞	1	2.6	∞	0.4	1.5	∞	0.5	1.3	8	0.5	3.1
F	∞	0.57	14.5	∞	4.4	5.9	∞	10	23.0	∞	7.3	11.7
V_4R	∞	0.62	8.8	∞	0.5	0.9	5	0.5	2.4	0.7	0.5	0.6
V_1	∞	0.75	20.2	2.4	0.3	1.1	1.6	0.3	1.0	0.8	0.5	0.6
V_2	∞	0.90	8.0	3	0.8	1.4	1.5	0.8	1.0	5.5	0.5	1.9
V_4	∞	2.14	2.6	3.7	1.0	1.6	∞	2	4.0	5.3	3.3	4.6
V_5	∞	3.14	6.3	16	1.3	6.0	∞	10	11.5	8.7	7.0	7.8
V_6	∞	4	8.1	∞	3.0	7.9	∞	24	32.0	∞	6.7	12.3

∞= Infinity.

(data from Gros, Gordon, and Miller, 1951, Pediatrics 8:349-361. Copyright American Academy of Pediatrics 1951)

TABLE 608

HEIGHT OF T-WAVE (mm) FOR INFANTS OF CHICAGO (IL)

Lead	1 wk. – 3 mo.			3 – 6 mo.			6 mo. – 1 yr.			1 – 5 yr.		
	Max.	Min.	Av.	Max.	Min.	Av.	Max.	Min.	Av.	Max.	Min.	Av.
1	3	-0.5	1.4	3	1	1.6	2	0.5	1.5	4	0.5	1.7
2	3	0.5	1.6	3	1	1.9	4	1	2.5	4	0.5	2.9
3	1	T	0.1	3	-0.3	1	2.5	-0.5	0.5	3	Ind.	0.9
R	-2	-0.5	-1	-3	-1	-2	-3	-0.5	-1.8	-4	-1	-2.6
L	2.5	T	0.9	3	Ind.	1.1	0.3	Ind.	0.2	2	Ind.	0.3
F	3	Ind.	0.8	2	0.5	1.2	4	Ind.	2.5	3	Ind.	2.3
V_4R	-2	-0.5	-1.1	-4	-1	-2.1	-2	-0.5	-1.1	-3	Ind.	-1.6
V_1	-4	-1	-1.6	-5	-1	-3.0	-3	-1	-2.1	-4	-1	-2.7
V_2	-4	-0.5	-1.2	-4	-1	-3.0	-4	-2.5	-3.2	-5	-1.2	-2.7
V_4	6	2	2.2	3	-1	1.8	2	1.5	1.7	5	-1.5	2.6
V_5	5	2	1	5	1	2.3	3	0.5	1.8	4.5	0.5	2.6
V_6	4	1	1.6	5	1	2.2	2	0.5	1.3	3.5	0.5	2.5

Ind. = Indiscernible; T = Tiny; – = Inverted.

(data from Gros, Gordon, and Miller, 1951, Pediatrics 8:349-361. Copyright American Academy of Pediatrics 1951)

TABLE 609

ELECTROCARDIOGRAPHIC DATA

Amplitude Lead V4 (direct writer data; mm)

Age	R Wave			S Wave		
	5%	Mean	95%	5%	Mean	95%
9 mo. – 2 yr.	8.7	17.9	28.3	3.1	11.4	19.9
2–5 yr.	7.6	18.3	30.7	3.3	11.0	18.1
6–13 yr.	8.9	18.4	--	2.1	10.7	21.4

Amplitude Lead V5 (direct writer data; mm)

Age	R Wave			S Wave		
	5%	Mean	95%	5%	Mean	95%
30 hr.	3.1	9.4	16.6	2.4	9.5	18.5
1 mo.	3.8	15.0	24.2	2.8	8.3	16.3
2–3 mo.	9.5	20.7	26.2	1.2	7.9	14.4
4–5 mo.	10.0	20.8	28.8	2.6	8.9	16.0
6–8 mo.	12.0	20.1	29.0	1.5	7.9	19.6
9 mo. – 2 yr.	7.3	17.4	28.4	0.6	5.4	10.5
2–5 yr.	9.4	21.5	33.3	0.6	4.3	8.9
6–13 yr.	12.4	22.0	33.0	0.0	4.0	9.2

Amplitude Lead V6 (direct writer data; mm)

Age	R Wave			S Wave		
	5%	Mean	95%	5%	Mean	95%
30 hr.	1.5	5.4	11.3	1.0	5.6	13.8
1 mo.	1.0	10.8	16.2	0.0	4.8	9.5
2–3 mo.	5.4	12.8	20.8	0.1	4.2	9.1
4–5 mo.	4.4	13.9	22.4	0.0	3.5	8.0
6–8 mo.	6.0	13.0	22.0	0.2	2.5	4.4
9 mo. – 2 yr.	5.7	12.1	20.0	0.3	2.3	5.2
2–5 yr.	6.4	14.4	22.1	0.0	1.5	3.7
6–13 yr.	7.7	15.7	23.3	0.0	1.4	4.1

% = percentile

TABLE 610

ELECTROCARDIOGRAPHIC DATA

Amplitude lead V4R (direct writer data; mm)

Age	R Wave			S Wave		
	5%	Mean	95%	5%	Mean	95%
30 hr.	4.0	8.6	14.2	0.2	3.8	13.0
1 mo.	3.3	6.3	8.5	0.8	1.8	4.6
2-3 mo.	1.1	5.1	10.1	0.0	3.4	9.3
4-5 mo.	2.4	5.2	7.5	0.3	3.5	6.7
6-8 mo.	1.3	4.4	7.1	0.2	3.9	11.7
9 mo.- 2 yr.	0.2	4.0	6.6	0.8	4.9	8.1
2-5 yr.	1.6	3.4	7.4	1.2	4.8	9.5
6-13 yr.	0.6	2.5	5.7	0.9	5.8	12.5

Amplitude lead V1 (direct writer data; mm)

Age	R Wave			S Wave		
	5%	Mean	95%	5%	Mean	95%
30 hr.	4.3	11.9	21.0	1.1	9.7	19.1
1 mo.	3.3	11.1	18.7	0.0	6.1	15.0
2-3 mo.	4.5	11.2	18.0	0.5	7.5	17.1
4-5 mo.	4.5	11.2	17.4	1.0	8.6	16.8
6-8 mo.	3.2	11.4	21.2	1.5	10.7	25.7
9 mo. - 2 yr.	2.5	9.7	15.6	2.0	8.5	17.2
2-5 yr.	2.1	7.5	13.9	2.1	10.9	21.6
6-13 yr.	1.1	5.3	10.7	3.8	12.6	22.3

Amplitude lead V2 (direct writer data; mm)

Age	R Wave			S Wave		
	5%	Mean	95%	5%	Mean	95%
9 mo. - 2 yr.	5.9	15.3	25.2	5.0	14.2	25.5
2-5 yr.	4.2	12.5	20.8	5.4	17.3	29.7
6-13 yr.	3.7	9.7	15.9	8.6	19.5	29.8

% = percentile

(data from Liebman and Plonsey, 1977, in Heart Disease in Infants, Children and Adolescents, 2nd edition, Moss, Adams, and Emmanouilides (eds.). Copyright 1977, The Williams & Wilkins Co., Baltimore)

TABLE   611

ELECTROCARDIOGRAPHIC DATA

Chest Lead T-Wave Voltages (direct writer data; mm)

Age	5%	Mean	95%	5%	Mean	95%
		V1			V5	
10 mo. - 2 yr.	-4.9	-3.1	-1.7	1.6	3.6	6.0
2 - 5 yr.	-4.9	-3.0	-1.4	2.5	4.9	7.6
6 - 13 yr.	-4.5	-2.4	-0.6	1.2	5.0	8.2
		V2			V6	
10 mo. - 2 yr.	-6.2	-3.0	-0.7	1.1	3.0	4.9
2 - 5 yr.	-5.1	-2.4	2.6	2.1	3.8	6.0
6 - 13 yr.	-4.0	0.5	5.0	1.5	3.6	6.0
		V4				
10 mo. - 2 yr.	-3.8	2.2	7.1			
2 - 5 yr.	0.6	3.6	8.0			
6 - 13 yr.	0.9	4.7	8.9			

R Wave in Limb Leads (direct writer data; mm)

Lead	Age	5%	Mean	95%
AVR	10 mo. - 2 yr.	0.1	2.0	3.3
	2 - 5 yr.	0	1.5	3.4
	6 - 13 yr.	0	1.3	3.5
AVL	10 mo. - 2 yr.	0.8	4.1	7.2
	2 - 5 yr.	0.3	3.3	7.4
	6 - 13 yr.	0.2	2.7	6.9
AVF	10 mo. - 2 yr.	2.7	8.7	16.8
	2 - 5 yr.	1.7	10.0	16.4
	6 - 13 yr.	2.9	10.0	16.4

% = percentile
AVR = augmented voltage to right arm
AVL = augmented voltage to left arm
AVF = augmented voltage to left leg

(data from Liebman and Plonsey, 1977, in Heart Disease in Infants, Children and
Adolescents, 2nd edition, Moss, Adams, and Emmanouilides (eds). Copyright 1977,
The Williams & Wilkins Co., Baltimore)

TABLE 612

MEAN FRONTAL, QRS, AND T AXES IN MATURE NEWLY BORN ILLINOIS
INFANTS

Age	QRS axis (frontal plane)					T axis (frontal plane)				
	Minimum	5%	Mean	95%	Maximum	Minimum	5%	Mean	95%	Maximum
0 - 24 hours	60	60	135	130	180	-20	0	70	140	180
1 - 7 days	60	80	125	160	180	-40	-40	25	80	100
8 - 30 days	0	60	110	160	180	-20	0	35	60	120

(data from Liebman, 1968; cited by Hastreiter and Abella, 1971, in Journal of Pediatrics
78:146-156)

TABLE 613

THE FRONTAL QRS AXIS ACCORDING TO AGE
IN ILLINOIS CHILDREN

Age	Percentile		
	10	Mean	90
30 hours	110	130	170
1 month	70	99	120
2-3 months	30	73	110
4-5 months	30	68	100
6-8 months	30	64	100
9-11 months	30	71	100
12-19 months	20	61	90
2-5 years	10	54	80
6-14 years	20	56	80

(data from Namin, pp. 60-120 in Gasul, B.M., Arulka, R.A.,
and Lev, M., eds., Heart Disease in Children: Diagnosis
and Treatment. Lippincott Co., Philadelphia, PA, 1966)

TABLE   614

AMPLITUDE OF THE R WAVE IN THE PRECORDIAL LEADS
EXPRESSED AS PERCENTAGE OF R + S
IN ILLINOIS CHILDREN

Age	Lead	Percentile		
		10	Mean	90
	V_4R			
30 hours		51.8	72.8	79.5
1 month		54.5	80.2	100
2- 3 months		40.0	65.0	92.2
4- 5 months		36.0	62.2	78.5
6- 8 months		33.1	58.7	85.2
9-11 months		29.0	58.9	90.0
12-19 months		24.5	52.4	78.5
2- 5 years		20.0	44.4	71.0
6-14 years		11.5	33.9	54.3
	V_1			
30 hours		43.5	59.6	62.5
1 month		49.5	66.6	89.5
2- 3 months		43.2	62.2	80.5
4- 5 months		49.5	59.3	77.0
6- 8 months		37.0	55.1	74.0
9-11 months		33.8	56.6	78.5
12-19 months		33.5	52.1	67.0
2- 5 years		23.5	41.5	59.5
6-14 years		11.8	28.4	46.3
	V_5			
30 hours		32.2	51.8	72.8
1 month		40.5	64.3	80.5
2- 3 months		54.2	70.1	85.4
4- 5 months		50.8	70.0	88.0
6- 8 months		56.3	73.8	88.1
9-11 months		55.3	73.3	84.0
12-19 months		54.5	82.1	87.0
2- 5 years		60.5	89.6	94.0
6-14 years		70.0	83.6	93.9
	V_6			
30 hours		25.8	54.1	89.5
1 month		40.5	68.5	85.5
2- 3 months		60.6	78.2	92.6
4- 5 months		58.0	81.6	100
6- 8 months		71.0	84.2	93.8
9-11 months		67.0	83.5	93.6
12-19 months		71.7	83.3	92.8
2- 5 years		76.3	89.3	98.5
6-14 years		77.8	89.8	100

(data from Namin, pp. 60-120 in Gasul, B.M., Arulka, R.A., and Lev,
M., eds., Heart Disease in Children: Diagnosis and Treatment.
Lippincott Co., Philadelphia, PA   1966)

TABLE   615

AMPLITUDE OF THE R WAVE IN THE PRECORDIAL LEADS*
IN ILLINOIS CHILDREN

Age	Lead	Percentile		
		10	Mean	90
	V_4R			
30 hours		5.0	8.7	13.0
1 month		4.0	6.3	8.0
2- 3 months		2.0	5.1	9.0
4- 5 months		3.0	5.2	7.0
6- 8 months		2.0	4.4	6.5
9-11 months		2.0	3.9	6.0
12-19 months		1.0	4.0	6.0
2- 5 years		2.0	3.4	6.5
6-14 years		1.0	2.5	5.0
	V_1			
30 hours		6.0	11.9	19.0
1 month		5.0	11.1	17.0
2- 3 months		6.0	11.2	16.5
4- 5 months		6.0	11.2	16.0
6- 8 months		5.0	11.4	19.0
9-11 months		5.0	10.8	16.0
12-19 months		5.0	11.8	18.0
2- 5 years		4.0	8.4	14.0
6-14 years		2.0	5.7	10.0
	V_5			
30 hours		4.5	9.4	15.0
1 month		3.8	15.0	22.2
2- 3 months		12.0	20.7	25.0
4- 5 months		12.0	20.8	27.0
6- 8 months		13.5	20.1	27.0
9-11 months		10.5	21.0	29.0
12-19 months		11.0	20.2	31.0
2- 5 years		13.0	24.4	36.5
6-14 years		16.0	25.0	36.0
	V_6			
30 hours		2.0	5.4	10.0
1 month		3.0	10.8	15.0
2- 3 months		7.0	12.8	19.0
4- 5 months		6.5	13.9	20.5
6- 8 months		7.0	13.0	20.0
9-11 months		6.0	13.8	22.5
12-19 months		7.0	13.1	21.0
2- 5 years		8.0	15.0	21.5
6-14 years		10.5	16.5	23.0

*Measurements are in millimeters or tenths of a millivolt.

(data from Namin, pp. 60-120 in Gasul, B.M., Arulka, R.A., and Lev,
M., eds., Heart Disease in Children: Diagnosis and Treatment.
Lippincott Co., Philadelphia, PA 1966)

TABLE 616

AMPLITUDE OF THE S WAVE IN THE PRECORDIAL LEADS*
IN ILLINOIS CHILDREN

Age	Lead	Percentile		
		10	Mean	90
	V_4R			
30 hours		1.0	3.8	11.0
1 month		1.0	1.8	4.0
2- 3 months		0.6	3.4	8.0
4- 5 months		1.0	3.5	6.0
6- 8 months		1.0	3.9	10.0
9-11 months		1.0	5.3	6.0
12-19 months		1.0	4.4	8.8
2- 5 years		2.0	4.8	8.5
6-14 years		2.0	5.8	11.0
	V_1			
30 hours		3.0	9.7	17.0
1 month		1.0	6.1	13.0
2- 3 months		2.0	7.5	15.0
4- 5 months		2.0	8.6	15.0
6- 8 months		2.5	10.7	22.0
9-11 months		2.5	9.4	18.0
12-19 months		5.0	10.8	17.0
2- 5 years		5.0	12.0	21.5
6-14 years		7.0	14.5	23.0
	V_5			
30 hours		4.0	9.5	16.5
1 month		4.0	8.3	14.5
2- 3 months		2.7	7.9	13.0
4- 5 months		4.0	8.9	15.0
6- 8 months		2.0	7.9	17.0
9-11 months		3.0	7.2	13.0
12-19 months		2.0	7.5	12.0
2- 5 years		2.0	5.2	9.5
6-14 years		1.0	5.1	10.0
	V_6			
30 hours		2.0	5.6	12.0
1 month		1.0	4.8	8.5
2- 3 months		1.0	4.2	8.0
4- 5 months		0	3.5	7.0
6- 8 months		0.5	2.5	4.0
9-11 months		0.5	3.0	5.0
12-19 months		1.0	2.9	6.0
2- 5 years		0	1.9	4.0
6-14 years		0	1.9	4.0

*Measurements are in millimeters or tenths of a millivolt.

(data from Namin, pp. 60-120 in Gasul, B.M., Arulka, R.A., and Lev,
M., eds., Heart Disease in Children: Diagnosis and Treatment.
Lippincott Co., Philadelphia, PA 1966)

TABLE 617

ELECTROCARDIOGRAPHIC DATA IN CHICAGO (IL) CHILDREN

Lead	5 or less years Average	6 to 10 years Average	11 to 15 years Average

Duration of P Wave (seconds)

Lead	5 or less years Average	6 to 10 years Average	11 to 15 years Average
I	0.07	0.08	0.08
II	0.07	0.08	0.08
III	0.06	0.07	0.07
aVR	0.07	0.08	0.10
aVL	0.05	0.06	0.06
aVF	0.06	0.07	0.07
V1	0.07	0.06	0.10
V2	0.06	0.07	0.07
V3	0.06	0.08	0.12
V4	0.07	0.08	0.11
V5	0.07	0.08	0.08
V6	0.07	0.07	0.07
V7	0.07	0.06	0.08
V3R	0.06	0.06	0.07
V4R	0.06	0.06	0.06

Duration of P-R Interval (seconds)

Lead	5 or less years Average	6 to 10 years Average	11 to 15 years Average
I	0.12	0.15	0.14
II	0.13	0.15	0.15
III	0.13	0.15	0.15
aVR	0.13	0.15	0.15
aVL	0.11	0.13	0.13
aVF	0.13	0.15	0.15
V1	0.13	0.14	0.15
V2	0.13	0.14	0.15
V3	0.13	0.15	0.15
V4	0.13	0.14	0.15
V5	0.13	0.14	0.15
V6	0.13	0.15	0.14
V7	0.13	0.14	0.14
V3R	0.13	0.14	0.15
V4R	0.13	0.14	0.14

(data from Switzer and Besoain, 1950, American Journal of Diseases of Children 79:449-466. Copyright 1950, American Medical Association)

TABLE 618

HEIGHT OF P WAVE
IN CHICAGO (IL) CHILDREN

Lead	5 or less years		6 to 10 years		11 to 15 years	
	Average	Usual	Average	Usual	Average	Usual
I	0.8	Positive	0.7	Positive	0.8	Positive
II	1.2	Positive	1.1	Positive	1.1	Positive
III	0.5	Positive	0.4	Positive	0.6	Positive
aVR	-1.0	Negative	-0.9	Negative	-1.9	Negative
aVL	0.4	Positive	0.3	Positive	0.3	Positive
aVF	1.0	Positive	0.8	Positive	0.8	Positive
V1	0.8	Positive	0.6	Positive	0.9	Positive
V2	0.9	Positive	0.9	Positive	1.0	Positive
V3	0.9	Positive	0.8	Positive	0.8	Positive
V4	0.7	Positive	0.6	Positive	0.8	Positive
V5	0.6	Positive	0.6	Positive	0.7	Positive
V6	0.6	Positive	0.5	Positive	0.5	Positive
V7	0.5	Positive	0.5	Positive	0.4	Positive
V3R	0.7	Positive	0.5	Positive	0.7	Positive
V4R	0.5	Positive	0.4	Positive	0.4	Positive

(data from Switzer and Besoain, 1950, American Journal of Diseases of Children 79:449-466.
Copyright 1950, American Medical Association)

TABLE 619

ELECTROCARDIOGRAPHIC DATA IN CHICAGO (IL) CHILDREN

Lead	5 or less years Average	6 to 10 years Average	11 to 15 years Average
	Height of R Wave (mm)		
I	5.2	5.7	6.1
II	11.2	10.2	11.2
III	8.0	7.0	5.9
aVR	1.4	1.5	1.2
aVL	2.7	2.6	2.4
aVF	10.9	8.1	8.7
V1	5.5	4.5	5.3
V2	11.1	10.3	9.9
V3	16.2	>12.7	>12.6
V4	>17.4	16.3	14.1
V5	16.0	14.6	>11.1
V6	15.1	>12.3	>12.6
V7	11.9	10.7	15.5
V3R	4.5	4.0	3.8
V4R	2.8	2.8	2.3
	Duration of T Wave (seconds)		
I	0.14	0.17	0.16
II	0.15	0.18	0.18
III	0.13	0.14	0.15
aVR	0.15	0.17	0.17
aVL	0.11	0.15	0.14
aVF	0.13	0.19	0.17
V1	0.13	0.14	0.13
V2	0.12	0.18	0.14
V3	0.15	0.21	0.20
V4	0.17	0.21	0.21
V5	0.17	0.19	0.18
V6	0.16	0.19	0.17
V7	0.16	0.18	0.16
V3R	0.14	0.14	0.14
V4R	0.13	0.13	0.13

The sign > indicates measurement was not accurate inasmuch as wave deflection exceeded the width of the recording film. Value equals the number of millimeters actually measurable.

(data from Switzer and Besoain, 1950, American Journal of Diseases of Children 79:449-466. Copyright 1950, American Medical Association)

TABLE 620

DEPTH OF Q WAVE (mm)
IN CHICAGO (IL) CHILDREN

Lead	5 or less years		6 to 10 years		11 to 15 years	
	Average	% with Q	Average	% with Q	Average	% with Q
I	0.4	50	0.3	28	0.5	54
II	1.3	70	0.3	33	0.6	46
III	1.9	80	0.9	61	0.6	50
aVR	1.7	30	4.4	72	2.4	37
aVL	0.1	20	0.3	33	0.5	50
aVF	1.3	90	0.5	33	0.7	54
V1	--	--	--	--	--	--
V2	--	--	--	--	--	--
V3	--	--	--	--	--	4
V4	1.4	60	0.4	39	0.8	46
V5	2.1	70	0.7	67	1.0	79
V6	1.7	70	0.6	67	1.1	83
V7	1.1	80	0.6	67	0.9	83
V3R	0.03	10	0.1	6	--	--
V4R	0.03	10	0.1	6	--	4

(data from Switzer and Besoain, 1950, American Journal of Diseases of
Children 79:449-466. Copyright 1950, American Medical Association)

TABLE 621

ELECTROCARDIOGRAPHIC DATA IN CHICAGO (IL) CHILDREN

Lead	5 or less years Average	6 to 10 years Average	11 to 15 years Average
	Depth of S Wave (mm)		
I	1.8	1.9	1.4
II	1.8	1.4	1.4
III	0.9	0.6	1.6
aVR	5.7	2.5	5.7
aVL	3.2	3.5	2.6
aVF	0.8	0.9	1.1
V1	6.2	6.8	11.4
V2	>12.2	>14.0	>15.5
V3	8.9	11.2	11.8
V4	4.8	7.1	7.8
V5	2.1	3.6	3.0
V6	1.7	> 2.5	1.5
V7	1.1	1.4	0.9
V3R	4.8	5.9	7.6
V4R	2.9	4.1	5.4
	Duration of QRS Complex (seconds)		
I	0.06	0.07	0.07
II	0.06	0.07	0.07
III	0.07	0.07	0.07
aVR	0.06	0.07	0.07
aVL	0.06	0.07	0.07
aVF	0.07	0.07	0.07
V1	0.07	0.08	0.08
V2	0.07	0.08	0.09
V3	0.08	0.08	0.08
V4	0.07	0.07	0.08
V5	0.07	0.07	0.08
V6	0.07	0.07	0.07
V7	0.07	0.07	0.07
V3R	0.07	0.07	0.08
V4R	0.07	0.07	0.08

The sign > indicates measurement was not accurate inasmuch as wave deflection exceeded the width of the recording film. Value equals the number of millimeters actually measurable.

(data from Switzer and Besoain, 1950, American Journal of Diseases of Children 79:449-466. Copyright 1950, American Medical Association)

TABLE 622

DURATION OF P WAVE IN A PRECORDIAL LEAD (seconds)
IN MICHIGAN CHILDREN

Age Group	.35	.40	.45	.50	.55	.60	.70	.80	.90	1.00	1.05
0 - 24 Hours	--	12	9	44	16	16	3	--	--	--	--
1 - 7 Days	4	25	42	4	18	7	--	--	--	--	--
8 - 30 Days	--	25	32	5	35	3	--	--	--	--	--
1 - 3 Months	--	34	32	10	19	5	--	--	--	--	--
3 - 6 Months	--	30	--	48	--	22	--	--	--	--	--
6 - 12 Months	--	3	3	7	13	67	7	--	--	--	--
1 - 3 Years	--	--	3	2	7	47	23	18	--	--	--
3 - 5 Years	--	2	--	4	--	32	28	28	6	--	--
5 - 8 Years	--	--	--	3	--	32	28	34	3	--	--
8 - 12 Years	--	--	--	1	--	16	30	40	11	2	--
12 - 16 Years	--	--	--	--	--	7	16	43	30	4	--

(data from Ziegler, Electrocardiographic Studies in Normal Infants and Children,
1951. Courtesy of Charles C Thomas, Publisher, Springfield, Illinois)

TABLE 623

DURATION OF PR INTERVAL IN STANDARD LEAD II (seconds)
IN MICHIGAN CHILDREN

Age Group		.06	.08	.10	.12	.14	.16	.18	.20	.22	.24	.26
0 - 24	Hours	--	25	63	12	--	--	--	--	--	--	--
1 - 7	Days	4	43	43	10	--	--	--	--	--	--	--
8 - 30	Days	--	38	60	2	--	--	--	--	--	--	--
1 - 3	Months	--	42	47	8	--	3	--	--	--	--	--
3 - 6	Months	--	26	52	22	--	--	--	--	--	--	--
6 - 12	Months	--	11	66	19	4	--	--	--	--	--	--
1 - 3	Years	--	11	37	38	11	3	--	--	--	--	--
3 - 5	Years	--	--	21	47	30	2	--	--	--	--	--
5 - 8	Years	--	--	8	41	38	11	2	--	--	--	--
8 - 12	Years	--	--	7	30	43	16	2	1	--	--	1
12 - 16	Years	--	--	1	24	36	34	4	1	--	--	--

(data from Ziegler, Electrocardiographic Studies in Normal Infants and Children,
1951. Courtesy of Charles C Thomas, Publisher, Springfield, Illinois)

TABLE   624

AMPLITUDE OF Q WAVES IN EACH EXTREMITY LEAD IN 1/10 OF A MILLIVOLT
IN MICHIGAN CHILDREN

LEADS

Age	1	2	3	aVR	aVL	aVF
0-24 hours	0.50	1.48	2.47	2.35	1.25	1.84
1 day-1 week	0.83	1.77	2.48	2.67	0.67	1.98
1 week-1 month	1.05	1.59	2.35	3.93	1.25	1.90
1-3 months	1.08	1.83	3.26	7.09	1.67	2.48
3-6 months	1.16	2.33	4.53	8.0	1.13	2.71
6 months-1 year	1.60	2.38	4.07	8.10	1.75	2.85
1-3 years	1.45	2.13	3.51	7.29	1.30	2.18
3-5 years	1.19	1.64	2.49	7.40	1.30	1.72
5-8 years	1.09	1.41	2.23	8.20	1.19	1.53
8-12 years	1.32	1.38	1.85	8.15	1.25	1.47
12-16 years	0.99	1.18	1.56	7.90	1.34	1.28

AMPLITUDE OF Q WAVES IN UNIPOLAR PRECORDIAL LEADS IN 1/10 OF A MILLIVOLT

LEADS

Age	V1	V2	V3	V4	V5	V6
0-24 hours	0	0	0	1.25	2.17	1.25
1 day-1 week	0	0	3.0	1.0	2.0	1.33
1 week-1 month	0	0	1.0	1.0	2.0	1.37
1-3 months	0	0	0.5	1.38	2.0	1.42
3-6 months	0	0	1.0	1.75	2.1	1.84
6 months-1 year	0	0	0	1.54	2.28	2.42
1-3 years	0	0	1.0	2.5	2.32	1.74
3-5 years	0	0	1.0	1.67	2.10	1.78
5-8 years	0	0	0.75	1.63	1.94	1.72
8-12 years	0	0	0.50	1.78	1.89	1.68
12-16 years	0	0	0	0	1.31	1.30

(data from Ziegler, Electrocardiographic Studies in Normal Infants and Children,
1951. Courtesy of Charles C Thomas, Publisher, Springfield, Illinois)

TABLE 625

DURATION OF QT INTERVAL IN STANDARD LEAD II (seconds)
IN MICHIGAN CHILDREN

Age Group		.16	.18	.20	.22	.24	.26	.28	.30	.34	.38	.42
0 - 24	Hours	--	--	--	--	6	19	22	41	12	--	--
1 - 7	Days	--	--	--	11	29	18	21	18	3	--	--
8 - 30	Days	--	--	12	15	50	20	3	--	--	--	--
1 - 3	Months	3	--	10	16	32	26	10	--	3	--	--
3 - 6	Months	--	--	9	9	30	35	13	4	--	--	--
6 - 12	Months	--	--	3	6	30	27	25	9	--	--	--
1 - 3	Years	--	--	--	--	--	--	--	--	23	52	25
3 - 5	Years	--	--	--	--	--	--	--	--	--	--	--
5 - 8	Years	--	--	--	--	--	1	11	59	28	1	--
8 - 12	Years	--	--	--	--	--	--	1	23	61	14	1
12 - 16	Years	--	--	--	--	--	--	--	33	41	23	3

(data from Ziegler, Electrocardiographic Studies in Normal Infants and Children,
1951. Courtesy of Charles C Thomas, Publisher, Springfield, Illinois)

TABLE 626

THE INCIDENCE AND AMPLITUDE (millivolts) OF RST SEGMENT DEVIATION IN THE
EXTREMITY LEADS OF THE ELECTROCARDIOGRAM OF NORMAL INFANTS AND CHILDREN
IN MICHIGAN

	Leads								
	1			2			3		
Age	Incidence	Min.	Max.	Incidence	Min.	Max.	Incidence	Min.	Max.
0 - 24 hours	3.13	-0.1	--	15.6	--	+0.1	25.0	--	+0.1
1 day - 1 week	14.3	-0.1	+0.1	42.8	--	+0.2	53.6	+0.1	+0.2
1 week - 1 mo.	20.0	-0.1	+0.1	17.5	--	+0.1	15.0	-0.2	+0.2
1 - 3 months	5.27	0.1	+0.1	5.27	--	+0.1	13.2	--	+0.1
3 - 6 months	--	--	--	26.1	-0.1	+0.1	26.1	-0.05	+0.1
6 mos. - 1 year	13.9	--	+0.1	13.9	--	+0.1	11.1	0.1	--
1 - 3 years	7.2	-0.1	+0.1	3.5	--	+0.2	8.8	-0.1	+0.1
3 - 5 years	3.78	-0.1	--	5.65	-0.1	+0.1	7.54	-0.05	+0.1
5 - 8 years	2.22	--	+0.1	2.22	-0.1	+0.1	4.44	-0.1	+0.05
8 - 12 years	3.4	-0.1	+0.05	3.4	-0.05	+0.2	1.14	--	+0.15
12 - 16 years	2.95	--	+0.1	8.8	--	+0.1	4.4	--	+0.1

	aVR			aVL			VF		
Age	Incidence	Min.	Max.	Incidence	Min.	Max.	Incidence	Min.	Max.
0 - 24 hours	9.4	-0.1	+0.1	0	--	--	0.4	--	+0.1
1 day - 1 week	25.0	-0.1	+0.1	10.7	-0.1	+0.1	32.4	--	+0.2
1 week - 1 mo.	0.5	-0.1	--	0.25	-0.1	--	10.0	--	+0.1
1 - 3 months	0	--	--	2.64	-0.1	--	7.9	--	+0.1
3 - 6 months	4.35	-0.1	--	0	--	--	21.8	-0.05	+0.1
6 mos. - 1 year	0	--	--	2.78	--	+0.1	8.15	--	+0.1
1 - 3 years	7.2	--	+0.2	1.78	-0.1	--	8.8	-0.1	+0.1
3 - 5 years	1.88	--	+0.1	1.88	-0.1	--	5.65	--	+0.1
5 - 8 years	0	--	--	1.11	--	+0.1	2.22	-0.1	--
8 - 12 years	1.14	--	+0.1	1.14	-0.05	--	3.4	--	+0.05
12 - 16 years	1.47	-0.05	--	1.47	--	+0.1	4.4	--	+0.1

THE INCIDENCE AND AMPLITUDE (millivolts) OF RST SEGMENT DEVIATION IN
THE PRECORDIAL LEADS OF THE ELECTROCARDIOGRAM OF NORMAL INFANTS AND CHILDREN

	V1			V2			V3		
Age	Incidence	Min.	Max.	Incidence	Min.	Max.	Incidence	Min.	Max.
0-24 hours	25.9	-0.1	+0.1	51.9	-0.2	+0.1	73.5	-0.05	+0.25
1 day - 1 week	50.0	-0.2	0	37.5	-0.2	+0.1	68.2	-0.2	+0.2
1 week - 1 month	41.7	-0.1	+0.1	25.0	-0.2	+0.1	50.0	-0.2	+0.2
1 - 3 months	23.1	-0.2	--	3.85	-0.1	--	62.6	-0.15	+0.2
3 - 6 months	50.0	-0.1	--	28.5	-0.1	+0.1	50.0	--	+0.1
6 mos. - 1 year	11.7	-0.1	+0.1	31.8	-0.1	+0.2	27.3	-0.05	+0.1
1 - 3 years	13.5	--	+0.2	43.4	--	+0.1	26.1	--	+0.1
3 - 5 years	10.7	-0.1	+0.1	57.1	-0.1	+0.1	32.2	--	+0.15
5 - 8 years	31.8	--	+0.2	68.1	--	+0.45	61.4	--	+0.25
8 - 12 years	33.3	--	+0.1	93.2	--	+0.2	89.0	--	+0.2
12 - 16 years	57.8	--	+0.2	95.5	--	+0.3	89.0	--	+0.25

TABLE 626   (continued)

THE INCIDENCE AND AMPLITUDE (millivolts) OF RST SEGMENT DEVIATION IN
THE PRECORDIAL LEADS OF THE ELECTROCARDIOGRAM OF NORMAL INFANTS AND CHILDREN
IN MICHIGAN

| | Leads | | | | | | | |
| | V4 | | | V5 | | | V6 | | |
Age	Inci-dence	Min.	Max.	Inci-dence	Min.	Max.	Inci-dence	Min.	Max.
0 - 24 hours	80.0	-0.1	+0.1	80.0	--	+0.2	7.4	--	+0.1
1 day - 1 week	91.0	--	+0.1	86.3	--	+0.2	12.5	--	+0.1
1 week - 1 month	75.0	--	+0.1	62.6	--	+0.2	8.35	--	+0.05
1 - 3 months	75.0	--	+0.1	37.5	--	+0.15	11.55	--	+0.1
3 - 6 months	75.0	--	+0.15	62.6	--	+0.2	21.5	--	+0.1
6 mos. - 1 year	31.8	--	+0.2	36.5	--	+0.2	27.1	--	+0.1
1 - 3 years	26.1	-0.1	+0.1	8.7	--	+0.1	4.45	--	+0.1
3 - 5 years	32.2	--	+0.2	17.9	--	+0.1	10.7	--	+0.1
5 - 8 years	36.4	--	+0.2	20.5	-0.1	+0.1	6.82	-0.1	+0.1
8 - 12 years	62.2	--	+0.2	37.8	--	+0.15	15.6	--	+0.1
12 - 16 years	66.6	--	+0.2	44.5	--	+0.1	20.0	--	+0.1

(data from Ziegler, Electrocardiographic Studies in Normal Infants and Children, 1951.
Courtesy of Charles C Thomas, Publisher, Springfield, Illinois)

TABLE   627

AMPLITUDE OF RST IN STANDARD LEAD I (in 1/10 millivolt)
IN MICHIGAN CHILDREN

Age Group	-1.0	-0.5	0	+0.5	+1.0
0 - 24 hours	3	--	97	--	--
1 - 7 days	7	--	86	--	7
8 - 30 days	5	3	80	5	7
1 - 3 months	3	--	94	--	3
3 - 6 months	--	--	100	--	--
6 - 12 months	--	--	85	6	9
1 - 3 years	2	2	92	--	4
3 - 5 years	2	2	96	--	--
5 - 8 years	--	--	98	1	1
8 - 12 years	2	--	97	1	--
12 - 16 years	--	--	96	2	2

(data from Ziegler, Electrocardiographic Studies in Normal
Infants and Children, 1951. Courtesy of Charles C Thomas,
Publisher, Springfield, Illinois)

TABLE 628

AMPLITUDE OF RST IN STANDARD LEAD II (in 1/10 millivolt)
IN MICHIGAN CHILDREN

Age Group	-1.0	-.05	0	+0.5	+1.0	+1.5	+2.0
0 - 24 Hours	--	--	84	3	13	--	--
1 - 7 Days	--	--	56	11	22	--	11
8 - 30 Days	--	--	81	3	13	3	--
1 - 3 Months	--	--	95	--	5	--	--
3 - 6 Months	4	--	75	4	17	--	--
6 - 12 Months	--	--	85	6	9	--	--
1 - 3 Years	--	--	96	--	2	--	2
3 - 5 Years	2	--	94	2	2	--	--
5 - 8 Years	1	--	98	--	1	--	--
8 - 12 Years	--	1	97	1	--	--	1
12 - 16 Years	--	--	91	--	9	--	--

(data from Ziegler, Electrocardiographic Studies in Normal Infants
and Children, 1951. Courtesy of Charles C Thomas, Publisher, Springfield,
Illinois)

TABLE 629

AMPLITUDE OF THE S WAVE IN THE STANDARD AND
UNIPOLAR EXTREMITY LEADS IN 1/10 OF A MILLIVOLT
IN MICHIGAN CHILDREN

LEADS

Age	1	2	3	aVR	aVL	aVF
0-24 hours	6.25	3.21	2.25	3.92	6.58	3.0
1 day-1 week	5.18	3.80	1.75	3.23	5.80	2.10
1 week-1 month	4.61	2.21	2.63	4.82	5.66	2.67
1-3 months	3.85	2.94	1.88	8.53	4.80	2.14
3-6 months	3.79	3.23	2.57	10.96	4.72	2.50
6 months-1 year	4.24	3.11	2.11	10.50	5.10	2.02
1-3 years	3.65	2.70	2.64	10.0	4.10	1.92
3-5 years	2.53	2.23	2.12	9.95	4.18	1.98
5-8 years	2.93	2.52	1.20	9.65	4.18	2.23
8-12 years	2.77	2.78	2.15	10.14	4.22	2.18
12-16 years	2.07	2.25	1.92	10.0	3.98	1.73

TABLE 630

AMPLITUDE OF THE S WAVE IN THE UNIPOLAR PRECORDIAL LEADS
IN 1/10 OF A MILLIVOLT

LEADS

Age	V1	V2	V3	V4	V5	V6
0-24 hours	10.0	22.0	26.4	23.0	12.0	4.5
1 day-1 week	11.1	21.7	18.7	13.3	7.6	3.2
1 week-1 month	7.0	17.1	14.1	9.7	5.9	2.6
1-3 months	6.9	13.0	14.9	11.0	5.3	2.0
3-6 months	7.3	17.0	16.0	8.5	4.8	1.7
6 months-1 year	8.1	17.9	13.9	11.0	3.3	2.0
1-3 years	11.9	21.1	14.7	7.32	3.1	1.0
3-5 years	13.0	22.8	13.1	5.9	2.8	0.6
5-8 years	13.9	23.9	16.8	8.2	3.2	0.9
8-12 years	15.2	23.1	15.1	7.8	3.0	1.0
12-16 years	14.5	23.7	12.6	6.0	3.0	1.4

(data from Ziegler, Electrocardiographic Studies in Normal Infants and Children,
1951. Courtesy of Charles C Thomas, Publisher, Springfield, Illinois)

TABLE 631

DURATION OF T WAVE IN A PRECORDIAL LEAD I (seconds)
IN MICHIGAN CHILDREN

Age Group	.07	.08	.09	.10	.11	.12	.13	.14	.15	.17	.19	.21
0 - 24 Hours	--	--	4	4	4	18	--	30	33	7	--	--
1 - 7 Days	--	--	--	5	5	20	10	10	30	10	10	--
8 - 30 Days	--	--	4	22	17	35	9	9	4	--	--	--
1 - 3 Months	3	6	3	20	3	26	6	20	10	3	--	--
3 - 6 Months	--	6	--	11	--	27	6	22	22	--	6	--
6 - 12 Months	--	--	--	--	4	23	14	32	18	9	--	--
1 - 3 Years	--	--	--	8	2	29	17	23	21	--	--	--
3 - 5 Years	--	--	--	2	--	11	9	30	32	9	5	2
5 - 8 Years	--	--	--	--	3	20	17	18	27	14	--	1
8 - 12 Years	--	--	1	--	--	13	2	13	32	28	11	--
12 - 16 Years	--	--	--	--	--	7	7	19	41	17	7	2

(data from Ziegler, Electrocardiographic Studies in Normal Infants and Children,
1951. Courtesy of Charles C Thomas, Publisher, Springfield, Illinois)

TABLE 632

AMPLITUDE OF THE T WAVES IN THE EXTREMITY LEADS IN 1/10 OF A MILLIVOLT
IN MICHIGAN CHILDREN

| Age | Leads | | | | | |
	1	2	3	aVR	aVL	aVF
0 - 24 hours	0.34	1.17	0.95	-0.42	0.05	0.91
1 day - 1 week	1.45	1.70	0.41	-1.45	1.02	0.98
1 week - 1 mo.	2.1	2.66	0.54	-2.36	0.86	1.65
1 - 3 months	2.24	3.05	0.895	-2.32	0.764	1.84
3 - 6 months	2.675	3.63	1.18	-2.91	0.91	2.07
6 mos. - 1 yr.	2.61	3.10	0.60	-2.85	1.24	1.84
1 - 3 years	2.88	3.00	0.126	-2.90	1.42	1.68
3 - 5 years	2.80	3.26	0.51	-2.93	1.36	2.0
5 - 8 years	2.69	3.32	0.605	-2.79	1.05	1.90
8 - 12 years	3.00	3.68	0.717	-2.17	1.33	2.10
12 - 16 years	2.65	3.48	0.89	-2.85	1.095	2.275

AMPLITUDE OF THE T WAVES IN THE PRECORDIAL LEADS IN 1/10 OF A MILLIVOLT

| Age | Leads | | | | | |
	V1	V2	V3	V4	V5	V6
0 - 24 hours	1.3	1.3	-0.4	-0.6	1.3	1.2
1 day - 1 week	-3.7	-4.5	-1.4	2.3	2.9	1.9
1 week - 1 mo.	-2.8	-2.1	-0.8	3.5	3.3	2.7
1 - 3 months	-2.8	-2.9	-0.5	3.4	3.6	2.5
3 - 6 months	-4.1	-5.2	-3.0	3.4	3.8	2.7
6 mos. - 1 yr.	-3.9	-4.6	-2.75	3.0	3.5	2.8
1 - 3 years	-3.6	-3.9	-1.46	4.5	4.15	3.15
3 - 5 years	-3.8	-4.2	0.33	4.6	4.67	3.6
5 - 8 years	-3.4	-1.4	1.84	5.85	5.68	4.04
8 - 12 years	-2.9	+1.2	2.93	5.7	5.91	4.4
12 - 16 years	-1.2	+6.0	5.6	7.5	5.91	3.98

(data from Ziegler, Electrocardiographic Studies in Normal Infants and Children,
1951. Courtesy of Charles C Thomas, Publisher, Springfield, Illinois)

TABLE 633

AMPLITUDE OF T WAVE IN STANDARD LEAD I (in 1/10 millivolt)
IN MICHIGAN CHILDREN

Age Group	-2.0	.1.0	+ -	0.0	+1.0	+2.0	+3.0	+4.0	+5.0	+6.0
1 - 7 Days	--	--	3	--	58	29	10	--	--	--
8 - 30 Days	6	--	--	--	34	39	21	--	--	--
1 - 3 Months	--	--	--	3	21	42	31	3	--	--
3 - 6 Months	--	--	--	--	13	39	35	4	9	--
6 - 12 Months	--	--	--	--	25	28	30	14	--	3
1 - 3 Years	--	--	--	--	16	32	25	18	5	4
3 - 5 Years	--	--	--	--	8	44	32	12	2	2
5 - 8 Years	--	--	--	--	7	43	33	14	3	--
8 - 12 Years	--	--	--	--	9	30	37	19	4	1
12 - 16 Years	--	--	--	--	14	41	32	11	1	1

(data from Ziegler, Electrocardiographic Studies in Normal Infants and Children,
1951. Courtesy of Charles C Thomas, Publisher, Springfield, Illinois)

TABLE 634

AMPLITUDE OF T WAVE IN STANDARD LEAD II (in 1/10 millivolt)
IN MICHIGAN CHILDREN

Age Group		+ -	+ - +	0 0	+1.0	+2.0	+3.0	+4.0	+5.0	+6.0	+7.0	+8.0
1 - 7	Days	3	--	--	57	29	11	--	--	--	--	--
8 - 30	Days	--	--	5	23	18	50	5	--	--	--	--
1 - 3	Months	--	--	3	13	19	23	37	5	--	--	--
3 - 6	Months	--	--	--	13	4	31	31	17	4	--	--
6 - 12	Months	--	--	--	8	28	33	14	14	3	--	--
1 - 3	Years	--	--	2	7	32	35	11	11	2	--	--
3 - 5	Years	--	--	--	10	26	32	16	14	2	--	--
5 - 8	Years	--	--	2	5	23	31	25	14	--	--	--
8 - 12	Years	--	1	--	5	17	25	29	17	5	1	--
12 - 16	Years	--	--	2	7	22	28	18	14	7	2	--

(data from Ziegler, Electrocardiographic Studies in Normal Infants and Children,
1951. Courtesy of Charles C Thomas, Publisher, Springfield, Illinois)

TABLE 635

THE INCIDENCE OF UPRIGHT, INVERTED, ISOELECTRIC, AND DIPHASIC (or multiphasic)
T WAVES IN THE EXTREMITY AND PRECORDIAL LEADS OF NORMAL INFANTS AND CHILDREN
IN MICHIGAN

Age		Leads 1	2	3	VR	VL	VF
0 - 24 hours	+	50	84.5	68.75	31.25	31.25	75
	-	31.25	3.125	6.25	53.1	37.5	6.25
	0	3.125	0	9.4	6.25	15.6	6.25
	+-	15.6	12.5	15.6	9.4	15.6	12.5
1 day - 1 wk.	+	92.7	92.7	53.6	0	82.2	68
	-	3.57	0	17.85	85.8	3.57	7.15
	0	0	3.57	10.7	0	10.7	7.15
	+-	3.57	3.57	17.85	14.3	3.57	17.85
1 wk. - 1 mo.	+	97.5	95	60	0	72.5	92.5
	-	2.5	0	20	100	5	0
	0	0	5	12.5	0	12.5	7.5
	+-	0	0	7.5	0	10	0
1 - 3 mos.	+	97.5	97.5	93.6	0	71	97.5
	-	0	0	3.2	95	15.8	2.5
	0	2.5	0	3.2	2.5	5.3	0
	+-	0	2.5	0	2.5	7.9	0
3 - 6 mos.	+	100	100	69.5	0	82.6	95.65
	-	0	0	13.1	100	4.35	0
	0	0	0	8.7	0	4.35	0
	+-	0	0	8.7	0	8.7	4.35
6 mos. - 1 yr.	+	100	100	52.8	0	75	97.2
	-	0	0	27.8	100	5.6	0
	0	0	0	0	0	16.6	0
	+-	0	0	19.4	0	2.8	2.8
1 - 3 yrs.	+	100	98	45.5	0	87.2	90
	-	0	0	31.5	100	2.0	4
	0	0	2	8.8	0	8.8	2
	+-	0	0	14.2	0	2.0	4
3 - 5 yrs.	+	100	100	53	0	88.5	88.5
	-	0	0	20.8	100	3.8	5.7
	0	0	0	13.1	0	5.7	0
	+-	0	0	13.1	0	2.0	0
5 - 8 yrs.	+	100	98	53.4	0	82	93.2
	-	0	0	17.8	100	3.33	3.33
	0	0	2	8.9	0	13.4	2.22
	+-	0	0	19.9	0	1.1	1.11
8 - 12 yrs.	+	100	99	60.3	0	91	93.2
	-	0	0	17	100	3.4	0
	0	0	0	10.2	0	2.27	2.27
	+-	0	1	12.5	0	3.43	4.5

TABLE 635 (continued)

THE INCIDENCE OF UPRIGHT, INVERTED, ISOELECTRIC, AND DIPHASIC (or multiphasic)
T WAVES IN THE EXTREMITY AND PRECORDIAL LEADS OF NORMAL INFANTS AND CHILDREN
IN MICHIGAN

Age		Leads					
		1	2	3	VR	VL	VR
12 - 16 yrs.	+	100	98.5	66.2	0	84	97
	–	0	0	10.3	98.5	6.0	0
	0	0	0	10.3	1.5	7.0	0
	+–	0	1.5	13.2	0	3.0	3.0

Age		Leads					
		V1	V2	V3	V4	V5	V6
0 - 24 hours	+	74.1	74.1	40.0	59.3	46.6	55.7
	–	7.41	3.7	40.0	22.2	13.4	14.8
	0	0	3.7	0	7.41	0	3.7
	+–	18.5	18.5	20.0	11.1	40.0	25.9
1 day - 1 wk.	+	18.75	18.75	27.3	56.2	100.0	87.5
	–	62.5	62.5	36.4	25	0	0
	0	0	6.25	0	6.24	0	6.25
	+0	18.75	12.5	36.4	12.5	0	6.25
1 wk. - 1 mo.	+	0	25	31.3	100	100.0	100
	–	83.5	33	25.0	0	0	0
	0	0	0	6.25	0	0	0
	+–	16.5	42	37.5	0	0	0
1 - 3 mos.	+	11.5	19.2	12.5	80.0	100.0	100
	–	77.0	61.6	12.5	7.7	0	0
	0	3.8	0	0	0	0	0
	+–	7.7	19.2	75.0	11.5	0	0
3 - 6 mos.	+	0	21.5	12.5	100	100.0	100
	–	100	64.5	12.5	0	0	0
	0	0	0	0	0	0	0
	+–	0	14	75.0	0	0	0
6 mos. 1 yr.	+	0	0	0	73	100	100
	–	100	91	57	4.3	0	0
	0	0	0	0	0	0	0
	+–	0	9	43	22.7	0	0
1 - 3 yrs.	+	0	4.3	15.4	92	100	100
	–	96	77.5	38.4	4	0	0
	0	0	0	0	0	0	0
	+–	4	18.2	46.2	4	0	0
3 - 5 yrs.	+	0	3.6	24	82	100	100
	–	93	50	21	7	0	0
	0	0	0	0	0	0	0
	+–	7	46.4	54	11	0	0

TABLE 635 (continued)

THE INCIDENCE OF UPRIGHT, INVERTED, ISOELECTRIC, AND DIPHASIC (or multiphasic)
T WAVES IN THE EXTREMITY AND PRECORDIAL LEADS OF NORMAL INFANTS AND CHILDREN
IN MICHIGAN

Age		V1	V2	V3	V4	V5	V6
5 - 8 yrs.	+	0	11.5	30.9	91	100	100
	-	88.5	20.3	14.3	0	0	0
	0	0	0	0	0	0	0
	+-	11.5	68.2	54.8	9	0	0
8 - 12 yrs.	+	0	33.3	56.5	86.5	100	100
	-	82	13.2	4.46	0	0	0
	0	0	0	0	0	0	0
	+-	18	53.5	39	13.5	0	0
12 - 16 years	+	31.2	75.5	96	98	100	100
	-	42.3	0	0	0	0	0
	0	2	0	0	0	0	0
	+-	3	24.5	4	2	0	0

(data from Ziegler, Electrocardiographic Studies in Normal Infants and Children,
1951. Courtesy of Charles C Thomas, Publisher, Springfield, Illinois)

TABLE 636

DURATION OF QRS INTERVAL IN A PRECORDIAL LEAD (seconds)
IN MICHIGAN CHILDREN

Age Group	.04	.05	.06	.07	.08	.09	.10
0 - 24 hours	--	7	66	4	19	4	--
1 - 7 days	14	29	47	3	7	--	--
8 - 30 days	12	23	52	10	3	--	--
1 - 3 months	--	5	82	--	13	--	--
3 - 6 months	--	--	48	26	26	--	--
6 - 12 months	--	9	48	28	15	--	--
1 - 3 years	--	5	59	24	12	--	--
3 - 5 years	--	--	34	18	46	2	--
5 - 8 years	--	3	22	39	31	--	--
8 - 12 years	--	2	20	28	48	2	--
12 - 16 years	3	3	32	42	15	4	1

(data from Ziegler, Electrocardiographic Studies in Normal Infants and Children,
1951. Courtesy of Charles C Thomas, Publisher, Springfield, Illinois)

TABLE 637

HEMOGLOBIN LEVELS (gm/100 ml) IN CHILDREN EXAMINED
IN PRESCHOOL NUTRITION SURVEY

| | | Transferrin saturation (Fe/TIBC x 100) | | | | | | | | |
| | | <10% | | 10-14% | | 15-19% | | 20-24% | | >25% | |
Age	Race	Mean	S.D.	Mean	S.D.	Mean	S.D.	Mean	S.D.	Mean	S.D.
12-23 mo.											
	Black	10.2	(1.8)	12.1	(1.0)	12.5	(0.8)	12.9	(1.1)	13.2	(1.0)
	White	10.6	(1.9)	12.1	(0.7)	12.3	(0.8)	12.4	(1.1)	13.0	(0.9)
24-47 mo.											
	Black	11.3	(1.1)	12.2	(0.9)	11.8	(0.8)	12.1	(0.9)	12.9	(1.1)
	White	11.6	(0.9)	12.4	(1.1)	12.6	(0.8)	12.5	(1.0)	12.8	(1.0)
48-71 mo.											
	Black	12.1	(1.0)	11.7	(1.0)	11.8	(1.0)	12.5	(1.0)	12.3	(1.1)
	White	12.4	(0.9)	12.6	(0.9)	12.8	(0.8)	13.0	(1.0)	13.0	(1.0)

Hemoglobin levels are given in relation to transferrin saturation.

(data from Owen, Lubin, and Garry, 1973, The Journal of Pediatrics 82:850-851)

TABLE 638

MEDIAN HEMOGLOBIN (gm/100 cc) VALUES FOR LEAN AND OBESE INDIVIDUALS
IN THE TEN-STATE NUTRITION SURVEY

Age (years)	White Boys		White Girls		Black Boys		Black Girls	
	Lean[a]	Obese[b]	Lean	Obese	Lean	Obese	Lean	Obese
1	11.4	11.4	11.7	11.8	10.4	9.4	9.5	10.0
2	11.4	12.2	11.9	11.9	11.1	10.6	10.4	11.0
3	12.1	11.7	12.1	11.5	10.9	11.8	11.1	11.5
4	12.0	12.7	12.1	12.3	11.4	11.6	11.4	11.9
5	12.3	12.5	12.3	12.5	11.4	11.4	11.7	11.4
6	12.4	12.6	12.3	12.3	11.4	11.5	11.8	12.0
7	12.6	12.7	12.7	12.8	11.8	11.8	11.6	12.0
8	12.6	13.0	12.8	13.1	12.0	11.8	11.4	11.8
9	12.8	13.3	12.8	13.4	12.1	11.9	11.8	11.8
10	12.8	13.3	12.5	13.2	11.8	12.2	12.1	12.0
11	13.1	13.1	12.7	13.4	12.0	12.2	12.5	12.2
12	13.1	13.5	12.9	13.4	12.1	12.1	12.0	12.1
13	13.4	13.7	12.9	13.3	12.1	12.4	12.5	12.5
14	13.6	14.4	13.5	13.0	12.8	12.8	12.1	12.2
15	13.6	14.3	13.7	13.4	13.2	14.0	12.0	12.3
16	14.2	14.9	13.7	13.5	13.6	14.1	12.5	12.5
17	14.8	15.1	13.3	13.7	13.7	14.2	12.3	12.8

[a] Below the 15th percentile for triceps fatfold for age, sex and race
[b] Above the 85th percentile for triceps fatfold for age, sex and race

(data from Garn and Clark, 1975)

TABLE 639

BLACK-WHITE COMPARISONS IN HEMOGLOBIN LEVELS
(gm/l00 ml) FOR CHILDREN IN THE TEN-STATE NUTRITION SURVEY

Age (years)	BOYS		GIRLS	
	Black Mdn	White Mdn	Black Mdn	White Mdn
1	10.7	11.1	10.1	11.7
2	10.8	12.0	10.8	11.9
3	11.4	12.0	11.4	11.9
4	11.4	12.2	11.8	12.3
5	11.4	12.5	11.5	12.4
6	11.6	12.5	11.8	12.4
7	11.8	12.6	11.8	12.7
8	11.8	12.8	11.8	12.8
9	12.0	12.8	11.8	12.9
10	12.1	13.0	12.0	12.8
11	12.1	13.1	12.1	13.0
12	12.1	13.3	12.1	13.0
13	12.5	13.4	12.1	13.3
14	12.8	13.9	12.1	13.3
15	13.1	14.2	12.2	13.3
16	13.6	14.6	12.1	13.4
17	13.8	14.9	12.3	13.5

Mdn = median

(data from Garn et al., 1975)

TABLE 640

BLACK-WHITE HEMOGLOBIN (gm/100 ml) DIFFERENCES
IN INCOME-MATCHED GROUPS OF CHILDREN
IN THE TEN-STATE NUTRITION SURVEY

| | Below-Poverty Group* | | | | Median-Income Group** | | | |
| | BOYS | | GIRLS | | BOYS | | GIRLS | |
Age (years)	Black Mdn	White Mdn	Black Mdn	White Mdn	Black Mdn	White Mdn	Black Mdn	White Mdn
1	10.8	10.1	9.8	11.1	10.1	11.4	10.2	11.8
2	11.0	11.4	10.6	11.5	10.5	12.2	10.8	11.9
3	11.2	11.9	11.6	12.4	11.3	12.0	11.1	11.5
4	11.4	12.1	11.8	12.2	10.8	12.2	11.4	12.1
5	11.4	12.4	11.4	12.2	11.4	12.3	11.8	13.2
6	11.7	12.3	11.8	12.2	11.1	12.5	11.5	12.5
7	11.8	12.4	11.8	12.6	11.8	12.4	11.5	12.7
8	11.8	12.8	11.8	12.6	11.8	12.8	11.8	12.6
9	11.8	12.8	11.8	13.1	12.0	12.9	11.8	12.7
10	12.0	13.0	12.1	12.8	12.1	12.9	11.9	12.9
11	12.0	13.0	12.1	12.9	12.5	13.2	11.8	12.8
12	12.1	12.9	12.1	12.8	12.1	13.2	12.1	13.3
13	12.2	13.3	12.3	13.0	12.5	13.1	11.9	13.4
14	12.6	13.9	12.1	13.1	12.8	14.0	11.8	13.3
15	13.0	13.7	12.1	13.1	13.2	14.1	12.1	13.3
16	13.4	14.2	12.1	13.4	13.7	14.6	12.2	13.4
17	13.6	14.7	12.3	13.3	13.1	14.8	11.8	13.7

*Per-capita income below $799.
**Per-capita income $2400 up
Mdn = median

(data from Garn et al., 1975)

TABLE 641

HEMOGLOBIN (gm/100 ml) FOR NAVAJO CHILDREN IN ARIZONA

Age	Mean	Range	Below Navajo Deficient	Percentage Below NNS Acceptable	NNS Deficient
BOYS
2 - 5 years	13.9	11.0-18.2	5	0	0
6 - 12 years	13.5	11.5-17.2	2	0	0
13 - 16 years	14.4	13.2-16.2	33	0	0
GIRLS
2 - 5 years	13.3	10.5-15.0	10	3	0
6 - 12 years	13.5	10.8-17.7	8	3	0
13 - 16 years	13.8	10.3-16.8	6	6	0

NNS - National Nutrition Survey

(data from Nutrition Survey of Lower Greasewood Chapter of Navajo Tribe, 1968-1969)

TABLE 642

HEMOGLOBIN VALUES (gm/100 cc) FOR RURAL CHILDREN
IN FLORIDA

Age	Girls		Boys	
(years)	Mean	S.D.	Mean	S.D.
6	10.00	1.88	10.00	2.07
7	10.44	2.20	10.29	1.91
8	10.58	1.86	10.58	2.00
9	10.44	1.93	10.44	2.18
10	10.44	1.97	10.73	2.29
11	10.73	1.68	10.58	2.11
12	10.87	1.88	9.86	1.84
13	11.02	1.88	10.73	1.95
14	11.16	1.69	11.02	1.86
15	10.58	1.56	11.16	1.78
16	10.73	1.73	11.02	2.49
17	11.74	2.22	11.45	2.07
18	10.87	1.73	10.87	2.50

(data from Abbott, Townsend, and Ahmann, 1945, American Journal of
Diseases of Children 69:346-349. Copyright 1945, American Medical
Association)

TABLE 643

PERCENTAGE DISTRIBUTION OF RURAL CHILDREN IN FLORIDA BY AGE ACCORDING TO HEMOGLOBIN VALUES

Age (years)	Sex	Normal Above 13.5 gm % of Total	Subnormal 11.4 to 13.6 gm % of Total	Anemic 9 to 11.4 gm % of Total	6.1 to 9 gm % of Total	3.6 to 6.1 gm % of Total
6	Girls	14.6	31.4	33.5	18.4	2.1
	Boys	18.0	30.3	25.5	21.6	4.7
7	Girls	20.9	35.3	25.5	16.3	2.0
	Boys	15.8	38.6	26.4	15.7	3.5
8	Girls	19.4	42.3	27.1	11.2	--
	Boys	20.3	39.1	27.4	10.6	2.6
9	Girls	21.0	37.1	28.7	12.5	0.8
	Boys	23.0	35.4	24.7	13.3	3.6
10	Girls	20.2	34.5	27.4	17.9	--
	Boys	32.3	27.3	24.2	14.2	2.0
11	Girls	20.0	46.3	23.7	10.0	--
	Boys	25.9	29.6	30.9	8.6	5.0
12	Girls	25.3	38.7	26.6	9.3	--
	Boys	11.0	34.1	32.9	18.3	8.7
13	Girls	32.8	29.5	29.5	8.2	--
	Boys	26.5	38.8	24.5	10.2	--
14	Girls	27.8	40.7	22.2	9.3	--
	Boys	30.9	36.4	25.5	7.2	--
15	Girls	20.0	48.0	20.0	12.0	--
	Boys	31.8	40.9	22.7	4.6	--
16	Girls	21.6	35.1	37.8	5.4	--
	Boys	37.5	37.5	12.5	4.2	8.3
17	Girls	61.5	--	15.4	23.1	--
	Boys	41.7	33.3	16.6	8.3	--
18	Girls	28.6	35.7	28.6	7.1	--
	Boys	33.3	33.3	24.9	--	8.4

(data from Abbott, Townsend, and Ahmann, 1945, American Journal of Diseases of Children 69:346-349. Copyright 1945, American Medical Association)

TABLE 644

HEMOGLOBIN CONCENTRATIONS (gm/dl) IN CHILDREN
OF BOGALUSA (LA)

	Age (years)							
	2.75	3.5	4.5	5.25	5.5	6.5	7.5	8.5
WHITE Percentiles (gm/dl)								
90	13.4	13.5	13.5	13.7	13.5	14.1	13.5	14.1
50	12.1	12.2	12.2	12.4	12.2	12.4	12.4	12.4
10	10.7	11.3	11.3	11.6	11.3	11.3	11.3	11.4
Percent Low*	12.1	3.9	3.0	0.0	6.2	10.2	10.6	10.0
Percent Deficient*	0.0	0.0	0.0	0.0	1.1	0.0	0.5	0.0
BLACK Percentiles (gm/dl)								
90	12.9	12.9	12.9	13.2	13.5	13.4	13.3	13.5
50	11.6	11.6	11.6	12.0	11.6	11.6	12.1	12.2
10	10.5	10.6	10.3	11.0	10.3	10.7	10.8	10.7
Percent Low*	10.3	9.4	15.1	2.0	19.2	36.4	25.9	25.9
Percent Deficient*	6.9	3.1	2.7	0.0	2.6	1.9	0.0	1.8

	Age (years)							
	9.5	10.5	11.5	12.5	13.5		14.5	
WHITE Percentiles (gm/dl)					Boys	Girls	Boys	Girls
90	13.9	14.1	14.4	14.4	14.4	14.1	15.6	14.7
50	12.4	12.9	12.9	12.9	13.4	12.9	14.1	13.5
10	11.6	12.0	11.6	11.6	12.2	12.1	12.2	12.2
Percent Low*	6.3	4.8	4.3	5.5	31.4	2.5	10.3	1.1
Percent Deficient*	0.6	0.0	0.0	0.0	5.6	0.0	5.2	0.0

TABLE 644 (continued)

HEMOGLOBIN CONCENTRATIONS (gm/dl) IN CHILDREN
OF BOGALUSA (LA)

Age (years)

	9.5	10.5	11.5	12.5	13.5		14.5	
BLACK Percentiles (gm/dl)					Boys		Girls	
90	13.2	13.5	13.5	13.8	13.5	13.5	14.4	14.1
50	12.2	12.2	12.2	12.3	12.4	12.2	12.9	12.4
10	11.0	11.0	11.0	11.3	11.3	11.1	11.3	11.2
Percent Low*	23.7	16.0	19.7	18.2	40.0	22.2	36.8	11.5
Percent Deficient*	0.0	0.8	0.0	0.6	25.3	0.0	17.6	1.6

*According to criteria of National Nutrition Survey (Pediatr. Res. 4: 103-106, 1970)

Age is at midpoint of interval.

(data from Frerichs, Webber, Srinivasan, et al., 1977, American Journal of Public Health 67:841-845)

TABLE 645

HEMATOCRIT VALUES (%) RELATED TO MATURITY STAGES
FOR ALABAMA CHILDREN

Stage

	1		2		3		4		5	
	Black	White	Black	White	Black	White	Black	White	Black	White
BOYS										
Mean	37.72	39.47	38.41	39.75	39.68	40.87	41.07	42.25	42.72	43.76
S.D.	2.53	2.35	2.45	3.01	2.36	2.63	2.70	2.54	3.15	2.67
GIRLS										
Mean	37.29	39.11	38.92	39.21	38.97	39.61	38.39	39.23	38.72	39.21
S.D.	2.62	2.98	3.22	2.14	3.69	2.61	3.49	2.35	2.81	3.01

Maturity stages based on secondary sex characteristics.

(data from Daniel, 1973, Pediatrics 52:388-394. Copyright American Academy of Pediatrics 1973)

TABLE 646

HEMATOCRIT VALUES (volume percent) FOR CHILDREN IN GEORGIA

Sex	Age (years)	Whites Mean	S.D.	Negroes Mean	S.D.
Males	15-24	46.07	2.77	46.51	2.65
Females	15-24	40.98	4.74	39.29	4.13

(data from McDonough, Hames, Garrison, et al., 1965, The relationship of hematocrit to cardiovascular states of health in the Negro and white population of Evans County, Georgia. Printed with permission from Journal of Chronic Diseases 18:243-257. Copyright 1965, Pergamon Press, Ltd.)

TABLE 647

HEMATOCRIT VALUES (%) RELATED TO AGE FOR CHILDREN IN ALABAMA

Age (years)	Boys Mean Black	White	S.D. Black	White	Girls Mean Black	White	S.D. Black	White
12	38.21	40.25	2.52	2.55	38.86	39.67	2.95	2.46
13	39.63	39.79	2.60	2.72	38.35	39.27	2.64	2.27
14	40.04	41.45	2.90	2.75	38.98	39.45	2.52	2.62
15	41.47	41.70	2.62	2.77	38.35	38.73	4.80	2.43
16	42.10	43.70	3.48	2.98	37.38	39.81	4.74	2.42
17 and older	43.88	44.39	2.98	2.35	38.91	39.33	2.53	3.19

(data from Daniel, 1973, Pediatrics 52:388-394. Copyright American Academy of Pediatrics 1973)

TABLE 648

BLOOD EOSINOPHIL COUNTS (n/mm^3)
IN NEW YORK CHILDREN

Age Group (years)	Mean ± S.E.	
	BOYS	GIRLS
1/4 - 2	267 ± 30	191 ± 20
2 - 4	244 ± 16	224 ± 31
4 - 6	266 ± 25	243 ± 21
6 - 8	305 ± 33	318 ± 32
8 - 10	332 ± 46	203 ± 28
10 - 12	205 ± 27	221 ± 29
> 12	204 ± 24	144 ± 13

(data from Cunningham, 1975, The Journal of Pediatrics 87:426-427)

TABLE 649

PERCENTAGE NORMOBLAST, GRANULOCYTES, LYMPHOCYTES, MONOCYTES, PLASMA CELLS
AND MEGAKARYOCYTES FROM STERNUM AND ILIUM IN CHILDREN FROM BIRTH TO 5 YEARS

Sites	Normoblasts %		Granulocytes %		Lymphocytes %	
	Avg.	Range	Avg.	Range	Avg.	Range
Sternum	22.9	4.9 - 36.4	53	39.1 - 67	19.1	15.8 - 24
Iliac crest	25.5	9.3 - 36.7	46	32.3 - 70.6	22.7	11.2 - 35.2

Sites	Monocytes %		Plasma Cells %		Megakaryocytes	
	Avg.	Range	Avg.	Range	Avg.	Range
Sternum	5.2	0.4 - 10.2	0.6	0.1 - 0.9	168	41 - 271
Iliac crest	5.1	0.4 - 8.5	0.3	0.0 - 0.9	163	45 - 279

(data from Sturgeon, 1951b, Pediatrics 7:642-650. Copyright American Academy of
Pediatrics 1951)

TABLE 650

RED CELL DIAMETER AND RETICULOCYTE VALUES
FOR 4-H CLUB HEALTH CHAMPIONS IN MINNESOTA

Age	Red Cell Diameter (μ)		Reticulocytes (percent)	
(years)	Mean	S.D.	Mean	S.D.
Boys				
13	7.87	0.17	0.94	0.27
14	7.84	0.19	0.98	0.37
15	7.91	0.19	1.03	0.46
16	7.86	0.20	1.07	0.31
17	7.89	0.17	0.86	0.41
Girls				
12	7.95	0.12	0.73	0.35
13	7.86	0.15	0.94	0.59
14	7.91	0.12	0.87	0.49
15	7.94	0.15	0.92	0.62
16	7.98	0.17	1.01	0.67
17	8.01	0.17	1.04	0.38

(data from Leichsenring, Norris, Lawson, et al., 1955, American Journal of Diseases
of Children 90:159-163. Copyright 1955, American Medical Association)

TABLE 651

M.E. AND FAT FROM TIBIAL, ILIAC CREST AND STERNAL ASPIRATIONS
MEASURED AS THICKNESSES OF EACH LAYER (mm)

	Site	Average	Range
M.E.	Tibia	4.7	1 to 11.0
	Ilium	4.1	1 to 7.0
	Sternum	3.8	2 to 7.0
Fat	Tibia	2.3	0 to 6.0
	Ilium	1.0	Trace to 2.0
	Sternum	1.0	Trace to 3.0

M.E. = layer of myeloid and erythroid cells

(data from Sturgeon, 1951a, Pediatrics 7:577-588. Copyright
American Academy of Pediatrics 1951)

TABLE 652

MEAN HEMATOCRIT IN RELATION TO ALTITUDE IN FULL-TERM INFANTS AT BIRTH (LAKE COUNTY 10,000 - 11,000 FEET ABOVE SEA LEVEL; DENVER 5,000 FEET ABOVE SEA LEVEL)

	Lake County		Denver	
	Cord	Capillary	Cord	Superficial Vein
Source	Blood	(4-5 days)	Blood	(30 min.-19 days)
Mean values	53.4	57.8	53.18	54.92

(data from Lichty, Ting, Bruns, et al., 1957, American Journal of Diseases of Children 93:666-678. Copyright 1957, American Medical Association)

TABLE 653

HEMATOCRIT (%) OF NAVAJO CHILDREN IN ARIZONA

			Percentage	
			Below NNS	NNS
Age	Mean	Range	Acceptable	Deficient
BOYS				
6 - 23 months	37	34-39	0	0
2 - 5 years	40	35-45	0	0
6 - 12 years	40	33-47	2	0
13 - 16 years	43	39-48	10	0
GIRLS				
6 - 23 months	37	33-41	0	0
2 - 5 years	39	32-43	3	0
6 - 12 years	40	29-48	4	1
13 - 16 years	41	35-45	3	0

NNS = National Nutrition Survey

(data from Nutrition Survey of Lower Greasewood Chapter Navajo Tribe, 1968-1969)

TABLE 654

HEMATOLOGIC DATA FOR INFANTS IN ROCHESTER (NY)

Age		Weight (kg)	Hemo-Globin (gm/100 ml)	Hema-tocrit (%)	Red blood cell count (x10^6)
1 day	Mean	3.14	17.2	52.3	4.90
	S.D.	--	--	--	--
2 days	Mean	3.18	17.4	53.2	4.92
	S.D.	--	--	--	--
3 days	Mean	3.05	16.8	50.1	4.65
	S.D.	--	--	--	--
4 days	Mean	3.28	15.9	46.8	4.57
	S.D.	--	--	--	--
5 days	Mean	3.48	16.4	49.3	4.77
	S.D.	--	--	--	--
4.0-5.9 weeks	Mean	4.47	11.6	32.5	3.52
	S.D.	--	--	--	--
6.0-6.9 weeks	Mean	4.53	11.1	33.0	3.24
	S.D.	--	--	--	--
9.0-11.0 weeks	Mean	5.23	10.3	30.4	3.20
	S.D.	--	--	--	--
16.0-19.9 weeks	Mean	6.60	10.6	31.3	3.62
	S.D.	--	--	--	--

Age		Plasma Volume (ml/gm)	Blood Volume (ml/gm)	Eryth-rocyte Volume (ml/gm)	Hemo-globin mass (gm/kg)
1 day	Mean	55.7	98.2	42.5	17.0
	S.D.	--	17.4	9.1	3.8
2 days	Mean	50.6	90.5	39.9	15.7
	S.D.	--	8.1	5.3	1.9
3 days	Mean	59.8	102.3	42.5	17.2
	S.D.	--	16.6	8.4	3.3
4 days	Mean	60.1	97.8	37.8	15.6
	S.D.	--	21.2	9.2	4.0
5 days	Mean	65.1	110.6	45.6	18.3
	S.D.	--	20.3	12.4	5.5

TABLE   654 (continued)

HEMATOLOGIC DATA FOR INFANTS IN ROCHESTER (NY)

Age		Plasma Volume (ml/gm)	Blood Volume (ml/gm)	Erythrocyte Volume (ml/gm)	Hemoglobin mass (gm/kg)
4.0-5.9 weeks	Mean	66.6	91.0	24.5	10.5
	S.D.	--	21.9	6.9	2.9
6.0-6.9 weeks	Mean	67.9	93.3	25.5	10.3
	S.D.	--	15.3	3.9	1.3
9.0-11.0 weeks	Mean	62.5	83.4	20.8	8.6
	S.D.	--	12.9	3.0	1.3
16.0-19.9 weeks	Mean	60.0	81.2	21.3	8.6
	S.D.	--	18.2	5.8	2.3

(data from Sisson, Lund, Whalen, et al., 1959, The Journal of Pediatrics 55:163-179)

TABLE 655

HEMOGLOBIN AND RED CELL VOLUME FOR
CHILDREN IN SAN FRANCISCO (CA)

Age (years)	Hemoglobin (gm/dl)		MCV (fl)	
	Median	Lower Limit	Median	Lower limit
0.5-2	12.5	11.0	77	70
2-5	12.5	11.0	79	73
5-9	13.0	11.5	81	75
9-12	13.5	12.0	83	76
12-14				
Female	13.5	12.0	85	77
Male	14.0	12.5	84	76
14-18				
Female	14.0	12.0	87	78
Male	15.0	13.0	86	77
18-49				
Female	14.0	12.0	90	80
Male	16.0	14.0	90	80

MCV = mean corpuscular volume.

(data from Dallman and Siimes, 1979, Journal of Pediatrics
94:26-31)

TABLE 656

HEMOGLOBIN, RED CELL COUNT, AND MEAN CORPUSCULAR HEMOGLOBIN VALUES
FOR 4-H CLUB HEALTH CHAMPIONS IN MINNESOTA

Age (years)	Hemoglobin (Gm/100 ml.)		Red Cell Count (Millions/cu. mm.)		Mean Corpuscular Hemoglobin (μγ)	
	Mean	S.D.	Mean	S.D.	Mean	S.D.
Boys						
13	13.8	0.75	4.44	0.25	31.2	1.03
14	14.2	0.76	4.62	0.26	30.8	0.86
15	14.2	0.88	4.67	0.31	30.4	1.20
16	15.1	0.76	4.85	0.36	31.1	1.14
17	14.9	0.73	4.88	0.41	30.7	1.34
Girls						
12	13.0	0.61	4.31	0.17	30.1	0.70
13	12.8	0.61	4.26	0.20	30.1	0.78
14	13.1	0.62	4.31	0.22	30.3	1.09
15	13.1	0.75	4.31	0.28	30.4	1.09
16	12.7	0.81	4.19	0.27	30.4	1.19
17	12.9	0.59	4.20	0.25	30.6	0.98

(data from Leichsenring, Norris, Lawson, et al., 1955, American Journal of Diseases
of Children 90:159-163. Copyright 1955, American Medical Association)

TABLE 657

HEMATOLOGIC FINDINGS IN BLACK AND WHITE CHILDREN
(mean age, 3 years)
IN NEW YORK CITY

Variables	Black	White
Hemoglobin (gm/dl)		
Mean	11.9	12.2
Hematocrit (%)		
Mean	35.7	36.9
Total leukocyte count/mm^3		
Mean	7,877	9,053
S.D.	1,936	1,843
Absolute neutrophil count/mm^3		
Mean	2,509	3,392
S.D.	1,176	1,234
Absolute lymphocyte count/mm^3		
Mean	5,149	5,254
S.D.	1,696	1,814

(data from Caramihai, Karayalcin, Aballi, et al., 1975, The Journal of
Pediatrics 86:252-254)

TABLE 658

RED BLOOD CELL COUNTS AND HEMOGLOBIN CONCENTRATIONS
FOR CHILDREN IN CINCINNATI (OH)

| Age Span in (Days) | Red Blood Cells | | | | | |
| | Volume % | | Count, (millions) | | MCV, Cu,μ | |
	Mean	S.D.	Mean	S.D.	Mean	S.D.
Cord Blood	52.3	5.3	4.638	0.495	113	6.1
< 2	58.2	6.6	5.305	0.555	110	5.6
2-3.9	54.5	6.3	5.060	0.603	108	6.8
4-9.9	54.9	6.2	5.216	0.628	106	5.0
10-29	46.2	7.2	4.640	0.730	100	5.3
30-60	36.5	4.8	3.874	0.501	94	5.6
61-90	33.1	2.6	3.766	0.344	88	4.9
91-151	35.3	3.1	4.332	0.511	82	5.7
152-212	34.8	2.5	4.483	0.412	78	5.2
213-273	35.6	3.2	4.713	0.503	76	5.8
274-334	34.9	2.8	4.834	0.440	73	7.8
335-394	34.5	4.1	4.758	0.515	73	7.7
395-455	34.6	3.7	4.700	0.542	74	7.6
456-516	34.7	3.4	4.904	0.517	71	8.5
517-577	34.8	2.9	4.868	0.450	72	8.7
578-638	35.9	3.1	4.807	0.496	75	7.3
639-820	35.8	3.0	4.777	0.429	76	7.3
821-1003	36.2	3.0	4.703	0.354	77	5.9
1004-1185	36.6	2.8	4.691	0.392	78	5.0
1186-1368	37.3	2.1	4.771	0.384	78	4.8
1369-1550	36.9	2.9	4.618	0.331	80	5.0
1551-2098	37.2	2.5	4.651	0.464	80	3.8
2099-3010	38.1	3.7	4.790	0.518	79	9.2
3011-4288	38.5	2.8	4.804	0.477	81	6.3
4289-5930	39.7	3.3	4.922	0.519	81	7.4

| Age Span in (Days) | Hemoglobin | | | | | |
| | In Whole Blood Gm. per 100 Cc. | | In Cells, MCHC, Gm. per 100 Cc. | | Per Cell, MCH | |
	Mean	S.D.	Mean	S.D.	Mean	S.D.
Cord Blood	17.1	1.78	32.6	1.41	36.9	2.16
< 2	19.4	2.14	33.5	1.29	36.6	1.89
2-3.9	18.5	2.16	33.9	1.32	36.7	2.67
4-9.9	18.8	2.13	34.3	1.43	36.2	2.42
10-29	15.9	2.33	34.5	1.60	34.4	2.07
30-60	12.7	1.64	34.9	1.11	32.8	1.98
61-90	11.4	0.96	34.6	1.27	30.4	1.63
91-151	11.9	1.02	33.7	1.43	27.7	2.23
152-212	11.6	0.98	33.5	1.59	26.1	2.23
213-273	11.7	1.32	32.8	1.72	24.9	2.66
274-334	11.3	1.35	32.4	2.12	23.7	3.57
335-394	11.1	1.71	32.3	2.32	23.5	3.65
395-455	11.1	1.80	31.8	2.69	23.8	3.74

TABLE   658   (continued)

RED BLOOD CELL COUNTS AND HEMOGLOBIN CONCENTRATIONS
FOR CHILDREN IN CINCINNATI (OH)

Age Span In (Days)	In Whole Blood Gm. per 100 Cc		In Cells, MCHC, Gm. per 100 cc		Per Cell, MCH	
	Mean	S.D.	Mean	S.D.	Mean	S.D.
456–516	11.0	1.58	31.7	2.20	22.8	3.90
517–577	11.2	1.97	31.9	2.83	23.2	4.38
578–638	11.7	1.49	32.6	2.14	24.6	3.58
639–820	11.7	1.47	32.6	2.16	24.7	3.59
821–1003	12.1	1.40	33.5	1.90	25.9	3.02
1004–1184	12.5	1.14	34.0	1.49	26.6	2.09
1186–1368	12.6	0.85	33.9	1.29	26.6	2.18
1369–1550	12.4	1.38	33.7	1.77	26.9	2.69
1551–2098	12.7	1.01	34.2	1.18	27.5	1.50
2099–3010	12.9	1.87	34.3	1.34	27.4	2.07
3011–4288	13.2	1.22	34.3	1.46	27.6	2.73
4289–5930	13.6	1.46	34.3	1.81	27.9	3.25

MCV = mean corpuscular volume
MCH = mean corpuscular hemoglobin
MCHC = mean corpuscular hemoglobin concentration

(data from Guest and Brown, 1957, American Journal of Diseases of Children
93:486-509. Copyright 1957, American Medical Association)

TABLE 659

HEMATOLOGIC DATA FOR INFANTS IN PORTLAND (OR)

| Age (days) | Red Blood Cells, Millions per Cu. Mm. | Hemoglobin Content | | Hemoglobin Coefficient | Color Index | Cell Volume | Volume Coefficient | Volume Index | Saturation Index |
		Percentage	Gm. per 100 Cc.						
				BOYS					
1	4.66	126.0	17.39	18.67	1.05	46.09	49.51	1.04	1.01
2	4.52	123.2	16.99	18.90	1.06	43.80	48.59	1.02	1.04
3	4.73	129.7	17.84	18.94	1.06	46.80	49.78	1.05	1.02
4	4.78	118.8	16.39	17.30	0.97	43.27	45.44	0.95	1.02
5	4.57	112.6	15.54	17.21	0.97	41.58	45.68	0.96	1.00
6	4.81	118.1	16.30	17.00	0.95	44.09	46.00	0.97	1.00
7	4.58	123.3	17.01	18.59	1.04	44.05	48.50	1.02	1.04
8	4.46	108.7	15.00	16.94	0.95	39.42	44.60	0.94	1.02
9	4.83	115.3	15.91	16.49	0.92	42.54	44.02	0.92	1.00
10	4.27	115.8	15.97	18.82	1.06	40.96	48.35	1.01	1.05
Average	4.63	119.2	16.44	17.87	1.00	43.30	47.03	0.99	1.02
				GIRLS					
1	4.65	123.3	17.01	18.44	1.03	45.60	49.68	1.04	0.99
2	4.71	127.4	17.58	18.64	1.04	46.54	49.32	1.04	1.01
3	4.95	125.2	17.28	17.52	0.98	48.80	49.38	1.04	0.95
4	4.56	126.4	17.45	19.08	1.07	44.78	49.01	1.03	1.04
5	4.56	111.5	15.39	16.98	0.95	42.40	46.88	0.98	0.97
6	4.53	115.8	15.99	17.86	1.00	43.46	48.44	1.02	0.98
7	4.07	102.2	14.10	17.60	0.99	38.92	48.55	1.02	0.97
8	4.52	116.1	16.02	17.75	1.00	44.13	48.94	1.03	0.97
9	4.58	112.2	15.48	16.95	0.95	40.64	44.28	0.93	1.03
10	4.28	107.8	14.88	17.44	0.98	39.46	46.19	0.97	1.01
Average	4.54	116.9	16.13	17.84	1.00	43.50	48.12	1.01	0.99

(data from Chuinard, Osgood, and Ellis, 1941, American Journal of Diseases of Children 62:1188-1196. Copyright 1941, American Medical Association)

TABLE 660

MEAN RESPIRATORY DATA FROM FULL-TERM BOSTON INFANTS

Measurement	Cc	Cc/kg	Cc/Sq.M	Rate/Min.
Tidal Volume	16.8	4.76	74.3	39
Minute Volume	642	182	2819	--

(data from Boutourline-Young and Smith, 1950, American Journal of Diseases of Children 80:753-766. Copyright 1950, American Medical Association)

TABLE 661

RESPIRATORY RATES (/min) IN NORMAL FULL-TERM BOSTON INFANTS

Age (hours)	Total	Room Air	Oxygen and Water Mist
0- 2	38.5±9.2	36 ±9.0	32 ±9.6
4- 6	40 ±9.5	37.5±7.5	45.5±9.5
8-10	42 ±9.6	38.5±8.7	37 ±9.6
12-16	43 ±9.3	39 ±7.7	48 ±9.0
20-24	43 ±6.7	40.5±6.1	45.5±7.3
28-32	43 ±5.7	40.5±5.8	44.5±6.9
36-40	43 ±8.5	40.5±7.8	46 ±7.9
48-56	43 ±5.6	40.5±7.6	45 ±7.6
64-72	43 ±6.0	40.5±7.8	44.5±6.2

(data from Haddad, Hsia, & Gellis, 1956, Pediatrics 17:204-213. Copyright American Academy of Pediatrics 1956)

TABLE 662

RESPIRATORY RATES (min) OF CHILD
RESEARCH COUNCIL CHILDREN (Denver, CO)

| Age | Respiratory Rate | | | |
| | BOYS | | GIRLS | |
(years)	Mean	S.D.	Mean	S.D.
0- 1	31	8	30	6
1- 2	26	4	27	4
2- 3	25	4	25	3
3- 4	24	3	24	3
4- 5	23	2	22	2
5- 6	22	2	21	2
6- 7	21	3	21	3
7- 8	20	3	20	2
8- 9	20	2	20	2
9-10	19	2	19	2
10-11	19	2	19	2
11-12	19	3	19	3
12-13	19	3	19	2
13-14	19	2	18	2
14-15	18	2	18	3
15-16	17	3	18	3
16-17	17	2	17	3
17-18	16	3	17	3

(data from Iliff and Lee, 1952)

BIOCHEMISTRY

TABLE 663

SERUM CHOLESTEROL (mg/100 ml; Tecumseh, MI)

Age	Boys		Girls	
(years)	Mean	S.D.	Mean	S.D.
4 - 9	177	30	178	27
10 - 14	172	28	176	31
15 - 19	166	31	173	31

(data from Johnson, Epstein, and Kjelsberg, 1965, printed with permission from Journal of Chronic Diseases 18:147-160, Copyright 1965, Pergamon Press, Ltd.)

TABLE 664

SERUM CHOLESTEROL LEVELS (mg/100 cc)
FOR CHILDREN IN NEW YORK CITY

Age Group	Males		Females	
(years)	Mean	S.D.	Mean	S.D.
3-7	179.8	6.48	209.0	7.13
8-12	180.4	3.98	196.4	4.40
13-17	175.5	5.10	182.9	4.77
18-22	185.2	11.76	192.6	8.78

(data from Adlersberg, Schaefer, Steinberg, et al., 1956, The Journal of the American Medical Association 162:619-622. Copyright 1956, American Medical Association)

TABLE 665

RATIO OF COMBINED TO FREE CHOLESTEROL IN SERUM
IN CHILDREN IN NEW YORK CITY

Age	Mean	S.D.
2 to 6 months	2.32	±0.21
7 to 12 months	2.46	±0.29
2d year	2.43	±0.53
3d year	2.63	±0.28
4th year	2.57	±0.31
5th year	2.47	±0.05
6th year	2.58	±0.20
7th year	2.66	±0.20
8th year	2.64	±0.05
9th year	2.59	±0.04
10th year	2.73	±0.22
11th year	2.63	±0.28
12th year	2.63	±0.21
13th year	2.56	±0.25

(data from Hodges, Sperry, and Andersen, 1943,
American Journal of Diseases of Children 65:858-
867. Copyright 1943, American Medical Association)

TABLE 666

TOTAL CHOLESTEROL CONTENT OF
SERUM (mg/100 cc) OF NORMAL CHILDREN
IN NEW YORK CITY

Age	Mean	S.D.
2 to 6 months	200.2	±49
7 to 12 months	206.6	±35.6
2d year	203.2	±38
3d year	202.4	±30.5
4th year	194.0	±29
5th year	212.6	±38
6th year	208.6	±39
7th year	204.3	±32
8th year	204.3	±34
9th year	220.4	±42
10th year	205.8	±34
11th year	206.2	±42
12th year	210.4	±42
13th year	204.3	±32

(data from Hodges, Sperry, and Andersen,
1943, American Journal of Diseases of Children
65:858-867. Copyright 1943, American Medical
Association)

TABLE 667

DISTRIBUTION OF PLASMA CHOLESTEROL AND TRIGLYCERIDE LEVELS
(mgm/deciliter) IN NONFASTING CHILDREN
IN CINCINNATI, OH

Age (years)	Cholesterol (Percentiles)					Triglyceride (Percentiles)				
	5	10	50	90	95	5	10	50	90	95
Boys										
White 6-11	118	124	154	189	192	32	36	58	88	115
12-17	104	117	146	178	195	44	47	73	124	147
Black 6-11	137	140	167	202	209	30	39	61	80	100
12-17	119	123	152	192	207	36	41	62	95	124
Girls										
White 6-11	117	128	158	195	196	29	34	68	108	135
12-17	119	124	158	190	204	45	48	69	108	129
Black 6-11	138	146	159	198	199	35	37	56	80	97
12-17	121	130	159	183	201	35	37	64	96	103

(data from de Groot, Morrison, Kelly, et al., 1977, Pediatrics 60:437-443. Copyright American Academy of Pediatrics 1977)

TABLE 668

PLASMA CHOLESTEROL LEVELS (mgm/deciliter)
IN FASTING WHITE CHILDREN
IN CINCINNATI, OH

Age (Years)	Boys Percentiles					Girls Percentiles				
	5	10	50	90	95	5	10	50	90	95
6	120	129	158	189	203	129	133	164	197	207
7	123	130	160	188	193	123	130	160	194	201
8	122	131	162	196	209	131	139	165	198	209
9	128	135	160	195	206	120	133	164	190	196
10	130	136	159	188	196	128	136	163	195	206
11	123	129	160	197	210	125	132	159	191	200
12	120	127	160	194	203	122	130	158	192	202
13	112	120	154	190	202	119	126	154	187	195
14	116	122	151	186	197	117	127	156	190	195
15	112	117	146	179	194	119	125	155	185	187
16	115	123	149	184	192	119	126	152	188	196
17	117	122	152	195	209	118	127	155	187	196

(data from de Groot, Morrison, Kelly, et al., 1977, Pediatrics 60:437-443. Copyright American Academy of Pediatrics 1977)

TABLE 669

MEAN SERUM CHOLESTEROL FOR PIMA INDIANS

Age (years)	Boys		Girls	
	Mean (mg/100 ml)	S.E.	Mean (mg/100 ml)	S.E.
Birth	86	0.9	88	0.4
5	144	2.5	135	3.0
6	145	2.4	147	2.7
7	145	2.8	155	2.8
8	150	3.4	153	4.4
9	153	3.5	152	3.2
10	158	3.5	149	3.8
11	152	5.2	151	3.7
12	150	3.1	146	5.0
13	146	3.7	153	4.2
14	146	3.5	147	4.9
15	141	3.2	150	4.0
16	145	4.0	147	4.8
17	151	5.5	161	5.4
18	160	5.0	167	6.2

(data from Savage, Hamman, Bartha, et al., 1976, Pediatrics 58:274-282. Copyright American Academy of Pediatrics 1976)

TABLE 670

PLASMA TRIGLYCERIDE LEVELS (mgm/deciliter)
IN FASTING WHITE CHILDREN
IN CINCINNATI, OH

Age (years)	Boys Percentiles					Girls Percentiles				
	5	10	50	90	95	5	10	50	90	95
6	33	35	52	81	96	34	38	55	82	94
7	31	35	52	80	105	34	38	55	91	114
8	31	34	50	86	104	33	37	56	93	104
9	30	33	51	93	112	33	37	59	94	119
10	31	35	52	89	104	35	39	59	91	113
11	30	34	50	83	101	37	41	67	108	124
12	36	41	65	112	135	41	51	79	120	135
13	37	41	71	129	162	48	51	80	125	142
14	37	43	70	121	143	44	48	75	118	141
15	44	49	76	124	165	44	50	76	129	154
16	36	48	74	136	163	39	44	66	105	114
17	42	48	74	128	174	41	44	65	94	108

(data from de Groot, Morrison, Kelly, et al., 1977, Pediatrics 60:437-443. Copyright
American Academy of Pediatrics 1977)

TABLE 671

PLASMA CHOLESTEROL AND TRIGLYCERIDE LEVELS (mgm/deciliter)
IN FASTING BLACK CHILDREN
IN CINCINNATI, OH

	Age (years)	Cholesterol (Percentiles)					Triglyceride (Percentiles)				
		5	10	50	90	95	5	10	50	90	95
Boys	6-11	130	136	164	199	214	32	34	49	75	88
	12-17	119	127	157	197	208	34	38	57	93	106
Girls	6-11	130	137	169	210	219	33	37	53	83	94
	12-17	123	131	163	202	213	40	44	63	101	114

(data from de Groot, Morrison, Kelly, et al., 1977, Pediatrics 60:437-443. Copyright
American Academy of Pediatrics 1977)

TABLE 672

SERUM PHOSPHOLIPID LEVELS (mg/100 cc)
LEVELS FOR CHILDREN IN NEW YORK CITY

Age Group (years)	Males		Females	
	Mean	S.D.	Mean	S.D.
3-7	227.1	8.35	261.9	7.96
8-12	233.2	4.94	241.7	5.10
13-17	220.6	5.98	235.5	6.54
18-22	217.0	13.40	243.7	9.35

(data from Adlersberg, Schaefer, Steinberg, et al., 1956, The Journal
of the American Medical Association 162:619-622. Copyright 1956,
American Medical Association)

TABLE 673

PLASMA PROTEIN (gm/100 cc) FOR NEWBORN INFANTS
DURING THE FIRST WEEK OF LIFE

Weight at Birth	Albumin		Globulin		Total Protein	
	Mean	S.D.	Mean	S.D.	Mean	S.D.
5 lb. 8 oz. to 6 lb. 7 oz.	4.7	0.55	1.69	0.55	6.3	0.66
6 lb. 8 oz. to 7 lb. 7 oz.	4.7	0.59	1.88	0.44	6.4	0.74
7 lb. 8 oz. to 8 lb. 7 oz.	4.8	0.89	1.67	0.54	6.4	0.41

(data from McMurray, Roe, and Sweet, 1948, American Journal of Diseases of Children 75:
265-278. Copyright 1948, American Medical Association)

TABLE 674

BLOOD PLASMA PROTEIN LEVELS IN NEWBORN TEXAS INFANTS

Wt. (gm)	Total Protein (gm./100 cc.)		Albumin (gm./100 cc.)		Globulin (gm./100 cc.)	
	Mean	S.D.	Mean	S.D.	Mean	S.D.
1,000 to 1,499	4,66	±0.50	3.14	±0.39	1.38	±0.22
1,500 to 1,999	5.19	±0.79	3.46	±0.56	1.70	±0.28
2,000 to 2,499	5.40	±0.67	3.50	±0.50	1.90	±0.33
2,500 to 2,999	5.77	±0.72	3.85	±0.73	1.90	±0.30
3,000 to 3,499	5.64	±0.80	3.76	±0.67	1.87	±0.45
3,500 to 3,999	5.49	±0.42	3.52	±0.44	1.87	±0.35
4,000 and above	5.57	±0.61	3.54	±0.41	1.87	±0.32
Total Infants 1,000 to 2,499 gm.	5.10	±0.71	3.40	±0.55	1.7	±0.33
Total Infants 2,500 gm. and above	5.6	±0.66	3.7	±0.58	1.9	±0.30

(data from Desmond and Sweet, 1949, Pediatrics 4:484-489. Copyright American
Academy of Pediatrics 1949)

TABLE 675

SERUM PROTEIN CONCENTRATION (grams per 100 cc)
IN RELATION TO TYPE OF ARTIFICIAL FEEDING
IN PHILADELPHIA CHILDREN

Group	Race	6 Months		1 Year	
		Mean	S.D.	Mean	S.D.
All	All White	6.46	±0.453	6.78	±0.447
	All Negro	6.53	±0.376	6.98	±0.377
Group I (irradiated evaporated milk)	White	6.43	±0.428	6.73	±0.400
	Negro	6.64	±0.436	7.03	±0.398
Group II (nonirradiated evaporated milk plus cod-liver oil)	White	6.45	±0.451	6.80	±0.394
	Negro	6.43	±0.380	6.85	±0.350
Group III (irradiated evaporated milk plus carotene)	White	6.58	±0.481	6.76	±0.601
	Negro	6.52	±0.315	7.03	±0.325
Group IV (irradiated evaporated milk plus carotene and yeast)	White	6.40	±0.489	6.85	±0.462
	Negro	6.49	±0.271	7.05	±0.403

Group	Race	2 Years		3 Years	
		Mean	S.D.	Mean	S.D.
All	All White	7.09	±0.437	7.08	±0.347
	All Negro	7.26	±0.380	7.28	±0.407
Group I (irradiated evaporated milk)	White	7.15	±0.422	7.09	±0.346
	Negro	7.27	±0.435	7.31	±0.347
Group II (nonirradiated evaporated plus cod-liver oil)	White	7.07	±0.518	7.07	±0.389
	Negro	7.25	±0.327	7.35	±0.343
Group III (irradiated evaporated milk plus carotene)	White	7.09	±0.381	7.15	±0.314
	Negro	7.34	±0.391	7.49	±0.364
Group IV (irradiated evaporated milk plus carotene and yeast)	White	7.00	±0.353	6.81	±0.215
	Negro	7.19	±0.372	7.02	±0.483

(data from Rhoads, Rapaport, Kennedy, et al., 1945, The Journal of Pediatrics 26: 415-454)

TABLE   676

AVERAGE BLOOD SUGAR VALUES (mg/100 cc)
IN INDIANA CHILDREN

Age (years)	Fasting	11 a.m.
2	79	87
3	82	79
4	80	86
5	88	86
6	87	81
7	85	85
8	88	92
9	91	84
10	91	87
11	88	84
12	93	92
13	93	85
14	91	89
15	87	84

(data from Rudesill and Henderson, 1941, American Journal
of Diseases of Children 61:108-115. Copyright 1941,
American Medical Association)

TABLE   677

BLOOD SUGAR LEVELS (mg/100 ml) OF PAPAGO SCHOOL
CHILDREN IN SOUTHWESTERN ARIZONA FOLLOWING A
1-HOUR GLUCOSE TOLERANCE TEST

Age (years)	Mean Level	S.D.
BOYS		
10	103.7	15.61
11	102.4	17.10
12	93.4	21.64
13	99.4	16.67
14	95.6	12.21
GIRLS		
10	108.6	18.40
11	96.1	29.95
12	99.1	20.07
13	101.7	20.03
14	96.9	15.15

(data from Adams et al., 1970)

TABLE 678

BLOOD ACID-BASE EQUILIBRIUM
FOR INFANTS IN NEW YORK CITY

Age (months)		3-6			6-9			9-12			12-15		
		Cry-ing	Quiet	Total	Cry-ing	Quiet	Total	Cry-ing	Quiet	Total	Cry-ing	Quiet	Total
pH	Mean	7.379	7.399	7.386	7.402	7.398	7.400	7.388	7.401	7.390	7.397	7.406	7.400
	S.D.	0.028	0.012	0.026	0.028	0.032	0.028	0.022	0.014	0.019	0.027	0.026	0.038
B.E.	Mean	-3.1	-2.2	-2.8	-3.9	-3.9	-3.9	-3.9	-3.4	-3.7	-3.2	-2.8	-3.0
(mEq/L)	S.D.	1.8	1.8	1.8	1.9	1.5	1.7	1.4	1.3	1.3	1.4	2.1	1.6
pCO_2	Mean	36.5	35.4	36.2	31.8	32.6	32.2	33.5	32.8	33.3	34.0	33.6	33.9
(mm Hg)	S.D.	3.1	3.4	3.2	3.6	2.9	3.3	5.3	2.8	4.3	3.8	2.6	3.4
TCO_2	Mean	21.9	22.3	22.0	20.1	20.4	20.2	20.3	20.6	20.4	20.9	21.5	21.1
mM/L	S.D.	1.6	2.4	1.9	2.2	1.6	1.9	2.4	1.6	2.1	1.4	2.1	1.6

Age (months)		15-18			18-21			21-24			Grand Total		
		Cry-ing	Quiet	Total	Cry-ing	Quiet	Total	Cry-ing	Quiet	Total	Cry-ing	Quiet	Total
pH	Mean	7.412	7.407	7.410	7.395	7.398	7.396	7.392	7.407	7.399	7.395	7.402	7.398
	S.D.	0.028	0.019	0.024	0.016	--	0.013	0.022	0.016	0.021	0.030	0.022	0.027
B.E.	Mean	-2.3	-2.9	-2.6	-3.2	-1.6	-2.7	-2.3	-2.9	-2.6	-3.2	-3.2	-3.2
(mEq/L)	S.D.	1.6	1.6	1.6	1.9	0.7	1.7	1.4	1.8	1.5	1.7	1.7	1.7
pCO_2	Mean	33.9	33.5	33.7	34.5	36.9	35.3	36.2	32.9	34.8	34.1	33.4	33.8
(mm Hg)	S.D.	3.4	1.5	2.7	5.5	2.0	4.3	2.8	2.9	2.7	4.0	2.8	3.7
TCO_2	Mean	21.8	21.4	21.6	21.3	22.6	21.7	22.4	21.0	21.8	21.1	21.1	21.1
(mM/L)	S.D.	1.9	1.5	1.7	1.6	1.9	1.8	1.1	2.1	1.6	2.0	1.8	1.9

B.E. = base excess
pCO_2 = CO_2 tension
TCO_2 = plasma total CO_2 content

(data from Albert and Winters, 1966, Pediatrics 37:728-732. Copyright American Academy of Pediatrics, 1966)

TABLE 679

SERUM ALKALINE PHOSPHATASE (international units)
IN RELATION TO AGE IN ADOLESCENTS OF BIRMINGHAM (AL)

Age (years)	BOYS Mean	BOYS S.D.	GIRLS Mean	GIRLS S.D.
9	--	--	90	24
10	--	--	90	19
11	--	--	103	23
12	99	38	91	21
13	100	45	78	29
14	102	36	52	19
15	103	39	38	16
16	75	28	31	10
17	64	46	28	6
18	42	14	--	--

(data from Bennett, Ward, and Daniel, 1976, The Journal
of Pediatrics 88:633-636)

TABLE 680

SERUM ALKALINE PHOSPHATASE (international units)
IN RELATION TO SEX MATURITY RATING (SMR)
IN WHITE AND BLACK ADOLESCENTS IN BIRMINGHAM (AL)

	SMR 1 Mean	± S.D.	SMR 2 Mean	± S.D.	SMR 3 Mean	± S.D.	SMR 4 Mean	± S.D.	SMR 5 Mean	± S.D.
BOYS										
White	72	13	77	18	101	27	75	29	58	27
Black	77	30	94	31	122	45	116	42	75	43
Combined	74	21	89	29	116	41	103	43	70	39
GIRLS										
White	--	--	89	20	76	26	33	11	38	17
Black	--	--	95	21	86	31	44	25	31	10
Combined	79	16	93	21	84	30	39	21	32	12

Sex maturity ratings adapted from Tanner (1962)

(data from Bennett, Ward, and Daniel, 1976, The Journal of Pediatrics 88:633-636)

TABLE 681

PLASMA ALKALINE PHOSPHATASE ACTIVITY
KA units/100 ml DURING THE ADOLESCENT
GROWTH SPURT IN ENGLISH CHILDREN

Age in relation to PHV		Mean	Mean -2SE	Mean +2SE
		BOYS		
Months	20	18.7	16.7	20.8
before	16	19.5	17.5	21.6
PHV	12	21.1	19.1	23.1
	8	21.5	19.6	23.4
	4	22.6	20.7	24.5
	PHV	24.3	22.4	26.2
Months	4	22.4	20.6	24.1
after	8	20.5	18.8	22.1
PHV	12	18.6	17.3	19.8
	16	16.1	15.1	17.2
	20	13.7	12.9	14.6
		GIRLS		
Months	20	13.0	10.5	15.5
before	16	11.7	9.9	13.4
PHV	12	15.7	12.8	18.6
	8	16.7	14.5	18.9
	4	17.5	15.3	19.7
	PHV	18.5	16.7	20.2
Months	4	18.0	16.2	19.7
after	8	15.2	13.3	17.1
PHV	12	12.7	11.1	14.4
	16	11.9	10.6	13.1
	20	10.1	8.6	11.6

PHV = age at peak height velocity.

(data from Round et al., 1979, Annals of Human Biology
6:129-136)

TABLE 682

BLOOD THIAMINE LEVELS IN ILLINOIS CHILDREN
AGED 11 TO 15 YEARS (γ/100 cc).

	Mean	S.D.
Boys	8.19	2.44
Girls	9.64	2.86

(data from Waisman, Coo, Richmond, et al.,
1951, American Journal of Diseases of
Children 82:555-560. Copyright 1951,
American Medical Association)

TABLE 683

SERUM MAGNESIUM LEVELS

Group	Mean	S.D.
Newborn	1.38	0.18
1-12 months	1.28	0.13
12-36 months	1.35	0.15
3-6 years	1.36	0.23
6-12 years	1.33	0.19
12-14 years	1.36	0.18

(Mean expressed in millequivalents per liter)

(data from Mays and Keele, 1961, American
Journal of Diseases of Children 102:623-624.
Copyright 1961, American Medical Association)

TABLE   684

SERUM LEVELS OF CALCIUM, INORGANIC PHOSPHORUS, PHOSPHATASE,
AND POTASSIUM IN UMBILICAL CORD BLOOD AND
IN FULL-TERM INFANTS IN BUFFALO, NY

	Full-Term Infants
Calcium, mg/100 cc.	
Mean of all determinations	9.06
S.D.	0.796
Mean of lowest value of each patient	8.87
S.D.	0.85
Inorganic phosphorus, mg/100 cc.	
Mean of all determinations	7.49
S.D.	1.76
Mean of highest value of each patient	8.35
S.D.	1.68
Phosphatase, B. U.	
Mean of all determinations	4.55
S.D.	1.38
Potassium, mEq/L.	
Mean of all determinations	4.48
S.D.	0.60

(data from Bruck and Weintraub, 1955, American Journal of Diseases
of Children 90:653-668. Copyright 1955, American Medical Association)

TABLE   685

SERUM CALCIUM AND INORGANIC PHOSPHORUS LEVEL
IN INFANTS OF BUFFALO, NY, ON DIFFERENT FEEDINGS

		Before feeding	Cow's milk	Human milk
Calcium	Mean	7.99	8.49	8.99
	S.D.	1.32	1.30	0.814
Phosphorus	Mean	7.29	8.27	6.82
	S.D.	1.87	1.68	1.04

(data from Bruck and Weintraub, 1955, American Journal of
Diseases of Children 90:653-668. Copyright 1955, American
Medical Association)

TABLE 686

SERUM CHEMICAL VALUES OF NORMAL IOWA
BREASTFED INFANTS (concentration per deciliter)

	Age (days)	Mean	S.D.
Age-related			
Total protein (g)	28	5.60	0.31
	56	5.62	0.44
	84	5.82	0.42
	112	6.04	0.44
Albumin (g)	28	3.76	0.30
	56	3.80	0.33
	84	4.02	0.33
	112	4.18	0.35
α_1 Globulin (g)	28	0.16	0.04
	56	0.18	0.05
	84	0.18	0.06
	112	0.19	0.05
α_2 Globulin (g)	28	0.49	0.09
	56	0.55	0.11
	84	0.60	0.11
	112	0.63	0.10
β Globulin (g)	28	0.58	0.11
	56	0.58	0.12
	84	0.60	0.10
	112	0.60	0.10
γ Globulin (g)	28	0.61	0.17
	56	0.51	0.16
	84	0.42	0.17
	112	0.43	0.18
Urea nitrogen (mg)	28	8.0	2.5
	56	6.6	2.1
	84	6.8	3.7
	112	6.8	3.7
Alkaline phosphatase[a]	28	40.1	12.6
	56	37.6	11.5
	84	34.2	10.4
	112	30.8	10.1
Age- and sex-related			
Phosphorus (mg)			
Boys	28	6.4	0.86
	56	6.2	0.74
	84	6.0	0.81
	112	6.0	0.95

TABLE   686 (continued)

SERUM CHEMICAL VALUES OF NORMAL IOWA
BREASTFED INFANTS (concentration per deciliter)

	Age (days)	Mean	S.D.
Girls	28	6.3	1.02
	56	5.8	0.87
	84	5.7	0.85
	112	5.8	1.00
Cholesterol (mg)			
Boys	28	140	23
	56	130	22
	84	137	25
	112	140	24
Girls	28	142	30
	56	140	39
	84	145	28
	112	147	39
Unrelated to age or sex			
Calcium (mg)		10.2	0.76
Magnesium (mg)		2.1	0.29

(data from Fomon et al., 1978)

TABLE 687

PLASMA AND ERYTHROCYTE CHEMISTRY IN NORMAL INFANTS
IN MEMPHIS (TN)

Variable		Normal mothers	Cord Blood	Babies, newborn	Babies, 24 hr.	Babies, 48 hr.	Babies, 72 hr.	Babies, 96 hr.	Babies 102 hr.
Plasma Na	Mean	134	134	134	140	139	141	140	141
	S.D.	±3.4	±4.2	±2.8	±4.4	±4.7	±5.0	±6.3	±3.6
K	Mean	3.8	5.0	4.1	4.6	4.3	4.4	4.6	4.6
	S.D.	±0.41	±1.01	±0.49	±0.40	±0.30	±0.37	±0.74	±0.42
Cl	Mean	162	109	105	109	105	105	105	106
	S.D.	±3.1	±6.3	±4.9	±8.2	±5.8	±6.6	±5.9	±3.8
Red Cell Na	Mean	13.4	13.0	12.5	15.5	16.9	16.2	14.9	17.0
	S.D.	±2.4	±1.8	±1.3	±2.8	±3.0	±2.6	±2.5	±3.7
K	Mean	97	98	97	100	100	100	102	105
	S.D.	±4.3	±4.5	±6.2	±4.0	±4.1	±4.7	±3.8	±4.1
Cl	Mean	61	63	63	50	56	57	58	50
	S.D.	±7.2	±7.0	±6.8	±4.5	±4.1	±3.8	±4.3	±5.6
Hematocrit	Mean	38.8	50.7	51.1	56.6	53.9	53.7	52.0	50.1
	S.D.	±5.9	±5.4	±5.4	±7.2	±4.6	±8.3	±5.5	±7.2
CO_2 Content	Mean	49.5	50.9	49.7	50.5	53.9	53.7	54.9	55.0
	S.D.	±5.3	±5.1	±7.5	±4.6	±6.0	±6.4	±5.2	±5.7
pH	Mean	7.47	7.39	7.33	7.43	7.43	7.42	7.42	7.39
	S.D.	±0.44	±0.16	±0.12	±0.62	±0.63	±0.81	±0.73	±0.24
NPN	Mean	25	32	31	38	33	34	33	33
	S.D.	±3.4	±3.6	±4.1	±4.8	±5.7	±5.8	±5.7	±5.0
Plasma Protein	Mean	6.8	6.3	6.2	6.1	5.7	6.0	5.5	5.8
	S.D.	±0.35	±0.62	±0.57	±0.70	±0.52	±0.51	±0.48	±0.35

Units of measurement are mEq/l (except NPN mg/100 cc), CO_2 (vols %), and plasma
protein (gm/100 cc)

NPN = non-protein nitrogen

(data from Overman, Etteldorf, Bass, et al., 1951, Pediatrics 7:565-576. Copyright
American Academy of Pediatrics 1951)

TABLE   688

PLASMA AND ERYTHROCYTE CHEMISTRY IN NORMAL INFANTS
IN MEMPHIS (TN)

Variable		Age					
		1 mo.	2-4 mo.	5-7 mo.	8-10 mo.	11-15 mo.	19-26 mo.
Plasma Na	Mean	141	142	141	140	139	141
	S.D.	±2.2	±4.3	±4.2	±2.4	±2.6	±2.1
K	Mean	5.0	4.7	4.5	4.5	4.6	4.2
	S.D.	±0.56	±0.40	±0.48	±0.26	±0.47	±0.39
Cl	Mean	105	106	105	103	104	104
	S.D.	±6.0	±1.6	±5.0	±1.9	±2.4	±2.1
Ca	Mean	5.4	5.3	5.4	5.3	5.4	5.1
	S.D.	±0.17	±0.24	±0.95	±0.23	±0.22	±0.22
Red Cell Na	Mean	19.0	19.8	19.8	19.4	19.7	18.7
	S.D.	±3.6	±3.8	±3.4	±3.9	±2.7	±2.6
K	Mean	98	100	99	105	107	101
	S.D.	±6.5	±7.3	±5.6	±6.0	±5.0	±6.4
Cl	Mean	61	65	62	65	65	68
	S.D.	±4.3	±4.3	±4.1	±5.2	±7.2	±4.5
Hematocrit	Mean	39.4	35.2	34.4	34.6	33.7	36.5
	S.D.	±3.6	±2.2	±2.5	±3.1	±2.7	±2.2
CO_2 Content	Mean	52.6	52.9	55.1	54.0	53.8	55.2
	S.D.	±7.7	±5.9	±4.7	±5.5	±5.3	±3.8
NPN	Mean	32	28	32	--	30	29
	S.D.	±5.4	±1.0	±4.8	--	±1.5	±1.5
Plasma Protein	Mean	5.6	6.0	6.3	6.5	6.7	6.8
	S.D.	±0.73	±0.24	±0.37	±0.52	±0.26	±0.52

Units of measurement are mEq/l (except NPN mg/100 cc), CO_2 (vols %), and plasma protein (gm/100 cc).

NPN = non-protein nitrogen

(data from Overman, Etteldorf, Bass, et al., 1951, Pediatrics 7:565-576. Copyright American Academy of Pediatrics 1951)

TABLE 689

SERUM CREATININE LEVELS (mg/100 cc) IN DUTCH
CHILDREN IN RELATION TO AGE AND BODY SURFACE AREA

	Age (years)								
	4	5	6	7	8	9	10	11	12
Boys									
Mean	0.362	0.415	0.429	0.452	0.486	0.482	0.549	0.560	0.598
S.D.	0.066	0.076	0.072	0.082	0.091	0.101	0.085	0.086	0.112
Girls									
Mean	0.382	0.400	0.432	0.439	0.479	0.497	0.500	0.562	0.587
S.D.	0.058	0.065	0.083	0.097	0.076	0.085	0.091	0.096	0.084
	Body surface area (cm^2)								
	--	6,575	7,450	8,325	9,200	10,075	10,950	11,824	>12,700
	<6,575	7,450	8,325	9,200	10,075	10,950	11,825	12,700	--
Boys									
Mean	0.430	0.389	0.417	0.453	0.504	0.525	0.542	0.572	0.570
S.D.	0.129	0.053	0.082	0.078	0.092	0.105	0.096	0.068	0.103
Girls									
Mean	0.390	0.384	0.423	0.443	0.506	0.481	0.545	0.507	0.592
S.D.	0.053	0.080	0.077	0.078	0.098	0.113	0.059	0.087	0.078

(data from Donckerwolcke et al., 1970)

TABLE 690

PLASMA PROTEINS (g/liter) AND
CALCIUM (mmol/liter)
DURING THE ADOLESCENT GROWTH
SPURT IN ENGLISH CHILDREN

	Range		Mean	
	Boys	Girls	Boys	Girls
Total protein	66-82	67-83	--	--
Albumin	43-54	42-54	--	--
Calcium	2.28	2.27	2.44	2.43
	2.60	2.59	--	--

(data from Round et al., 1979, Annals of Human Biology 6:
129-136)

TABLE   691

PLASMA INORGANIC PHOSPHORUS (mmol/l) DURING THE
ADOLESCENT GROWTH SPURT IN ENGLISH CHILDREN

Age in relation to PHV		Mean	Mean -2SE	Mean +2SE
BOYS				
Months before PHV	20	1.32	1.26	1.37
	16	1.31	1.25	1.37
	12	1.32	1.26	1.38
	8	1.33	1.28	1.38
	4	1.37	1.33	1.41
	PHV	1.37	1.34	1.41
Months after PHV	4	1.35	1.32	1.38
	8	1.29	1.26	1.32
	12	1.20	1.17	1.23
	16	1.19	1.15	1.24
	20	1.18	1.14	1.22
GIRLS				
Months before PHV	20	1.31	1.21	1.42
	16	1.31	1.24	1.39
	12	1.26	1.21	1.30
	8	1.27	1.22	1.31
	4	1.29	1.24	1.34
	PHV	1.28	1.24	1.33
Months after PHV	4	1.26	1.21	1.31
	8	1.23	1.18	1.29
	12	1.20	1.13	1.26
	16	1.19	1.12	1.27
	20	1.18	1.10	1.25

PHV = age at peak height velocity.

(data from Round et al., 1979, Annals of Human Biology
6:129-136)

TABLE 692

MEAN BLOOD GAS AND pH DETERMINATIONS DURING THE FIRST DAY FOR INFANTS IN ANN ARBOR, MI

Type of Analgesia or Anesthesia	pH Birth Mean	After Birth Number of hours	After Birth Mean	CO_2 Content (mEq./L) Birth Mean	After Birth Number of hours	After Birth Mean
Inhalation...............	7.34	4	7.41	19.1	4	19.4
Caudal or saddle.........	7.31	4	7.40	18.4	4	19.2
Caudal or saddle.........	7.34	8	7.41	21.0	8	20.8
Caudal or saddle.........	7.34	12	7.45	19.6	12	19.3
Caudal or saddle.........	7.36	24	7.43	20.9	24	20.6
Inhalation...............	7.32	24	7.45	19.2	24	19.8

Type of Analgesia or Anesthesis	CO_2 Tension (Mm. Hg) Birth Mean	After Birth Number of hours	After Birth Mean	% O_2 Saturation Birth Mean	After Birth Number of hours	After Birth Mean
Inhalation...............	33	4	30	83	4	90
Caudal or saddle.........	34	4	30	84	4	92
Caudal or saddle.........	37	8	32	83	8	91
Caudal or saddle.........	33	12	27	86	12	96
Caudal or saddle.........	35	24	30	85	24	91
Inhalation...............	36	24	28	85	24	92

(data from Graham and Wilson, 1954, American Journal of Diseases of Children 87:287-297.
Copyright 1954, American Medical Association)

TABLE 693

AGE- AND SEX-SPECIFIC CHEMICAL CONSTITUENTS IN SERUM OF TORONTO CHILDREN

Chemical Constituent	Age (years)	Sex	Percentile Estimates in mg/dl		Percentile Estimates in SI Units		Units
			5	95	5	95	
Inorganic phosphorus	4-8	F	3.9	5.4	1.28	1.74	mmol/l
	9-13	F	3.5	5.1	1.13	1.64	mmol/l
	14-20	F	2.8	4.3	0.90	1.39	mmol/l
	4-9	M	3.8	5.4	1.23	1.75	mmol/l
	10-15	M	3.4	5.3	1.10	1.71	mmol/l
	16-20	M	2.5	4.4	0.81	1.43	mmol/l
Creatinine	4-5	F	0.4	0.7	35	62	μmol/l
	6-9	F	0.5	0.8	44	71	μmol/l
	10-14	F	0.6	0.9	53	80	μmol/l
	15-18	F	0.7	1.0	62	88	μmol/l
	19-20	F	0.8	1.1	71	97	μmol/l
	4-6	M	0.4	0.7	35	62	μmol/l
	7-9	M	0.5	0.8	44	71	μmol/l
	10-11	M	0.6	0.9	53	80	μmol/l
	12-14	M	0.7	1.0	62	88	μmol/l
	15-17	M	0.8	1.1	71	97	μmol/l
	18-20	M	0.9	1.2	80	106	μmol/l

SI = standard international.

(data from Cherian and Hill, 1978, reproduced with permission from the American Journal of Clinical Pathology, Volume 69, pages 24-31, 1978)

TABLE 694

NON-SEX-RELATED CHEMICAL CONSTITUENTS IN
SERUM OF TORONTO BOYS AND GIRLS COMBINED

Chemical Constituent	Units	Age (years)	Percentile Estimates in Present Units		Percentile Estimate in SI Units		Units
			5	95	5	95	
Sodium	mEq/l	4-20	136	142	136	142	mmol/l
Potassium	mEq/l	4-20	3.8	5.0	3.8	5.0	nmol/l
Chloride	mEq/l	4-20	100	109	100	109	mmol/l
Urea nitrogen	mg/dl	4-20	9	18	3.20	6.55	mmol/l
Magnesium	mEq/l	4-20	1.54	1.86	0.77	0.93	mmol/l
Calcium	mg/dl	4-20	9.6	10.9	2.40	2.72	mmol/l
Total protein	g/dl	4-11	6.6	7.9	--	--	--
Total protein	g/dl	12-20	6.8	8.4	--	--	--
Albumin	g/dl	4-20	3.8	5.0	--	--	--
Alpha-1-globulin	g/dl	4-20	0.1	0.3	--	--	--
Alpha-2-globulin	g/dl	4-20	0.6	1.1	--	--	--
Beta-globulin	g/dl	4-20	0.8	1.2	--	--	--
Gamma-globulin	g/dl	4-20	0.6	1.4	--	--	--

SI = standard international.

(data from Cherian and Hill, 1978, reproduced with permission from the American Journal of Clinical Pathology, Volume 69, pages 24-31, 1978)

DIET

TABLE   695

PERCENT ON DIFFERENT TYPES OF FEEDING BY AGE AT
DISCHARGE FOR U.S. INFANTS

Age at discharge	Breast only	Breast and bottle	Bottle only
Under 8 days	41	28	31
8 days and over	35	25	40

(data from Bain, 1948, Pediatrics 2:313-320. Copyright
American Academy of Pediatrics 1948)

TABLE 696

PERCENTAGE OF MOTHERS WHO BREASTFED THEIR FIRST
BABY BY EDUCATION AND RACE: MOTHERS, AGED 15-54
INCLUDED IN THE NATIONAL FERTILITY SURVEY OF 1965

Education in Years	Race		
	White	Black	Other
0-6 years	74	77	--
7-9 years	56	74	64
10-11 years	46	53	75
12 years	44	41	57
13-15 years	54	36	--
16 or more	60	61	--
Total	50	58	64

(data from Hirschman and Sweet, 1974)

TABLE 697

PERCENTAGE OF MOTHERS WHO BREASTFED THEIR FIRST
BABY BY OCCUPATION AND EDUCATION: MOTHERS, AGE 15-54,
IN 1965 NATIONAL FERTILITY SURVEY

Occupation	Education in Years					
	0-6	7-9	10-11	12	13-15	16+
Never worked	79	66	44	46	63	52
White collar	--	60	50	51	58	63
Clerical	--	53	45	40	43	52
Blue collar	76	49	43	38	58	--
Service	68	60	55	47	66	--
Farm	--	59	--	--	--	--
Total	74	58	47	44	54	60

(data from Hirschman and Sweet, 1974)

TABLE 698

PERCENTAGE OF MOTHERS WHO BREASTFED THEIR FIRST BABY
BY RELIGION AND REGION OF RESIDENCE: MOTHERS,
AGE 15-54, IN 1965 NATIONAL FERTILITY SURVEY

Region of Residence	Religion			
	Protestant	Catholic	Jewish	Other/None
Northeast	44	37	24	54
Midwest	50	45	--	58
South	49	45	--	44
West	55	58	34	63
Total	53	44	27	57

(data from Hirschman and Sweet, 1974)

TABLE 699

PERCENTAGE OF MOTHERS WHO BREASTFED THEIR
FIRST BABY BY REGION OF
RESIDENCE AND FARM RESIDENCE:
MOTHERS, AGE 15-54, IN 1965
NATIONAL FERTILITY SURVEY

Region of Residence	Farm	Nonfarm
Northeast	55	39
Midwest	59	48
South	69	54
West	66	55
Total	64	49

(data from Hirschman and Sweet, 1974)

TABLE 700

PERCENTAGE OF MOTHERS WHO BREASTFED THEIR FIRST BABY
BY ETHNIC ORIGIN AND RELIGION: MOTHERS,
AGE 15-54, IN 1965 NATIONAL FERTILITY SURVEY

Ethnic Origin	Protestant	Catholic	Jewish	Other/None
Black	60	41	--	--
Latin American	40	59	--	--
English	54	44	--	58
German	52	46	--	68
Slavic	57	53	26	42
Irish	51	31	--	--
French	53	34	--	--
Italian	42	38	--	--
Others, D.K.	52	46	--	56
Total	53	44	27	57

D.K. = don't know

(data from Hirschman and Sweet, 1974)

TABLE 701

PERCENTAGE OF MOTHERS WHO BREASTFED THEIR FIRST BABY BY MOTHER'S BIRTH COHORT
AND YEAR OF BIRTH OF INFANT: MOTHERS, AGE 15-54, IN 1965 NATIONAL FERTILITY SURVEY

Mother's Birth Cohort	Mother's Age in 1965	Year of Birth of First Baby							
		1926-1930	1931-1935	1936-1940	1941-1945	1946-1950	1951-1955	1956-1960	1961-1965
1911-15	50-54	81	72	71	57	55	--	--	--
1916-20	45-49	--	70	82	65	40	36	--	--
1921-25	40-44	--	--	75	66	52	48	31	26
1926-30	35-39	--	--	--	64	50	47	29	16
1931-35	30-34	--	--	--	--	60	47	39	42
1936-40	25-29	--	--	--	--	--	46	35	34
1941-45	20-24	--	--	--	--	--	--	38	34
1946-50	15-19	--	--	--	--	--	--	--	21
Total	--	81	72	72	65	50	46	36	32

(data from Hirschman and Sweet, 1974)

TABLE 702

PER CENT OF INFANTS RECEIVING SPECIFIED FEEDING AT TIME OF DISCHARGE
FROM HOSPITALS IN 1946 AND 1956 (BY GEOGRAPHIC REGIONS)

United States	Breast Only		Breast and Bottle		Bottle Only	
	1946	1956	1946	1956	1946	1956
Regions:						
Northeast	23	12	16	9	61	79
East and Central	36	20	30	15	34	65
Southeast	55	27	27	16	18	57
Southwest	47	27	35	23	18	50
Mountain and Plains	44	26	28	17	28	57
Pacific	31	25	29	19	40	56

(data from Meyer, 1958, Pediatrics 22:116-121. Copyright American Academy of
 of Pediatrics 1958)

TABLE 703

PERCENT OF CURRENTLY MARRIED WOMEN WHO BREAST FED THEIR FIRST
OR SECOND CHILD, BY BIRTH COHORT OF MOTHER AND YEAR OF BIRTH OF CHILD:
UNITED STATES, 1973 AND 1965

Birth cohort of mother and year of birth of child	First child		Second child	
	1973	1965	1973	1965
All women	38.4	50.5	26.8	37.6
Birth cohort of mother				
1951-59	25.6	--	16.6	--
1946-50	25.7	21.8	19.8	13.8
1941-45	37.7	34.8	26.1	24.1
1936-40	43.2	36.6	26.8	25.9
1931-35	49.6	45.9	31.4	30.8
1926-30	47.1	47.5	32.6	33.4
1921-25	--	57.8	--	40.0
1916-20	--	65.7	--	49.4
1911-15	--	68.3	--	62.8
Year of birth of child				
1971-73	29.1	--	23.6	--
1966-70	28.3	--	22.7	--
1961-65	38.0	32.3	24.6	22.7
1956-60	43.1	36.0	28.3	27.5
1951-55	48.8	46.6	34.4	33.2
1946-50	58.9	50.5	55.0	41.3
1941-45	--	64.5	--	53.7
1936-40	--	77.4	--	65.9
1931-35	--	72.0	--	75.7

(data from Hirschman and Hendershot, 1979)

TABLE   704

PERCENT OF EVER-MARRIED WOMEN 15-44 YEARS OF AGE WHO BREAST FED THEIR
FIRST OR SECOND CHILD, BY DURATION OF BREAST FEEDING, BIRTH COHORT
OF MOTHER, AND YEAR OF BIRTH OF CHILD: UNITED STATES, 1973

Birth cohort of mother and Year of birth of child	Duration of breast feeding			Duration of breast feeding		
	All durations	Less than 3 months	3 months or more	All durations	Less than 3 months	3 months or more
	Percent breast feeding first child			Percent breast feeding second child		
All women	38.6	25.6	13.0	27.0	17.0	10.0
Birth cohort of mother						
1951-59	25.1	21.1	4.0	17.8	13.7	4.1
1946-50	25.4	18.7	6.7	19.4	14.2	5.2
1941-45	37.8	25.4	12.4	26.2	17.1	9.1
1936-40	43.2	27.6	15.6	26.7	16.1	10.6
1931-35	50.0	31.3	18.7	31.9	19.8	12.1
1929-30	49.6	29.4	20.2	34.4	18.8	15.6
Year of birth of child						
1971-73	28.7	21.6	7.1	23.5	16.4	7.1
1966-70	27.9	19.7	8.2	22.1	15.2	6.9
1961-65	37.5	25.2	12.3	24.7	15.4	9.3
1956-60	42.9	29.5	13.4	29.1	18.5	10.6
1951-55	49.8	31.7	18.1	34.2	19.7	14.5
1950 or before	59.9	27.9	32.0	56.7	27.1	29.6

(data from Hirschman and Hendershot, 1979)

TABLE   705

PERCENTAGE OF INFANTS STARTED ON BREAST FEEDINGS, DISCONTINUED IN
HOSPITAL AND DISCHARGED ON BREAST FEEDING, FROM NEW HAVEN HOSPITAL
(University Service) BY YEAR FROM 1942 THROUGH 1951

Year	Percent Started on Breast	Number Stopped in Hospital Expressed as Percent of Number Started	Percent of Total Babies Discharged on Breast
1942	81.9	14.5	70.0
1943	75.6	20.8	59.9
1944	60.1	26.5	44.1
1945	56.6	18.8	46.0
1946	48.9	14.1	42.0
1947	53.6	13.8	46.2
1948	51.8	9.0	47.2
1949	50.3	11.3	44.7
1950	48.7	7.0	45.3
1951	47.2	5.4	44.7

(data from Jackson, Wilkin, and Auerbach, 1956, Pediatrics 17:700-
713. Copyright American Academy of Pediatrics 1956)

TABLE   706

PERIOD AT WHICH BREAST FEEDING WAS DISCONTINUED,
1947 THROUGH 1949 IN A CONNECTICUT HOSPITAL

Rooming-in Mothers*

Stopped Breast Feeding	1947 Cumulative %	1948 Cumulative %	1949 Cumulative %
1-2 wk.	14.1	22.1	21.8
3 wk.-2 mo.	46.7	50.0	56.4
3-4 mo.	69.6	66.3	75.6
5-6 mo.	80.7	78.5	85.0
7-9 mo.	94.1	94.2	95.3
10-12 mo.	97.8	97.7	98.7
after 12 mo.	100.0	100.0	100.0
Total	100.0	100.0	100.0

Nursery Mothers

1-2 wk.	31.5	32.9	35.0
3 wk.-2 mo.	71.5	63.6	80.0
3-4 mo.	81.7	79.7	92.5
5-6 mo.	89.4	89.5	95.0
7-9 mo.	95.7	95.1	100.0
10-12 mo.	99.6	98.6	--
after 12 mo.	100.0	100.0	--
Total	100.0	100.0	100.0

*Mothers rooming in same unit as newborns for duration
of hospital stay. Infants of "Nursery Mothers" cared
for in neo-natal nursery.

(data from Jackson, Wilkin, and Auerbach, 1956, Pediatrics
17:700-713. Copyright American Academy of Pediatrics 1956)

TABLE 707

MEAN DURATION (months) OF BREAST FEEDING
BY EDUCATION AND PERSONAL-SOCIAL FACTORS
IN A CONNECTICUT HOSPITAL

Group	0 through High School 3		College Graduate and Postgraduate	
	Mean	S.D.	Mean	S.D.
Total	2.9	0.23	3.9	0.13
University Service	2.9	0.23	4.4	0.27
Private	0	--	3.3	0.26
Primiparas	2.8	0.30	3.8	0.25
Multiparas	3.0	0.38	4.0	0.30
Negro	5.0	0.68	--	--
University Service, White	2.4	0.22	4.3	0.27

(data from Jackson, Wilkin, and Auerbach, 1956, Pediatrics 17:700-713.
Copyright American Academy of Pediatrics 1956)

TABLE 708

SIGNIFICANT EPISODES OF ILLNESS PER 100 NEW JERSEY INFANTS
REGARDLESS OF FEEDING MODE AT ONSET OF ILLNESS

Age (months)	Breast-fed	Limited Breast	Artificial
1-2	0.7 (1)	3.8 (3)	11.8 (34)
3-4	5.9 (8)	11.3 (9)	16.0 (46)
5-6	7.4 (10)	20.0 (16)	20.5 (59)
7-8	18.5 (25)	16.3 (13)	21.5 (62)
9-10	14.1 (19)	22.5 (18)	21.2 (61)
11-12	11.9 (16)	20.0 (16)	19.8 (57)
First Year	58.5 (79)	93.8 (75)	110.8 (319)

(data from Cunningham, 1979, The Journal of Pediatrics 95:685-689)

TABLE   709

OVERALL ENERGY EXPENDITURE OF IGLOOLIK COMMUNITY OF ESKIMOS

Group	Daily Expenditure (KCal x 10^{-3})	Yearly Expenditure (KCal x 10^{-6})
Boys		
0- 5 years	0.64	11.8
5-10 years	1.39	23.8
10-15 years	2.30	28.5
Girls		
0- 5 years	0.64	14.4
5-10 years	1.39	22.8
10-15 years	1.84	18.1

(data from Godin and Shephard, 1973)

TABLE 710

AVERAGE HOURS BETWEEN FEEDINGS AT NIGHT IN RELATION TO AGE SOLIDS ACCEPTED
IN COLORADO INFANTS

| Age solids accepted (months) | Age in months | | | | | | | | | Age at reaching 8 hour span | |
	¼	1	1¼	2	2¼	3	3¼	4	5	Median (months)	Range (months)
½ to 1	4.7	5.7	5.8	7.3	8.2	10.1	10.1	10.4	10.3	1.6	0.1-4.0
1 to 1½	4.9	5.1	6.0	8.5	9.2	10.5	10.8	11.0	11.5	1.3	0.1-15.0
1½ to 2	4.8	5.7	6.3	7.6	8.3	9.2	9.8	10.4	11.1	1.5	0.1-15.0
2 to 2½	4.2	5.7	6.6	8.0	9.2	9.9	10.1	10.2	11.3	1.0	0.3-2.3
2½ to 3	4.4	4.8	5.4	7.4	7.6	9.2	9.9	10.3	11.1	1.1	0.1-4.0
>3	4.7	5.7	6.3	8.1	8.5	8.9	9.7	10.6	11.2	1.2	0.1-3.0

(data from Beal, 1969, The Journal of Pediatrics 75:690-692)

TABLE 711

MEAN RETENTION OF NITROGEN IN RELATION TO INTAKE
(Data as milligrams per kilogram daily)

| Intake range | One-year-olds | | Two-year-olds | | Three-year-olds | |
	Intake	Retention	Intake	Retention	Intake	Retention
300-349	--	--	--	--	--	--
350-399	--	--	--	--	--	--
400-424	--	--	--	--	--	--
425-449	441	81	439	56	439	46
450-474	463	85	466	66	--	--
475-499	485	91	484	90	--	--
500-524	511	118	510	105	--	--
525-549	535	112	535	83	538	124
550-574	562	119	565	88	559	106
575-599	584	128	591	89	586	105
600-624	613	126	612	116	--	--
625-649	--	--	638	105	--	--
650-699	678	182	674	146	--	--
700-749	728	217	--	--	--	--

(data from Stearns, Newman, McKinley, et al., 1958, Annals of the New York Academy of
Sciences 69:857-868)

TABLE 712

MEAN RETENTION OF NITROGEN IN RELATION TO INTAKE
(Data as milligrams per kilogram daily)

Intake range	4- to 7-year-olds		7- to 11-year-olds	
	Intake	Retention	Intake	Retention
250-299	--	--	273	23
300-324	309	40	--	--
325-349	--	--	339	21
350-374	--	--	365	33
375-399	--	--	385	37
400-424	--	--	410	53
425-449	--	--	437	67
450-474	--	--	463	67
475-499	487	83	488	73
500-524	515	90	508	105
525-549	537	73	537	104
550-574	558	89	--	--
575-599	585	116	--	--
600-649	619	105	620	101

(data from Stearns, Newman, McKinley, et al., 1958, Annals of the
New York Academy of Sciences 69:857-868)

TABLE   713

MEAN HEMOGLOBIN CONCENTRATION (in grams per 100 cc)
IN RELATION TO TYPE OF ARTIFICIAL FEEDING
IN PHILADELPHIA CHILDREN

Group	Race	6 Months Mean	6 Months S.D.	1 Year Mean	1 Year S.D.	2 Years Mean	2 Years S.D.
Group I (irradiated evaporated milk)	White	12.7	+1.26	13.2	+0.83	12.7	+0.95
	Negro	13.0	+0.93	12.9	+1.14	12.4	+1.00
Group II (nonirradiated evaporated milk plus cod-liver oil)	White	12.8	+1.36	13.2	+1.09	12.4	+0.98
	Negro	12.4	+1.39	12.9	+1.05	12.5	+1.12
Group III (irradiated evaporated milk plus carotene)	White	13.1	+1.08	13.4	+1.01	12.5	+0.85
	Negro	12.9	+1.15	13.1	+1.09	12.1	+0.91
Group IV (irradiated evaporated milk plus carotene and yeast)	White	13.5	+0.86	13.2	+1.20	12.6	+0.89
	Negro	13.4	+0.67	13.2	+1.13	12.0	+0.85

Group	Race	3 Years Mean	3 Years S.D.	4 Years Mean	4 Years S.D.
Group I (irradiated evaporated milk)	White	12.2	+0.74	12.9	+0.72
	Negro	12.2	+1.00	12.3	+0.84
Group II (nonirradiated evaporated milk plus cod-liver oil)	White	12.4	+0.69	12.9	+0.75
	Negro	12.4	+0.79	12.7	+0.74
Group III (irradiated evaporated milk plus carotene)	White	12.6	+0.74	13.0	+0.90
	Negro	12.1	+0.87	12.6	+0.85
Group IV (irradiated evaporated milk plus carotene and yeast)	White	12.6	+0.75	12.7	+0.92
	Negro	12.5	+0.91	12.2	+0.87

(data from Rhoads, Rapaport, Kennedy, et al., 1945, The Journal of Pediatrics 26:415-454)

MISCELLANEOUS

    Otoscopy

    Skin

    Sleep

    Temperature

    Urine

    Saliva

TABLE 714

AUDITORY THRESHOLDS OF OHIO CHILDREN 6 YEARS OLD

Frequency (Hz)	Boys		Girls	
	Mean	S.D.	Mean	S.D.
Right ear				
500	4.44	7.17	3.33	7.06
1000	3.84	8.30	2.48	6.84
2000	0.05	7.42	-0.13	5.96
4000	2.00	7.47	3.79	6.15
6000	4.00	8.43	3.50	8.14
Left ear				
500	5.13	9.03	3.14	8.73
1000	2.52	8.45	-0.09	8.06
2000	2.69	10.73	1.25	7.15
4000	4.97	10.10	6.00	9.56
6000	6.06	8.37	6.00	11.80
Better ear				
500	2.00	6.72	1.41	5.89
1000	0.65	6.90	-0.90	5.97
2000	-2.22	6.71	-1.53	5.70
4000	-0.32	6.79	2.76	6.24
6000	1.68	7.49	1.71	7.81
Worse ear				
500	8.28	8.26	5.62	9.24
1000	6.32	8.95	4.17	8.24
2000	5.09	10.05	3.00	6.65
4000	7.58	9.15	7.36	8.95
6000	9.00	7.77	8.38	11.11
Left-right differences				
500	-0.41	9.20	0.38	7.92
1000	-1.61	8.57	-1.91	8.12
2000	2.51	11.29	1.25	6.43
4000	2.97	11.72	2.09	5.77
6000	1.67	9.35	3.05	10.89

(data from Roche, Himes, Siervogel, et al., 1979)

TABLE 715

AUDITORY THRESHOLDS OF OHIO CHILDREN 7 YEARS OLD

Frequency (Hz)	Boys		Girls	
	Mean	S.D.	Mean	S.D.
Right ear				
500	3.57	8.51	3.35	7.01
1000	2.97	9.16	1.23	7.96
2000	-0.47	7.43	-0.26	6.30
4000	3.32	7.22	3.68	8.81
6000	3.00	8.34	2.32	8.92
Left ear				
500	1.03	7.16	0.67	5.32
1000	0.95	8.35	-0.43	5.64
2000	-0.11	8.29	0.71	4.88
4000	3.00	7.49	0.50	6.75
6000	3.89	9.79	0.50	7.46
Better ear				
500	0.16	7.36	0.26	5.88
1000	-0.26	8.06	-1.10	6.34
2000	-2.68	7.07	-1.81	4.80
4000	0.21	7.36	-0.39	6.35
6000	0.58	8.66	-0.52	6.71
Worse ear				
500	4.56	7.95	4.22	6.36
1000	4.22	8.98	2.14	7.30
2000	2.11	7.89	2.43	5.74
4000	6.11	6.03	5.00	8.75
6000	6.38	8.55	3.64	9.29
Left-right differences				
500	-2.39	6.50	-2.81	5.64
1000	-1.68	6.17	-1.50	5.41
2000	0.37	6.32	0.71	5.37
4000	-0.32	8.40	-3.71	6.92
6000	0.92	7.19	-2.00	5.91

(data from Roche, Himes, Siervogel, et al., 1979)

TABLE 716

AUDITORY THRESHOLDS OF OHIO CHILDREN 8 YEARS OLD

Frequency (Hz)	Boys Mean	Boys S.D.	Girls Mean	Girls S.D.
Right ear				
500	1.33	8.04	3.59	6.82
1000	0.33	6.80	0.86	6.49
2000	-0.24	7.98	-1.31	7.45
4000	1.71	6.56	2.91	8.45
6000	2.62	9.09	0.69	9.63
Left ear				
500	-0.29	7.98	0.18	7.29
1000	0.29	6.21	-2.06	7.34
2000	-0.62	7.71	-2.29	5.31
4000	0.48	6.17	-0.82	6.64
6000	2.76	9.32	-0.65	8.75
Better ear				
500	-1.33	7.41	-1.12	6.90
1000	-1.62	6.53	-2.51	6.92
2000	-2.95	7.65	-4.29	4.63
4000	-1.14	5.86	-2.69	6.67
6000	-0.14	7.53	-2.63	6.99
Worse ear				
500	2.38	8.23	4.88	6.29
1000	2.24	5.88	1.41	6.67
2000	2.10	7.18	0.76	7.10
4000	3.33	6.00	4.94	6.98
6000	5.52	9.82	2.76	10.36
Left-right differences				
500	-1.62	4.76	-3.41	6.63
1000	-0.05	5.38	-3.06	4.60
2000	-0.38	6.96	-1.12	7.41
4000	-1.24	6.10	-4.00	9.07
6000	0.14	8.24	-1.47	7.82

(data from Roche, Himes, Siervogel, et al., 1979)

TABLE 717

AUDITORY THRESHOLDS OF OHIO CHILDREN 9 YEARS OLD

Frequency	Boys		Girls	
(Hz)	Mean	S.D.	Mean	S.D.
Right ear				
500	1.84	9.62	-0.33	5.71
1000	0.97	8.39	-1.53	4.97
2000	0.92	8.12	-1.73	5.65
4000	1.73	6.53	0.93	7.44
6000	2.05	8.08	1.87	8.07
Left ear				
500	0.34	8.66	-0.69	5.93
1000	-0.23	8.23	-1.21	6.85
2000	-1.06	8.19	-2.83	4.91
4000	-0.67	6.22	0.29	5.52
6000	3.94	10.05	0.93	7.84
Better ear				
500	-0.65	8.47	-1.93	5.50
1000	-1.68	7.67	-3.40	5.23
2000	-1.95	7.82	-4.60	4.04
4000	-1.89	5.60	-1.40	5.54
6000	-0.16	8.37	-0.73	7.47
Worse ear				
500	2.97	9.55	1.15	5.72
1000	2.57	8.44	0.79	5.90
2000	1.89	8.14	0.14	5.40
4000	3.06	6.37	2.79	6.91
6000	6.29	8.71	3.71	7.83
Left-right differences				
500	-1.77	4.70	-0.38	4.31
1000	-1.26	6.06	0.29	5.75
2000	-1.94	4.82	-1.10	6.36
4000	-2.50	6.26	-0.43	6.07
6000	1.89	7.87	-0.71	6.72

(data from Roche, Himes, Siervogel, et al., 1979)

TABLE 718

AUDITORY THRESHOLDS OF OHIO CHILDREN 10 YEARS OLD

Frequency (Hz)	Boys		Girls	
	Mean	S.D.	Mean	S.D.
Right ear				
500	2.35	7.80	-1.31	5.35
1000	0.25	7.31	-1.03	5.55
2000	1.15	6.56	-0.28	6.41
4000	2.50	6.84	1.24	7.38
6000	1.55	6.48	1.10	7.16
Left ear				
500	0.05	7.89	-1.45	5.93
1000	-0.67	6.98	-2.69	7.84
2000	-1.23	6.35	-4.07	5.79
4000	1.69	6.74	-0.36	7.66
6000	3.85	7.94	-0.14	8.33
Better ear				
500	-0.70	7.90	-3.72	5.36
1000	-2.10	6.26	-4.90	5.89
2000	-2.20	6.21	-5.10	4.80
4000	-1.30	5.95	-2.34	7.11
6000	-0.15	6.77	-3.10	7.36
Worse ear				
500	3.35	7.39	0.97	4.86
1000	1.74	7.49	1.13	6.31
2000	2.21	6.15	0.76	6.42
4000	5.59	5.72	3.36	6.84
6000	5.59	6.68	4.07	6.38
Left-right differences				
500	-2.16	5.38	-0.14	5.78
1000	-0.92	4.98	-1.66	7.50
2000	-2.62	4.92	-3.79	6.01
4000	-0.87	8.72	-2.07	6.52
6000	2.36	7.37	-1.24	9.26

(data from Roche, Himes, Siervogel, et al., 1979)

TABLE 719

AUDITORY THRESHOLDS OF OHIO CHILDREN 11 YEARS OLD

FREQUENCY	BOYS		GIRLS	
(Hz)	MEAN	S.D.	MEAN	S.D.
Right Ear				
500	0.06	6.92	-1.67	4.93
1000	-0.22	6.23	-2.67	5.59
2000	-2.17	6.00	-2.00	7.93
4000	-0.22	6.10	0.60	7.54
6000	0.56	8.49	0.33	8.12
Left Ear				
500	-0.39	6.60	-1.27	5.79
1000	-0.44	5.64	-2.40	4.91
2000	-5.56	5.06	-3.47	6.56
4000	-2.00	6.91	1.00	9.21
6000	0.33	7.86	3.33	8.67
Better Ear				
500	-1.44	6.57	-2.80	5.19
1000	-1.94	5.73	-3.80	4.68
2000	-5.94	5.14	-5.20	5.29
4000	-3.22	5.96	-2.27	7.23
6000	-2.67	7.19	-1.47	6.97
Worse Ear				
500	1.14	6.71	-0.13	5.22
1000	1.28	5.70	-1.27	5.50
2000	-1.78	5.67	-0.27	8.15
4000	1.00	6.48	3.87	8.37
6000	3.56	7.90	5.13	8.64
Left-Right Differences				
500	-0.80	4.09	0.40	3.30
1000	-0.22	4.16	0.27	3.96
2000	-3.39	5.17	-1.47	7.50
4000	-1.78	5.12	0.40	7.87
6000	-0.22	8.79	3.00	8.03

(data from Roche, Himes, Siervogel, et al., 1979)

TABLE 720

AUDITORY THRESHOLDS OF OHIO CHILDREN 12 YEARS OLD

FREQUENCY	BOYS		GIRLS	
(Hz)	MEAN	S.D.	MEAN	S.D.
Right Ear				
500	-0.53	4.81	-0.11	5.55
1000	-0.88	4.49	-0.56	6.00
2000	-1.29	4.37	-1.50	7.03
4000	-0.06	5.49	-0.22	7.85
6000	0.35	6.56	-0.44	8.09
Left Ear				
500	-1.27	8.98	1.54	12.22
1000	-1.94	4.49	1.54	13.98
2000	-3.82	6.64	-2.17	10.93
4000	-2.24	7.26	0.74	11.63
6000	-0.24	7.79	2.63	13.13
Better Ear				
500	-3.12	3.94	-2.17	5.77
1000	-3.12	3.76	-2.72	6.22
2000	-5.35	4.56	-4.89	6.77
4000	-3.59	6.28	-3.33	7.04
6000	-1.94	6.54	-2.89	8.45
Worse Ear				
500	1.39	8.84	3.60	11.46
1000	0.36	4.54	3.77	13.16
2000	0.36	5.35	1.31	10.16
4000	1.39	5.71	3.94	10.98
6000	2.12	7.24	5.14	11.73
Left-Right Differences				
500	-0.73	9.26	1.66	11.09
1000	-1.21	4.55	2.06	13.27
2000	-2.48	7.05	-0.51	10.26
4000	-2.06	6.43	0.86	11.61
6000	-0.55	6.27	2.91	10.13

(data from Roche, Himes, Siervogel, et al., 1979)

TABLE 721

AUDITORY THRESHOLDS OF OHIO CHILDREN 13 YEARS OLD

FREQUENCY	BOYS		GIRLS	
(Hz)	MEAN	S.D.	MEAN	S.D.
Right Ear				
500	-2.58	5.62	-0.25	7.13
1000	-2.79	3.93	-1.40	7.02
2000	-4.21	6.25	-3.40	6.67
4000	-2.58	7.20	-0.77	7.59
6000	-2.16	8.32	0.07	9.17
Left Ear				
500	-3.19	5.78	-1.65	8.82
1000	-3.68	5.36	-2.11	11.27
2000	-5.21	6.21	-2.64	11.10
4000	-2.74	7.06	0.28	11.29
6000	-2.58	7.31	-0.67	9.79
Better Ear				
500	-4.16	5.33	-3.30	7.04
1000	-4.84	4.64	-5.05	5.88
2000	-6.32	5.78	-5.93	6.09
4000	-5.16	6.06	-3.44	6.78
6000	-5.16	6.98	-3.68	8.36
Worse Ear				
500	-1.57	5.78	1.40	8.31
1000	-1.57	4.17	1.54	10.95
2000	-3.11	6.29	-0.07	10.64
4000	-0.16	7.22	2.95	10.90
6000	0.42	7.61	3.09	9.34
Left-Right Differences				
500	-0.59	3.68	-1.40	6.07
1000	-0.86	4.44	-0.70	11.02
2000	-1.00	4.96	0.96	11.13
4000	-0.16	6.70	1.05	10.30
6000	-0.42	7.41	-0.74	9.12

(data from Roche, Himes, Siervogel, et al., 1979)

TABLE 722

AUDITORY THRESHOLDS OF OHIO CHILDREN 14 YEARS OLD

FREQUENCY (Hz)	BOYS		GIRLS	
	MEAN	S.D.	MEAN	S.D.
Right Ear				
500	-1.96	6.10	-2.19	6.11
1000	-3.07	4.79	-3.45	5.68
2000	-2.79	6.62	-3.95	6.00
4000	-0.39	7.38	-1.07	6.47
6000	-1.04	7.54	-0.52	8.45
Left Ear				
500	-3.04	5.95	-2.79	6.32
1000	-4.61	5.69	-3.56	10.52
2000	-2.96	6.67	-4.49	9.31
4000	-0.32	7.00	-0.14	11.19
6000	-0.32	8.32	-1.05	8.46
Better Ear				
500	-4.21	5.85	-4.19	5.75
1000	-5.43	5.21	-6.16	4.59
2000	-4.79	5.90	-6.82	5.05
4000	-2.50	6.61	-3.78	6.11
6000	-3.46	7.00	-3.53	7.37
Worse Ear				
500	-0.79	5.73	-0.79	6.22
1000	-2.25	4.93	-0.85	10.37
2000	-0.96	6.79	-1.62	9.14
4000	1.79	7.10	2.58	10.48
6000	2.11	7.85	1.96	8.58
Left-Right Differences				
500	-1.07	4.58	-0.60	4.46
1000	-1.54	3.98	-0.11	11.11
2000	-0.18	5.34	-0.55	9.00
4000	0.07	5.73	0.93	10.89
6000	0.71	7.23	-0.53	7.33

(data from Roche, Himes, Siervogel, et al., 1979)

TABLE 723

AUDITORY THRESHOLDS OF OHIO CHILDREN 15 YEARS OLD

FREQUENCY	BOYS		GIRLS	
(Hz)	MEAN	S.D.	MEAN	S.D.
Right Ear				
500	-2.46	6.10	-3.70	6.02
1000	-2.46	5.86	-3.78	5.73
2000	-2.98	6.40	-4.41	5.63
4000	0.52	7.56	-1.90	6.69
6000	-1.11	9.58	-1.81	8.46
Left Ear				
500	-3.26	5.87	-4.63	6.23
1000	-2.89	6.22	-3.75	10.58
2000	-2.80	7.25	-3.64	10.02
4000	0.25	7.80	-0.79	10.57
6000	-0.22	9.27	-1.32	8.29
Better Ear				
500	-5.02	5.34	-6.03	5.24
1000	-5.08	5.12	-6.77	4.55
2000	-5.60	5.64	-6.66	4.91
4000	-1.88	7.40	-4.45	6.16
6000	-3.54	7.92	-3.92	7.11
Worse Ear				
500	-0.71	5.84	-2.30	6.39
1000	-0.28	5.93	-0.77	10.28
2000	-0.18	6.85	-1.40	9.71
4000	2.65	7.27	1.75	9.98
6000	2.22	9.93	0.79	8.86
Left-Right Differences				
500	-0.80	5.54	-0.93	4.73
1000	-0.43	6.28	0.03	11.27
2000	0.18	7.96	0.77	9.82
4000	-0.28	6.21	1.11	9.74
6000	0.89	7.67	0.49	6.50

(data from Roche, Himes, Siervogel, et al., 1979)

TABLE 724

AUDITORY THRESHOLDS OF OHIO CHILDREN 16 YEARS OLD

FREQUENCY	BOYS		GIRLS	
(Hz)	MEAN	S.D.	MEAN	S.D.
Right Ear				
500	-1.72	5.41	-4.03	6.39
1000	-2.31	5.32	-4.22	6.02
2000	-1.53	5.91	-4.70	6.15
4000	1.13	7.62	-2.06	6.75
6000	1.75	7.74	-3.11	7.92
Left Ear				
500	-3.69	6.04	-5.43	5.67
1000	-3.03	6.17	-6.29	4.71
2000	-1.75	7.25	-6.16	5.22
4000	1.94	7.87	-1.05	9.97
6000	1.97	9.49	-1.52	8.70
Better Ear				
500	-4.63	5.06	-6.95	4.43
1000	-4.34	5.40	-7.43	3.84
2000	-3.94	5.78	-7.84	4.02
4000	-1.50	7.02	-4.95	5.81
6000	-1.31	7.79	-5.43	6.48
Worse Ear				
500	-0.78	5.88	-2.51	6.66
1000	-1.00	5.65	-3.08	6.02
2000	0.66	6.59	-3.02	6.18
4000	4.56	7.24	1.84	9.39
6000	5.03	8.29	0.79	8.83
Left-Right Differences				
500	-1.97	4.68	-1.40	6.87
1000	-0.72	4.18	-2.06	6.24
2000	-0.22	5.87	-1.46	7.23
4000	0.81	7.50	1.02	11.48
6000	0.22	8.02	1.59	8.32

(data from Roche, Himes, Siervogel, et al., 1979)

TABLE 725

AUDITORY THRESHOLDS OF OHIO CHILDREN 17 YEARS OLD

FREQUENCY	BOYS		GIRLS	
(Hz)	MEAN	S.D.	MEAN	S.D.
Right Ear				
500	-0.91	12.83	-4.59	7.78
1000	-1.78	13.02	-3.94	9.14
2000	-2.04	12.65	-4.35	6.54
4000	1.43	15.25	-4.35	6.83
6000	2.61	16.03	-1.12	10.68
Left Ear				
500	-2.96	13.08	-5.76	6.26
1000	-1.70	13.45	-5.88	4.43
2000	-2.00	14.03	-5.71	6.84
4000	1.70	14.92	0.12	10.35
6000	2.43	16.44	0.29	8.49
Better Ear				
500	-4.17	12.89	-7.35	5.32
1000	-3.35	13.13	-7.29	4.06
2000	-4.65	12.99	-7.71	3.91
4000	-1.30	14.66	-5.76	5.82
6000	-0.91	14.65	-4.18	7.34
Worse Ear				
500	0.30	12.70	-3.00	7.90
1000	-0.13	13.15	-2.53	8.77
2000	0.61	13.19	-2.35	7.78
4000	4.43	14.94	1.53	10.14
6000	5.96	16.99	3.35	10.19
Left-Right Differences				
500	-2.04	5.29	-1.18	7.02
1000	0.09	4.38	-1.94	9.58
2000	0.04	6.81	-1.35	8.37
4000	0.26	7.50	4.47	10.92
6000	-0.17	9.41	1.41	10.85

(data from Roche, Himes, Siervogel, et al., 1979)

TABLE 726

SIX MONTH INCREMENTS IN AUDITORY THRESHOLDS
IN OHIO CHILDREN 7 YEARS OLD

FREQUENCY	BOYS		GIRLS	
(Hz)	MEAN	S.D.	MEAN	S.D.
Right Ear				
500	0.70	5.81	-0.95	6.54
1000	1.10	7.24	0.00	5.66
2000	-0.82	7.35	-1.05	6.58
4000	-0.18	8.70	-1.05	4.82
6000	-0.45	6.47	-0.53	8.24
Left Ear				
500	-1.70	7.00	-2.43	10.87
1000	2.29	8.16	-0.25	9.60
2000	-0.18	5.92	-1.13	7.97
4000	0.00	8.12	-3.20	10.39
6000	-2.90	7.85	-2.53	11.94

(data from Roche, Himes, Siervogel, et al., 1979)

TABLE 727

SIX MONTH INCREMENTS IN AUDITORY THRESHOLDS
IN OHIO CHILDREN 8 YEARS OLD

FREQUENCY	BOYS		GIRLS	
(Hz)	MEAN	S.D.	MEAN	S.D.
Right Ear				
500	-2.08	8.24	-0.92	8.56
1000	-2.32	7.34	-1.70	8.03
2000	-0.16	5.47	-0.89	10.13
4000	-1.20	7.02	-1.70	7.29
6000	-0.72	6.16	-2.44	8.64
Left Ear				
500	-0.88	6.91	-1.28	9.91
1000	-2.56	5.31	-2.72	8.14
2000	-1.04	6.19	-1.28	5.47
4000	-3.36	6.99	-2.64	7.91
6000	-1.68	7.18	-1.44	7.99

(data from Roche, Himes, Siervogel, et al., 1979)

TABLE 728

SIX MONTH INCREMENTS IN AUDITORY THRESHOLDS
IN OHIO CHILDREN 9 YEARS OLD

FREQUENCY	BOYS		GIRLS	
(Hz)	MEAN	S.D.	MEAN	S.D.
Right Ear				
500	-0.08	7.79	-0.32	6.44
1000	-0.67	6.97	1.00	5.09
2000	1.00	6.04	-1.00	4.66
4000	1.42	8.18	-0.90	7.80
6000	-1.08	6.07	-0.10	9.37
Left Ear				
500	-0.18	6.11	-0.12	6.73
1000	-0.09	6.28	3.78	6.36
2000	-1.91	4.80	0.53	6.03
4000	0.26	7.94	1.67	6.52
6000	0.27	9.04	1.44	7.38

(data from Roche, Himes, Siervogel, et al., 1979)

TABLE 729

SIX MONTH INCREMENTS IN AUDITORY THRESHOLDS
IN OHIO CHILDREN 10 YEARS OLD

FREQUENCY	BOYS		GIRLS	
(Hz)	MEAN	S.D.	MEAN	S.D.
Right Ear				
500	0.16	4.54	0.10	7.06
1000	0.40	4.12	1.33	7.11
2000	-0.80	4.28	-0.57	7.65
4000	-1.12	7.03	-0.95	8.80
6000	-1.60	7.30	-1.81	7.92
Left Ear				
500	-0.95	7.53	0.42	9.39
1000	-0.08	5.12	-2.50	11.07
2000	-0.33	5.80	-0.10	7.58
4000	0.75	6.07	-1.16	8.20
6000	0.17	9.00	-4.00	8.58

(data from Roche, Himes, Siervogel, et al., 1979)

TABLE 730

SIX MONTH INCREMENTS IN AUDITORY THRESHOLDS
IN OHIO CHILDREN 11 YEARS OLD

FREQUENCY	BOYS		GIRLS	
(Hz)	MEAN	S.D.	MEAN	S.D.
Right Ear				
500	-1.08	4.93	0.78	5.74
1000	-0.08	4.45	-1.13	8.20
2000	-0.80	7.07	0.87	7.00
4000	-1.12	6.08	-0.52	6.72
6000	-1.12	10.28	-1.57	7.58
Left Ear				
500	-0.83	5.72	-0.78	6.29
1000	1.04	5.04	-1.57	6.35
2000	-1.36	4.07	0.70	7.74
4000	-2.72	6.37	-0.87	9.87
6000	-1.28	7.21	2.70	8.88

(data from Roche, Himes, Siervogel, et al., 1979)

TABLE 731

SIX MONTH INCREMENTS IN AUDITORY THRESHOLDS
IN OHIO CHILDREN 12 YEARS OLD

FREQUENCY	BOYS		GIRLS	
(Hz)	MEAN	S.D.	MEAN	S.D.
Right Ear				
500	-2.08	5.49	0.91	6.81
1000	-1.76	5.11	1.48	5.33
2000	1.28	5.88	1.13	5.75
4000	-1.28	6.73	0.87	8.50
6000	-1.28	7.35	2.26	9.27
Left Ear				
500	-1.50	5.96	2.36	18.64
1000	-1.00	4.13	4.55	18.94
2000	1.08	6.10	1.09	10.97
4000	-1.58	7.64	2.55	11.99
6000	-2.00	6.78	1.45	13.37

(data from Roche, Himes, Siervogel, et al., 1979)

TABLE 732

SIX MONTH INCREMENTS IN AUDITORY THRESHOLDS
IN OHIO CHILDREN 13 YEARS OLD

FREQUENCY	BOYS		GIRLS	
(Hz)	MEAN	S.D.	MEAN	S.D.
Right Ear				
500	-1.30	5.80	-0.39	7.28
1000	0.17	4.30	-1.22	5.84
2000	-2.17	4.51	-1.72	4.81
4000	-1.74	9.23	-0.61	6.06
6000	-2.78	7.90	-0.33	9.09
Left Ear				
500	-0.36	7.50	-1.89	9.10
1000	-1.73	6.66	-1.56	8.70
2000	-0.52	6.88	-1.37	5.63
4000	1.04	9.12	0.06	9.24
6000	-0.96	8.20	-1.17	10.24

(data from Roche, Himes, Siervogel, et al., 1979)

TABLE 733

SIX MONTH INCREMENTS IN AUDITORY THRESHOLDS
IN OHIO CHILDREN 14 YEARS OLD

FREQUENCY	BOYS		GIRLS	
(Hz)	MEAN	S.D.	MEAN	S.D.
Right Ear				
500	-0.16	6.77	-0.94	6.48
1000	-1.03	6.10	-1.84	6.03
2000	-0.22	5.01	-1.76	5.59
4000	0.92	7.10	-1.76	6.36
6000	0.27	6.13	0.53	8.16
Left Ear				
500	-0.22	7.16	-0.41	5.77
1000	-0.43	7.27	-0.73	5.24
2000	0.27	5.19	-0.04	4.41
4000	0.76	7.46	-1.06	6.52
6000	1.03	7.21	-0.61	8.76

(data from Roche, Himes, Siervogel, et al., 1979)

TABLE 734

SIX MONTH INCREMENTS IN AUDITORY THRESHOLDS
IN OHIO CHILDREN 15 YEARS OLD

FREQUENCY	BOYS		GIRLS	
(Hz)	MEAN	S.D.	MEAN	S.D.
Right Ear				
500	-0.67	6.83	-0.53	7.11
1000	0.31	6.44	-0.19	5.93
2000	0.27	5.09	0.34	4.65
4000	-0.31	5.92	-0.49	7.17
6000	-0.55	6.08	-0.34	8.49
Left Ear				
500	-0.86	6.70	-1.02	5.76
1000	0.47	6.42	0.34	4.94
2000	0.47	5.95	0.60	3.73
4000	-0.59	6.91	-1.32	7.51
6000	0.55	8.11	0.13	9.70

(data from Roche, Himes, Siervogel, et al., 1979)

TABLE 735

SIX MONTH INCREMENTS IN AUDITORY THRESHOLDS
IN OHIO CHILDREN 16 YEARS OLD

FREQUENCY	BOYS		GIRLS	
(Hz)	MEAN	S.D.	MEAN	S.D.
Right Ear				
500	0.20	5.26	-1.23	4.65
1000	-0.16	6.01	-1.32	5.30
2000	0.24	5.13	-1.66	4.50
4000	-0.37	6.24	-2.49	5.50
6000	1.02	6.62	-2.38	7.39
Left Ear				
500	-0.49	6.23	-0.85	5.00
1000	0.04	6.76	-1.49	4.62
2000	0.69	5.25	-1.23	4.64
4000	0.33	7.22	-0.30	9.08
6000	0.37	8.08	-1.79	6.24

(data from Roche, Himes, Siervogel, et al., 1979)

TABLE 736

SIX MONTH INCREMENTS IN AUDITORY THRESHOLDS
IN OHIO CHILDREN 17 YEARS OLD

FREQUENCY	BOYS		GIRLS	
(Hz)	MEAN	S.D.	MEAN	S.D.
Right Ear				
500	-1.69	6.33	-2.30	4.32
1000	-1.44	5.66	-0.20	4.85
2000	-0.81	4.06	-2.60	6.16
4000	-2.19	5.60	-4.00	7.62
6000	-3.56	6.85	-1.30	8.24
Left Ear				
500	-0.56	4.32	-1.00	6.07
1000	-1.13	4.54	-0.90	4.70
2000	-0.75	6.92	-1.60	5.53
4000	-1.94	8.42	-0.50	6.08
6000	-2.25	10.63	-1.20	8.04

(data from Roche, Himes, Siervogel, et al., 1979)

TABLE 737

SPONTANEOUS VISIBILITY OF TYMPANIC MEMBRANE BY AGE
FOR CHILDREN IN NEW ORLEANS (LA)

	Age (days)*					
	1	2	3	4	5	6
Tympanic membrane and canals visible	0	3	19	78	49	42
Tympanic membrane and canals not visible	14	77	95	76	43	28

*Day of birth not counted; data are for numbers of cases

(data from McLellan and Webb, 1961, The Journal of Pediatrics 58:523-527)

TABLE 738

SKIN pH READINGS IN INFANTS
(mostly Puerto Rican or Negro) IN NEW YORK STATE

Age Group	Skin Area Tested	Number of Infants in Each pH Range					
		3-3.9	4-4.9	5-5.9	6-6.9	7-7.9	8.0
I (1-48 hr.)	Shoulder	3	23	17	21	21	2
	Axilla	3	16	9	21	33	4
	Abdomen	0	3	9	29	39	2
II (3-6 days)	Shoulder	21	40	13	10	1	0
	Axilla	7	38	20	16	2	0
	Abdomen	2	31	18	22	7	2
III (7-30 days)	Shoulder	16	54	3	2	1	0
	Axilla	8	41	22	3	1	0
	Abdomen	2	47	15	6	2	1

MEAN SKIN pH BY AGE GROUPS

Site	I	II	III
Shoulder			
Mean	5.8	4.7	4.4
S.D.	1.27	0.96	0.67
Axilla			
Mean	6.3	5.0	4.7
S.D.	1.2	1.03	0.7
Abdomen			
Mean	6.8	5.6	4.9
S.D.	0.83	1.06	0.79

(data from Behrendt and Green, 1958, American Journal of Diseases of Children 95:35-41. Copyright 1958, American Medical Association)

TABLE 739

SKIN pH IN BOYS
IN NEW YORK STATE

Maturity Group	Skin Area Tested	Number of Boys in Each pH Range							
		2-2.9	3-3.9	4-4.9	5-5.9	6-6.9	7-7.9	8-8.9	9-9.9
I	Axillary vault	0	9	73	18	3	0	0	0
	Axillary fossa	0	20	71	11	1	0	0	0
	Shoulder	10	42	33	4	0	0	0	0
II	Axillary vault	0	6	28	26	35	18	1	0
	Axillary fossa	2	14	75	16	6	1	0	0
	Shoulder	9	33	44	10	0	0	0	0
III	Axillary vault	0	0	4	14	48	35	15	0
	Axillary fossa	0	6	57	30	17	6	0	0
	Shoulder	5	25	59	20	0	0	0	0
IV	Axillary vault	0	0	1	2	22	60	47	0
	Axillary fossa	0	10	65	33	19	4	1	0
	Shoulder	10	34	66	19	0	0	0	0
V	Axillary vault	0	0	0	0	6	15	13	3
	Axillary fossa	0	2	14	13	7	1	0	0
	Shoulder	0	10	25	2	0	0	0	0

VALUES OF SKIN pH FOR THREE TEST AREAS

Site	Maturity Group				
	I	II	III	IV	V
Axillary vault					
Mean	4.6	5.7	6.8	7.6	7.9
S.D.	0.56	1.15	0.93	0.69	0.71
Axillary fossa					
Mean	4.4	4.6	5.1	5.1	5.3
S.D.	0.56	0.73	0.89	0.93	0.93
Shoulder					
Mean	3.8	4.0	4.3	4.2	4.2
S.D.	0.7	0.76	0.69	0.82	0.48

INFLUENCE OF RACE -- AXILLARY VAULT pH

Race	Mean in Each Maturity Group				
	I	II	III	IV	V
White	4.59	5.80	6.77	7.57	7.84
Negro	4.83	6.01	7.24	7.60	8.03
Puerto Rican	4.52	5.19	6.44	7.48	7.78

Group I = prepubescent; Group II = early pubescent; Group III = midpubescence;
Group IV = later pubecent; Group V = adult.

TABLE 740

MEAN SKIN pH OF GIRLS IN NEW YORK STATE BY MATURITY GROUPS
(Colorimetric Measurements)

| Maturity | Site | Skin pH | |
		Mean	S.D.
I	Shoulder	4.3	1.06
	Axillary vault	4.8	1.18
	Axillary fossa	4.5	0.94
II	Shoulder	4.3	0.60
	Axillary vault	6.1	1.21
	Axillary fossa	4.7	0.87
III	Shoulder	4.3	0.79
	Axillary vault	7.5	0.78
	Axillary fossa	4.9	1.05
IV	Shoulder	5.1	0.79
	Axillary vault	7.1	0.67
	Axillary fossa	6.1	1.14

I - prepubescent; II - early pubescence; III - later pubescence
but pre-menarche; IV - post menarche.

(data from Behrendt, Green, and Carol, 1964, American Journal of
Diseases of Children 108:37-43. Copyright 1964, American Medical
Association)

TABLE   741

MEDIAN RADIOGRAPHIC THICKNESS OF SKIN OF
FOREARM (mm) AT 15 YEARS IN RELATION TO WEIGHT (lb)
IN MISSOURI CHILDREN

Weight	BOYS	GIRLS
60	1.29	1.07
80	1.34	1.12
100	1.39	1.15
120	1.42	1.18
140	1.45	1.21
160	1.48	1.23
180	1.51	1.25

(data from Bliznak and Staple, 1975, Radiology 118:
55-60)

TABLE   742

VALUES FOR SKIN REFLECTANCE
IN WHITE OHIO CHILDREN

Age Range (years)	Inner Arm		Chest		Forehead		Left Areola		Scrotum	
	Mean	S.D.	Mean	S.D.	Mean	S.D.	Mean	S.D.	Mean	S.D.
0-4	29.9	3.9	29.2	3.7	27.6	3.7	24.2	4.1	15.8	4.2
4½-10	27.3	3.7	24.8	4.3	24.4	3.2	21.0	4.3	16.3	2.8
4½-10	27.0	3.7	26.4	4.8	25.7	4.4	23.3	3.4	--	-
12½-16	27.8	5.2	24.9	6.7	22.9	5.1	18.8	4.6	16.2	4.1
8½-15½	28.8	2.8	31.2	5.1	26.3	4.2	19.4	3.6	--	-

(data from Garn, Selby, and Crawford, 1956)

TABLE 743

CHANGE IN SKIN COLOR OF NEGRO INFANTS IN TENNESSEE BY COLOR AT BIRTH

Color Change by 2½-3 Yr.	Color at Birth						
	3	4	5	6	7	8	Total
No change	--	5	3	4	1	--	13
1 shade lighter	2	--	1	4	1	1	9
2 shades lighter	--	--	--	--	1	--	1
1 shade darker	--	5	8	3	--	--	16
2 shades darker	--	2	4	2	--	--	8
3 shades darker	--	--	1	--	--	--	1
Total	2	12	17	13	3	1	48

Changes of specific skin color between birth
and 36 months by percentage

Change in Specific Skin Color	Age of Child	
	30 Months	36 Months
No change	34.8	20.0
Lighter	13.0	28.0
Darker	52.2	52.0

(data from Horton and Crump, 1959, Archives of
Dermatology 80:421-426. Copyright 1959, American
Medical Association)

TABLE 744

SKIN RESISTANCE TO ELASTICITY AND SWEAT CHLORIDE CONTENT
IN TENNESSEE CHILDREN (means and 95 percent confidence limits)

	White	Negro
Sweat chloride (mEq./L.)	17.2 (14.6 to 21.0)	11.3 (9.4 to 13.6)
Skin resistance (ohms x 10^3)	228 (214 to 243)	365 (318 to 419)

(data from Batson, Young, and Shepard, 1962, The Journal of Pediatrics 60: 716-720)

TABLE 745

MORPHOLOGICAL CLASSIFICATION AND MEASUREMENTS OF HAIR ROOTS AND HAIR SHAFTS
IN FULL-TERM NEWLY BORN INFANTS IN BALTIMORE (MD)

Hair	Type	Mean ± S.D. (μ)	Range (μ)
Hair roots	Anagen*	123 ± 32	60 to 200
	Telogen	87 ± 21	20 to 160
Hair Shafts**	--	31 ± 6	15 to 48

*Includes four hair roots in catagen phase.

**All the available hair shafts from 63 newborns are included in the measurement

although only 714 (70%) had identifiable hair roots.

(data from Saadat, Khan, Gutberlet, et al., 1976, Pediatrics 57:960-962. Copyright American Academy of Pediatrics 1976)

TABLE   746

RATES OF FINGERNAIL GROWTH IN INDIVIDUAL NORMAL INFANTS IN NEW
YORK CITY DURING THE FIRST WEEK AFTER BIRTH

Infant	Nail Growth (mm/day)
Al	.122
Bi	.118
Wo	.115
Mu	.113
La	.113
Br	.111
Ri	.111
Mo	.110
Or	.105
Th	.102
Average	.112

(data from Sibinga, 1959, Pediatrics 24:225-233. Copyright
American Academy of Pediatrics 1959)

TABLE   747

AVERAGE NUMBER OF HOURS OF SLEEP, THE AVERAGE LONGEST SLEEP
PERIOD, AND THE AVERAGE NUMBER OF FEEDINGS FOR EACH DAY
IN NEWLY BORN INFANTS IN CALIFORNIA

	First Day	Second Day	Third Day
Average number of hours of sleep	17.0	16.5	16.3
Average length of longest sleep period	4.9	4.3	4.2
Average number of feedings	4.9	6.0	6.2

(data from Parmelee, Schulz, and Disbrow, 1961, The Journal
of Pediatrics 58:241-250)

TABLE   748

PERCENTAGE OF TIME SPENT IN EACH SLEEP STAGE DURING EACH
THIRD OF THE NIGHT IN 2-YEAR-OLD FLORIDA CHILDREN

Sleep Stage	Thirds I	Thirds II	Thirds III
0	18.1	33.0	48.9
1	38.5	37.2	24.3
REM*	16.4	35.4	48.2
2	30.7	34.5	34.7
4	66.2	25.3	8.5

*REM = rapid eye movement

(data from Kohler, Coddington, and Agnew, 1968, The Journal
of Pediatrics 72:228-233)

TABLE   749

SEQUENCE OF SLEEP STAGE CHANGES IN TERMS OF PERCENT OF TIME
ONE STAGE WAS FOLLOWED BY ANOTHER IN 2-YEAR-OLD FLORIDA CHILDREN

Following stage	This sleep stage 0	This sleep stage 1	This sleep stage 2	This sleep stage 4
0	--	7.3	4.6	1.6
1	97.8	--	53.6	6.4
2	2.2	92.7	--	92.0
4	0.0	0.0	41.8	--

(data from Kohler, Coddington, and Agnew, 1968, The Journal of
Pediatrics 72:228-233)

TABLE 750

DISTRIBUTION OF SLEEP STAGE LENGTHS FROM RECORDS REDUCED TO
483 MINUTES IN LENGTH (boys 8 to 11 years)
IN FLORIDA BOYS

Duration (minutes)	Sleep Stage					
	0	1	2	3	4	REM
80-84	--	--	1	--	1	--
75-79	--	--	1	--	1	--
70-74	--	3	5	--	1	1
65-69	--	3	1	--	2	6
60-64	1	4	9	--	3	3
55-59	0	4	7	--	2	5
50-54	0	11	7	--	11	5
45-49	0	16	19	--	12	15
40-44	1	15	25	--	12	19
35-39	0	25	36	--	11	20
30-34	0	22	33	--	17	25
25-29	0	24	50	--	22	18
20-24	1	28	61	--	26	29
15-19	1	34	56	5	19	32
10-14	1	43	59	21	27	26
5-9	6	53	134	113	44	9
1-4	81	206	151	294	36	4

The stages were defined from electroencephalographic and electro-oculographic data.

REM = rapid eye movement stage.

(data from Ross, Agnew, Williams, et al., 1968, Pediatrics 42:324-335. Copyright
American Academy of Pediatrics 1968)

TABLE   751

PERCENT OF TIME ONE SLEEP STAGE FOLLOWED ANOTHER
IN BOYS AGED 8 TO 11 YEARS IN FLORIDA

	This sleep stage				
	0	1	2	3	4
Was followed by					
this sleep stage					
0	0.0	10.2	5.7	1.1	1.2
1	80.9	0.0	53.4	2.5	3.2
2	19.1	89.8	0.0	39.1	27.2
3	0.0	0.0	40.9	0.0	68.3
4	0.0	0.0	0.0	57.2	0.0

Stages defined from electroencephalographic and electroculographic data

(data from Ross, Agnew, Williams, et al., 1968, Pediatrics 42:324-335. Copyright
American Academy of Pediatrics 1968)

TABLE 752

DISTRIBUTION OF SLEEP STAGE AMOUNTS WITH RESPECT TO THIRDS
OF THE SLEEP PERIOD IN BOYS AGED 8 TO 11 YEARS IN FLORIDA

Sleep stage	1	2	3
0	26.5	14.5	59.0
1	41.0	33.9	25.1
REM	6.0	33.6	60.4
2	26.4	40.8	32.8
3	55.7	26.5	17.8
4	77.5	18.2	4.3

The stages were defined from electroencephalographic and
electroculographic data

REM = rapid eye movement stage

(data from Ross, Agnew, Williams, et al., 1968, Pediatrics
42:324-335. Copyright American Academy of Pediatrics 1968)

TABLE 753

MEAN ORAL TEMPERATURE (^{O}F) in
CALIFORNIA CHILDREN

Age (years)	Girls		Boys	
	Mean	S.D.	Mean	S.D.
11.5	98.01	.35	97.86	.44
12.0	97.87	.41	97.89	.37
12.5	97.84	.33	97.95	.37
13.0	97.95	.23	97.83	.20
13.5	97.98	.35	97.90	.37
14.0	98.00	.40	97.73	.46
14.5	97.98	.34	97.79	.40
15.0	98.03	.46	97.57	.44
15.5	97.96	.27	97.59	.48
16.0	97.88	.43	97.46	.51
16.5	97.80	.39	97.39	.51
17.0	97.79	.43	97.44	.47
17.5	97.75	.37	97.42	.58

(data from Eichorn and McKee, 1953)

TABLE 754

BODY TEMPERATURES (°F) OF CHILD
RESEARCH COUNCIL CHILDREN (Denver, CO)

| Age | Body Temperature | | | |
| | BOYS | | GIRLS | |
(years)	Mean	S.D.	Mean	S.D.
0- 1	R99.1	0.7	R99.1	0.4
1- 2	R99.1	0.5	R98.9	0.5
2- 3	R99.0	0.4	R98.8	0.4
3- 4	R98.9	0.4	R98.8	0.4
3- 4	98.7	0.5	98.7	0.5
4- 5	98.6	0.5	98.5	0.5
5- 6	98.5	0.4	98.5	0.4
6- 7	98.4	0.4	98.5	0.4
7- 8	98.3	0.4	98.4	0.4
8- 9	98.3	0.4	98.3	0.4
9-10	98.1	0.5	98.2	0.4
10-11	98.0	0.5	98.1	0.4
11-12	98.0	0.4	98.0	0.5
12-13	97.8	0.4	97.9	0.4
13-14	97.7	0.4	97.9	0.5
14-15	97.6	0.4	97.9	0.6
15-16	97.4	0.4	97.9	0.4
16-17	97.3	0.5	97.8	0.5
17-18	97.2	0.4	97.9	0.5

R = temperatures measured rectally

(data from Iliff and Lee, 1952)

TABLE 755

URINARY OUTPUT (cc) IN INFANCY IN RELATION
TO INTAKE AND VOLUME AT ONE VOIDING

Age (Days)	Milk (cc)	Urine (cc)	Percent of intake returned as output
1	28.3	21	74.0
2	67.0	22	46.5
3	146.9	37	25.0
4	243.2	62	25.5
5	311.5	99	31.7
6	364.0	115	31.6
7	364.8	144	39.5
8	382.5	142	37.1
9	412.5	162	39.2

VOLUME AT ONE VOIDING (cc)

| | Age (years) | | | | |
	1	2	3	4	5
Maximum Mean	30.0	34.0	34.0	38.0	32.0
Minimum Mean	1.5	1.0	2.0	3.5	2.0
Average Mean	9.2	10.2	12.5	13.5	11.7

(data from J. Thompson, 1944, Archives of Disease in Childhood 19:169-177)

TABLE 756

24-HOUR CREATININE EXCRETION

Age (months)	Stature (cm)	24-Hour Urinary Creatinine (mg)	Age (months)	Stature (cm)	24-Hour Urinary Creatinine (mg)
0	50	35.50	39	98.0	263.62
1	53.5	44.94	40	98.6	268.19
2	56.9	55.19	41	99.2	272.80
3	60.4	66.44	42	99.8	277.44
4	62.4	72.38	43	100.4	281.12
5	64.4	78.57	44	101.0	283.81
6	66.4	84.99	45	101.6	287.53
7	68.0	90.44	46	102.2	290.25
8	69.6	96.05	47	102.8	292.98
9	71.2	101.82	48	103.4	295.72
10	72.5	107.30	49	104.0	299.52
11	73.8	112.91	50	104.5	303.05
12	75.2	118.82	51	105.1	305.84
13	76.3	123.61	52	105.6	308.35
14	77.4	128.48	53	106.2	311.17
15	78.5	132.66	54	106.7	313.70
16	79.6	137.71	55	107.1	318.09
17	80.7	142.84	56	107.4	322.20
18	81.8	147.24	57	107.7	325.25
19	82.8	156.49	58	108.0	329.40
20	83.7	159.03	59	108.4	333.87
21	84.7	165.16	60	108.7	336.97
22	85.6	171.20	63	111.0	359.64
23	86.6	177.53	66	112.2	379.24
24	87.5	183.75	69	113.2	384.88
25	88.5	189.39	72	114.1	390.22
26	89.0	194.03	72	114.5	391.59
27	89.8	198.46	75	115.9	399.86
28	90.5	203.62	78	117.2	407.86
29	91.5	209.54	81	118.6	431.70
30	92.1	213.67	84	120.0	456.00
31	92.8	219.94	87	121.5	477.50
32	93.5	226.27	90	123.0	499.38
33	94.1	231.49	93	124.5	527.88
34	94.8	237.96	96	126.0	556.92
35	95.5	244.48	99	127.5	586.50
36	96.2	250.12	102	129.0	616.62
37	96.8	256.52	105	130.5	--
38	97.4	259.08	108	132.0	--

Based on the data of Stuart and Stevenson (1954), Stearns et al. (1958), and Daniels and Hejinian (1929) for U.S. children.

(data collated by Viteri and Alvarado, 1970, Pediatrics 46:696-706. Copyright American Academy of Pediatrics 1962)

TABLE 757

MEAN DAILY URINARY CREATININE PER KILOGRAM OF BODY WEIGHT
IN IOWA CHILDREN

Age (years)	Sex	Mean	S.D.
1	MF	13.5	3.1
2	MF	16.3	6.2
3	MF	17.4	2.5
4	M	18.3	2.0
5	M	20.9	2.2
6	M	20.9	2.0
7	M	22.3	2.1
8	M	22.9	2.0
9	M	24.7	1.9
10	M	24.0	2.1

(data from Stearns, Newman, McKinley, et al., 1958, Annals
of the New York Academy of Sciences 69:857-868)

TABLE   758

CREATININE COEFFICIENT (gm/24 hours/kg body weight)
AND CREATININE EXCRETION (gm/24 hours) OF
MINNESOTA CHILDREN

Age (years)	Creatinine Coefficient		Creatinine	
	Mean	S.D.	Mean	S.D.
Boys				
12½–14½	25.3	4.7	1.27	0.32
14½–16½	25.5	3.2	1.69	0.26
16½–18½	27.0	3.0	1.85	0.19
Girls				
12½–14½	23.0	6.1	1.04	0.28
14½–16½	21.7	3.3	1.15	0.19
16½–18½	21.3	1.3	1.20	0.16

(data from Novak, 1963, Annals of New York Academy of Sciences 110:545-577)

TABLE   759

CREATININE EXCRETION (gm/day) IN RELATION
TO STATURE, WEIGHT AND SURFACE AREA
IN OHIO CHILDREN

Age (months)	Sex	A Creatinine	B Creatinine	C Stature (cm)	D Weight (kg)	E Surface area (m^2)
48 to 71	M	0.376	0.589	108	17.9	0.73
	F	0.346	0.535	107	17.2	0.71
72 to 95	M	0.501	0.635	123	23.6	0.90
	F	0.536	0.735	123	23.0	0.89
96 to 119	M	0.688	0.840	134	30.0	1.07
	F	0.593	0.759	134	28.1	1.04
120 to 143	M	0.826	0.934	144	36.0	1.21
	F	0.797	0.991	146	36.1	1.22
144 to 167	M	1.06	1.17	154	45.9	1.41
	F	0.946	1.08	155	43.9	1.39
168 to 191	M	1.51	1.64	170	60.0	1.70
	F	1.17	1.28	161	54.2	1.58
192 to 215	M	1.87	1.98	175	71.1	1.87
	F	1.19	1.26	162	56.7	1.60

Age (months)	Sex	A X 10^3 D	B X 10^3 D	A X 10^4 C	B X 10^4 C	A X 10^2 E	B X 10^2 E
48 to 71	M	21.0	32.9	34.8	54.5	51.5	80.0
	F	20.1	31.1	32.3	50.0	48.7	75.4
72 to 95	M	21.2	26.9	40.8	51.6	55.9	70.6
	F	23.3	32.0	43.6	59.8	60.4	82.6
96 to 119	M	22.9	28.0	51.4	62.6	64.2	78.4
	F	21.1	27.0	44.1	56.5	57.0	72.8
120 to 143	M	22.9	25.9	57.4	64.9	68.4	77.1
	F	22.1	27.4	54.5	68.0	65.2	81.1
144 to 167	M	23.1	25.5	68.9	76.0	75.1	83.0
	F	21.5	24.6	61.1	69.9	68.1	77.8
168 to 191	M	25.2	27.3	88.9	96.4	88.9	96.4
	F	21.6	23.6	72.6	79.6	74.0	81.0

TABLE   759   (continued)

CREATININE EXCRETION (gm/day) IN RELATION
TO STATURE, WEIGHT AND SURFACE AREA
IN OHIO CHILDREN

Age (months)	Sex	$\dfrac{F}{A \ X \ 10^3}$ $\overline{D}$	$\dfrac{G}{B \ X \ 10^3}$ $\overline{D}$	$\dfrac{H}{A \ X \ 10^4}$ $\overline{C}$	$\dfrac{I}{B \ X \ 10^4}$ $\overline{C}$	$\dfrac{J}{A \ X \ 10^2}$ $\overline{E}$	$\dfrac{K}{B \ X \ 10^2}$ $\overline{E}$
192 to 215	M	26.3	27.8	107.0	113.0	100.0	106.0
	F	21.0	22.2	73.5	77.9	74.6	78.6

a = creatinine total was calculated by multiplying the creatine by 0.862
b = surface area was calculated by the formula of Dubois and Dubois (Arch. Int.
     Med. 15:868-881, 1915)

(data from Clark, Thompson, Beck, et al., 1951, American Journal of Diseases
of Children 81:774-783. Copyright 1951, American Medical Association)

TABLE   760

DAILY CREATININE EXCRETION AND
CREATININE EXCRETION/WEIGHT
IN OHIO CHILDREN

Age (years)	Weight (kg) Mean	S.D.	Creatinine (gm) Mean	S.D.	Creatinine/Weight Ratio Mean	S.D.
			BOYS			
3 - 4.9	16.64	1.37	.33	.07	20.9	5.7
5 - 6.9	21.27	3.28	.49	.13	23.3	7.3
7 - 8.9	26.18	2.82	.64	.21	24.4	7.9
9 -10.9	32.74	4.89	.78	.20	23.9	5.1
11 -12.9	41.88	9.40	.98	.30	23.4	4.5
13 -14.9	54.68	12.64	1.33	.32	25.6	5.2
15 up	68.87	9.48	1.85	.30	27.0	3.4
			GIRLS			
3 - 4.9	16.38	1.88	.31	.07	18.9	4.4
5 - 6.9	20.59	2.95	.46	.13	21.9	4.3
7 - 8.9	24.31	3.18	.59	.13	24.4	5.7
9 -10.9	30.62	4.68	.77	.20	25.5	7.1
11 -12.9	40.47	7.80	.97	.27	24.2	5.6
13 -14.9	47.63	8.94	1.11	.22	23.6	3.7
15 up	59.20	13.07	1.26	.16	21.9	4.2

(data from Reynolds and Clark, 1947)

TABLE 761

URINARY EXCRETION OF CREATINE AND CREATININE (gm/day)
IN RELATION TO AGE IN OHIO CHILDREN

Age		Creatine		Creatinine	
(months)	Sex	Mean	S.D.	Mean	S.D.
48 to 71	M	0.247	0.23	0.376	0.10
	F	0.219	0.21	0.346	0.10
72 to 95	M	0.156	0.10	0.501	0.10
	F	0.231	0.16	0.536	0.12
96 to 119	M	0.176	0.12	0.688	0.11
	F	0.193	0.10	0.593	0.11
120 to 143	M	0.125	0.10	0.826	0.19
	F	0.225	0.24	0.797	0.17
144 to 167	M	0.129	0.08	1.06	0.29
	F	0.150	0.14	0.946	0.19
168 to 191	M	0.147	0.20	1.51	0.32
	F	0.131	0.12	1.17	0.18
192 to 215	M	0.129	0.18	1.87	0.32
	F	0.086	0.10	1.19	0.18

(data from Clark, Thompson, Beck, et al., 1951, American Journal
of Diseases of Children 81:774-783. Copyright 1951, American
Medical Association)

TABLE   762

URINARY TOTAL HYDROXYPROLINE AND CREATININE
EXCRETION (mg/24 hr) IN ENGLISH CHILDREN

Age	Boys		Girls	
(years)	Mean	S.D.	Mean	S.D.
Total hydroxyproline:				
11	68.6	32.5	54.7	27.6
12	78.0	29.4	69.1	30.2
13	93.2	43.9	51.1	25.8
14	110.0	48.0	38.0	17.7
15	106.4	41.6	31.1	12.9
16	81.1	32.0	31.7	13.2
17 and 18	55.1	24.0	27.1	14.5
Creatinine:				
11	446.0	211.0	303.0	147.0
12	521.0	227.0	403.0	273.0
13	688.0	296.0	326.0	176.0
14	676.0	364.0	425.0	356.0
15	856.0	311.0	426.0	173.0
16	829.0	337.0	552.0	237.0
17 and 18	852.0	378.0	599.0	270.0

(data from Zorab, 1969)

TABLE 763

TOTAL HYDROXYPROLINE EXCRETION (mg/24 hours)
AND HEIGHT VELOCITY IN ADOLESCENT ENGLISH BOYS

Years from Peak Velocity	Mean	S.D.
Total Hydroxyproline		
- 1 yr. 11 mth.	69.28	25.92
- 1 yr. 6 mth.	90.75	30.66
- 1 yr. 3 mth.	83.20	--
- 11 mth.	98.81	35.46
- 6 mth.	118.00	41.08
- 3 mth.	117.67	--
Peak Height Velocity		--
+ 1 mth.	130.12	46.82
+ 6 mth.	146.67	--
+ 9 mth.	113.50	--
+ 1 yr. 1 mth.	96.57	17.57
+ 1 yr. 6 mth.	89.00	--
Stature Growth Rate (cm/yr.)		
- 2 yr.	4.47	1.12
- 1 yr.	5.87	1.27
Peak Height Velocity	9.50	1.02
+ 1 yr.	5.61	1.60
+ 2 yr.	2.79	1.27

(data from Zorab, Clark, Harrison, et al., 1970,
Archives of Disease in Childhood 45:763-765,
B.M.A. House, London)

TABLE   764

PAROTID SALIVARY FLOW RATE (ml/min) WITH VARIOUS STIMULANTS
IN TEXAS CHILDREN

Parotid Fluid Variables	Peppermint		Cherry		Grape	
	Mean	S.D.	Mean	S.D.	Mean	S.D.
Flow rate (ml/min)	0.329	0.177	0.612	0.275	0.993	0.421
Acid phosphatase (units/100 ml)	0.23	0.09	0.20	0.09	0.20	0.12
Flow rate (ml/min)	0.275	0.153	0.477	0.234	0.929	0.458
Ammonia (mg/100 ml)	0.21	0.14	0.11	0.10	0.06	0.03
Flow rate (ml/min)	0.339	0.183	0.610	0.275	0.986	0.417
Amylase (unites x 10^3/100 ml)	176.	89.2	239.	119.9	401.	206.2
Flow rate (ml/min)	0.315	0.183	0.570	0.285	0.878	0.406
Bicarbonate (mEq/liter)	10.9	5.56	19.8	7.85	29.8	9.19
Flow rate (ml/min)	0.343	0.130	0.556	0.173	0.830	0.235
Calcium (mg/100 ml)	2.81	1.61	2.73	1.34	3.26	2.18
Flow rate (ml/min)	0.325	0.119	0.565	0.170	0.920	0.303
Chloride (mEq/liter)	17.7	10.52	25.8	12.73	33.3	13.44
Flow rate (ml/min)	0.419	0.179	0.693	0.258	1.093	0.361
Creatinine (mg/100 ml)	0.15	0.04	0.14	0.05	0.13	0.05
Flow rate (ml/min)	0.374	0.115	0.623	0.206	0.896	0.156
Fluoride (mg/liter)	0.020	0.005	0.019	0.005	0.020	0.005
Flow rate (ml/min)	0.211	0.83	0.367	0.126	0.623	0.225
Glucose (mg/100 ml)	0.29	0.12	0.19	0.08	0.15	0.06
Flow rate (ml/min)	0.327	0.100	0.491	0.114	0.676	0.140
IgA (mg/100 ml)	12.6	5.49	8.6	2.57	5.9	2.00
Flow rate (ml/min)	0.343	0.130	0.556	0.173	0.830	0.235
Magnesium (mg/100 ml)	0.07	0.03	0.07	0.02	0.05	0.03
Flow rate (ml/min)	0.419	0.179	0.683	0.258	1.093	0.361
Nonprotein nitrogen (mg/100 ml)	15.6	4.1	13.0	4.0	11.5	3.4
Total nitrogen (mg/100 ml)	45.8	18.7	50.9	18.8	69.6	29.7
Flow rate (ml/min)	0.305	0.174	0.561	0.278	0.878	0.400
Urea nitrogen (mg/100 ml)	9.0	2.45	7.7	1.84	7.5	1.83
Flow rate (ml/min)	0.333	0.189	0.619	0.298	1.019	0.458
Osmolality (mOsm/kg)	57.1	18.44	87.2	28.25	127.2	34.95
Flow rate (ml/min)	0.350	0.196	0.631	0.302	1.015	0.457
pH	7.08	0.30	7.43	0.21	7.67	0.18
Flow rate (ml/min)	0.343	0.130	0.556	0.173	0.830	0.235

TABLE 764   (continued)

PAROTID SALIVARY FLOW RATE (ml/min) WITH VARIOUS STIMULANTS
IN TEXAS CHILDREN

Parotid Fluid Variables	Peppermint		Cherry		Grape	
	Mean	S.D.	Mean	S.D.	Mean	S.D.
Phosphorus (mg/100 ml)	11.8	2.62	10.5	2.41	9.7	1.56
Flow rate (ml/min)	0.316	0.114	0.569	0.168	0.966	0.338
Potassium (mEq/liter)	15.0	3.56	15.9	3.66	16.0	2.65
Flow rate (ml/min)	0.337	0.164	0.586	0.244	0.950	0.389
Total protein (mg/100 ml)	146.	55.5	205.	78.5	319.	122.8
Flow rate (ml/min)	0.204	0.128	0.379	0.200	0.621	0.279
Sialic acid (mg/100 ml)	0.74	0.50	0.73	0.43	1.05	0.53
Flow rate (ml/min)	0.316	0.114	0.569	0.168	0.966	0.338
Sodium (mEq/liter)	15.4	9.96	33.2	18.23	54.9	16.91
Flow rate (ml/min)	0.363	0.176	0.634	0.284	0.943	0.402
Total solids (%)	0.49	0.13	0.66	0.19	0.92	0.26
Flow rate (ml/min)	0.323	0.171	0.577	0.256	0.943	0.402
Specific gravity	1.0024	0.0009	1.0036	0.0012	1.0061	0.0030
Flow rate (ml/min)	0.333	0.168	0.633	0.291	0.962	0.389
Free 17-OHCS (μg/100 ml)	2.48	1.08	2.20	0.81	2.06	0.74
Flow rate (ml/min)	0.153	0.047	0.253	0.069	0.428	0.119
Thiocyanate (mg/100 ml)	5.47	3.23	3.81	1.83	3.20	1.77
Flow rate (ml/min)	0.329	0.177	0.612	0.275	0.993	0.421
Uric acid (mg/100 ml)	3.82	1.09	3.15	0.91	2.85	0.80
Flow rate (ml/min)	0.323	0.170	0.577	0.267	0.943	0.402
Viscosity (centipoises)	1.16	0.14	1.18	0.12	1.20	0.10

(data from Shannon, 1973)

TABLE   765

RATE OF SALIVARY FLOW (cc/hour)
IN CALIFORNIA CHILDREN

Age Group (years)	Mean	S.D.
5-9	15	9
10-14	19	15
15-19	21	17

(data from Becks and Wainwright, 1943)

TABLE   766

MEAN SECRETORY RATE OF THE PAROTID GLAND
FOR CHILDREN IN NEW YORK CITY

Age (years)	Rate in 0.01 cc per Five Minutes
3-4	74.2
4-5	52.6
5-6	24.1
6-7	15.8
7-8	12.8
8-9	15.7
9-10	8.9
10-11	11.2
11-14	10.6

(data from Lourie, 1943, American Journal of
Diseases of Children 65:455-479. Copyright
1943, American Medical Association)

TABLE   767

CALCIUM CONTENT OF SALIVA IN CALIFORNIA CHILDREN

Age Group (years)	Total calcium (mg %)		Total calcium (mg/hr)	
	Mean	S.D.	Mean	S.D.
5-9	4.92	0.88	0.72	0.44
10-14	5.23	0.78	0.98	0.82
15-19	5.69	1.03	1.15	1.02

(data from Becks, 1943

TABLE   768

INORGANIC PHOSPHORUS CONTENT OF SALIVA IN CALIFORNIA CHILDREN

Age Group (years)	Inorganic phosphorus (mg %)		(Inorganic phosphorus (mg/hr)	
	Mean	S.D.	Mean	S.D.
5-9	13.8	3.9	2.08	1.09
10-14	14.7	4.1	2.76	1.55
15-19	15.4	4.9	2.91	1.94

(data from Wainwright, 1943)

TABLE 769

MEAN SALIVA ANALYSES FOR CHILDREN IN OREGON

County	Age (years)	Sex	pH Mean	Starch Hydrolyzing Time Mean (sec- onds)	Buffer Capacity Mean	Ammonia Nitrogen Mean (mg%)	Lacto- Bacillus Count Mean	Snyder Test Mean
Coast Region								
Clatsop	14	F	7.25	91	0.78	4.05	3.17	2.42
	14	M	7.25	38	0.78	4.71	3.60	2.35
	15	F	7.20	40	0.66	3.48	3.06	2.31
	15	M	7.40	39	1.23	4.04	2.46	1.92
	16	F	7.25	85	0.66	3.83	2.62	1.54
	16	M	7.14	126	0.78	4.52	3.06	2.11
Coos	14	F	7.41	68	0.71	3.06	3.79	2.21
	14	M	7.47	65	0.88	3.94	3.62	2.00
	15	F	7.28	70	0.58	4.09	4.37	1.43
	15	M	7.40	56	1.01	4.28	3.98	1.69
	16	F	7.42	48	0.77	4.10	3.35	1.12
	16	M	7.41	73	1.07	4.24	3.95	2.18
Central Oregon Region								
Deschutes	14	F	7.40	68	0.58	4.69	2.84	2.00
	14	M	7.56	59	0.74	5.03	2.90	1.70
	15	F	7.48	112	0.51	4.65	3.06	1.53
	16	F	7.29	97	0.53	4.13	3.00	2.00
	16	M	7.41	59	0.95	4.89	2.92	2.23
Klamath	14	F	7.42	90	0.57	5.90	2.75	1.64
	14	M	7.36	48	0.79	6.68	2.65	1.65
	15	F	7.34	62	0.57	5.56	2.53	1.70
	15	M	7.42	56	0.85	5.96	3.18	1.97
	16	F	7.22	85	0.60	6.03	2.64	1.84
	16	M	7.42	59	0.84	5.91	3.24	2.24

(data from Sullivan and Storvick, 1950)

TABLE 770

SALIVA ANALYSES FOR CHILDREN IN OREGON

Region	County	Mean and Range	pH	Starch hydro-lyzing time (seconds)	Buffer ca-pacity	Ammonia nitrogen mg/100 ml	Lacto-bacillus counts*	Snyder Test
Coast	Clatsop	Mean	7.25	70	0.83	4.15	3.01	2.11
		Range	6.60-7.70	15-900	0.1-2.7	0-9.99	0 - 7	0 - 4
	Coos	Mean	7.40	64	0.84	3.94	3.86	1.80
		Range	6.55-7.93	15-600	0.1-3.6	0-9.99	0 - 7	0 - 4
Central Oregon	Deschutes	Mean	7.42	80	0.63	4.70	3.09	1.87
		Range	6.68-7.97	30-900	0.1-1.6	0-9.99	0 - 7	0 - 4
	Klamath	Mean	7.37	66	0.70	6.02	2.94	1.82
		Range	6.52-8.23	15-1350	0.2-1.7	0-9.99	0 - 7	0 - 4

*The lactobacillus counts were coded and grouped according to the following classification:

```
0 ----------- negative          ⎞
1 ----------- 1-999             ⎟
2 ----------- 1,000-9,999       ⎟
3 ----------- 10,000-99,999     ⎟
4 ----------- 100,000-249,999   ⎟
5 ----------- 250,000-499,999   ⎬   per ml. of saliva
6 ----------- 500,000-999,999   ⎟
7 ----------- 1,000,000-1,499,999 ⎟
8 ----------- 1,500,000-1,999,999 ⎟
9 ----------- more than 2,000,000 ⎠
```


(data from Sullivan and Storvick, 1950)

SUPPLEMENTAL INFORMATION

    Published sources of raw data

    Relevant sources of data not included

Published Sources of Raw Data

(The names of authors supplying
serial data are underlined)

EXERCISE PHYSIOLOGY

Energy expenditure: Spady, 1980

Metabolic rate: Webster et al., 1941; Clagett and Hathaway, 1941; Garn and Clark, 1953

CENTER OF GRAVITY

Center of gravity: Palmer, 1944; Paŕízková, 1976

BODY COMPOSITION

^{40}K: Maresh and Groome, 1966; Cheek, 1968

Total body sodium: Cheek, 1968

Total body water: Edelman, 1952; Owen et al., 1962; Cheek et al., 1966; Flynn et al., 1967; Young et al., 1968; Cheek, 1968

Extracellular water: Owen et al., 1962; Cheek, 1968

Intracellular water: Cheek, 1968

BONE

Physical properties of infant cortical bone: Hirsch and Evans, 1965

Volume of marrow: Sturgeon, 1951a

FAT

Body density of newly born: Friis-Hanson, 1961, 1963

Skinfold thicknesses in premature infants: Kornfeld, 1957

ORGAN SIZE

Heart

Size of silhouette: Schwarz, 1946

Surface area: Young et al., 1968

Respiratory

Trachea, length: Coldiron, 1968; Butz, 1968

Trachea, cross-sectional area: Butz, 1968

Trachea, angle of bifurcation: Alavi et al., 1970

BRAIN

Blood flow in infants: Leahy et al., 1979

PERFORMANCE

    Standing long jump: Berg, 1968

    Strength: Tuddenham and Snyder, 1954

HEMATOLOGY

    Blood volume in infants: Brines et al., 1941; Sisson et al., 1959

    Plasma volume in infants: Brines et al., 1941; Sisson et al., 1959

    Blood volume at birth: De Marsh et al., 1942

    Hemoglobin: Gottfried et al., 1954; Sisson et al., 1959; Beal et al., 1962

    Mean corpuscular hemoglobin concentration: Beal et al., 1962

    Hematocrit in the newly born: Gottfried et al., 1954

    Hematocrit in infants: Brines et al., 1941; Sisson et al., 1959

    Red cell volume in infants: Brines et al., 1941; Sisson et al., 1959

    Number of red blood cells in the newly born: Gottfried et al., 1954

    Differential blood counts: Medoff and Barbero, 1950

    Marrow histology: Sturgeon, 1951 b,c

    Coagulation factors in infants: Owen and Hurn, 1953

BLOOD CHEMISTRY

    Ph in the newly born: Graham and Wilson, 1954

    Acid-base balance: Burnard and James, 1961; Weisbrot et al., 1958

    CO_2 in the newly born: Gottfried et al., 1954; Graham and Wilson, 1954

    O_2 saturation in the newly born: Graham and Wilson, 1954

    O_2 content in the newly born: Graham and Wilson, 1954

    Chloride in the newly born: Gottfried et al., 1954

    Sodium in the newly born: Gottfried et al., 1954

    Plasma pepsinogen: Grayzel et al., 1958

    Galactose: Hartman et al., 1953

    Gamma globulin: Trevorrow, 1970

ELECTROCARDIOGRAM

    Children less than 5 years: Switzer and Besoain, 1950

DIETS IN INFANCY

    Consumption of formula and strained foods: Fomon et al., 1975

    Iron intake: Beal et al., 1962

ENDOCRINES

    Growth hormone during sleep: Underwood et al., 1971

    Luteinizing hormone: Lee et al., 1976

    Follicle stimulating hormone: Lee et al., 1976

    Dehydroepiandrosterone: Lee et al., 1976

    Estradiol: Lee et al., 1976

    Androstenedione: Lee et al., 1976

    17-hydroxyprogesterone: Lee et al., 1976

    Testosterone: Lee et al., 1976

    Luteinizing hormone during sleep: Lee et al., 1976

    Somatomedin levels in relation to pubescence: Rothenberg et al., 1977

    Thyroidal radioiodine uptake: Fisher et al., 1962

    17-OHCS in urine: Cheek, 1968

SLEEP

    Percent total sleep by stages: Kohler et al., 1968

    Sleep patterns in infants: Ross et al., 1968

URINE

    Glomerular filtration rate in newly born: Guignard et al., 1975

    Creatinine excretion: Young et al., 1968; Cheek, 1968

    Pepsinogen: Grayzel et al., 1958

    Bilirubin in newly born: Halbrecht and Brzoza, 1950

    Protein in newly born: Halbrecht and Brzoza, 1950

    Renal clearance: Rubin et al., 1949

    Hydroxyproline excretion: Cheek, 1968

    Hydroxyproline/creatinine ratio: Cheek, 1968

RESPIRATORY

    Rate: Howard and Bauer, 1953; Kreiger, 1963

Rate, waking and asleep in infants: Hoppenbrowers et al, 1978

Functional residual capacity in infants: Kreiger, 1963

Specific compliance in infants: Kreiger, 1963

Tidal volume in infants: Kreiger, 1963; Boutourline-Young and Smith, 1950; Howard and
     Bauer, 1953

Pressure changes in infants: Kreiger, 1963

Resistance (inspiratory, expiratory) in infants: Kreiger, 1963

Minute volume in infants: Boutourline-Young and Smith, 1950; Howard and Bauer, 1953;
     Kreiger, 1963

Elastic work in infants: Kreiger, 1963

Breath work in infants: Kreiger, 1963

Nasal resistance in infants: Polgar and Kong, 1965

EXCRETION IN STOOLS

Fat by infants: Fomon et al., 1970

Pigment: Tat et al., 1943

Urobilinogen: Tat et al., 1943

FINGERNAILS

Growth in infants: Sibinga, 1959

Relevant Sources of Reference Data That Could Not
Be Included Because of Copyright Restrictions

Baker, G. L., 1969, Nutritional survey of Northern Eskimo infants and children, Am. J. Clin. Nutr., 22:612.

Beal, V. A., 1969, Breast- and formula-feeding of infants, J. Am. Diet. Assoc., 55:31.

Clarke, H. H., 1971, "Physical and Motor Tests in the Medford Boy's Growth Study," Prentice-Hall, Englewood Cliffs.

Clarke, R. P., Merrow, S. B., Morse, E. H., and Keyser, D. E., 1970, Interrelationships between plasma lipids, physical measurements, and body fatness of adolescents in Burlington, Vermont, Am. J. Clin. Nutr., 23:754.

Currarino, G., 1965, Roentgenographic estimation of kidney size in normal individuals with emphasis on children, Am. J. Roentgenol., 93:464.

Futrell, M. T., Kilgore, L. T., and Windham, F., 1971, Nutritional status of Negro preschool children in Mississippi, J. Am. Diet. Assoc., 59:218.

Garn, S. M., Smith, N. J., and Clark, D. C., 1975, The magnitude and the implications of apparent race differences in hemoglobin values, Am. J. Clin. Nutr., 28:563.

Giammona, S. T., 1971, Evaluation of pulmonary function in children, Pediatr. Clin. North Am., 16:285.

Hampton, M. C., Huenemann, R. L., Shapiro, L. R., Mitchell, B. W., and Behnke, A. R., 1966, A longitudinal study of gross body composition and body conformation and their association with food and activity in a teen-age population. Anthropometric evaluation of body build, Am. J. Clin. Nutr., 19:422.

Hampton, M. C., Huenemann, R. L., Shapiro, L. R., and Mitchell, B. W., 1967, Caloric and nutrient intakes of teen-agers, J. Am. Diet. Assoc., 50:385.

Himes, J. H., Roche, A. F., and Siervogel, R. M., 1979, Compressibility of skinfolds and the measurement of subcutaneous fatness, Am. J. Clin. Nutr., 32:1734.

Huenemann, R. L., Shapiro, L. R., Hampton, M. C., and Mitchell, B. W., 1966, A longitudinal study of gross body composition and body conformation and their association with food and activity in a teen-age population. Views of teen-age subjects on body conformation, food and activity, Am. J. Clin. Nutr., 18:325.

Huenemann, R. L., Shapiro, L. R., Hampton, M. C., and Mitchell, B. W., 1968, Food and eating practices of teen-agers, J. Am. Diet. Assoc., 53:17.

Jeníček, M., and Demirjian, A., 1972, Triceps and subscapular skin-fold thickness in French-Canadian school-age children in Montreal, Am. J. Clin. Nutr., 25:576.

Johnston, F. E., McKigney, J. I., Hopwood, S., and Smelker, J., 1978, Physical growth and development of urban native Americans: A study in urbanization and its implications for nutritional status. Am. J. Clin. Nutr., 31:1017.

Lantz, E. L., and Wood, P., 1958, Nutrition of New Mexican, Spanish-American and "Anglo" adolescents, J. Am. Diet. Assoc., 34:145.

Malina, R. M., 1971, Skinfolds in American Negro and white children, J. Am. Diet. Assoc., 59:34.

Mazess, R. B., and Mather, W., 1974, Bone mineral content of North Alaskan Eskimos, Am. J. Clin. Nutr., 27:916.

Merrow, S. B., 1967, Triceps skinfold thickness of Vermont adolescents, Am. J. Clin. Nutr., 20:978.

Montoye, H. J., Epstein, F. H., and Kjelsberg, M. O., 1965, The measurement of body fatness. A study in a total community, Am. J. Clin. Nutr., 16:417.

Pomerance, H. H., 1979, "Growth Standards in Children," Harper and Row, Hagerstown.

Silverman, F. N., 1957, Roentgen standards for size of the pituitary fossa from infancy through adolescence, Am. J. Roentgenol., 78:451.

Wenberg, B. G., Boedeker, M. T., and Schuck, C., 1965, Nutritive value of diets in Indian boarding schools in the Dakotas, J. Am. Diet. Assoc., 46:96.

Young, C. M., Bogan, A. D., Roe, D. A., and Lutwak, L., 1968, Body composition of pre-adolescent and adolescent girls. IV. Total body water and creatinine excretion, J. Am. Diet. Assoc., 53:579.

Young, C. M., Sipin, S. S., and Roe, D. A., 1968, Body composition of pre-adolescent and adolescent girls. I. Density and skinfold measurements, J. Am. Diet. Assoc., 53:25.

Zavaleta, A., and Malina, R. M., 1980, Growth, fatness, and leanness in Mexican-American children, Am. J. Clin. Nutr., 33:2008.

ANNOTATED BIBLIOGRAPHY

ANNOTATED BIBLIOGRAPHY

The following citations are presented in one of two styles. One style provides a brief sample description and a list of the variables included in the tables of this volume. The other citations provide bibliographic data only and are included because they are mentioned in some tables or are listed in the footnotes to some tables, usually in relation to methodology, or are included in the table of raw data sources.

"AAHPER Youth Fitness Test Manual," 1965, American Alliance for Health, Physical Education and Recreation, Washington, D.C. Data from a nationally representative sample of 9,200 children drawn by the Survey Research Center, University of Michigan. (Tables of softball throw and sit-ups)

"AAHPER Youth Fitness Test Manual," 1976, American Alliance for Health, Physical Education and Recreation, Washington, D.C. Data from a nationally representative sample of U.S. youth: 3,808 boys and 3,669 girls. (Tables of 50-yard dash, 600-yard run, 9-minute and 1-mile run, 12-minute and 1.5-mile run, pull-up, flexed-arm hang, sit-up, shuttle run, and standing long jump)

Abbott, O. D., Townsend, R. O., and Ahmann, C. F., 1945, Hemoglobin values for 2,205 rural school children in Florida, Am. J. Dis. Child., 69:346. (Tables of hemoglobin values by age)

Abraham, S., Lowenstein, F. W., and O'Connell, D. E., 1975, "Preliminary Findings of the First Health and Nutrition Examination Survey, United States, 1971-1972. Anthropometric and Clinical Findings," DHEW Publ. No. (HRA) 75-1229, U.S. Govt. Print. Office, Washington. Data from the first half of a projected nationally representative sample (N = 10,126) of those aged 1 to 74 years. (Tables of triceps and subscapular skinfold thicknesses by age and race and in relation to income level)

Abuid, J., Klein, A. H., Foley, T. P., Jr., and Larsen, P. R., 1974, Total and free triiodothyronine and thyroxine in early infancy, J. Clin. Endocrinol. Metab., 39:263.

Adams, F. H., Linde, L. M., and Miyake, H., 1961, The physical working capacity of normal school children. I. California, Pediatrics, 28:55. Data from 243 children aged 6 to 14 years. (Tables of stature, weight, surface area, blood pressure, vital capacity, and working capacity)

Adams, M. S., Brown, K. S., Iba, B. Y., and Niswander, J. D., 1970, Health of Papago Indian children, Public Health Rep., 85:1047. Data from 920 Papago Indians aged 5 to 15 years living in southwestern Arizona. (Tables of blood sugar after glucose tolerance test and visual acuity)

Adlersberg, D., Schaefer, L. E., Steinberg, A. G., and Wang, C-I., 1956, Age, sex, serum lipids, and coronary atherosclerosis, J. Am. Med. Assoc., 162:619. Data from 134 boys and 144 girls; all were normal and 98% were white. (Tables of serum cholesterol levels and serum phospholipids)

Ahlgren, J. G. A., Ingervall, B. F., and Thilander, B. L., 1973, Muscle activity in normal and postnormal occlusion, Am. J. Orthod., 64:445. Data from 15 boys aged 9 to 11 years with normal occlusion examined in a school dental service in Gothenburg (Sweden). (Table of voltage amplitude in chewing and duration of activity)

Alavi, S. M., Keats, T. E., and O'Brien, W. M., 1970, The angle of tracheal bifurcation: its normal mensuration, Am. J. Roentgenol., 108:546.

Albert, M. S., and Winters, R. W., 1966, Acid-base equilibrium of blood in normal infants, Pediatrics, 37:728. Data from 139 normal infants in New York City aged 3 to 24 months. (Table of pH, B.E. (base excess), pCO_2 (CO_2 tension), and TCO_2 (plasma total CO_2 content)

Albritton, E. C. (ed), 1954, "Standard Values in Nutrition and Metabolism," W. B. Saunders Co., Philadelphia. Collated data from the literature. (Tables of basal metabolic rates)

Alderman, R. B., 1959, Age and sex differences in PWC_{170} of Canadian school children, Res. Quart., 40:1. Data from 48 boys and 51 girls aged 10 or 14 years examined twice, one year apart. (Table of physical work capacity)

Alimurung, M. M., Joseph, L. G., Nadas, A. S., and Massell, B. F., 1951, The unipolar precordial and extremity electrocardiogram in normal infants and children, Circulation, 4:420.

Angsusingha, K., Kenny, F. M., Nankin, H. R., and Taylor, F. H., 1974, Unconjugated estrone, estradiol and FSH and LH in prepubertal and pubertal males and females, J. Clin. Endocrinol. Metab., 39:63.

Åstrand, P.-O., and Rhyming, I., 1954, A nomagram for calculation of aerobic capacity (physical fitness) from pulse rate during submaximal work, J. Appl. Physiol., 7:218.

August, G. P., Grumbach, M. M., and Kaplan, S. L., 1972, Hormonal changes in puberty. III. Correlations of plasma testosterone, LH, FSH, testicular size, and bone age with male pubertal development, J. Clin. Endocrinol. Metab., 34:319. Data from 51 healthy California boys aged 5 to 16 years. (Table of LH, FSH, testosterone, testicular volume index, and bone age in relation to sexual development)

Avruskin, T. W., Tang, S. C., Shenkman, L., Mitsuma, T., and Hollander, C. S., 1973, Serum triiodothyronine concentrations in infancy, childhood, adolescence and pediatric thyroid disorders, J. Clin. Endocrinol. Metab., 37:235.

Bailey, D. A., Shephard, R. J., and Mirwald, R. L., 1974, A current view of Canadian cardio-respiratory fitness, Can. Med. Assoc. J., 111:25. Data from 111 boys and 145 girls aged 15 to 19 years in Saskatoon (Sask.). (Table of stature, weight, and maximal oxygen uptake/weight)

Bailey, D. A., Ross, W. D., Mirwald, R. L., and Weese, C., 1978, Size dissociation of maximal aerobic power during growth in boys, Med. Sport, 11:140. Serial data from 51 boys in Saskatoon (Sask.) aged 8 to 15 years. (Table of stature, weight, maximal oxygen uptake, maximal oxygen uptake/weight, treadmill performance, and heart rate at maximal oxygen uptake)

Bain, K., 1948, The incidence of breast feeding in hospitals in the United States, Pediatrics, 2:313. Data from 2,513 hospitals. (Table of numbers of infants by feeding at time of discharge)

Baker, P. T., Hunt, E. E., Jr., and Sen, T., 1958, The growth and interrelations of skinfolds and brachial tissues in man, Am. J. Phys. Anthropol., 16:39. Data from 160 Boston (MA) children. (Tables of skinfolds at arm, forearm, waist, back and calf, and areas and relative areas of brachial components [marrow, compact bone, muscle, fat])

Bar-Or, O., Shephard, R. J., and Allen, C. L., 1971, Cardiac output of 10- to 13-year-old boys and girls during submaximal exercise, J. Appl. Physiol., 30:219. Data from 56 Toronto children. (Table of oxygen consumption and circulatory responses)

Batson, R., Young, W. C., and Shepard, F. M., 1962, Observations on skin resistance to electricity and sweat chloride content. A preliminary report, J. Pediatr., 60:716. Data from 229 normal children in Nashville (TN; 150 white, 79 Negro). (Tables of sweat chloride content and skin resistance to elasticity)

Bayley, N., 1965, Comparisons of mental and motor test scores for ages 1-5 months by sex, birth order, race, geographical location, and education of parents, Child Dev., 36:379. Data from 1,409 infants in 10 metropolitan areas of the U.S. (Tables of mental and motor development scores by age, by parity, and by race)

Bayley, N., 1969, "Manual for the Bayley Scales of Infant Development," Psychological Corp., New York. Data from 1,262 children aged 2 to 30 months stratified according to the 1960 U.S. census. (Table of ages at which psychomotor levels are reached)

Beal, V. A., 1969, Termination of night feeding in infancy, J. Pediatr., 75:690. Data from 95 infants examined serially at the Child Research Council, Denver (CO). (Table of average hours between feedings at night in relation to age and age at which solids were accepted)

Beal, V. A., Meyers, A. J., and McCammon, R. W., 1962, Iron intake, hemoglobin and physical growth during the first two years of life, Pediatrics, 30:518.

Becks, H., 1943, Human saliva. XIV. Total calcium content of resting saliva of 650 healthy individuals, J. Dent. Res., 22:397. Data from 179 San Francisco (CA) children. (Table of total calcium percent and per hour)

Becks, H., and Wainwright, W. W., 1943, Human saliva. XIII. Rate and flow of resting saliva of healthy individuals, J. Dent. Res., 22:391. Data from 239 San Francisco (CA) children. (Table of rate of saliva flow)

Behrendt, H., and Green, M., 1955, The relationship of skin pH pattern to sexual maturation in boys, Am. J. Dis. Child., 90:164. Data from 502 boys aged 4 weeks to 22 years. Most were at the New York State Training School for Boys in Orange County. All were healthy or convalescing from an illness. None had a known endocrine disorder. (Table of skin pH in relation to maturity group and race)

Behrendt, H., and Green, M., 1958, Skin pH pattern in the newborn infant, Am. J. Dis. Child., 95:35. Data from 222 New York State infants aged 1 hour to 40 days (60 Negro, 98 Puerto Rican; the rest Caucasian). They were mostly full-term babies and none had skin eruptions. (Table of skin pH in relation to age)

Behrendt, H., Green., M., and Carol, B., 1964, Relation of skin pH pattern to sexual maturation in girls, Am. J. Dis. Child., 108:37. Data from 417 New York State girls aged 1 month to 17 years (68 Puerto Rican, 145 Negro). They were healthy and free of illnesses known to have effects on sweat glands. (Table of skin pH by maturity group)

Bennett, D. L., Ward, M. S., and Daniel, W. A., Jr., 1976, The relationship of serum alkaline phosphatase concentrations to sex maturity ratings in adolescents, J. Pediatr., 88:633. Data from 510 children aged 12 to 19 years enrolled in a children and youth program for low income families in Birmingham (AL). (Tables of serum alkaline phosphatase in relation to sexual maturity, age, and race)

Berg, S. J., 1968, "Relationship Between Selected Body Measurements and Success in the Standing Broad Jump," M.S. Thesis, Washington State University Department of Physical Education.

Bidlingmaier, F., Wagner-Barnack, M., Butenandt, O., and Knorr, D., 1973, Plasma estrogens in childhood and puberty under physiologic and pathologic conditions, Pediatr. Res., 7: 901.

Bliznak, J., and Staple, T. W., 1975, Roentgenographic measurement of skin thickness in normal individuals, Radiology, 118:55. Data from 418 patients (177 white male, 40 black male, 141 white female, and 60 black female). The number per age group was not reported. (Table of skin thickness on the medial aspect of the proximal or middle third of the forearm in relation to weight)

Blizzard, R. M., Penny, R., Foley, T. P., Jr., Baghdassarian, A., Johanson, A., and Yen, S. S. C., 1972, The pituitary-gonadal inter-relationships in relation to puberty, in: "Gonadotropins," B. B. Saxena, C. G. Beling, and H. M. Gandy, eds., Wiley, New York.

Boileau, R. A., Bowen, A., Heyward, V. H., and Massey, B. H., 1977, Maximal aerobic capacity on the treadmill and bicycle ergometer of boys 11-14 years of age, J. Sports Med. Phys. Fitness, 17:153. Data from 21 Illinois children in a sports fitness school. (Table of stature, weight, maximal oxygen uptake, and maximal oxygen uptake/weight)

Boltshauser, E., and Hoare, R. D., 1976, Radiographic measurements of the normal spinal cord in childhood, Neuroradiology, 10:235. Data from English children aged 1 month to 15 years free of conditions which influence the size of the spinal cord or the subarachnoid space. (Table of spinal cord size/subarachnoid space width)

Bouchard, C., Malina, R. M., Hollmann, W., and Leblanc, C., 1977, Submaximal working capacity, heart size and body size in boys 8-18 years, Eur. J. Appl. Physiol., 36:115. Data from 237 boys in Cologne (W. Germany). (Tables of stature, weight, heart dimensions, heart volume, heart volume in relation to weight, physical work capacity, and maximum oxygen consumption)

Boutourline-Young, H. J., and Smith, C. A., 1950, Respiration of full term and of premature infants, Am. J. Dis. Child., 80:753. Data from 17 full-term infants in Boston (MA). (Table of tidal volume and minute volume)

Brines, J. K., Gibson, J. G., and Kunkel, P., 1941, The blood volume in normal infants and children, J. Pediatr., 18:447.

Brown, R. C., Jr., Unpublished, "A Study of the Physical Performance of 7,000 Boys and Girls in the New York Metropolitan Area," Data from 3,000 boys and 4,000 girls aged 6 to 18 years attending one of three schools in the suburban areas of New York City. Race was not reported. (Tables of Kraus Weber test, push-ups, standing long jump, agility run, and chins)

Bruck, E., and Weintraub, D. H., 1955, Serum calcium and phosphorus in premature and full-term infants. A longitudinal study in the first three weeks of life, Am. J. Dis. Child., 90:653. Serial data from 21 full-term infants. (Tables of serum calcium and inorganic phosphorus, phosphatase and potassium in relation to feeding)

Burley, L. R., Dobell, H. C., and Farrell, B. J., 1961, Relations of power, speed, flexibility, and certain anthropometric measures of Junior High School girls, Res. Quart., 32: 443. Data from 100 girls in Albuquerque (NM). (Table of flexibility of ankle, knee, hip, thigh, wrist, elbow, shoulder; and performance on long jump, basketball throw, and 50-yard dash)

Burmeister, L. F., Flatt, A. E., and Weiss, M. W., 1974, "Size and Strength Development of the Hand in Elementary School Children," Iowa State Services for Crippled Children, Iowa City. Data from 1,741 children in grades K to 6. Approximately 81% were white, 5% were black, and 7% were Spanish. (Table of hand pinch strength)

Burmeister, W., and Bingert, A., 1967, Die quantitativen Veründerungen der menschlichner Zellmasse zwischen dem 8 und 19 Lebensfahr, Klin. Wochenschr., 45:409. Data from 2,252 boys and 1,935 girls living in Hamburg (W. Germany). (Tables of weight, stature, lean body mass, and total body fat)

Burnard, E. D., and James, L. S., 1961, Radiographic heart size in apparently healthy newborn infants: clinical and biochemical correlations, Pediatrics, 27:727. Data in the first 6 hours of life (majority in first 2 hours) from 109 full-term infants in New York City chosen at random. (Table of reduction in transverse diameter)

Burr, I. M., Sizonenko, P. C., Kaplan, S. E., and Grumbach, M. M., 1970, Hormonal changes in puberty. 1. Correlation of serum luteinizing hormone and follicle stimulating hormone with stages of puberty, testicular size and bone age in normal boys, Pediatr. Res., 4: 25. Data from 106 normal California boys. (Tables of luteinizing hormone, follicle stimulating hormone, and LH/FSH and testicular volume index in relation to pubertal stage, chronological age, bone age, and testicular volume index)

Butz, R. O., Jr., 1968, Length and cross-section growth patterns in the human trachea, Pediatrics, 42:336.

CAHPER, n.d., "The Physical Fitness Performance and Work Capacity of Canadian Adults Aged 18 to 44 years," Canadian Association for Health, Physical Education and Recreation, Ottawa. Data from 229 individuals aged 18 to 19 years selected from each of the 10 provinces. (Tables of grip strength, flexibility, standing long jump, speed sit-ups, physical work capacity, physical work capacity in relation to weight, and skinfold thicknesses at the subscapular, triceps, abdominal, and suprailiac sites)

CAHPER, 1966, "Fitness Performance Test Manual for Boys and Girls 7 to 17 Years of Age," Canadian Association for Health, Physical Education and Recreation, Ottawa. Data from 4,665 boys and 4,537 girls. (Tables of speed sit-ups, standing long jump, shuttle run, flexed arm hang, 50-yard run, and 300-yard run)

CAHPER, 1968, "The Physical Work Capacity of Canadian Children Aged 7 to 17," Canadian Association for Health, Physical Education and Recreation, Ottawa. Data from 2,017 children in 175 randomly selected schools throughout Canada. Also, the children within each school were selected randomly. (Tables of physical work capacity and physical work capacity by weight)

Caramihai, E., Karayalcin, G., Aballi, A. J., and Lanzkowsky, P., 1975, Leukocyte count differences in healthy white and black children 1 to 5 years of age, J. Pediatr., 86: 252. Data from 150 black and 150 white children from well baby clinics in Jamaica (NY) after excluding those with infections or positive sickle cell preparations. (Table of hemoglobin, hematocrit, leukocyte, neutrophil and lymphocyte counts in relation to race)

Carron, A. V., and Bailey, D. A., 1973, A longitudinal examination of speed of reaction and speed of movement in young boys 7 to 13 years, Hum. Biol., 45:663. Data from 146 boys enrolled in the Saskatchewan Child Growth and Development Study. Each was examined serially for 7 years. (Table of hand and total body reaction time and hand movement time)

Carron, A. V., and Bailey, D. A., 1974, Strength development in boys from 10 through 16 years, Monogr. Soc. Res. Child Dev., 39 (4, Serial No. 157). Data from 207 boys enrolled in the Saskatchewan Child Growth and Development Study. Strength was tested annually from 10 through 16 years of age. (Tables of status and increment values for composite strength, upper strength, and lower strength in relation to stature and weight)

Carter, J. E. L., 1972, "The Heath-Carter Somatotype Method," Department of Physical Education, San Diego State College, San Diego.

Castelli, W. A., Ramirez, P. C., and Nasjleti, C. E., 1973, Linear growth study of the pharyngeal cavity, J. Dent. Res., 52:1245. Data from 60 randomly selected standardized lateral view cephalograms of children aged 6 to 15 years examined in the Department of Orthodontics, University of Michigan School of Dentistry. (Table of pharyngeal length and depth and size of laryngeal inlet)

Cheek, D. B., 1961, ECF volume: Its structure and measurement and the influences of age and disease, J. Pediatr., 58:103.

Cheek, D. B., 1968, "Human Growth: Body Composition, Cell Growth, Energy and Intelligence,"
    Lea and Febiger, Philadelphia. Data from 4,158 normal children in Baltimore (MD).
    (Tables of lean body mass, percent body fat, and total body fat)

Cheek, D. B., Mellits, D., and Elliott, D., 1966, Body water, height and weight during growth
    in normal children, Am. J. Dis. Child., 112:312.

Cheldelin, L. V., Davis, P. C., Jr., and Grant, W. W., 1975, Normal values for transillumina-
    tion of skull using a new light source, J. Pediatr., 87:937. Data from 300 infants seen
    in a well-baby clinic at William Beaumont Army Medical Center (TX). (Table of trans-
    illumination distance)

Cherian, A. G., and Hill, J. G., 1978, Percentile estimates of reference values for fourteen
    chemical constituents in sera of children and adolescents, Am. J. Clin. Pathol., 69:24.
    Data from 1,062 healthy children aged 4 to 20 years. These children lived in Toronto
    (Ont.) and about 90 percent of them were Caucasian. (Tables of serum, inorganic phos-
    phorus, creatinine, sodium, potassium, chloride, urea nitrogen, magnesium, calcium,
    total protein, albumin, Alpha-1-globulin, Alpha-2-globulin, Beta-globulin, and Gamma-
    globulin)

Chuinard, E. G., Osgood, E. E., and Ellis, D. M., 1941, Hematologic standards for healthy
    newborn infants. Erythrocyte count, hemoglobin content, cell volume, color index,
    volume index and saturation index, Am. J. Dis. Child., 62:1188. Data from 196 healthy
    white full-term infants of low socioeconomic level. (Table of red cell count, hemo-
    globin, hemoglobin coefficient, color index, cell volume, volume coefficient, volume
    index, and saturation index)

Clagett, D. D., and Hathaway, M. L., 1941, Basal metabolism of normal infants from three to
    fifteen months of age with special reference to twins, Am. J. Dis. Child., 62:967.

Clark, L. C., Thompson, H. L., Beck, E. I., and Jacobson, W., 1951, Excretion of creatine
    and creatinine by children, Am. J. Dis. Child., 81:774. Mixed longitudinal data from
    Ohio children aged 3 to 18 years enrolled in the Fels Longitudinal Study. (Table of
    urinary excretion of creatine and creatinine in relation to age, stature, weight, and
    body surface area)

Clarke, H. H., and Degutis, E. W., 1962, Comparison of skeletal age and various physical and
    motor factors with the pubescent development of 10, 13, and 16 year old boys, Res.
    Quart., 33:356. Data from 237 Caucasian boys in the Medford (OR) public schools.
    (Tables of lung capacity, cable tension test results, and standing long jump distances
    in relation to puberty status)

Cleveland, R. H., 1979, Symmetry of bronchial angles in children, Radiology, 133:89. Data
    from 50 children aged from birth to 18 years, chosen at random, excluding any with
    evidence or history of lung, heart or mediastinal disease and those with chest wall or
    spinal deformity. (Table of bronchial angles and lateral differences in bronchial
    angles)

Cohen, A. M., and Vig, P. S., 1976, A serial growth study of the tongue and intermaxillary
    space, Angle Orthod., 46:332. Serial data from 50 children, average age ranging from
    4 to 19 years, who were born in King's Hospital, London (25 boys and 25 girls). This
    study retrospectively used the first 50 individuals in Leighton's series [Leighton, B.
    C., M.D.S. Thesis, University of London, 1960]. (Tables of tongue shadow area and
    ratio of tongue shadow area to intermaxillary space area)

Coldiron, J. S., 1968, Estimation of nasotracheal tube length in neonates, Pediatrics, 41:
    823.

Coon, V., Donato, G., Houser, C., and Bleck, E. E., 1975, Normal ranges of hip motion in
    infants six weeks, three months and six months of age, Clin. Orthop., 110:256. Data
    from 84 normal California infants, 44 of whom were examined at two ages. (Table of hip
    flexion contracture, internal rotation and external rotation)

Corcoran, J. M., Eastman, C. J., Carter, J. N., and Lazarus, L., 1977, Circulating thyroid
    hormone levels in children, Arch. Dis. Childh., 52:716.

Cron, G. W., and Pronko, N. H., 1957, Development of the sense of balance in school children,
    J. Educ. Res., 51:33. Data from 322 boys and 178 girls in Wichita (KS). (Table of
    balance scores)

Cumming, G. R., 1967, Current levels of fitness, Can. Med. Assoc. J., 96:868. Data from boys
    and girls aged 6 to 18 years in Winnipeg (Man.), sample size not given. (Table of
    maximal oxygen uptake/weight)

Cumming, G. R., and Friesen, W., 1967, Bicycle ergometer measurement of maximal oxygen uptake in children, Can. J. Physiol. Pharmacol., 45:937. Data from 20 boys aged 11 to 15 years in Winnipeg (Man). (Table of stature, weight, maximal oxygen uptake/weight)

Cunningham, A. S., 1975, Eosinophil counts: age and sex differences, J. Pediatr., 87:426. Data from 1,289 children examined at scheduled visits to a pediatric clinic in Cooperstown (NY). All were free of acute illness, 4 percent had asthma, none was known to have invasive parasites. (Table of eosinophil counts)

Cunningham, A. S., 1979, Morbidity in breast-fed and artificially fed infants, II, J. Pediatr., 95:685. Data from 724 infants born at a rural hospital in New York State. Those with birth weights below the 5th percentile or with serious neonatal problems were excluded. Of the infants, 505 were seen regularly in a clinic for the first year of life. (Table of prevalence of illness in relation to feeding)

Cunningham, D. A., and Eynon, R. B., 1973, The working capacity of young competitive swimmers 10-16 years of age, Med. Sci. Sports, 5:227. Data from competitive swimmers (24 boys and 19 girls) in Ontario aged 11 to 15 years. (Table of stature, weight, maximal oxygen uptake, maximal oxygen uptake/weight)

Cunningham, D. A., Telford, P., and Swart, G. T., 1976, The cardiopulmonary capacities of young hockey players: age 10, Med. Sci. Sports, 8:23. Data from 15 members of a competitive ice hockey team aged about 10 years. (Table of stature, weight, maximal oxygen uptake, maximal oxygen uptake/weight)

Cunningham, D. A., MacFarlane-van Waterschoot, B., Paterson, D. H., Lefcoe, M., and Sangal, S. P., 1977, Reliability and reproducibility of maximal oxygen uptake measurement in children, Med. Sci. Sports, 9:104. Data from 66 members of a competitive ice hockey team aged about 10 years. (Table of stature, weight, maximal oxygen uptake, maximal oxygen uptake/weight)

Cureton, K. J., Boileau, R. A., and Lohman, T. G., 1975, A comparison of densitometric, potassium-40 and skinfold estimates of body composition in prepubescent boys, Hum. Biol. 47:321. Data from 49 boys aged 8 to 11 years. (Tables of stature, weight, body density total body potassium, sum of 10 skinfolds, and percent body fat)

Cureton, T. K., and Barry, A. J., 1964, Improving the physical fitness of youth. A report of research in the Sports-Fitness School of the University of Illinois, Monogr. Soc. Res. Child Dev., 29 (4, Serial No. 95). Data from 707 boys aged 7 to 15 years. (Tables of dynamometer strength (right and left), back, left leg extension, total strength, strength/weight, 60-yard dash, 440-yard run, 600-yard run, running long jump, running high jump, standing long jump, shot-put, chin-ups, dips, floor push-ups, vertical jump, visual reaction time, auditory reaction time, combined visual plus auditory reaction time, vital capacity, agility run, balance beam, endurance hops, and trunk flexibility)

Cycle II of the Health Examination Survey by the National Center for Health Statistics, unpublished results. This survey was conducted in 1963-1965 and included 7,119 children aged 6 to 11 years who were representative of the noninstitutionalized population of the U.S. (Table of estimated upper arm muscle circumference)

Cycle III of the Health Examination Survey by the National Center for Health Statistics, unpublished results. This survey was conducted in 1966-1970 and included 6,768 youths aged 12 through 17.9 years who were representative of the noninstitutionalized youths in the U.S. (Tables of estimated upper arm muscle circumference)

Dallman, P. R., and Siimes, M. A., 1979, Percentile curves for hemoglobin and red cell volume in infancy and childhood, J. Pediatr., 94:26. Data from 8,847 white children aged 5 to 16 years enrolled in the Kaiser Permanente Out-patient Clinic in San Francisco (CA), and from 210 white children aged 10 months to 7 years seen at the Moffitt Hospital Outpatient Clinic in San Francisco (CA). Data were obtained also from 777 Finnish children aged 2 to 15 years (about 50% from Helsinki and about 50% from rural areas) and from 238 Finnish children examined serially at 6, 9, and 12 months. Those likely to have hemoglobin deficiency or thalassemia minor were excluded from all groups. (Table of hemoglobin and mean corpuscular volume)

Daniel, W. A., Jr., 1973, Hematocrit: maturity relationship in adolescence, Pediatrics, 52: 388. Data from 1,000 boys and 1,007 girls aged 11 to 20 years who were attending the Adolescent Unit of the University of Alabama Medical Center and whose families' annual incomes were less than $4,000. Those who were seriously or chronically ill, who had hematologic disorders, or who were seriously injured were excluded. (Tables of hematocrit in relation to pubertal stages, race, and age)

Daniels, J., and Hejinian, L. M., 1929, Growth in infants from the standpoint of physical measurements and nitrogen metabolism. I. Creatinine, Am. J. Dis. Child., 37:1128.

Daniels, J., and Oldridge, N., 1971, Changes in oxygen consumption of young boys during growth and running training, Med. Sci. Sports, 3:161. Data from 14 children in Albany (WI) studied serially. (Table of stature, weight, maximal oxygen uptake, and maximal oxygen uptake/weight)

De Groot, I., Morrison, J. A., Kelly, K. A., Rauh, J. L., Mellies, M. J., Edwards, B. K., and Glueck, C. J., 1977, Lipids in school children 6 to 17 years of age: upper normal limits, Pediatrics, 60:437. Data from 7,337 children in Cincinnati (OH). They excluded American Indians, Orientals, and those receiving oral contraceptives, antidiabetic, antihypertensive, uric acid-lowering or lipid-lowering drug regimens, and those whose pregnancy status was unknown. (Tables of plasma cholesterol and triglycerides in relation to fasting and in relation to race)

De Marsh, Q. B., Windle, W. F., and Alt, H. L., 1942, Blood volume of newborn infant in relation to early and late clamping of umbilical cord, Am. J. Dis. Child., 63:1123.

de Peretti, E., and Forest, M. G., 1976, Unconjugated dehydroepiandrosterone plasma levels in normal subjects from birth to adolescence in humans: The use of a sensitive radio-immunoassay, J. Clin. Endocrinol. Metab., 43:982.

Desmond, M. M., and Sweet, L. K., 1949, Relation of plasma proteins to birth weight, multiple births and edema in the newborn, Pediatrics, 4:484. Data from newborn infants in Washington (DC), 88 of whom were normal full-term, 270 premature, 32 twin, and 10 with edema. (Table of plasma protein levels)

Dickman, M. L., Schmidt, C. W., and Gardner, R. M., 1971, Spirometric standards for normal children and adolescents (ages 5 through 18 years), Am. Rev. Respir. Dis., 104:680. Data from 482 healthy children in Utah. (Table of forced vital capacity and expiratory volume in relation to stature)

Diem, K., and Lentner, C. (eds), 1972, "Scientific Tables," Ciba-Geigy Ltd., Basel. Collated data from the literature for a wide range of physical and biochemical measures. (Tables of BMR values)

Dinucci, J. M., and Shows, D. A., 1977, A comparison of the motor performance of black and Caucasian girls age 6-8, Res. Quart., 48:680. Data from 90 girls (45 black, 45 Caucasian) in Louisiana. This was a random sample of public and private school children. (Tables of vertical jump and reach, standing long jump, pull-ups, push-ups, bent arm hang, grip strength, leg lift; shoulder, hip, and trunk strength; flexibility of wrist, trunk, hip, leg, neck, and arm; sit and reach, balance, 6-second run, 10- and 50-yard dash, shuttle run, dodging run, 300- and 600-yard run)

Donaldson, S. W., and Tompsett, A. C., Jr., 1952, Tracheal diameter in the normal newborn infant, Am. J. Roentgenol., 67:785. Data from 350 Michigan infants screened at birth. (Table of anteroposterior diameter of trachea)

Donckerwolcke, R. A. M. G., Sander, P. C., Van Stekelenburg, G. J., Stoop, J. W., and Tiddens, H. A. W. M., 1970, Serum creatinine values in healthy children, Acta Paediatr. Scand., 59:399. Data from 408 healthy Dutch children aged 4 to 13 years. (Table of serum creatinine in relation to age and body surface area)

Ducharme, J.-R., Forest, M. G., de Peretti, E., Sempé, M., Collu, R., and Bertrand, J., 1976, Plasma adrenal and gonadal sex steroids in human pubertal development, J. Clin. Endocrinol. Metab., 42:468. Data from 109 French children aged 8 to 16 years. (Table of plasma steroid levels in relation to age)

Eckert, H. M., and Eichorn, D. H., 1977, Developmental variability in reaction time, Child Dev., 48:452. Serial data from white children in the Berkeley and Oakland (CA) Growth Studies (105 boys and 115 girls). (Table of reaction time)

Edelman, I. S., 1952, Exchange of water between blood and tissues. Characteristics of deuterium oxide equilibration in body water, Am. J. Physiol., 171:279.

Edelman, I. S., Haley, H. B., Schloerb, P. R., Sheldon, D. B., Friis-Hansen, B. J., Stoll, G., and Moore, F. D., 1952, Further observations on total body water. I. Normal values throughout the life span, Surg. Gynecol. Obstet., 95:1. Data from 41 Boston (MA) children. (Table of weight, surface area, and total body water)

Eichorn, D. H., and McKee, J. P., 1953, Oral temperature and subcutaneous fat during adolescence, Child Dev., 24:235. Serial data from 103 white children enrolled in the Oakland (CA) Growth Study aged 11.5 to 17.5 years. (Table of oral temperature)

Eklöf, O., and Ringertz, H., 1976, Kidney size in children. A method of assessment, Acta Radiol. [Diagn.], 17:617. Data from about 150 Swedish children. (Table of kidney length in relation to Ll-L3 length)

Elliott, D. A., and Cheek, D. B., 1968, Muscle electrolyte patterns during growth, in: "Human Growth: Body Composition, Cell Growth, Energy, and Intelligence," D. B. Cheek, ed., Lea and Febiger, Philadelphia. Data from 44 normal infants and children in Baltimore (MD). (Table of muscle electrolyte values)

Ellis, J. D., Carron, A. V., and Bailey, D. A., 1975, Physical performance in boys from 10 through 16 years, Hum. Biol., 47:263. Serial data from 106 boys enrolled in the Saskatchewan Child Growth and Development Study. (Tables of performance on standing long jump, flexed arm hang, bent knee sit-ups, and increments in these)

Espenschade, A., Dable, R. R., and Schoendube, R., 1953, Dynamic balance in adolescent boys, Res. Quart., 24:270. Data from 363 boys aged 12.5 to 18 years. (Table of beam walking scores in relation to age and maturity)

Faiman, C., and Winter, J. S. D., 1974, Gonadotropins and sex hormone patterns in puberty: Clinical data, in: "Control of the Onset of Puberty," M. M. Grumbach, G. D. Grave, and F. E. Mayer, eds., Wiley, New York.

Faust, M. S., 1977, Somatic development of adolescent girls, Monogr. Soc. Res. Child Dev., 42 (1, Serial No. 169). Serial data from 97 white girls enrolled in the Guidance Study or the Berkeley (CA) Growth Study. (Tables of stages of breast, pubic, and axillary hair development in relation to puberty, and ages at transition of secondary sex characteristics)

Ferguson, A. D., Cutter, F. F., and Scott, R. B., 1956, Growth and development of Negro infants. VI. Relationship of certain environmental factors to neuromuscular development during the first year of life, J. Pediatr., 48:308. Data from 708 healthy full-term infants in Washington (DC), 511 of them from well-baby clinics for low income groups and 197 from a private pediatric practice for middle income groups. (Table of age of first walking in relation to the education and parity of the mother and the weight of the infant)

Ferris, A. G., Beal, V. A., Laus, M. J., and Hosmer, D. W., 1979, The effect of feeding on fat deposition in early infancy, Pediatrics, 64:397. Serial data from 92 female infants in western Massachusetts; race not stated. (Tables of suprailiac, midaxillary, triceps, and subscapular skinfold thicknesses)

Ferris, B. G., Jr., and Smith, C. W., 1953, Maximum breathing capacity and vital capacity in female children and adolescents, Pediatrics, 12:341. Data from 466 white girls attending private schools near Boston (MA) aged 5 years 1 month to 18 years 2 months. (Tables of vital capacity and maximum breathing capacity and of these variables in relation to weight, surface area, and stature)

Ferris, B. G., Jr., Whittenberger, J. L., and Gallagher, J. R., 1952, Maximum breathing capacity and vital capacity of male children and adolescents, Pediatrics, 9:659. Data from 161 boys aged 5 to 18 years attending private schools near Boston (MA). (Tables of vital capacity and maximum breathing capacity and of these variables in relation to weight, surface area, and stature)

Fisher, D. A., 1973, Advances in the laboratory diagnosis of thyroid disease. Part 1., J. Pediatr., 82:1.

Fisher, D. A., Oddie, T. H., and Burroughs, J. C., 1962, Thyroidal radioiodine uptake rate measurement in infants, Am. J. Dis. Child., 103:738.

Flatau, E., Josefsberg, Z., Reisner, S. H., Bialik, O., and Laron, Z., 1975, Penile size in the newborn infant, J. Pediatr., 87:663. Data from 100 healthy Israeli infants with birth weights appropriate for gestational age (mean 279 ± 7 days). (Table of penis length, width, and circumference; testicular volume, recumbent length, weight, and head circumference)

Fleisch, A., 1951, Le métabolisme basal standard et sa détermination au moyen du "Metabolo-calculator," Helv. Med. Acta, 18:23. Collated data from 17 reports. (Tables of basal metabolic rates)

Fleishman, E. A., 1964, "The Structure and Measurement of Physical Fitness," Prentice Hall, Inc., Englewood Cliffs (NJ). Data from more than 20,000 children in cities with populations over 25,000 in seven geographic areas of the U.S. (northeast, mid-Atlantic, southeast, midwest, southwest, far west, and Rocky Mountain). For most tests, about 2,000 students were examined in the age range 15 to 18 years. (Tables of hold half-sit-up, dodge run, 50-yard dash, long jump, 600-yard walk-run, pull-ups, leg lifts, grip strength, softball throw, shuttle run, dynamic and extent flexibility)

Flint, M. M., Drinkwater, B. L., Wells, C. L., and Horvath, S. M., 1977, Validity of estimating body fat of females: effect of age and fitness, Hum. Biol., 49:559. Data from 14 girls in Santa Barbara (CA) aged 12 years. (Table of stature, weight, maximal oxygen uptake, and maximal oxygen uptake/weight)

Flynn, M. A., Hanna, F. M., and Lutz, R. N., 1967, Estimation of body water compartment of preschool children. I. Normal children, Am. J. Clin. Nutr., 20:1125.

Flynn, M. A., Murthy, Y., Clark, J., Comfort, G., Chase, G., and Bentley, A. E. T., 1970, Body composition of Negro and white children, Arch. Environ. Health, 20:604. Data from 32 Negro and 34 white children aged 4 to 6 years living in or near Columbia (MO). (Table of stature, weight, triceps skinfold thickness, waist circumference, upper arm circumference, bicristal diameter, total body potassium, lean body mass, total body fat, and percent body fat)

Fomon, S. J., 1967, Body composition of the male infant during the first year of life, Pediatrics, 40:863. Data at birth from analyses of six male stillborn infants. Data at older ages obtained from literature. (Table of total body water, total body fat, total body protein, and fat-free body mass)

Fomon, S. J., 1974, "Infant Nutrition," 2nd edition, W. B. Saunders Co., Philadelphia. Data from white children of above average economic status living in or near Iowa City (IA) and from the literature. (Tables of increase in weight, water, protein, fat, and "other" and fat-free body mass from birth to 3 years in a male reference infant)

Fomon, S. J., Filer, L. J., Jr., Thomas, L. N., and Rogers, R. R., 1970, Growth and serum chemical values of normal breastfed infants, Acta Paediatr. Scand., Suppl. 202.

Fomon, S. J., Filer, L. J., Jr., Thomas, L. N., Anderson, T. A., and Nelson, S. E., 1975, Influence of formula concentration on caloric intake and growth of normal infants, Acta Paediatr. Scand., 64:172.

Fomon, S. J., Ziegler, E. E., Filer, L. J., Jr., Anderson, T. A., Edwards, B. B., and Nelson, S. E., 1978, Growth and serum chemical values of normal breastfed infants, Acta Paediatr. Scand., Suppl. 273. Serial data from 233 normal breastfed infants in Iowa City (IA). (Tables of total protein, albumin, α_1 globulin, α_2 globulin, β globulin, γ globulin, urea nitrogen, alkaline phosphatase, phosphorus, cholesterol, calcium, and magnesium in relation to age and sex)

Forbes, G. B., 1972, Growth of lean body mass in man, Growth, 36:325. Data from 577 normal children aged 8 to 18 years in Rochester (NY). (Table of stature, weight, lean body mass, percent body fat, and total body fat)

Forbes, G. B., and Amirhakimi, G. H., 1970, Skinfold thickness and body fat in children, Hum. Biol., 42:401. Data from healthy children (293 boys and 179 girls) living in or near Rochester (NY), aged 7.5 to 18 years. (Tables of stature, weight, total body fat, triceps and subscapular skinfold thicknesses, and average of six skinfolds)

Forest, M. G., Sizonenko, P. C., Cathiard, A. M., and Bertrand, J., 1974, Hypophyso-gonadal function in humans during the first year of life. 1. Evidence for testicular activity in early infancy, J. Clin. Invest., 53:819.

Francis, C. C., 1948, Growth of the human pituitary fossa, Hum. Biol., 20:1. Serial data from 400 white children and 391 Negro children included in the Brush and Bolton studies (Cleveland, OH). (Tables of pituitary fossa length and depth)

Frankenburg, W. K., and Dodds, J. B., 1967, The Denver Developmental Screening Test, J. Pediatr., 71:181. Data from 1,036 children in Denver (CO) aged 2 weeks to 6.4 years. Children with a high risk of developmental abnormalities were excluded; otherwise they were somewhat representative of the Denver community. (Table of gross motor abilities and fine motor abilities in relation to age)

Frerichs, R. R., Webber, L. S., Srinivasan, S. R., and Berenson, G. S., 1977, Hemoglobin levels in children from a biracial Southern community, Am. J. Public Health, 67:841. Data from 4,238 children in Bogalusa (LA) aged 2.5 through 14 years (2,610 white and 1,628 black). (Table of hemoglobin concentration in relation to age, race, and socio-economic status)

Friedenberg, M. J., Walz, B. J., McAlister, W. H., Locksmith, J. P., and Gallagher, T. L., 1965, Roentgen size of normal kidneys: computer analysis of 1,286 cases, Radiology, 84:1022. Data from 332 out-patients in St. Louis (MO) whose history, physical examination and routine laboratory findings disclosed no evidence of renal disease. (Tables of renal index, and lateral differences in renal index and length)

Friis-Hansen, B., 1957, Changes in body water compartments during growth, Acta Paediatr., 46, Suppl. 110.

Friis-Hansen, B., 1961, Body water compartments in children: changes during growth and related changes in body composition, Pediatrics, 28:169.

Friis-Hansen, B., 1963, The body density of newborn infants, Acta Paediatr., 52:513.

Friis-Hansen, B., 1971, Body composition during growth, Pediatrics, 47:264.

Frisancho, A. R., Garn, S. M., and Ascoli, W., 1970, Subperiosteal and endosteal bone apposition during adolescence, Hum. Biol., 42:639. Data from 1,151 serial radiographs of white children from southwestern Ohio enrolled in the Fels Longitudinal Study. (Table of metacarpal II total area, medullary area, cortical area, and percent cortical area)

Fry, P., Howard, J. E., and Logan, B. C., 1975, Body weight and skinfold thickness in Black, Mexican-American, and white infants, Nutrition Reports International, 11:155. Data from a total of 50 black, Mexican-American, and white infants. (Table of triceps and subscapular skinfold thicknesses and arm circumference in relation to race)

Gamble, J. L., 1946, Physiological information gained from studies on the life raft ration, Harvey Lecture Series, 42:247.

Gampel, B., 1965, The relation of skinfold thickness in the neonate to sex, length of gestation, size at birth and maternal skinfold, Hum. Biol., 37:29. Data from 419 live births in London (England); race not stated. They were measured within 24 hours of birth. (Tables of triceps and subscapular skinfold thicknesses, recumbent length and head circumference in relation to birth weight)

Garn, S. M., 1952, Changes in areolar size during the steroid growth phase, Child Dev., 23:55. Data from 147 white children aged 9 to 20 years examined at the Forsyth Dental Infirmary for Children, Boston (MA). (Table of areolar diameters in relation to breast development and age)

Garn, S. M., 1958, Fat, body size and growth in the new born, Hum. Biol., 30:265. Data from 146 children and 95 mothers (Caucasian) from southwestern Ohio. They were all full-term single births. (Table of fat thickness over the 11th rib)

Garn, S. M., and Clark, D. C., 1975, Hemoglobin and fatness, Ecology of Food and Nutrition, 4:131. Data from children aged 1 to 18 years included in the Ten-State Nutrition Survey of 1968-1970 (492 white males, 440 white females, 467 black males, and 481 black females). (Table of hemoglobin values in lean and obese individuals in relation to race and age)

Garn, S. M., and Clark, L. C., Jr., 1953, The sex difference in basal metabolic rate, Child Dev., 24:215. Data from 145 white children in southwestern Ohio enrolled in the Fels Longitudinal Study. (Table of stature, weight, basal oxygen consumption, and of basal oxygen consumption by weight for boys and girls paired by calf muscle width and caloric intake)

Garn, S. M., Greaney, G. R., and Young, R. W., 1956, Fat thickness and growth progress during infancy, Hum. Biol., 28:232. Data from more than 300 white southwestern Ohio infants examined serially from birth to 12 months in the Fels Longitudinal Study. (Tables of status and increments in medial plus lateral calf fat thickness)

Garn, S. M., Selby, S., and Crawford, M. R., 1956, Skin reflectance studies in children and adults, Am. J. Phys. Anthropol., 14:101. Data from 112 children in southwestern Ohio. They were all white, American-born, and free from known endocrinopathy or abnormality of the pigmentary system. The measurements were made from May through October using a Photovolt Model 610 Colorimeter. (Table of skin reflectance values on inner arm, chest, forehead, left areola, and scrotum)

Garn, S. M., Smith, N. J., and Clark, D. C., 1975, Lifelong differences in hemoglobin levels between blacks and whites, J. Natl. Med. Assoc., 67:91. Data from the Ten-State Nutrition Survey of 1968-1970 including about 30,000 individuals aged 1 to 18 years. (Tables of hemoglobin levels by age and race and matched for income)

Garn, S. M., Poznanski, A. K., and Larson, K., 1976, Metacarpal lengths, cortical diameters and areas from the 10-State Nutrition Survey including: estimated skeletal weights, weight, and stature for whites, blacks, and Mexican-Americans, in: "Proceedings of First Workshop on Bone Morphometry," Z. F. G. Jaworski, ed., University of Ottawa Press. Data from 9,053 children aged 1 to 17 years. (Tables of estimated skeletal weight in relation to income and race, metacarpal II length, total medullary and cortical widths, total area, medullary area, cortical area, and percent cortical area in relation to race)

Garner, L. D., and Kotwal, N. S., 1973, Correlation study of incisive biting forces with age, sex and anterior occlusion, J. Dent. Res., 52:698. Data from 45 boys and 45 girls aged 10 to 18 years. These were normal white children in Indiana without apparent neuromuscular defects, with eight permanent teeth erupted and no badly damaged or heavily restored incisors. (Table of incisor biting force)

Gatewood, O. M. B., Glasser, R. L., and Vanhoutte, J. J., 1965, Roentgen evaluation of renal size in pediatric age groups, Am. J. Dis. Child., 110:162. Data from 130 normal intravenous pyelograms of children in Baltimore (MD) aged 16 hours to 15 years. (Table of renal size and angulation)

Gilliam, T. B., Sady, S., Thorland, W. G., and Weltman, A. L., 1977, Comparison of peak performance measures in children ages 6 to 8, 9 to 10, and 11 to 13 years, Res. Quart., 48:695. Data from 63 Michigan children enrolled in a summer fitness program. (Table of maximal oxygen uptake and maximal oxygen uptake/weight)

Glassow, R. B., and Kruse, P., 1960, Motor performance of girls age 6 to 14 years, Res. Quart., 31:426. Data from 125 elementary school girls in Wisconsin. (Table of ability to run, jump, and throw)

Godin, G., and Shephard, R. J., 1973, Activity patterns of the Canadian Eskimo, in: "Polar Human Biology. The Proceedings of the SCAR/IUPS/IUBS Symposium on Human Biology and Medicine in the Antarctic," O. G. Edholm and E. K. E. Gunderson, eds., Heinemann Medical Books, London. Data from 266 children aged from birth to 15 years in the Igloolik region of the Eastern Canadian Arctic. (Table of overall daily and yearly energy expensiture)

Goldring, D., Londe, S., Sivakoff, M., Hernandez, A., Britton, C., and Choi, S., 1977, Blood pressure in a high school population. I. Standards for blood pressure and the relation of age, sex, weight, height and race to blood pressure in children 14 to 18 years of age, J. Pediatr., 91:884. Data from 6,838 school children in the St. Louis (MO) metropolitan area aged 14 to 18 years. Of the group, 30.4% were black and 69.5% were white. (Table of systolic and diastolic blood pressure)

Gottfried, S. P., Bogin, M., and Levyckyer, N. V., 1954, Blood and electrolyte studies on normal newborn full-term infants, Am. J. Dis. Child., 87:543.

Govatos, L. A., 1966, Sex differences in children's motor performances, in: "Collected Papers, The Eleventh Inter-Institutional Seminar in Child Development," Education Dept., Henry Ford Museum and Greenfield Village, Dearborn (MI). Serial data from 23 boys and 25 girls of high I.Q. and above average socioeconomic status, examined in Grades 2 through 6. (Tables of grip strength, vital capacity, ball bounce, standing long jump, basketball throw, horizontal ladder, shuttle run, balance beam, and turn under bar in relation to season)

Graham, B. D., and Wilson, J. L., 1954, Chemical control of respiration in newborn infants, Am. J. Dis. Child., 87:287. Data from 64 full-term normal infants. (Table of blood pH, CO_2 content, CO_2 tension and percent oxygen saturation)

Grayzel, H. G., Elkan, B., Schneck, L., Garza, S. L., and Cavies, A. P., 1958, Plasma and urine pepsinogen (uropepsin) values in normal infants and children, Am. J. Dis. Child., 96:666.

Greulich, W. W., and Pyle, S. I., 1959, "Radiographic Atlas of Skeletal Development of the Hand and Wrist," Stanford University Press (CA).

Greulich, W. W., Dorfman, R. I., Catchpole, H. R., Solomon, C. I., and Culotta, C. S., 1942, Somatic and endocrine studies of puberal and adolescent boys, Monogr. Soc. Res. Child Dev., VII (3, Serial No. 33).

Gros, G., Gordon, A., and Miller, R., 1951, Electrocardiographic patterns of normal children from birth to five years of age, Pediatrics, 8:349. Data from 104 infants and children in Chicago (IL) who were normal by physical examination. (Tables of height and duration of P-wave, duration of P-R interval and QRS complex, depth of Q-wave, height of R-wave, depth of S-wave, R/S ratio, and height of T-wave)

Gruenwald, P., and Minh, H. N., 1960, Evaluation of body and organ weights in perinatal pathology. I. Normal standards derived from autopsies, Am. J. Clin. Pathol., 34:247. Data from more than 3,000 autopsies performed in several eastern U.S. cities. (Table of weights of heart, lung, spleen, liver, adrenals, kidneys, thymus, and brain in relation to total body weight and gestational age)

Guest, G. M., and Brown, E. W., 1957, Erythrocytes and hemoglobin of the blood in infancy and childhood. III. Factors in variability, statistical studies, Am. J. Dis. Child., 93: 486. Data from 1,568 white infants and children in Cincinnati (OH) examined between 1932 and 1942. (Table of hemoglobin, mean corpuscular hemoglobin, packed red cell volume, and mean corpuscular volume)

Guignard, J. P., Torrado, A., Da Cunha, O., and Gautier, E., 1975, Glomerular filtration rate in the first three weeks of life, J. Pediatr., 87:268.

Gupta, D., Attanasio, A., and Roaf, S., 1975, Plasma estrogen and androgen concentrations in children during adolescence, J. Clin. Endocrinol. Metab., 40:636.

Gutin, B., Trinidad, A., Norton, C., Giles, E., Giles, A., and Stewart, K., 1978, Morphological and physiological factors related to endurance performance of 11-to-12 year-old girls, Res. Quart., 49:44. Data from 33 black girls (5 West Indian, 28 American black) aged 12 years in low income families in New York City. (Table of stature, weight, maximal oxygen uptake, and maximal oxygen uptake/weight)

Haas, S. S., Epps, C. H., Jr., and Adams, J. P., 1973, Normal ranges of hip motion in the newborn, Clin. Orthop., 91:114. Data from 400 newborn babies (200 black, 200 white) in Washington (DC). (Table of range of hip motion in relation to race)

Haddad, H. M., Hsia, D. Y-Y., and Gellis, S. S., 1956, Studies on respiratory rate in the newborn: its use in the evaluation of respiratory distress in infants of diabetic mothers, Pediatrics, 17:204. Data from 55 normal full-term infants born in the Beth Israel Hospital, Boston (MA). (Table of respiratory rates)

Halbrecht, I., and Brzoza, H., 1950, Evaluation of hepatic function in newborn infants by means of chemical study of cord blood, Am. J. Dis. Child., 79:988.

Hamill, P. V. V., Palmer, A., and Drizd, T., 1978, "Forced Vital Capacity of Children 6-11 Years. United States," DHEW Publ. No. (PHS) 78-1651, U.S. Govt. Print. Office, Washington. Data from 7,119 children aged 6 to 11 years who constituted a nationally representative sample. (Table of forced vital capacity by race)

Handelman, C. S., and Osborne, G., 1976, Growth of the nasopharynx and adenoid development from one to eighteen years, Angle Orthod., 46:243. Serial data from 12 white children studied at the Child Research Council, Denver (CO). (Tables of nasopharyngeal size and shape and increments in these measures)

Hanson, M. R., 1965, "Motor Performance Testing of Elementary School Age Children," Ph.D. Thesis, University of Washington, Seattle (WA). Data from 22 schools in Minnesota with physical education programs at least 3.5 days/week. Included were 2,870 children in grades 1 through 6. (Tables of performance in overhand throw, wall pass, soccer punt, soccer wall volley, volley ball serve, pitching accuracy, potato race, full-hang pull-ups, jump and reach, rope skipping, sit-ups, Hanson shoulder test, Bass balance test, 50-yard dash, standing long jump, and 600-yard walk-run)

Hardin, D. H., and Ramirez, J., 1972, Elementary school performance norms, Texas Assoc. Health, Phys. Educ., Recreation J., 40:8. Data from 1,130 children aged 6 to 11 years in schools of El Paso (TX). (Table of performance in 50-yard dash, standing long jump, and softball throw)

Harlan, W. R., Grillo, G. P., Cornoni-Huntley, J., and Leaverton, P. E., 1979, Secondary sex characteristics of boys 12 to 17 years of age. The U.S. Health Examination Survey, J. Pediatr., 95:293. Data from a nationally representative sample of 6,768 boys. (Table of genital stages in relation to pubic hair stages within age groups, and gynecomastia in relation to genital stage and age)

Harrison, J. C. E., 1958, "The Construction of Cable-tension Strength Test Norms for Boys Seven, Nine, Twelve, and Fifteen Years of Age," M.S. Thesis, University of Oregon. Data from 373 white boys enrolled in the Medford (OR) Growth Study. (Table of cable tension strength test results for elbow, shoulder, trunk, hip, knee, and ankle)

Hartmann, A. F., Grunwaldt, E., and James, D. H., Jr., 1953, Blood galactose in infants and children, J. Pediatr., 43:1.

Hastreiter, A. R., and Abella, J. B., 1971, The electrocardiogram in the newborn period. 1. The normal infant, J. Pediatr., 78:146. Data from 11 mature newly born infants. (Table of amplitude of R, S, and T waves, QRS axis, and T axis in frontal plane)

Hayek, A., Maloof, F., and Crawford, J. D., 1973, Thyrotropin behavior in thyroid disorders of childhood, Pediatr. Res., 7:28.

Hays, G. C., and Mullard, J. E., 1974, Normal free thyroxine index values in infancy and childhood, J. Clin. Endocrinol. Metab., 39:958.

Henry, F. M., 1961, Stimulus complexity, movement complexity, age, and sex in relation to reaction latency and speed in limb movements, Res. Quart., 32:353. Data from 60 boys in Berkeley (CA). (Table of reaction time and movement time of hand-wrist)

Hindley, C. B., Filliozat, A. M., Blackenberg, C., Nicolet-Meister, D., and Sand, E. A., 1966, Differences in age of walking in five European longitudinal samples, Hum. Biol., 38:364. Serial data from studies in Brussels, London, Paris, Stockholm, and Zurich. A description of these studies is available in Courrier, 30, Special Issue, 1980. (Tables of age at first walking by social class)

Hirsch, C., and Evans, F. G., 1965, Studies on some physical properties of infant component bone, Acta Orthop. Scand., 35:303.

Hirschman, C., and Hendershot, G. E., 1979, "Trends in Breast Feeding Among American Mothers," DHEW Publ. No. (PHS) 79-1979, U.S. Govt. Print. Office, Washington. Data from national U.S. surveys made in 1965 and 1973. (Tables of percentage of mothers who breast-fed their children, in relation to birth cohort of mother, year of birth of child, and parity. Duration of breast feeding given also)

Hirschman, C., and Sweet, J. A., 1974, Social background and breast feeding among American mothers, Soc. Biol., 21:39. Data from 4,918 women included in The National Fertility Study of 1965. (Tables of percentages of mothers breast feeding in relation to birth cohort, ethnic origin, region of residence, farm and non-farm, year of birth of infant, religion, education, and occupation)

Hodges, R. G., Sperry, W. M., and Andersen, D. H., 1943, Serum cholesterol values for infants and children, Am. J. Dis. Child., 65:858. Data from 417 children in New York City. (Tables of total serum cholesterol and ratio of combined to free serum cholesterol)

Holinger, P. H., 1975, The esophagus, in: "Textbook of Pediatrics," V. C. Vaughn and R. J. McKay, eds., W. B. Saunders Co., Philadelphia. Source of data not given. (Table of esophageal length)

Holliday, M. A., 1971, Metabolic rate and organ size during growth from infancy to maturity and during late gestation and early infancy, Pediatrics, 47:169.

Holliday, M. A., 1978, Body composition and energy needs during growth, in: "Human Growth," Vol. II, F. Falkner and J. M. Tanner, eds., Plenum Publishing Corp., New York. Collated data. (Table of stature, weight, total organ weight, muscle mass, body fat, and extracellular body fluid)

Homer, M. J., 1978, The hilar height ratio, Radiology, 129:11. Data from anteroposterior radiographs of 40 individuals aged 15 to 30 years living in Boston (MA). On each side a line was drawn parallel to the thoracic spine from the apex of the lung to the dome of the diaphragm. A perpendicular from each line was drawn through the midpoint of the corresponding hilus. The ratio is $\frac{\text{apex to hilus}}{\text{hilus to diaphragm}}$. (Table of right and left hilar height ratios)

Honzik, M. P., and McKee, J. P., 1962, The sex difference in thumb-sucking, J. Pediatr., 61:726. Data from white children examined in the Institute of Human Development, Berkeley (CA). (Table of prevalence of thumb-sucking)

Hoppenbrouwers, T., Harper, R. M., Hodgman, J. E., Sterman, M. B., and McGinty, D. J., 1978, Polygraphic studies of normal infants during the first six months of life. II. Respiratory rate and variability as a function of state, Pediatr. Res., 12:120.

Hopper, B. R., and Yen, S. S. C., 1975, Circulating concentrations of dehydroepiandrosterone and dehydroepiandrosterone sulfate during puberty, J. Clin. Endocrinol. Metab., 40:458.

Horton, C. P., and Crump, E. P., 1959, Growth and development. VI. Changes in skin color of fifty-one Negro infants from birth through three years of age, as related to skin color of parents, socioeconomic status, and developmental quotient, Arch. Dermatol., 80:421. Serial data from 51 infants in Nashville (TN). (Table of change in skin color with age)

Howard, P. J., and Bauer, A. R., 1953, Quantitative variations in the respiration of the newborn infant, Am. J. Dis. Child., 86:284.

Hughes, I. A., and Winter, J. S. D., 1976, The application of a serum 17 OH-progesterone radioimmunoassay to the diagnosis and management of congenital adrenal hyperplasia, J. Pediatr., 88:766.

Hunsicker, P. and Reiff, G., 1980, Comparison of youth fitness tests, 1958-65-75, in: "Proceedings of the First National Conference on Fitness and Sports for All," U.S. Govt. Print. Office, Washington. (Tables of pull-ups, sit-ups, flexed arm hang, shuttle run, standing long jump, 50-yard dash, and 600-yard run)

Hupprich, F. L., and Sigerseth, P. O., 1950, The specificity of flexibility in girls, Res. Quart., 21:25. Data from 300 white girls in Eugene (OR) aged 6 to 18 years. (Table of flexibility of hip, trunk, neck, wrist, ankle, elbow, shoulder, and knee)

Hutchinson-Smith, B., 1973, Skinfold thickness in infancy in relation to birthweight, Dev. Med. Child Neurol., 15:628. Serial data from a general sample of 200 English infants. (Tables of skinfolds at biceps, triceps, subscapular and suprailiac in relation to birthweight)

Hutinger, P. W., 1959, Differences in speed between American Negro and white children in performance of the 35-yard dash, Res. Quart., 30:366. Data from 792 children in the fourth to sixth grades in five public elementary schools in Kansas City (MO). (Table of time for 35-yard dash)

Iliff, A., and Lee, V. A., 1952, Pulse rate, respiratory rate, and body temperature of children between two months and eighteen years of age, Child Dev., 23:237. Serial data from 197 white children enrolled in the Child Research Council Study, Denver (CO). (Tables of respiratory rates and rectal temperatures)

Jackson, C., and Jackson, C. L., 1950, "Bronchoesophagology," W. B. Saunders Co., Philadelphia. Source of data not given. (Table of esophageal length)

Jackson, E. B., Wilkin, L. C., and Auerbach, H., 1956, Statistical report on incidence and duration of breast feeding in relation to personal-social and hospital maternity factors, Pediatrics, 17:700. Data from the Grace-New Haven Community Hospital (CT). (Tables of number of babies started on breast and breast feeding at discharge, period at which breast feeding was discontinued, and duration by education, parity and ethnic group of mother)

Jackson, R. L., Einstein, R. A. J., Blau, A., and Kelly, H. G., 1944, Angle of clearance of the left ventricle as an index to cardiac size. Modified technique for its determination and range of values for normal children, Am. J. Dis. Child., 68:157. Data from 102 normal Iowa children. (Table of angles of clearance)

Jéniček, M., and Demirjian, A., 1974, Age at menarche in French Canadian urban girls, Ann. Hum. Biol., 1:339. Data from a random sample of 1,002 girls aged 10.5 to 15.0 in Montreal schools. (Tables of age at menarche in relation to family income, occupation, education, and body frame)

Jenner, M. R., Kelch, R. P., Kaplan, S. L., and Grumbach, M. M., 1972, Hormonal changes in puberty. IV. Plasma estradiol, LH, and FSH in prepubertal children, pubertal females, and in precocious puberty, premature thelarche, hypogonadism, and in a child with a feminizing ovarian tumor, J. Clin. Endocrinol. Metab., 34:521. Data from 26 prepubertal children and 76 pubertal girls in California. (Table of estradiol, LH, and FSH in relation to pubertal stage and menstrual cycle phase)

Johnson, B. C., Epstein, F. H., and Kjelsberg, M. O., 1965, Distributions and femilial studies of blood pressure and serum cholesterol levels in a total community--Tecumseh, Michigan, J. Chronic Dis., 18:147. Data from 3,002 children from birth to 18 years. (Tables of systolic and diastolic [fourth and fifth phase] blood pressure and serum cholesterol)

Johnson, R. D., 1962, Measurements of achievement in fundamental skills of elementary school children, Res. Quart., 33:94. Data from 308 Minnesota children in grades 1 through 6. (Tables of kick, pass and catch, jump and reach, zig zag, and batting scores)

Johnston, F. E., and Malina, R. M., 1966, Age changes in the composition of the upper arm in Philadelphia children, Hum. Biol., 38:1. Data from 363 clinically-normal white Philadelphia (PA) children aged 6 to 16 years. (Tables of arm diameter, total muscle and fat diameter, relative fat and muscle, total and relative bone, total marrow space, and cortex)

Johnston, F. E., Hamill, P. V. V., and Lemeshow, S., 1973, "Skinfold Thickness of Children 6-11 Years, United States," DHEW Publ. No. (HSM) 73-1602, U.S. Govt. Print. Office, Washington. Data from 7,119 children aged 6 to 11 years who constituted a nationally representative sample. (Tables of triceps, subscapular and mid-axillary skinfold thicknesses in relation to age and race)

Johnston, F. E., Hamill, P. V. V., and Lemeshow, S., 1974, "Skinfold Thickness of Youths 12-17 Years, United States," DHEW Publ. No. (HRA) 74-1614, U.S. Govt. Print. Office, Washington. Data from 6,727 youths aged 12-17 years who constituted a nationally representative sample. (Tables of triceps, subscapular, mid-axillary, suprailiac, and medial calf skinfold thicknesses in relation to age and race)

Jones, D. V., and Work, C. E., 1961, Volume of a swallow, Am. J. Dis. Child., 102:427. Data from 10 Ohio children. (Table of swallow volume)

Jones, H. E., 1947, Sex differences in physical abilities, Hum. Biol., 19:12. Data from the Adolescent Growth Study participants; this study was conducted at the University of California (Berkeley). Sample size not given. (Table of strength and weight in terms of adult stature, rate of maturing, and dextrality index in relation to age)

Jones, H. E., 1949, "Motor Performance and Growth. A Developmental Study of Static Dynamometric Strength," University of California Press, Berkeley. Serial data from 183 white children aged 11 to 18 years. (Tables of strength in relation to age and maturity, grip strength in relation to maturing, thrust, pull and total strength, lateral differences, seasonal differences, weight, stature, and somatotype)

Kahane, J. C., 1978, A morphological study of the human prepubertal and pubertal larynx, Am. J. Anat., 151:11. Data from 20 specimens from white children (10 boys and 10 girls) aged 9 to 19 years. They had not had long term endotracheal intubation, local diseases or diseases affecting growth. (Tables of laryngeal cartilage shape and size and vocal cord length and anterior cricothyroid distance by pubertal group and in relation to adult values)

Kendall, H. O., and Kendall, F. P., 1948, Normal flexibility according to age groups, J. Bone Joint Surg., 30A:690. Data from 5,115 children in Baltimore (MD) aged 5 to 22 years. (Tables of limitation of touching fingers to toes and limitation of touching forehead to knees when sitting with legs extended)

Kenny, F. M., Gancayco, G. P., Heald, F. P., and Hung, W., 1966, Cortisol production rate in adolescent males in different stages of sexual maturation, J. Clin. Endocrinol. Metab., 26:1232. Data from 16 boys in Pittsburgh (PA). (Tables of cortisol production rate and urinary 17-OCHS in relation to surface area, body weight, creatinine excretion, and pubertal status)

Keogh, J., 1965, "Motor Performance of Elementary School Children," mimeographed, Dept. Physical Education, University of California, Los Angeles. Data from 266 boys and 243 girls aged 5 to 9 years. These children were living in Santa Monica (CA). (Tables of performance on 30-yard dash, shuttle run, standing long jump, hurdle standing jump, ball throw, accuracy throw, 50-foot hop, mat hop, beam balance and walk, side step, cable jump, and grip strength)

Kidd, L., Levison, H., Gemmel, P., Aharon, A., and Swyer, P. R., 1966, Limb blood flow in the normal and sick newborn, Am. J. Dis. Child., 112:402. Data from 15 normal full-term infants in Toronto (Ont.). (Table of calf systolic blood pressure, blood flow and resistance)

Klemm, T., Bamzer, D. H., and Schneider, V., 1976, Bone mineral content of the growing skeleton, Am. J. Roentgenol., 126:1283. Data from 137 German children aged 3 to 16 years using absorptiometry. (Tables of bone mineral content of the os calcis, cortical width and thickness, stature, and weight)

Knuttgen, H. G., 1967, Aerobic capacity of adolescents, J. Appl. Physiol., 22:655. Data from Boston (MA) children aged 15 to 18 years (95 boys and 95 girls). (Tables of maximal oxygen uptake, maximal oxygen uptake/weight, stature, and weight)

Kohler, W. C., Coddington, R. D., and Agnew, H. W., Jr., 1968, Sleep patterns in 2-year-old children, J. Pediatr., 72:228. Data from 16 healthy Florida children. (Table of percent of total sleep in each sleep stage and sequence of stages)

Kondo, S., and Eto, M., 1975, Physical growth studies on Japanese-American children in comparison with native Japanese, in: "Comparative Studies on Human Adaptability of Japanese, Caucasians and Japanese Americans," S. M. Horvath, S. Kondo, H. Matsui, and H. Yoshimura, eds., University of Tokyo Press, Tokyo. Data from 956 Japanese-American children in Los Angelese. (Tables of skinfold thicknesses at triceps, subscapular, and abdominal sites)

Kopetzky, M. T., Maselli, R., and Ellis, E. F., 1974, Pulmonary function studies with simple equipment in 323 normal children, J. Allergy Clin. Immunol., 53:1. Data from 323 normal children in Denver (CO) aged 4 to 18 years. (Table of forced vital capacity, forced expiratory volume, maximum voluntary ventilation, maximum expiratory flow rate, and peak expiratory flow rate)

Kornfeld, W., 1957, Typical and atypical changes in the soft tissue distribution during childhood, Hum. Biol., 29:62. Data from 2,216 healthy white children in New Jersey, predominantly from middle class families. About two-thirds were of Jewish descent, the remainder being of Anglo-Saxon, German, Scandinavian, or Slavic origin. (Tables of skinfold thicknesses of cheeks, anterior chest, subscapular, and paraumbilical sites)

Korth-Schutz, S., Levine, L. S., and New, M. I., 1976, Serum androgens in normal prepubertal and pubertal children and in children with precocious adrenarche, J. Clin. Endocrinol. Metab., 42:117.

Krahenbuhl, G. S., Pangrazi, R. P., Burkett, L. N., Schneider, M. J., and Petersen, G., 1977 Field estimation of VO$_2$ max in third grade Arizona children eight years of age, Med. Sc. Sports, 9:37. Data from 20 boys and 18 girls. (Tables of stature, weight, maximal oxygen uptake, maximal oxygen uptake/weight, maximum heart rate, time for running 549 meters, 1207 meters, and 1609 meters)

Kramer, J. D., and Lurie, P. R., 1964, Maximal exercise tests in children. Evaluation of maximal exercise tests as an index of cardiovascular fitness in children with special consideration of the recovery pulse curve, Am. J. Dis. Child., 108:283. Data from 26 Indianapolis boys aged less than 13 years who were in physical fitness classes. Also they reported data for 15 trained athletes aged 15 years. (Table of maximal oxygen uptake, maximal oxygen uptake/weight, stature, and weight)

Krieger, I., 1963, Studies on mechanics of respiration in infancy, Am. J. Dis. Child., 105:4

Krogman, W. M., 1970, Growth of head, face, trunk, and limbs in Philadelphia white and Negro children of elementary and high school age, Monogr. Soc. Res. Child Dev., 35 (3, Serial No. 136). Data from 942 white and 988 Negro children in Philadelphia (PA) who were examined annually. (Table of auricular height)

Krogman, W. M., 1971, "The Manual and Oral Strengths of American White and Negro Children, Ages 3-6 Years," mimeographed, Philadelphia Center for Research in Child Growth. Data from 800 children aged 3 to 6 years in Philadelphia (PA) and the Berkeley-Oakland area of California. The white group in the California sample includes 10% who were Mexican-American or Oriental (mostly Chinese). (Tables of grip strength, bite strength, wrist turning strength, thumb opposability strength, and push strength in relation to race and learning)

Kuhns, L. R., Berger, P. E., Roloff, D. W., Poznanski, A. K., and Holt, J. F., 1974, Fat thickness in the newborn infant of a diabetic mother, Radiology, 111:665. Data from 147 "normal" infants varying in gestational age from <30 to <41 weeks. These Michigan infants had respiratory problems at birth after a normal gestation and delivery. They were all within 2 s.d. of mean weight for gestational age and none had significant hypoglycemia. (Table of skinfold thickness at 10th rib)

Laga, E. M., Driscoll, S. G., and Munro, H. N., 1972a, Comparison of placentas from two socioeconomic groups. I. Morphometry, Pediatrics, 50:24. Data from consecutive deliveries in Boston (MA) and Guatemala City. (Tables of trimmed placental weight, percent parenchyma, microscopic components of parenchyma, peripheral and stem mass of villi, trophoblasts, fibroblasts and connective tissue, and surface area)

Laga, E. M., Driscoll, S. G., and Munro, H. N., 1972b, Comparison of placentas from two socioeconomic groups. II. Biochemical characteristics, Pediatrics, 50:33. Data from consecutive births in Boston (MA) and Guatemala City. (Tables of DNA, RNA, RNA/DNA, protein, protein/DNA, alkaline phosphatase, polysome/monosome ratio, membrane-bound ribosomes/total ribosomes (%), and cell-free ^{14}C-leucine incorporation (10 ^{3}Xcpm/ placenta)

Larivière, G., Lavallee, H., and Shephard, J., 1974, Correlations between field tests of performance and laboratory measurements of fitness, Acta Paediatr. Belg., 28 (Suppl.): 19. Data from 20 boys and 14 girls aged about 10 years in Quebec (Canada). (Table of stature, weight, maximal oxygen uptake, and maximal oxygen uptake/weight)

Leahy, F. A. N., Sankaran, K., Cates, D., MacCallum, M., and Rigatoo, H., 1979, Quantitative noninvasive method to measure cerebral blood flow in newborn infants, Pediatrics, 64:277

Lee, P. A., and Migeon, C. J., 1975, Puberty in boys: Correlation of plasma levels of gonadotrophins (LH, FSH), androgens (testosterone, androstenedione, dehydroepiandrosterone and its sulfate), estrogens (estrone and estradiol) and progestins (progesterone and 17-hydroxyprogesterone), J. Clin. Endocrinol. Metab., 41:556. Data from Ann Arbor (MI) children: 11 boys studied throughout pubescence and 43 boys examined at various stages during pubescence. (Table of serum steroid levels in relation to age)

Lee, P. A., Xenakis, T., Winer, J., and Matsenbaugh, S., 1976, Puberty in girls: correlation of serum levels of gonadotropins, prolactin, androgens, estrogens, and progestins with physical changes, J. Clin. Endocrinol. Metab., 43:775. Data from 27 girls in Ann Arbor

(MI). (Table of serum hormone concentrations in relation to age)

Lee, V. A., and Iliff, A., 1956, The energy metabolism of infants and young children during postprandial sleep, Pediatrics, 18:739. Serial data from 78 white children enrolled in the Child Research Council Study, Denver (CO). (Tables of metabolic rates, fed and asleep, in relation to surface area, weight, and stature)

Leichsenring, J. M., Norris, L. M., Lawson, S. A., and Halbert, M. L., 1955, Blood cell values for healthy adolescents, Am. J. Dis. Child., 90:159. Data from 110 boys and 130 girls aged 12 to 21 years chosen by physicians in Minnesota as being unusually healthy. (Tables of hemoglobin, red cell count and mean corpuscular hemoglobin, red cell diameter, and percentage reticulocytes)

Leighton, J. R., 1956, Flexibility characteristics of males ten to eighteen years of age, Arch. Phys. Med. Rehab., 37:494. Data from 400 school boys in Ontario (OR). (Table of flexibility of neck, shoulder, elbow, wrist, hip, knee, ankle, and trunk)

Lemoh, J. N., and Brooke, O. G., 1979, Frequency and weight of normal stools in infancy, Arch. Dis. Childh., 54:719. Data from 55 healthy English infants aged from 3 days to 2 years. None were breastfed. (Table of frequency, weight, and water content of stools)

Lewis, R. C., Duval, A. M., and Iliff, A., 1943, Standards for the basal metabolism of children from 2 to 15 years of age, inclusive, J. Pediatr., 23:1. Data from 1,007 determinations in boys and 713 in girls. These white children were living in Denver (CO) and were enrolled in the Child Research Council studies. (Tables of calories/hour per m^2 and per kg and per cm of stature, surface area, and weight)

Lichty, J. A., Ting, R. Y., Bruns, P. D., and Dyar, E., 1957, Studies of babies born at high altitude. 1. Relation of altitude to birth weight, Am. J. Dis. Child., 93:666. Data from 180 Colorado mothers with 615 newborns living at an altitude of 10,000 to 11,000 feet. (Table of hematocrit in relation to altitude)

Liebman, J., 1968, Electrocardiography, in: "Heart Disease in Infants, Children, and Adolescents," A. J. Moss and F. H. Adams, eds., Williams and Wilkins Co., Baltimore.

Liebman, J., and Plonsey, R., 1977, Electrocardiography, in: "Heart Disease in Infants, Children and Adolescents," 2nd edition, A. J. Moss, F. H. Adams, and G. C. Emmanouilides, eds., Williams and Wilkins Co., Baltimore. Data obtained by combining the reports of Namin and Miller (1966) and Alimurung et al. (1951). (Tables of amplitude of R and S waves, T wave voltage, and R wave in limb leads)

Lohman, T. G., Boileau, R. A., and Massey, B. H., 1975, Prediction of lean body mass in young boys from skinfold thickness and body weight, Hum. Biol., 47:245. Data from 162 Illinois boys aged 6.3 to 12.5 years enrolled in a sports fitness program. Of these, 32 were black, 4 oriental and 126 white. (Table of stature, weight, lean body mass, percent body fat, and triceps and subscapular skinfolds)

Lombard, O. M., 1950, Breadth of bone and muscle by age and sex in childhood. Studies based on measurements derived from several roentgenograms of the calf of the leg, Child Dev., 21:229. Serial data from about 177 white Boston (MA) children. (Tables of breadth of bone plus muscle, and breadth of skin plus subcutaneous tissue)

Londe, S., 1968, Blood pressure standards for normal children as determined under office conditions, Clin. Pediatr., 7:400. Data from 1,588 predominantly white children aged 3 to 15 years examined in the St. Louis (MO) Labor Health Institute. (Tables of blood pressure: systolic, and diastolic fifth phase)

Lourie, R. S., 1943, Rate of secretion of the parotid glands in normal children. A measurement of function of the autonomic nervous system, Am. J. Dis. Child., 65:455. Data from 51 children in New York City aged 3 to 14 years. (Table of rate of secretion)

Maaser, R., 1972, Die Ultraschallmessung der subcutanen Fettgewebsdicke zur Beurteilung des Ernährungszustandes von Kindern, Z. Kindeheilkd., 112:321. Data from 1,200 healthy children in Dortmund (W. Germany) aged 3 to 14 years measured with ultrasound. (Table of triceps, subscapular, and suprailiac subcutaneous fat thicknesses)

Maaser, R., Stolley, H., and Droese, W., 1972, Die Hautfettfaltenmessung mit dem Caliper. II. Standardwerte der subcutanen Fettgewebsdicke, 2-14 Jahriger gesunder Kinder, Monatsschr. Kinderheilkd., 120:350. Data from 3,400 German children aged 2 to 14 years. (Table of triceps, subscapular, and suprailiac skinfold thicknesses)

MacMahon, B., 1973, "Age at Menarche," DHEW Publ. No. (HRA) 74-1615, U.S. Govt. Print. Office, Washington. Data from three nationally representative U.S. samples of individuals aged 6 to 74 years. (Tables of distributions of menstruating and non-menstruating girls, age of menarche in relation to race, income, and size of place of residence, geographic location, and reported age of menarche in relation to skinfold thicknesses)

Maksud, M. G., Coutta, K. D., and Hamilton, L. H., 1971, Oxygen uptake, ventilation, and heart rate: Study in Negro children during strenuous exercise, Arch. Environ. Health, 23:23. Data from 47 black elementary inner city school children in Milwaukee (WI) aged 9 to 11 years. (Table of maximal oxygen uptake, maximal oxygen uptake/weight, stature, and weight)

Malina, R. M., Unpublished a, Data from 1,448 black and 1,396 white Philadelphia (PA) school children aged 6 to 13 years. The white sample was from a middle to upper middle socio-economic level whereas the Negro sample was from a lower socioeconomic level. (Tables of weight, stature, skinfolds at triceps, subscapular, and midaxillary sites, manual dexterity, standing long jump, 35-yard dash, softball throw, hand grip, push and pull strength in relation to age and sex)

Malina, R. M., Unpublished b, Data from 300 boys and 299 girls aged 8 to 18 years in Phila-delphia (PA) of whom 69% were white and 31% black. (Tables of response to fixed work loads on bicycle ergometer)

Malina, R. M., 1972, Weight, height and limb circumferences in American white and Negro children: Longitudinal observations over a one year period, J. Trop. Pediatr., 13:280. Data from 830 children aged 6 through 13 years. The white sample was from a middle to upper middle socioeconomic level whereas the Negro sample was from a lower socioeconomic level; all the children lived in Philadelphia (PA). (Tables of estimated arm muscle circumference)

Malina, R. M., and Johnston, F. E., 1967, Relations between bone, muscle and fat widths in the upper arms and calves of boys and girls studied cross-sectionally at ages 6 to 16 years, Hum. Biol., 39:211. Data from 356 clinically normal children studied in the Philadelphia Center for Research in Child Growth (PA). (Tables of total calf width, total muscle, total bone, medial fat, lateral fat, total tibia, tibial cortex, and tibial medullary cavity)

Malina, R. M., Hamill, P. V. V., and Lemeshow, S., 1973, "Selected Body Measurements of Children 6-11 Years. United States," DHEW Publ. No. (HSM) 73-1605. Data from 7,119 children aged 6 to 11 years who constituted a nationally representative sample. (Tables of popliteal height, knee height, buttock-popliteal length, buttock-knee length, seat breadth, and thigh clearance)

Maresh, M., and Groome, D. S., 1966, Potassium-40: Serial determinations in infants, Pediatrics, 38:642.

Maroney, M., and Rantz, L. A., 1950, Electrocardiogram in 679 healthy infants and children, Pediatrics, 5:396. Data from San Francisco (CA) children aged 6 months to 9 years (309 white, 338 Negro, 32 other). (Table of QRS interval, PR interval, and variations in T and P)

Marshall, W. A., and Tanner, J. M., 1969, Variations in pattern of pubertal changes in girls, Arch. Dis. Childh., 44:291.

Martin, W. E., 1955, "Children's Body Measurements for Planning and Equipping Schools," DHEW, Office of Education, Spec. Publ. No. 4, U.S. Govt. Print. Office, Washington. Data from 3,318 children aged 5 to 21 years from 10 schools in southern Michigan in the area between Ann Arbor and Detroit. Blacks formed 11.7% of the sample; 0.7% were other non-white children. (Tables of stature; eye, shoulder, and elbow height, wrist height, fist carrying height, finger tip height, trochanteric height [heights from floor]; knee joint to sole of foot, maximum upward reach, total arm length [arc], upper arm length [arc], shoulder to wrist [arc], wrist to finger tip [arc], arm span, shoulder to opposit finger tip, back to finger tip, manubrium to finger tip, buttocks to abdomen depth, sitting height; eye, shoulder, lumbar and elbow heights above seat; top of thigh to seat, top of knee to seat, top of knee to sole of foot, back length; ischium, buttocks, and abdomen to front of knee; maximum upward reach with arm at 45°; height of downward reach with arm at 45°; seat front to front of knee; height of hand and arm straight for-ward; maximum space for forward reach; buttocks to seat front, buttocks to sole of foot, lumbar support to front of knee; and ischium to seat front and to lumbar back support)

Mathews, D. K., Shaw, V., and Woods, J. B., 1959, Hip flexibility of elementary school boys as related to body segments, Res. Quart., 3:297. Data from 158 boys in Pullman (WA). (Table of performance on tests of flexibility)

Mayberry, W. E., Gharib, H., Bilstad, J. M., and Sizemore, G. W., 1971, Radioimmunoassay for human thyrotropin. Clinical value in patients with normal and abnormal thyroid function, Ann. Int. Med., 74:471.

Mays, J. E., Jr., and Keele, D. K., 1961, Serum magnesium levels in healthy children and in various disease states, Am. J. Dis. Child., 102:623. Data from 234 normal Oklahoma children. (Table of serum magnesium levels)

Mazess, R. B., and Cameron, J. R., 1972, Growth of bone in school children: comparison of radiographic morphometry and photon absorptiometry, Growth, 36:77. Data from 322 white children aged 6 to 14 years from Middleton (WI). (Tables of bone mineral, bone width, and mineral/width for mid-radius and distal radius and ulna and midshaft of humerus; cortical thickness and cortical area and cortical area/total area for distal radius, mid-radius and total radius)

Mazess, R. B., and Mather, W. E., 1975, Bone mineral content in Canadian Eskimos, Hum. Biol., 47:45. Data from 177 children measured using direct photon absorptiometry. (Tables of bone mineral, bone width, and mineral/width in mid-radius, distal radius, mid-ulna, and distal ulna)

McCammon, R., Unpublished data from the Child Research Council, Denver (CO). Serial data recorded from 334 white children of middle socioeconomic status living in or near Denver. These unpublished data are more complete than those in McCammon, 1970. (Tables of skinfold thicknesses at forearm, biceps, triceps, subscapular, pectoral, mid-axillary, paraumbilical, suprailiac, medial thigh and medial calf sites; radiographic subcutaneous fat thicknesses at forearm, deltoid, hip, thigh, and calf sites; and cross-sectional area of fat in upper arm)

McCammon, R. W., 1970, "Human Growth and Development," Charles C Thomas, Springfield (IL). Data from 334 white children studied serially at the Child Research Council, Denver (CO). (Tables of heart width/internal chest width, muscle widths and fat widths at maximum forearm width, mid-thigh and maximum calf width; bone widths at mid-humerus, mid-radius, mid-thigh, maximum forearm, and maximum calf; subcutaneous fat thickness at maximum forearm, mid-thigh, maximum calf, deltoid insertion, and maximum hip bulge)

McCue, C. M., Miller, W. W., Mauck, H. P., Jr., Robertson, L., and Parr, E. L., 1979, Adolescent blood pressure in Richmond, Virginia, schools, Va. Med., 106:210. Data from 2,594 children in the ninth and tenth grades of the seven major public high schools in Richmond. About 80% of the children were black. Also included were 572 students in the ninth, tenth and twelfth grades from three private schools; this group included 2% blacks. (Table of systolic and fourth phase diastolic blood pressures by sex, age, and race)

McDonough, J. R., Hames, C. G., Garrison, G. E., Stulb, S. C., Lichtman, M. A., and Hefelfinger, D. C., 1965, The relationship of hematocrit to cardiovascular states of health in the Negro and white population of Evans County, Georgia, J. Chronic Dis., 18:243. Data from 218 white and 159 Negro individuals aged 15 to 24 years. (Table of hematocrit by age)

McLellan, M. S., and Webb, C. H., 1961, Ear studies in the newborn infant. II. Age of spontaneous visibility of the auditory canal and tympanic membrane, and the appearance of these structures in healthy newborn infants, J. Pediatr., 58:523. Data from 524 examinations of 250 healthy full-term Louisiana infants, most of whom were Negro. (Tables of prevalence of spontaneous visibility)

McMurray, L.-G., Roe, J. H., and Sweet, L. K., 1948, Plasma protein studies on normal newborn and premature infants. I. Plasma protein values for normal full term and normal premature infants. II. Use of concentrated normal human serum albumin in treatment of premature infants, Am. J. Dis. Child., 75:265. Data from 46 normal full term infants. (Table of albumin, globulin, and total protein in relation to birth weight)

Medoff, H. S., and Barbero, G. J., 1950, Total blood eosinophil counts in the newborn period, Pediatrics, 6:737.

Mergen, D. C., and Jacobs, R. M., 1970, The size of the nasopharynx associated with normal occlusion and Class II malocclusion, Angle Orthod., 40:342. Data from 20 Caucasian girls with normal occlusions attending the orthodontic clinic of the University of Iowa College of Dentistry. (Table of nasopharyngeal area and depth, facial convexity, and midface depth)

Metz, K. F., and Alexander, J. F., 1970, An investigation of the relationship between maximum aerobic work capacity and physical fitness in twelve-to-fifteen-year old boys, Res.

Quart., 41:75. Data from 60 boys and 30 girls in Minnesota. (Table of stature, weight maximal oxygen uptake, and maximal oxygen uptake/weight)

Meyer, H. F., 1958, Breast feeding in the United States: extent and possible trend. A survey of 1,904 hospitals with two and a quarter million births in 1956, _Pediatrics_, 22:116. Data from all geographic regions of the U.S. (Table of percent of infants receiving specified feeding on discharge by geographic region)

Michael, E. D., Jr., and Katch, F. I., 1968, Prediction of body density from skinfold and girth measurements of 17-year-old boys, _J. Appl. Physiol._, 25:747. Data from 48 white California boys. (Table of stature, weight, underwater weight, body density, lean body weight, percent body fat, functional residual capacity, expiratory reserve capacity, residual volume, vital capacity, circumferences of shoulder, chest, buttocks, abdomen, thigh, knee, calf, ankle, upper arm, forearm and wrist, and skinfold thicknesses at triceps, chest, iliac, abdomen, scapula, and thigh)

Milne, C., Seefeldt, V., and Reuschlein, P., 1976, Relationship between grade, sex, race, and motor performance in young children, _Res. Quart._, 47:726. Data from 553 Michigan children in kindergarten, grade 1, or grade 2. Of these children, 417 were white and 136 were black. (Table of agility, shuttle run, standing long jump, 30-yard dash, 400-foot run, and Well's sit and reach flexibility test)

Montoye, H. J., 1978, 1970, "An Introduction to Measurement in Physical Education," Allyn & Bacon, Boston [originally published in 1970 by Phi Epsilon Kappa Fraternity, Indianapolis]. (Tables of subscapular, suprailiac, paraumbilical, and triceps skinfolds, and sum of these skinfolds)

Montoye, H. J., and Lamphear, D. E., 1977, Grip and arm strength in males and females, age 10 to 69, _Res. Quart._, 48:109. Data from 999 children in Tecumseh (MI). (Tables of sum of grip strengths, grip strength/body weight, arm strength, arm strength/weight, strength index, strength index/body weight, and relative strength index)

Morse, M., Schultz, F. W., and Cassels, D. E., 1949, Relation of age to physiological responses of the older boy (10-17 years) to exercise, _J. Appl. Physiol._, 1:683. Data from 110 Chicago boys who were neither extremely thin nor obese. (Table of maximal oxygen uptake)

Moss, A. J., and Adams, F. H., 1962, "Problems of Blood Pressure in Childhood," Charles C Thomas, Springfield (IL). Data from 1,418 white children aged from birth to 18 years. (Tables of systolic and fourth and fifth phase diastolic blood pressure at the wrist and ankle and upper extremity)

Nagle, F. J., Hagberg, J., and Kamel, S., 1977, Maximal O_2 uptake of boys and girls - ages 14-17, _Eur. J. Appl. Physiol._, 36:75. Data from 120 white high school boys in the midwest part of the U.S. (Table of stature, weight, maximal oxygen uptake, and maximal oxygen uptake/weight)

Namin, E. P., 1966, Pediatric electrocardiography and vectorcardiography, _in_: "Heart Disease in Children: Diagnosis and Treatment," B. M. Gasul, R. A. Arulka, and M. Lev, eds., Lippincott Co., Philadelphia. Source of data unclear but presumably Illinois children. (Tables of frontal QRS axis, amplitude of R wave, amplitude of R wave as a percentage of R plus S wave, and amplitude of S wave)

Namin, E. P., and Miller, R. A., 1966, The normal electrocardiogram and vectorcardiogram in children, _in_: "Electrocardiography in Infants and Children," D. E. Cassels and R. F. Ziegler, eds., Grune & Stratton, New York. Electrocardiographic R and S wave amplitude data from 437 children aged 30 hours to 14 years; standard Sanborn or Cambridge units were used. (Used by Liebman and Plonsey, 1977, in tables of heart rate, P-R interval, and maximum QRS duration)

Nankin, H. R., Sperling, M., Kenny, F. M., Drash, A. L., and Troen, P., 1975, Correlation between sexual maturation and serum gonadotropins: Comparison of black and white youngsters, _Am. J. Med. Sci._, 268:139.

Novak, L. P., 1963, Age and sex differences in body density and creatinine excretion of high school children, _Ann. NY Acad. Sci._, 110:545. Data from 57 boys and 53 girls aged 12.5 to 18.5 years. They were attending the University of Minnesota High School and appeared to be of high socioeconomic status. (Tables of body density, creatinine coefficient, and creatinine excretion in relation to age)

Novak, L. P., 1966, Total body water and solids in six- to seven-year-old children: differences between the sexes, _Pediatrics_, 38:483. Data from 64 Minnesota children of heterogeneous socioeconomic background who seemed to be healthy. (Table of total body water, percent body water, total body solids, and percent body solids)

Novak, L. P., Hamamoto, K., Orvis, A. L., and Burke, E. C., 1970, Total body potassium in infants. Determination by whole-body counting of radioactive potassium (^{40}K), Am. J. Dis. Child., 119:419. Data from 64 healthy white full-term Minnesota infants examined at an average age of 32 days. (Tables of biceps, triceps, forearm, subscapular, thigh and calf skinfold thicknesses; bicristal, elbow, wrist, knee and ankle diameters; limb diameters corrected for fat thickness at upper arm, forearm, thigh, and calf)

Novak, L. P., Tauxe, W. N., and Orvis, A. L., 1973, Estimation of total body potassium in normal adolescents by whole-body counting: age and sex differences, Med. Sci. Sports, 5:147. Data from 111 boys and 100 girls in Minnesota; all were white. (Tables of stature, weight, total body potassium, total body potassium/weight, muscle diameter and muscle volume of upper arm and calf, and corrected limb diameters)

Noyes, F. R., and Grood, E. S., 1976, The strength of the anterior cruciate ligament in humans and rhesus monkeys, J. Bone Joint Surg., 58A:1074. Data from six human specimens at ages 16 to 26 years. (Tables of biomechanical properties)

Nutrition Survey of Lower Greasewood Chapter Navajo Tribe, 1968-1969, conducted by the Department of Community Medicine, University of Pittsburgh; The National Nutrition Survey; and the United States Public Health Service, Indian Health Service. Data from 168 boys and 161 girls aged from birth to 14 years. (Tables of triceps skinfold, hematocrit, and hemoglobin)

Oberhausen, E., Burmeister, W., and Huycke, E. J., 1966, Das Wachstum des Kaliumbestandes in Menschen gemessen mit dem Ganzkörperzähler, Ann. Pediatr., 205:381. Data from 2,261 boys and 1,785 girls aged 8 to 19 years living in Hamburg (W. Germany). (Tables of total body potassium)

Oddie, T. H., and Fisher, D. A., 1967, Protein-bound iodine level during childhood and adolescence, J. Clin. Endocrinol. Metab., 27:89. Data from 279 euthyroid children aged 0.1 to 20 years. (Tables of serum thyroxine iodine levels)

O'Halloran, M. T., and Webster, H. L., 1972, Thyroid function assays in infants, J. Pediatr., 81:916.

Oppel, W. C., Harper, P. A., and Rider, R. V., 1968, The age of attaining bladder control, Pediatrics, 42:614. Data from 859 children born in Baltimore (MD). Of these, 48% were Negro and the rest Caucasian. (Table of percentage of bed-wetters and duration of relapse)

Overman, R. B., Etteldorf, J. N., Bass, A. C., and Horn, G. B., 1951, Plasma and erythrocyte chemistry of the normal infant from birth to two years of age, Pediatrics, 7:565. Data from 188 normal infants in Memphis (TN) and their mothers and from cord blood. (Tables of sodium, potassium, and chloride in plasma and red cells, hematocrit, plasma CO_2, pH, non-protein nitrogen, and protein)

Owen, C. A., Jr., and Hurn, M. M., 1953, Changes in blood coagulation factors during the first week of life, J. Pediatr., 42:424.

Owen, G. M., Jensen, R. L., and Fomon, S. J., 1962, Sex-related differences in total body water and exchangeable chloride during infancy, J. Pediatr., 60:858.

Owen, G. M., Lubin, A. H., and Garry, R. J., 1973, Hemoglobin levels according to age, race, and transferrin saturation in preschool children of comparable socioeconomic status, J. Pediatr., 82:850. Data from 162 white and 249 black children aged 12 to 71 months examined in the Ten-State Nutrition Survey, 1968-1970. (Table of hemoglobin in relation to transferrin saturation, age, and race)

Owen, G. M., Kram, K. M., Garry, P. J., Lowe, J. E., and Lubin, A. H., 1974, A study of nutritional status of preschool children in the United States, 1968-1970, Pediatrics, 53:597. Data from 3,441 children aged 1 to 6 years; approximately 80% were white, 15% Negro, and 4% Spanish-American. The group constituted 65% of randomly selected children in 74 communities. (Tables of lateral thoracic skinfold thickness in relation to age, socioeconomic status, and race)

Palmer, C. E., 1944, Studies of the center of gravity in the human body, Child Dev., 15:99. Data from 1,172 normal white individuals in Minneapolis (MN) aged from birth to 20 years. (Tables of position of standing center of gravity in relation to stature)

Pařízková, J., 1976, Growth and growth velocity of lean body mass and fat in adolescent boys, Pediatr. Res., 10:647. Serial data from 40 Czechoslovakian boys from 10 to 17 years of age. (Table of stature, weight, bone age, body density, lean body mass, and percent body fat)

Parmelee, A. H., Schulz, H. R., and Disbrow, M. A., 1961, Sleep patterns of the newborn, J. Pediatr., 58:241. Data from 100 consecutive births at the Medical Center of the University of California at Los Angeles (CA). Race was not reported. (Tables of total hours of sleep, average period, and number of feedings)

Penny, R., Guyda, H., Baghdassarian, A., Johanson, A. J., and Blizzard, R. M., 1970, Correlation of serum follicular stimulating hormone (FSH) and luteinizing hormone (LH) as measured by radioimmunoassay in disorders of sexual development, J. Clin. Invest., 49: 1847.

Penny, R., Goldstein, I. P., and Frasier, S. D., 1978, Gonadotrophin excretion and body composition, Pediatrics, 61:294. Data from 282 California children aged 3 to 16.9 years. Race was not reported. They were all between the 10th and 97th percentiles for stature and weight. (Tables of total body water, percent body water, percent body fat, total body fat, and gonadotrophin excretion in relation to chronological age and pubertal stage)

Pett, L. B., and Ogilvie, G. F., 1957, The report on Canadian average weights, heights and skinfolds, Can. Bull. Nutr., 5:1. Data from a nationally representative sample. (Tables of triceps skinfold thickness)

Piscopo, J., 1962, Skinfold and other anthropometrical measurements of preadolescent boys from three ethnic groups, Res. Quart., 33:255. Data from 647 Italian, Jewish or Negro boys in Boston (MA). (Table of skinfold thickness at abdomen, chest, lateral and posterior arm and scapula; stature, weight, biiliac diameter, and the circumferences of chest, upper arm, and thigh)

Polgar, G., and Kong, G. P., 1965, The nasal resistance of newborn infants, J. Pediatr., 67: 557.

Ramsey, G. V., 1950, Sexual growth of Negro and white boys, Hum. Biol., 22:146. Data from 286 white boys in a midwest city enrolled in the seventh, eighth, or ninth grade. They were fairly representative of the city socioeconomically. The 37 Negro boys were all those in the seventh through the ninth grade of a small eastern town. This group was middle class in relation to the general Negro population in the northeastern United States. (Table of age of first ejaculation, voice change, and first appearance of pubic hair)

Rarick, L., and Thompson, J. A. J., 1956, Roentgenographic measures of leg muscle size and ankle extensor strength of seven-year-old children, Res. Quart., 27:321. Data from 51 American-born white children in Wisconsin aged 70 to 84 months. (Table of muscle area, total area, bone area, muscle breadth in the calf, calf circumference, stature and weight)

Rauh, J. L., and Schumsky, D. A., 1968, An evaluation of triceps skinfold measures from urban school children, Hum. Biol., 40:363. Data from 1,573 students in Cincinnati (OH) aged 6 through 17 years. They were representative of the local population in stature and weight; about 35% were Negro. (Tables of triceps skinfold thickness in relation to race)

Reiquam, C. W., Allen, R. P., and Akers, D. R., 1965, Normal and abnormal small bowel lengths. An analysis of 389 autopsy cases in infants and children, Am. J. Dis. Child., 109:447. Data from 389 autopsies in Denver (CO); race not stated. (Table of small bowel length in relation to recumbent length)

Reynolds, E. L., 1948, Distribution of tissue components in the female leg from birth to maturity, Anat. Rec., 100:621. Serial data from 190 white girls in southwestern Ohio. These girls were of middle socioeconomic status. (Table of comparisons between calf circumference and radiographic calf diameter, absolute and relative breadths of fat, muscle and bone in calf; total calf breadth, and tissue breadths relative to adult, and measured and calculated calf circumferences)

Reynolds, E. L., 1949, The fat/bone index as a sex-differentiating character in man, Hum. Biol., 21:199. Data from 297 white children aged 7.5 to 16.5 years. These southwestern Ohio children were enrolled in the Fels Longitudinal Study and were of middle socioeconomic status. (Table of fat width/bone width in the calf)

Reynolds, E. L., 1951, The distribution of subcutaneous fat in childhood and adolescence, Monogr. Soc. Res. Child Dev., 15 (2, Serial No. 50). Data from 176 white children in southwestern Ohio examined annually from 1 to 11 years. These children were of middle socioeconomic status. (Tables of calf, trochanteric, waist, chest, forearm, and deltoid fat thicknesses, increments in these, and relative fat thicknesses)

Reynolds, E. L., and Clark, L. C., 1947, Creatinine excretion, growth progress and body
    structure in normal children, Child Dev., 18:155. Data from 400 sets of observations
    in 155 white children in southwestern Ohio enrolled in the Fels Longitudinal Study.
    These children were of middle socioeconomic status. (Tables of weight, creatinine ex-
    cretion, and creatinine/weight ratio)

Reynolds, E. L., and Grote, P., 1948, Sex differences in the distribution of tissue compon-
    ents in the human leg from birth to maturity, Anat. Rec., 102:45. Data from south-
    western Ohio white children enrolled in the Fels Longitudinal Study; the number at each
    age is 13 to 30 within each sex. These children were of middle socioeconomic status.
    (Tables of calf widths: total, muscle, fat, and bone)

Reynolds, E. L., and Wines, J. V., 1948, Individual differences in physical changes asso-
    ciated with adolescence in girls, Am. J. Dis. Child., 75:329. Serial data from 49 white
    girls in southwestern Ohio examined in the Fels Longitudinal Study. These girls were
    of middle socioeconomic status. (Tables of breast development stages and pubic hair
    stages in relation to age)

Reynolds, E. L., and Wines, J. V., 1951, Physical changes associated with adolescence in
    boys, Am. J. Dis. Child., 82:529. Data from 59 white boys in southwestern Ohio examined
    annually from 9 to 21 years. These boys were from families of middle socioeconomic
    status. (Tables of penis size, pubic hair and genitalia stages in relation to chrono-
    logical age)

Rhoads, T. F., Rapoport, M., Kennedy, R., and Stokes, J., Jr., 1945, Studies on the growth
    and development of male children receiving evaporated milk. II. Physical growth, denti-
    tion, and intelligence of white and Negro children through the first four years as in-
    fluenced by vitamin supplements, J. Pediatr., 26:415. Serial data from 233 children
    seen in the Outpatient Department of the Children's Hospital of Philadelphia. These
    boys had birth weights of 5 pounds or more. They were all followed from about 42 days
    of age to at least 2 years. At 2 years of age, there were 58% white and 42% Negro. The
    whites were predominantly of Irish, British, or German origin. (Tables of hemoglobin
    and serum protein concentrations in relation to age, type of feeding, and race; and
    age of first standing and walking)

Rich, G. Q., III, 1957, Muscular fatigue curves of boys and girls, Res. Quart., 31:485.
    Data from 200 children in Albany (CA) aged 8 to 17 years. (Table of stature, weight,
    and strength)

Roberts, K. E., and Schoellkopf, J. A., 1951a, Eating, sleeping, and elimination practices
    of a group of two-and-one-half-year-old children. IV. Elimination practices: bowel,
    Am. J. Dis. Child., 82:137. Data from 783 children studied in the Well Baby Clinic of
    the Rochester (MN) Child Health Institute. (Table of frequency and variability of
    bowel movements)

Roberts, K. E., and Schoellkopf, J. A., 1951b, Eating, sleeping, and elimination practices
    of a group of two-and-one-half-year-old children. V. Elimination practices: bladder,
    Am. J. Dis. Child., 82:144. Data from 775 children studied in the Well Baby Clinic of
    the Rochester (MN) Child Health Institute in which mothers were advised to postpone a
    definite training program "until near the beginning of the third year." (Table of
    frequency of daytime urination, prevalence of dryness and wetness, frequency of lack of
    urinary control, and frequency of daytime urination)

Roche, A. F., Unpublished, Serial data from white children living in southwestern Ohio.
    These children are enrolled in the Fels Longitudinal Study and are from families of
    middle socioeconomic status. (Tables of skinfold thickness at triceps and subscapular
    sites, and radiographic fat thickness over the tenth rib)

Roche, A. F., and Barkla, D. H., 1965, The level of the larynx during childhood, Ann. Otol.
    Rhinol. Laryngol., 74:645. Serial data from 16 white Melbourne (Australia) children
    of British ancestry. (Table of levels of the body of the hyoid, tip of epiglottis and
    arytenoid in relation to cervical vertebrae)

Roche, A. F., French, N. Y., and Davila, G. H., 1971, Areolar size during pubescence, Hum.
    Biol., 43:210. Data from serial measurements of 144 Melbourne children aged 9 to 15
    years. These children were of British ancestry and their families were approximately
    representative of the Melbourne community socioeconomically. (Tables of areolar dia-
    meters in relation to chronological age, peak height velocity, age at menarche, and
    pubic hair stages)

Roche, A. F., Eichorn, D., McCammon, R. W., Reed, R. B., Valadian, I., Himes, J. H., Kent, R. L., Jr., and Siervogel, R. M., 1979, The natural history of blood pressure. Final report on Contract NOl-HV-42985. Unpublished. Serial data from 1,912 children included in the Berkeley, Denver, Fels, Guidance, Harvard, and Oakland Growth Studies. (Tables of pulse rate, basal pulse rate, pulse rate after recovery from exercise; basal systolic, seated systolic, and standing systolic blood pressures; difference between standing systolic and diastolic fifth phase; difference between basal systolic and diastolic, fourth and fifth phases; systolic after recovery from exercise; basal diastolic, fourth and fifth phases; standing diastolic fifth phase; seated diastolic, fourth and fifth phases; supine diastolic, fourth phase; diastolic fourth and fifth phases after recovery from exercise; difference between seated systolic and fourth phase diastolic; difference between seated systolic and fifth phase diastolic; difference between supine systolic and fourth phase diastolic; difference between supine systolic and fifth phase diastolic; and difference between systolic and fifth phase diastolic after exercise)

Roche, A. F., Himes, J. H., Siervogel, R. M., and Johnson, D. L., 1979, "Longitudinal Study of Human Hearing: Its Relationship to Noise and Other Factors. II. Results from the First Three Years," AMRL-TR-79-103, Wright Patterson Air Force Base, OH. Serial data from 258 children (aged 6 to 18 years) living in southwestern Ohio. All except 13 of the children are white. (Tables of auditory thresholds and increments in auditory thresholds)

Rodahl, K., Åstrand, P.-O., Birkhead, N. C., Hettinger, T., Issekutz, B., Jones, D. M., and Weaver, R., 1961, Physical work capacity. A study of some children and young adults in the United States, Arch. Environ. Health, 2:499. Data from 601 children aged 8 to 18 years in Philadelphia (PA). Of these, 31% were non-white. (Table of stature, weight, muscle strength, manual dexterity, pulse response, and maximal oxygen uptake)

Rogowski, P., Siersbaek-Nielsen, K., and Mølholm-Hansen, J., 1974, Seasonal variation in neonatal thyroid function, J. Clin. Endocrinol. Metab., 39:919. Data from 246 full term infants born in Copenhagen (Denmark) after uncomplicated pregnancies. The blood samples were obtained on the second or third day of life. (Tables of T_3 and T_4 and TSH in relation to age, weight, and season)

Root, A. W., 1973, Endocrinology of puberty. I. Normal sexual maturation, J. Pediatr., 83:1. Data compiled from the recent literature. (Tables of estrogen and androgen levels in plasma and urine, luteinizing and follicle stimulating hormones in relation to age and pubertal stage)

Ross, J. J., Agnew, H. W., Jr., Williams, R. L., and Webb, W. B., 1968, Sleep patterns in pre-adolescent children: an EEG-EDG study, Pediatrics, 42:324. Data from 18 healthy Florida boys aged 3 to 11 years. Race was not stated. (Tables of sleep stage patterns and stages by part of sleep period)

Rothenberg, L. H., Hintz, R., and Van Camp, M., 1977, Assessment of physical maturation and somatomedin levels during puberty, Am. J. Orthod., 71:666.

Round, J. M., Butcher, S., and Steele, R., 1979, Changes in plasma inorganic phosphorus and alkaline phosphatase activity during the adolescent growth spurt, Ann. Hum. Biol., 6:129. Serial data from 44 boys and 23 girls in England. (Tables of plasma alkaline phosphatase, protein, albumin, and calcium in relation to peak height velocity)

Rubin, M. I., Bruck, E., and Rapoport, M., 1949, Maturation of renal function in childhood: clearance studies, J. Clin. Invest., 28:1144.

Rubinstein, H. A., Butler, V. P., Jr., and Werner, S. C., 1973, Progressive decrease in serum triiodothyronine concentrations with human aging: Radioimmunoassay following extractions of serum, J. Clin. Endocrinol. Metab., 37:247.

Rudesill, C. L., and Henderson, R. A., 1941, Normal blood sugar values in children, Am. J. Dis. Child., 61:108. Data from 144 children in Indianapolis (IN) examined when fasting. (Table of blood sugar values)

Rutledge, M. M., Clark, J., Woodruff, C., Krause, G., and Flynn, M. A., 1976, A longitudinal study of total body potassium in normal breast-fed and bottle-fed infants, Pediatr. Res., 10:114. Data from 84 Missouri infants each measured three times during the first year of life. They were of middle socioeconomic status and were admitted to the study regardless of race. The racial distribution was not reported. (Table of recumbent length, weight, and total body potassium)

Saadat, M., Khan, M. A., Gutberlet, R. L., and Heald, F. P., 1976, Measurements of hair in normal newborns, Pediatrics, 5:960. Data from 63 healthy infants in Baltimore (MD) with birthweights not less than 2500 gm (28 boys, 35 girls; 26 Caucasians, 37 Negroid). They were divided to anagen (active cellular proliferation) and telogen (resting) phases. The few in a catagen phase (transitional between anagen and telogen) were included with the anagen phase. (Table of hair root and shaft diameter)

Safrit, M. J., Unpublished, The physical performance of inner city children in Milwaukee (WI). Data from 60 boys and 60 girls. Table of 50-yard run, shuttle run, 600-yard run-walk, standing jump, softball throw, flexed arm hang, pull-ups, and sit-ups.

Savage, P. J., Hamman, R. F., Bartha, G., Dippe, S. E., Miller, N., and Bennett, P. H., 1976, Serum cholesterol levels in American (Pima) Indian children and adolescents, Pediatrics, 58:274. Data from 1,915 individuals from birth to 18 years. (Table of serum cholesterol)

Schoenberg, J. B., Beck, G. J., and Bouhuys, A., 1978, Growth and decay of pulmonary function in healthy blacks and whites, Respir. Physiol., 33:367. Data from children in a rural town and an urban district in Connecticut and in a semi-rural town in South Carolina. The children included 1,001 whites and 1,022 blacks aged 7 to 19 years. (Tables of stature, weight, forced vital capacity, forced expiratory volume, and peak expiratory flow)

Schonfeld, W. A., and Beebe, G. W., 1942, Normal growth and variation in the male genitalia from birth to maturity, J. Urol., 48:759. Data from 1,500 normal males in New York City. (Table of penis length and circumference and testis volume)

Schulz, D. M., Giordano, D. A., and Schulz, D. H., 1962, Weights of organs of fetuses and infants, Arch. Pathol., 74:244. Data from 983 autopsy records after rejecting multiple births, multiple anomalies, congenital heart disease, infants of diabetic mothers, severe maceration, erythroblastosis, leukemia and related diseases of the reticuloendothelial system and those conditions leading to extreme emaciation. (Tables of recumbent length, body weight, weight of brain, thymus, heart, both lungs, liver, spleen, pancreas, both adrenals and both kidneys in relation to gestational and chronological age)

Schutte, J. E., 1979, "Growth and Body Composition of Lower and Middle Income Adolescent Black Males," Ph.D. Thesis, Southern Methodist University, Dallas. Data from 203 boys in Dallas County (TX) aged 10 to 18 years. (Tables of subscapular, chest, umbilical, suprailiac, triceps, biceps, forearm, thigh, and calf skinfolds; upper arm, forearm, thigh, and calf muscle diameters; total body water, percent body water, total body fat, and minimum and maximum areolar diameters)

Schutte, J. E., 1980, Growth differences between lower and middle income black male adolescents, Hum. Biol., 52:193. Data from 203 individuals in Texas. (Table of lean body mass, total body fat, and percent body fat)

Schwarz, G. S., 1946, Determination of frontal plane area from the product of the long and short diameters of the cardiac silhouette, Radiology, 47:360.

Scott, R. B., Ferguson, A. D., Jenkins, M. E., and Cutter, F. F., 1955, Growth and development of Negro infants. V. Neuromuscular patterns of behavior during the first year of life, Pediatrics, 16:24. Data from 708 full-term infants from contrasting socioeconomic backgrounds. (Tables of ages at which tasks could be performed)

Seely, J. E., Guzman, C. A., and Becklake, M. R., 1974, Heart and lung function at rest and during exercise in adolescence, J. Appl. Physiol., 36:34. Data from lower and middle class high school children in Montreal (Quebec). The group included 110 boys and 59 girls aged 13 to 18 years; about 10% of these were non-white. (Table of stature, weight, and maximal oxygen uptake/weight)

Seils, L. G., 1951, The relationship between measures of physical growth and gross-motor performance of primary-grade school children, Res. Quart., 22:244. Data from 269 boys and 224 girls in grades 1, 2, or 3 of the public schools of four Massachusetts communities. (Tables of stature, weight, motor skill performance in running, balance, agility, jumping, throwing, striking, and catching)

Shannon, I. L., 1973, Reference table for human parotid saliva collected at varying levels of exogenous stimulation, J. Dent. Res., 55:1157. Data from 534 fasting males aged 17 to 25 years. Flow was elicited by flavored drops. (Table of parotid saliva flow rate)

Sheldon, W. H., Lewis, N. D. C., and Tenny, A. M., 1969, Psychotic patterns and physical constitution, in: "Schizophrenia, Current Concepts and Research," D. V. Siva Sankar, ed., PJD Publications, Hicksville, NY.

Shephard, R. J., 1974, Work physiology and activity patterns of circumpolar Eskimos and Ainu, Hum. Biol., 46:263. Data from 76 individuals aged 9 to 18 years. (Table of heart rate, gas exchange ratio and ventilation, total body water, percent body fat, and energy expenditure at maximal oxygen uptake)

Shephard, R. J., Jones, G., and Brown, J. R., 1968, Some observations on the fitness of a Canadian population, Can. Med. Assoc. J., 98:977. Data from 193 men enrolled in a technical institute, of whom 176 were white, 12 Negroid, 4 Asian, and 1 was a North American Indian. (Table of grip strength, forced vital capacity, forced expiratory volume aerobic power, and relative aerobic power)

Shephard, R. J., Allen, C., Bar-Or, O., Davies, C. T. M., Degré, S., Hedman, R., Ishii, K., Kaneko, M., La Cour, J. R., de Prampero, P. E., and Seliger, V., 1969, The working capacity of Toronto schoolchildren, Can. Med. Assoc. J., 100:560, 705. Data from 57 boys and 72 girls in Toronto (Ont.) aged 11 to 13 years. (Table of stature, weight, maximal oxygen uptake, and maximal oxygen uptake/weight)

Shephard, R. J., Hatcher, J., and Rode, A., 1973, On the body composition of the Eskimo, Eur. J. Appl. Physiol., 32:3. Data from 20 juvenile Eskimos living in a Canadian arctic area. (Table of stature, weight, average skinfold thickness, total body water, body solids, body fat, and fat free solids)

Shephard, R. J., Lavallée, H., Larivière, G., Rajic, M., Brisson, G. R., Beaucage, C., Jéquier, J.-C., and La Barre, R., 1974, La capacité physique des enfants Canadiens: une comparison entre les enfants Canadiens-français, Canadiens-anglais et Esquimaux. 1. Consommation maximale d'oxygène et débit cardiaque, Union Med. Can., 103:1767. Data from urban Quebec children (26 boys and 19 girls) and from rural Quebec children (17 boys and 13 girls) aged 9 to 12 years, and from 14 Point Hope Eskimos aged 11 to 18 years. (Table of maximal oxygen uptake and maximal oxygen uptake/weight)

Sherry, S. N., and Kramer, I., 1955, The time of passage of the first stool and first urine by the newborn infant, J. Pediatr., 46:158. Data from 500 full-term infants weighing more than 2,000 gm at birth. (Tables of time of passage of first stool and first voiding)

Shock, N. W., 1942, Standard values for basal oxygen consumption in adolescents, Am. J. Dis. Child., 64:19. Serial data from 100 children enrolled in the Oakland (CA) Growth Study. They were chosen from five elementary schools on the basis of the parents' willingness to participate and the likelihood of long-term residence in the area. (Tables of basal metabolic rate in relation to surface area; basal oxygen consumption in relation to surface area and in relation to weight)

Siassi, B., Hodgman, J. E., Cabal, L., and Hon, E. H., 1979, Cardiac and respiratory activity in relation to gestation and sleep status in newborn infants, Pediatr. Res., 13:1163. Data from 32 newly born infants free from cardiopulmonary disease. They were studied during the first six days of life. (Table of heart rate, and respiratory rate in relation to sleep status)

Sibinga, M. S., 1959, Observations on growth of fingernails in health and disease, Pediatrics, 24:225. Data from 13 premature and 10 normal infants in New York City. (Table of rate of fingernail growth)

Sigerseth, P. O., 1970, Flexibility, in: "An Introduction to Measurement in Physical Education. Vol. 4, Physical Fitness," H. J. Montoye, ed., Phi Epsilon Kappa Fraternity, Indianapolis (IN). Copyright 1978, Allyn & Bacon, Inc., Boston. Data from about 100 children (50 boys and 50 girls) in each age group. (Tables of joint flexibility)

Simmons, K., 1944, The Brush Foundation Study of Child Growth and Development. II. Physical growth and development, Monogr. Soc. Res. Child Dev., 9 (1, Serial No. 37). Data from serial examinations of 999 children aged 3 months to 19 years. These children were selected from the Cleveland (OH) population to represent a group free of gross physical and mental defects. All were white, most were of North European ancestry, and the families were above average both economically and educationally. (Tables of knee height and acromial height)

Siri, W. E., 1956, Gross composition of the body, in: "Advances in Biological and Medical Physics," Vol. IV, J. H. Lawrence and C. A. Tobias, eds., Academic Press, New York.

Sisson, T. R. C., Lund, C. J., Whalen, L. E., and Telek, A., 1959, The blood volume of infants. 1. The full-term infant in the first year of life, J. Pediatr., 55:163. Data from 126 full-term apparently normal infants who had 223 plasma volume determinations. The parity and age of the mothers varied as did the ancestral histories. (Table of weight, hemoglobin, hematocrit, red cell count, plasma volume, blood volume, erythrocyte volume, and hemoglobin mass)

Sizonenko, P. C., and Paunier, L., 1975, Hormonal changes in puberty, III: Correlation of plasma dehydroepiandrosterone, testosterone, FSH and LH with stages of puberty and bone age in normal boys and girls and in patients with Addison's disease or hypogonadism or with premature or late adrenarche, J. Clin. Endocrinol. Metab., 41:894. Data from 104 normal boys and 123 normal girls in Switzerland. (Tables of dehydroepiandrosterone, testosterone, FSH and LH in plasma in relation to bone age, chronological age and stage of puberty, and stages of pubic and axillary hair and testicular volume index)

Slaughter, M. H., and Lohman, T. G., 1977, Relationship of body composition to somatotype in boys ages 7 to 12 years, Res. Quart., 48:750. Data from 45 boys (43 white, 2 black) aged 7.25 to 12.59 years. There were no apparent nutritional deficiencies on dietary survey. (Table of stature, weight, percent body fat, lean body mass, triceps, sub-scapular, suprailiac, and calf skinfold thicknesses, circumferences at biceps and calf, elbow width, knee width, and somatotypes)

Slaughter, M. H., Lohman, T. G., and Boileau, R. A., 1978, Relationship of anthropometric dimensions to lean body mass in children, Ann. Hum. Biol., 5:469. Data from 163 boys and 44 girls in Illinois aged 7 to 12 years. Of these children, 203 were white and 4 were black. All were participants in a sports fitness summer program. (Table of weight, stature, total body potassium, lean body mass, percent body fat, midaxillary and para-umbilical skinfold thicknesses, forearm and chest circumferences)

Snyder, M. L., 1940, Simple colorimetric method for estimation of relative numbers of lacto-bacilli in saliva, J. Dent. Res., 19:349.

Snyder, R. G., Spencer, M. L., Owings, C. L., and Schneider, L. W., 1975, "Anthropometry of U.S. Infants and Children," Paper No. 750423, SAE Automotive Engineering Congress and Exposition, Detroit, Michigan. [Also published by Consumer Protection Safety Commission, Report No. UM-HSRI-Bl-75-5.] Data from 4,027 infants and children in schools, day-care centers, nurseries, and clinics in eight states. (Tables of crotch height, buttock-knee length, minimum hand clearance, inside grip diameter, outside grip diameter, buttocks depth, hip depth, standing center of gravity as percent of stature, seated center of gravity as percent of sitting height, and sitting-midshoulder height)

Sorsby, A., Benjamin, B., and Sheridan, M., 1961, "Refraction and its Components During the Growth of the Eye from the Age of Three," Medical Research Council Special Report Series No. 301, Her Majesty's Stationery Office, Norwich, England. Data from 1,530 English children aged 3 to 15 years. (Table of ocular refraction and corneal power in vertical meridian, rate of axial elongation in relation to rate of change in ocular refraction and in corneal power and in lens power, axial length/lens power, depth of anterior chamber, equivalent power of lens, total power of eye, axial length, mean ocular refraction, stature, and weight)

Spady, D. W., 1980, Daily energy expenditure of healthy, free ranging children, Am. J. Clin. Nutr., 33:766.

Stearns, G., Newman, K. J., McKinley, J. B., and Jeans, P. C., 1958, The protein requirements of children from one to ten years of age, Ann. NY Acad. Sci., 69:857. Serial data from 458 studies of 51 children aged 1 to 4 years and 481 studies of 67 children aged 4 to 11 years. (Tables of daily urinary creatinine in relation to body weight and retention of nitrogen in relation to intake)

Stewart, K. J., and Gutin, B., 1976, Effects of physical training on cardiorespiratory fit-ness in children, Res. Quart., 47:110. Data from 24 boys aged 10 to 12 years attending private schools in New York City. (Table of maximal oxygen uptake/weight)

Stine, O. C., Hepner, R., and Greenstreet, R., 1975, Correlation of blood pressure with skinfold thickness and protein levels, Am. J. Dis. Child., 129:905. Data from 920 elementary school children in Baltimore (MD) in low-income urban families (711 black, 209 white). (Tables of triceps skinfold thickness and systolic and diastolic blood pressure in relation to race)

Strobel, C. T., Byrne, W. J., Ament, M. E., and Euler, A. R., 1979, Correlation of esophageal lengths in children with height: application to the Tuttle test without prior esophageal manometry, J. Pediatr., 94:81. Data from 119 patients aged 3 months to 19.5 years. (Table of esophageal length)

Strong, W. B., Spencer, D., Miller, M. D., and Salehbhai, M., 1978, The physical working capacity of healthy black children, Am. J. Dis. Child., 132:244. Data from 170 children aged 7 to 14 years in Augusta (GA). Ages are given as at last birthday. They were participants in a recreational or sports program. None had evidence of respiratory disease. (Tables of physical work capacity and heart rate at maximum exercise and total work load in relation to age and surface area)

Stuart, H. C., and Dwinell, P. H., 1942, The growth of bone, muscle and overlying tissues in children six to ten years of age as revealed by studies of roentgenographs of the leg area, Child Dev., 13:196. Data from about 130 white children studied serially from 6 to 10 years of age. These children were enrolled in the Harvard School of Public Health Growth Study. (Tables of muscle width, subcutaneous tissue width, bone area, muscle area, area of skin plus subcutaneous fat, and relative areas of bone, muscle and skin plus subcutaneous fat)

Stuart, H. C., and Sobel, E. H., 1946, The thickness of the skin and subcutaneous tissue by age and sex in childhood, J. Pediatr., 28:637. Data from about 130 white children enrolled in the Harvard School of Public Health Growth Study. (Table of skin plus subcutaneous fat)

Stuart, H. C., and Stevenson, S. S., 1954, Physical growth and development, in: "Textbook of Pediatrics," edition 6, W. E. Nelson, ed., W. B. Saunders Co., Philadelphia.

Sturgeon, P., 1951a, Volumetric and microscopic pattern of bone marrow in normal infants and children. I. Volumetric pattern, Pediatrics, 7:577. Data from 72 aspirations from 50 normal infants and children. (Table of amount of fat and myeloid plus erythoid cells in tibia, ilium and sternum)

Sturgeon, 1951b, Volumetric and microscopic pattern of bone marrow in normal infants and children. II. Cytologic pattern, Pediatrics, 7:642. Data from 72 aspirations of 50 normal infants and children. (Table of mean percentages of cell types from sternum and iliac crest)

Sturgeon, P., 1951c, Volumetric and microscopic pattern of bone marrow in normal infants and children. III. Histologic pattern, Pediatrics, 7:774.

Sullivan, J. H., and Storvick, C. A., 1950, Statistical interpretation of salivary analyses on 555 school children in two geographic regions in Oregon, J. Dent. Res., 29:173. Data from native-reared school children. (Tables of pH, starch hydrolysing time, buffer capacity, ammonia nitrogen, lactobacillus count and Snyder test of saliva in relation to place of residence in Oregon and age)

Swearingen, J. J., and Young, J. W., 1965, "Determination of Centers of Gravity of Children, Sitting and Standing," Report AM-65-23, Office of Aviation Medicine, Civil Aeronautics Research Institute, Oklahoma City (OK). Data from 1,200 Oklahoma children aged 5 to 18 years. (Tables of thigh clearance height, sitting height, buttock-knee length, knee height, buttock depth, center of gravity from floor as percent of stature, and center of gravity sitting and standing)

Switzer, J. L., and Besoain, M., 1950, Electrocardiograms of normal children with special reference to the aV link leads and chest leads, Am. J. Dis. Child., 79:449. Data from 52 normal children (35 boys and 17 girls) aged 1.5 to 15 years. (Tables of shift of electrical axis, electrical position, transition zone, heart rate, depth of Q and S waves, height of P and R waves, duration of QRS complex, P-R interval and P and T waves)

Tanner, J. M., 1962, "Growth at Adolescence," 2nd edition, Blackwell Scientific Publications, Oxford.

Tat, R. J., Greenwalt, T. J., and Dameshek, W., 1943, Output of bile pigment by newborn infants and by older infants and children, Am. J. Dis. Child., 65:558.

Ten-State Nutrition Survey 1968-1970, 1972, DHEW Publ. No. (HSM) 72-8130, U.S. Govt. Print. Office, Washington. Data from 10 states plus New York City. The sample was drawn from low income areas but those examined are not all from low income groups [includes 7,337 white, 7,779 black, and 1,431 Mexican-American children aged 1 to 18 years]. (Tables of triceps skinfold thickness by ethnic group in Texas and Louisiana)

Thompson, J., 1944, Observations in the urine of the new-born infant, Arch. Dis. Childh., 19:169. Data from 200 normal Scottish infants. (Table of intake of milk and output of urine and amount at each voiding in relation to age)

Thompson, M. E., and Dove, C. C., 1942, A comparison of physical achievement of Anglo and Spanish American boys in Junior High School, Res. Quart., 13:341. Data from 65 boys in Las Cruces (NM), age not given. (Table of performance in baseball throw, base running, chinning, 60-yard dash, jump and reach, and shot-put, in relation to race)

Trevorrow, V. E., 1970, Gamma globulin studied longitudinally between the ages of 4 and 20 years, Hum. Biol., 42:598.

Trotter, M., 1971, The density of bones in the young skeleton, Growth, 35:221. Data from 143 skeletons from birth to 22 years representing both sexes and American white and Negro groups. (Table of density of parts of humerus, radius, femur, and tibia)

Trotter, M., and Hixon, B. B., 1974, Sequential changes in weight, density and percentage ash weight of human skeletons from an early fetal period through old age, Anat. Rec., 179:1. Data from 144 skeletons from birth to 22 years. These represented each sex and white and Negro groups. (Tables of bone density for humerus, radius, femur, and tibia)

Tuddenham, R. D., and Snyder, M. M., 1954, "Physical Growth of California Boys and Girls from Birth to Eighteen Years," University of California Publications in Child Development, Vol. 1, No. 2, University of California Press, Berkeley. Serial data from 137 white children enrolled in the Guidance Study in Berkeley (CA). (Tables of grip strength)

Underwood, L. E., Azumi, K., Voina, S. J., and Van Wyk, J. J., 1971, Growth hormone levels during sleep in normal and growth hormone deficient children, Pediatrics, 48:946.

Usher, R., and McLean, F., 1969, Intrauterine growth of live-born Caucasian infants at sea level: standards obtained from measurements in 7 dimensions of infants born between 25 and 44 weeks of gestation, J. Pediatr., 74:901. Data from 300 Caucasian infants with varying socioeconomic backgrounds and national origins who were born alive in Montreal and measured within 36 hours of birth. Infants were excluded who had major congenital abnormalities, erythroblastosis or marked fetal malnutrition or who were born to diabetic mothers. (Table of paraumbilical skinfold thickness in relation to both gestational age and birthweight)

Vincent, M. F., 1968, Motor performance of girls from twelve through eighteen years of age, Res. Quart., 39:1094. Data from 300 girls in Georgia enrolled in physical education. (Table of performance in ball bounce, jump rope, high jump, ball throw, accuracy throw, side step, distance throw, and base run)

Viteri, F. E., and Alvarado, J., 1970, The creatinine height index: its use in the estimation of the degree of protein depletion and repletion in protein calorie malnourished children, Pediatrics, 46:696. Data from normal children reported by Stuart and Stevenson (1954), Daniels and Hejinian (1929), and Stearns et al. (1958). (Table of stature and 24-hour urinary creatinine)

Voors, A. W., Foster, T. A., Frerichs, R. R., Webber, L. S., and Berenson, G. S., 1976, Studies of blood pressures in children, ages 5-14 years, in a total biracial community. The Bogalusa Heart Study, Circulation, 54:319. Data from 3,524 children aged 5 through 14 years. In the study group, 37% were black and 63% were white. (Table of systolic and diastolic blood pressure)

Wainwright, W. W., 1943, Human saliva. XV. Inorganic phosphorus content of resting saliva of 650 healthy individuals, J. Dent. Res., 22:403. Data from 179 children. (Table of inorganic phosphorus content percent and mg/hour)

Waisman, H. A., Coo, T. S., Richmond, J. B., and Williams, S. J., 1951, Blood level of thiamine in normal children, Am. J. Dis. Child., 82:555. Data from 54 healthy non-institutionalized children aged 8 to 17 years and from 214 presumably healthy children aged 6 to 20 years living in an institution where they were grouped in cottages and offered identical diets from a central kitchen. (Table of blood thiamine levels)

Warner, W. L., Meeker, M., and Eells, K., 1949, "Social Class in America: A Manual of Procedure for the Measurement of Social Status," Science Research Associate, Chicago.

Webster, B., Harrington, H., and Wright, L. M., 1941, The standard metabolism of adolescence, J. Pediatr., 19:347.

Weisbrot, I. M., James, L. S., Prince, C. E., Holaday, D. A., and Apgar, V., 1958, Acid-base homeostasis of the newborn infant during the first 24 hours of life, J. Pediatr., 52:395.

Widdowson, E. M., and Dickerson, J. W. T., 1964, "Chemical Composition of the Body in Minimal Metabolism," Vol. 11A, C. L. Comar and F. Bronner, eds., Academic Press, New York.

Wieland, R. G., Chen, J. C., Zorn, E. M., and Hallberg, M. C., 1971, Correlation of growth, pubertal staging, growth hormone, gonadotropins, and testosterone levels during the pubertal growth spurt in males, J. Pediatr., 79:999. Data from 49 boys at a correctional school near Cleveland (OH). Race was not stated. (Table of stature, serum testosterone, human growth hormone, luteinizing hormone, and follicle stimulating hormone, in relation to pubertal stage)

Wilmore, J. H., and McNamara, J. J., 1974, Prevalence of coronary heart disease risk factors in boys, 8 to 12 years of age, J. Pediatr., 84:527. Data from 94 boys. (Table of stature, weight, lean body weight, total body fat, percent body fat, specific gravity, vital capacity, forced expiratory volume, residual volume, total lung capacity, plasma cholesterol, plasma triglycerides, blood pressure, and work capacity)

Wilmore, J. H., and Sigerseth, P. O., 1967, Physical work capacity of young girls, 7-13 years of age, J. Appl. Physiol., 22:923. (Table of stature, weight, maximal oxygen uptake, and maximal oxygen uptake/weight)

Winter, J. S. D., 1978, Prepubertal and pubertal endocrinology, in: "Human Growth," Vol. II, F. Falkner and J. M. Tanner, eds., Plenum Publishing Corp., New York. (Tables of thyroid hormone in serum, and gonadotrophins and sex steroids by age and by pubertal stage)

Winter, J. S. D., and Faiman, C., 1972, Pituitary-gonadal relations in male children and adolescents, Pediatr. Res., 6:126.

Winter, J. S. D., and Faiman, C., 1973, Pituitary-gonadal relations in female children and adolescents, Pediatr. Res., 7:948.

Winter, J. S. D., Faiman, C., Hobson, W. C., Prasad, A. V., and Reyes, F. I., 1975, Pituitary gonadal relations in infancy. I. Patterns of serum gonadotrophin concentrations from birth to four years of age in man and chimpanzee, J. Clin. Endocrinol. Metab., 40:545. Data from mixed cord sera of 55 infants at birth and from 198 children aged 5 days to 4 years in Winnipeg (Man.). (Table of serum FSH and LH)

Winter, J. S. D., Hughes, I. A., Reyes, F. I., and Faiman, C., 1976, Pituitary-gonadal relations in infancy. 2. Patterns of serum gonadal steroid concentrations in man from birth to two years of age, J. Clin. Endocrinol. Metab., 42:679.

Woodard, H. Q., 1964, The composition of human cortical bone. Effect of age and of some abnormalities, Clin. Orthop., 37:187. Data from 16 children aged 2 to 16 years. (Table of chemical composition of bone)

Yen, S. S. C., Vicic, W. J., and Kearchner, D. V., 1969, Gonadotrophin levels in puberty. I. Serum luteinizing hormone, J. Clin. Endocrinol. Metab., 29:382. Data from 175 children aged 8 through 15 years. (Tables of luteinizing hormone in relation to age and menarche)

Young, C. M., Bogan, A. D., Roe, D. A., and Lutwak, L., 1968, Body composition of pre-adolescent and adolescent girls. IV. Total body water and creatinine excretion, J. Am. Dietet. Assoc., 53:579.

Younoszai, M. K., and Mueller, S., 1975, Clinical assessment of liver size in normal children Clin. Pediatr., 14:378. Data from 105 healthy children aged 5 to 12 years in Iowa City (IA). (Table of stature, weight, and liver height and extension below costal margin)

Zachmann, M., Prader, A., Kind, H. P., Häfliger, H., and Budliger, H., 1974, Testicular volume during adolescence: cross-sectional and longitudinal studies, Helv. Paediatr. Acta, 29:61. Data from 2,113 Swiss boys aged 12.5 to 17 years. (Tables of testicular volume in relation to age and pubic hair stage)

Zavaleta, A. N., 1976, "Densitometric Estimates of Body Composition in Mexican American Boys," Ph.D. Dissertation, University of Texas at Austin. Data from 95 boys aged 9 to 14.9 years living in a lower socioeconomic neighborhood. (Tables of arm circumference, estimated arm muscle circumference, calf circumference, estimated calf muscle circumference; triceps, subscapular, midaxillary, suprailiac, medial calf, biceps, thigh, juxtanipple, and abdominal skinfold thicknesses; grip strength, and body density)

Ziegler, R. F., 1951, "Electrocardiographic Studies in Normal Infants and Children," Charles C Thomas, Springfield (IL). Data from 510 normal children aged from birth to 16 years. (Tables of duration of P wave, PR, QT and QRS intervals, amplitude of Q and S waves and RST segment deviation, incidence of RST segment deviation, duration and amplitude of T wave, and incidence of variations in T wave)

Zorab, P. A., 1969, Normal creatinine and hydroxyproline excretion in young persons, Lancet, ii:1164. Data from 673 boys and 430 girls from 10 English boarding schools whose diet for the day excluded soup, fish, meat, ice cream, gelatin-containing foods, and sweets. (Table of urinary hydroxyproline and creatinine excretion in relation to age)

Zorab, P. A., Clark, S., Harrison, A., and Seel, J. R., 1970, Hydroxyproline excretion and height velocity in adolescent boys, Arch. Dis. Childh., 45:763. Data from boarding school boys aged 11 to 18 years on a collagen free diet at the time of the urinary examination. (Table of both total hydroxyproline and growth rate in stature in relation to peak height velocity)

Zuti, W. B., and Corbin, C. B., 1977, Physical fitness norms for college freshmen, Res. Quart., 48:499. Data from 3,000 subjects aged 17.6 to 19.5 years tested in the Exercise Physiology Laboratory at Kansas State University. Race not stated. (Table of grip, leg and back strength, trunk flexibility, estimated $\dot{V}O_2$ maximum, stature, weight, triceps skinfold, and percent body fat)

INDEX